D1608552

Theory and Practice of CONTEMPORARY PHARMACEUTICS

Theory and Practice of CONTEMPORARY PHARMACEUTICS

EDITED BY

Tapash K. Ghosh
Bhaskara R. Jasti

CRC PRESS

Boca Raton London New York Washington, D.C.

Samford University Library

Library of Congress Cataloging-in-Publication Data

Theory and practice of contemporary pharmaceutics/edited by Tapash K. Ghosh and Bhaskara R. Jasti
p. cm.
Includes bibliographical references and index.
ISBN 0-415-28863-0 (alk. paper)
1. Drugs—Dosage forms. 2. Pharmacy. 3. Pharmaceutical technology.
[DNLM: 1. Technology, Pharmaceutical. 2. Organizational Case Studies. QV 778 G427t 2003] I. Ghosh, Tapash K. II. Jasti, Bhaskara R. III. Title.

RS200.T465 2004
615′.4—dc21 2002152232

This book contains information obtained from authentic and highly regarded sources. Reprinted material is quoted with permission, and sources are indicated. A wide variety of references are listed. Reasonable efforts have been made to publish reliable data and information, but the author and the publisher cannot assume responsibility for the validity of all materials or for the consequences of their use.

Neither this book nor any part may be reproduced or transmitted in any form or by any means, electronic or mechanical, including photocopying, microfilming, and recording, or by any information storage or retrieval system, without prior permission in writing from the publisher.

All rights reserved. Authorization to photocopy items for internal or personal use, or the personal or internal use of specific clients, may be granted by CRC Press LLC, provided that $1.50 per page photocopied is paid directly to Copyright Clearance Center, 222 Rosewood Drive, Danvers, MA 01923 USA. The fee code for users of the Transactional Reporting Service is ISBN 0-415-28863-0/05/$0.00+$1.50. The fee is subject to change without notice. For organizations that have been granted a photocopy license by the CCC, a separate system of payment has been arranged.

The consent of CRC Press LLC does not extend to copying for general distribution, for promotion, for creating new works, or for resale. Specific permission must be obtained in writing from CRC Press LLC for such copying.

Direct all inquiries to CRC Press LLC, 2000 N.W. Corporate Blvd., Boca Raton, Florida 33431.

Trademark Notice: Product or corporate names may be trademarks or registered trademarks, and are used only for identification and explanation, without intent to infringe.

Visit the CRC Press Web site at www.crcpress.com

© 2005 by CRC Press LLC

No claim to original U.S. Government works
International Standard Book Number 0-415-28863-0
Library of Congress Card Number 2002152232
Printed in the United States of America 1 2 3 4 5 6 7 8 9 0
Printed on acid-free paper

RS
200
.T465
2005

Preface

Defining pharmaceutics is not an easy task, due to its multidisciplinary nature. It may be defined as that branch of pharmaceutical sciences that deals with the optimization of drug therapy through (a) investigations of physical and chemical properties of drug molecules; (b) design, fabrication, and evaluation of drug delivery systems; and (c) monitoring how drug products are absorbed, distributed, metabolized, and excreted in the body. Traditionally, different components of pharmaceutics are presented under different course titles in undergraduate pharmacy curriculum. Physical-chemical principles underlying development of a successful dosage form belongs to physical pharmacy; the art and science of formulation and manufacture of dosage forms are described under dosage form design; the study of the factors influencing the bioavailability of a drug in humans and animals and the use of this information to optimize pharmacological and therapeutic activity of drug products are covered under biopharmaceutics; and the study of the kinetics of absorption, distribution, metabolism, and excretion (ADME) of drugs and their corresponding pharmacologic, therapeutic, or toxic responses in humans and animals are taught under pharmacokinetics. Though all the above four disciplines can be categorized under pharmaceutics, due to the extensive amount of information being presented to modern pharmacy students to prepare them for subsequent pharmacy practice courses, biopharmaceutics and pharmacokinetics are offered as separate courses in all undergraduate pharmacy curricula.

Conventionally, most undergraduate pharmacy curricula offer physical pharmacy and dosage form design as two-semester sequence courses, but the courses are not tied up properly to link principles of physical pharmacy with the practice of dosage form development. Therefore, students struggle through courses, spending hours studying and memorizing, only to forget almost everything after the examinations. Under the new pharmacy curriculum, all programs have been or will be entry level Pharm. D. programs very soon. The demands and expectations from practicing pharmacists under the new curriculum in delivering quality pharmaceutical care has also increased substantially. To keep up with that, several institutions are offering pharmaceutics as two-semester sequence courses as Pharmaceutics I and Pharmaceutics II in which theory and application are interspersed with each other. Unfortunately, the books currently available do not address this concept. Therefore, an undergraduate text combining these two components is badly needed.

Like most other professional fields, the curriculum in the undergraduate pharmacy program has changed dramatically over the past few years. With the current guideline given by ACPE (Accrediation Council for Pharmacy Education), major emphasis is being placed on problem-based learning and critical thinking. Most textbooks currently on the market do not address those issues. The last two concepts have been introduced in the textbook through assignments of simpler case studies at the end of each chapter. With the aid of practical examples, demonstrations, and case studies, this book has combined enough basic physical science background and information to prepare future practicing pharmacists for the competitive professional market.

The book is written by people who are in the practice of pharmacy. Apart from explaining the subject matter in a much simpler way, the text includes ample examples, cases, and problems. References have been added at the end of each chapter for further reading. Each chapter has been divided into the following sections:

- Basic science
- Self-evaluation (tutorial)
- Practice numerical problems (homework)
- Food for thought (case studies)

Tutorials have been designed to have the students evaluate their understanding of the subject matter in a logical and step-wise manner, whereas homework deals with in-depth mathematical treatment of a given problem. At the end of each chapter case studies have been fabricated to present a real-life scenario that will challenge students to analyze a problem critically and come up with a plausible solution utilizing the subject matter learned from the chapter. The organization of each chapter may vary based on the subject matter. Some chapters require more graphic representations, whereas others require more mathematical treatment. Answers to the questions in these sections have been included for the students.

This book consists of 17 chapters divided into 4 lecture modules with diagrams, illustrations, practical applications, and case studies. Two modules should be taught each semester in a two-semester pharmaceutics sequence. The modules are designed to assure proper flow of course materials from beginning to end. By the time the students finish both courses, they will have a good understanding of the entire dosage form development process, from identification of the drug candidate to approval of the product.

We believe this book offers a wealth of up-to-date information organized in a logical sequence corresponding to the art and science required for future formulators and dispensing pharmacists. Authors have been selected from academia, industry, and regulatory agencies for their experience and expertise in their selected areas to objectively present a balanced view of the science and its application. Their insights will prove useful to pharmacy students as well as practicing pharmacists involved in the development or dispensation of existing and next-generation biotechnology-based drug products. Therefore, this simplified, colorful, and user-friendly book will present pharmaceutics in a way that has never been presented before.

Tapash K. Ghosh, Ph.D.
Bhaskara R. Jasti, Ph.D.

Dedication

The book is dedicated to our beloved children:
Roshni Ghosh, Shoham Ghosh, Sowmya Jasti and Sravya Jasti

The Editors

Tapash K. Ghosh, Ph.D., is employed by the Office of Clinical Pharmacology and Biopharmaceutics at CDER, FDA. Before joining the FDA, he held faculty positions in the division of pharmaceutical sciences at the Massachusetts College of Pharmacy in Boston and Howard University in Washington, DC. He also worked in various pharmaceutical industries. He is the author of numerous scientific publications and the principal editor of other scientific books.

Bhaskara R. Jasti, Ph.D., is an associate professor in the Department of Pharmaceutics and Medicinal Chemistry, TJL School of Pharmacy, University of the Pacific. Dr. Jasti teaches an integrated curriculum covering basic pharmaceutics and biopharmaceutics courses at the University of the Pacific. Following Ph.D. and postdoctoral training, Dr. Jasti's first academic appointment was at Wayne State University in the Departments of Pharmacy and Internal Medicine. His current research interests are identifying the barriers for drug delivery and targeted delivery of drugs to cancer cells. Dr. Jasti has published more than 30 papers and presented more than 50 papers at various national and international meetings.

Contributors

William Abraham
Monsanto
St. Louis, MO

Emmanuel O. Akala
Howard University
Washington DC

Harisha Alturi
University of Missouri-Kansas City
Kansas City, MO

Hemant Alur
University of Missouri-Kansas City
Kansas City, MO

Edward Dennis Bashaw
Food and Drug Administration
Rockville, MD

Melgardt M. De Villiers
University of Louisiana at Monroe
Monroe, LA

Surajit Dey
University of Missouri-Kansas City
Kansas City, MO

Clapton S. Dias
PRA International
Lenexa, KS

Manjori Ganguly
UNMC College of Pharmacy
Omaha, NE

Tapash K. Ghosh
The Food and Drug Administration
Rockville, MD

Xin Guo
University of the Pacific
Stockton, CA

Emily Ha
University of the Pacific
Stockton, CA

Jeffrey A. Hughes
University of Florida
Gainesville, FL

Blisse Jain
University of Missouri-Kansas City
Kansas City, MO

Sunil S. Jambhekar
Massachusetts College of Pharmacy
Boston, MA

Bhaskara R. Jasti
University of the Pacific
Stockton, CA

Thomas P. Johnston
University of Missouri-Kansas City
Kansas City, MO

Feirong Kang
North Dakota State University
Fargo, ND

William M. Kolling
The University of Louisiana at Monroe
Monroe, LA

Uday B. Kompella
University of Nebraska
Omaha, NE

Xiaoling Li
University of the Pacific
Stockton, CA

Carl J. Malanga
West Virginia University
Morgantown, WV

Ashim K. Mitra
University of Missouri-Kansas City
Kansas City, MO

Vien Nguyen
University of Florida
Gainesville, FL

Jignesh Patel
University of Missouri-Kansas City
Kansas City, MO

Adam M. Persky
University of North Carolina
Chapel Hill, NC

Laszlo Prokai
Center for Drug Discovery
Gainesville, FL

Sumeet K. Rastogi
North Dakota State University
Fargo, ND

Yon Rojanasakul
West Virginia University
Morgantown, WV

Nikhil R. Shitut
North Dakota State University
Fargo, ND

Jagdish Singh
North Dakota State University
Fargo, ND

Somnath Singh
North Dakota State University
Fargo, ND

Craig K. Svensson
University of Iowa
Iowa City, IA

Giridhar S. Tirucherai
University of Missouri-Kansas City
Kansas City, MO

Contents

Module I:

Module II:

Module III:

Module IV:

Module I

1 Introduction: Methods of Data Representation, Interpretation, and Analysis

*Craig K. Svensson and Tapash K. Ghosh**

CONTENTS

Learning Objectives

After finishing this chapter, students will have a thorough knowledge of

- The meaning and use of significant figures
- Basic rules in the manipulation of exponents and logarithms
- How to determine the slope of a line
- Methods for summarizing an array of data, including calculation of mean, standard deviation, and coefficient of variation
- The meaning of precision, accuracy, and bias
- How to determine the best-fit line for a set of data

* No official support or endorsement of this article by the FDA is intended or should be inferred.

0-4152-8863-0/05/$0.00+$1.50
© 2005 by CRC Press LLC

I. INTRODUCTION

If you pick up a container of almost any medication, you will find an expiration date somewhere on the label or embedded in the container itself. But how do pharmaceutical companies determine how long a medicine is stable in its container? This is just one of many problems that can only be answered using quantitative techniques. The student of pharmaceutics must, therefore, be familiar with key mathematical concepts. The purpose of this introductory chapter is to review those concepts in the context of the pharmaceutical sciences.

II. SIGNIFICANT FIGURES

While serving as a consultant on a clinical pharmacokinetics service in a teaching hospital, one of the authors was asked to mediate a conflict between the pharmacy department and a fellow in pulmonary medicine. The division of pulmonary medicine had obtained a new computer program for calculating dosage regimens of theophylline, a drug used in the treatment of asthma. The fellow was disturbed that the pharmacy department was not following his medication orders explicitly and was unable to understand why. Further investigation revealed that the medical fellow had ordered an intravenous solution that required 407.562 mg of theophylline in 1000 ml of fluid, whereas the pharmacy department compounded an intravenous solution containing 400 mg of theophylline per 1000 ml. In the view of the pulmonary fellow, this represented an unacceptable deviation from his medication order. Resolution of the conflict required patient explanation of the concept of significant figures and the realistic precision one can expect in the measurement and delivery of intravenous fluids. Such episodes arise as a consequence of the ready access to computational devices (calculators and personal computers) that are able to provide results with an expansive number of consecutive figures to the right of a decimal point, irrespective of the precision of the actual measurement.

Any measurement, weight or volume, is made with the recognition of a certain level of error. This means that our measurements are actually *approximations*. If a pharmacist pours 100 ml of a suspension into a graduated cylinder, it is unlikely that the cylinder contains exactly 100 ml. Error is introduced based on the pharmacist's line of vision and the precision of the glassware. Both the instrument and the operator impact the level of precision of any measurement. This needs to be taken into account when performing pharmaceutical calculations. For example, if a graduated cylinder is calibrated with markings of 1 ml, it would be inaccurate to claim to have used it to measure 42.5 ml. Similarly, a balance with a limit of sensitivity of 1 mg cannot be used to weigh 891.7 mg of powder.

In essence, only digits that have practical meaning are *significant* figures. Sometimes zeros merely identify the location of the decimal point, while in other instances they are of practical significance. Only in the latter case would zeros be included in the sum of significant figures.

The rule for significant figures is quite simple: ***When performing a mathematical operation with two or more numbers, the answer should contain no more decimal places than the lowest approximate number used in the operation.*** For example, the following volumes were measured and mixed together: 10.2 ml, 15.65 ml, 1.28 ml, and 1.79 ml. What is the final volume of the mixture, expressed with the appropriate number of significant figures? First, the raw sum should be determined:

	10.2 ml
	15.65 ml
	1.28 ml
	1.79 ml
Raw Sum:	28.92 ml

Though the raw sum is 28.92 ml, since the value with the lowest number of significant figures is 10.2 ml, the sum is properly expressed as 28.9 ml. Using the raw sum to express the volume of the mixture would misrepresent the level of accuracy with which the volume is actually known.

The rule is applied in a similar fashion when values are multiplied or divided, with the recognition that not all values may be approximations. For example, if a pharmacist combines three unit dose packages of a liquid that are 2.5 ml each, the total volume obtained is 7.5 ml. Initially, one might assume that the answer should be expressed as 8 ml, since the lowest number of digits in the product (3 × 2.5 ml) is one. The number of containers, however, is not an approximate number. What is approximate is the volume in each container. Therefore, the number of significant figures properly expressed in the answer is based on the fewest digits in the approximate number in the product (which is two in the volume value of 2.5 ml).

III. EXPONENTS AND LOGARITHMS

We are all familiar with the use of acronyms, words formed from the initial letters of a series of words. Acronyms simplify notation and reduce the volume of words necessary to communicate. It is much easier to say (or write) *NHL* than *National Hockey League*. Mathematically, exponents serve the same function. Very large numbers are difficult to manipulate and take considerable space to write. The use of exponents facilitates scientific notation and computation.

In the use of exponents, a number is expressed as an integer multiplied by 10 with the appropriate exponent. For example,

$$456 = 4.56 \times 10^2$$

The exponent represents the number of times the number must be multiplied by 10. In the example above,

$$456 = 4.56 \times 10^2 = 4.56 \times 10 \times 10$$

Expressing large numbers in this fashion eliminates the need to count the number of digits to the left or right of the decimal place, as that is readily expressed in the exponent. This is particularly helpful when multiplying or otherwise manipulating two or more large numbers. It is more difficult to determine

$$30{,}000 \times 200{,}000$$

in comparison to

$$(3 \times 10^4) \times (2 \times 10^5)$$

To manipulate exponents in this manner requires knowledge of several important laws of exponents. These are summarized below:

Laws of Exponents	**Example**
$A^x \times A^y = A^{(x+y)}$	$10^3 \times 10^6 = 10^9$
$\left(A^x\right)^y = A^{xy}$	$\left(10^3\right)^6 = 10^{18}$
$\dfrac{A^x}{A^y} = A^{x-y}$	$\dfrac{10^3}{10^6} = 10^{-3}$
$\dfrac{1}{A^x} = A^{-x}$	$\dfrac{1}{10^3} = 10^{-3}$
$\sqrt[x]{A} = A^{1/x}$	$\sqrt[3]{10} = 10^{1/3}$

Another means by which large numbers may be expressed is *logarithmic notation*. In this method of notation, a number is expressed solely as a power of 10. A series of numbers can thereby be manipulated with simple operations using only the exponents. The logarithm of a number is the power to which 10 must be raised in order to equal that number. Therefore, if

$$A = b^x$$

then

$$\log_b A = x$$

Common logarithms utilize base 10 and are, by convention, written without notation of the base. Hence,

$$\log 1 = \log 10^0 = 0$$

$$\log 10 = \log 10^1 = 1$$

$$\log 100 = \log 10^2 = 2$$

$$\log 1000 = \log 10^3 = 3$$

$$\log 10{,}000 = \log 10^4 = 4$$

Logarithms for multiples of 10 are integers and readily determined by inspection. Values such as 1425 are more difficult and will be composed of two components: an integral (called the *characteristic*) and a fraction (called the *mantissa*). The mantissa provides the numerals that determine the figures that comprise the antilogarithm, and the characteristic identifies the position of the decimal point. For example, in the logarithm 3.5119, 0.5119 is the mantissa and 3 is the characteristic. Using a table of logarithms or a hand-held calculator, one can determine that the mantissa 0.5119 corresponds to the sequence of figures 325. The characteristic, 3, indicates that the number belongs in the thousands and should be identified as 3250. As with exponentials, students should be familiar with important laws of logarithms.

Laws of Logarithms

$$\log xy = \log x + \log y$$

$$\log \frac{x}{y} = \log x - \log y$$

$$\log x^{a} = a \log x$$

$$-\log \frac{x}{y} = +\log \frac{y}{x}$$

While logarithms to the base 10 are more familiar, the use of logarithmic transformation was originally introduced by John Napier over 300 years ago using e as the base. This system, sometimes referred to as the *Napier* system, uses the irrational number 2.718282... as the base and are known more commonly as *natural* logarithms (ln). Sometimes it becomes necessary to convert between natural and common logarithms. This may be accomplished using the following equation:

$$\ln A = 2.303 \log \mathrm{A}$$

Moreover, students will find the following relationship useful in handling exponential equations:

$$\ln e^{-x} = -x$$

IV. GRAPHICAL REPRESENTATION OF DATA

It has often been said that a picture is worth a thousand words. This is certainly true when it comes to representing scientific data. Summarizing data in a manner that allows rapid assimilation of key relationships is often best accomplished through graphical representation of data. This means allows one to identify how a *dependent* variable changes as a function of an *independent* variable. For example, if one wanted to know the stability of a particular drug in an intravenous solution, a preparation would be compounded and sampled over a period of time. The samples would then be assayed for drug content. In this setting, drug concentration represents a dependent variable and time represents the independent variable. How does one determine how concentration changes as a function of time? The data could be represented in tabular form (Table 1.1).

TABLE 1.1
Stability of Hypothetical Drug in Intravenous Solution

Time (hours)	Concentration (mg/L)
0	100
1	90
2	80
4	60
6	40

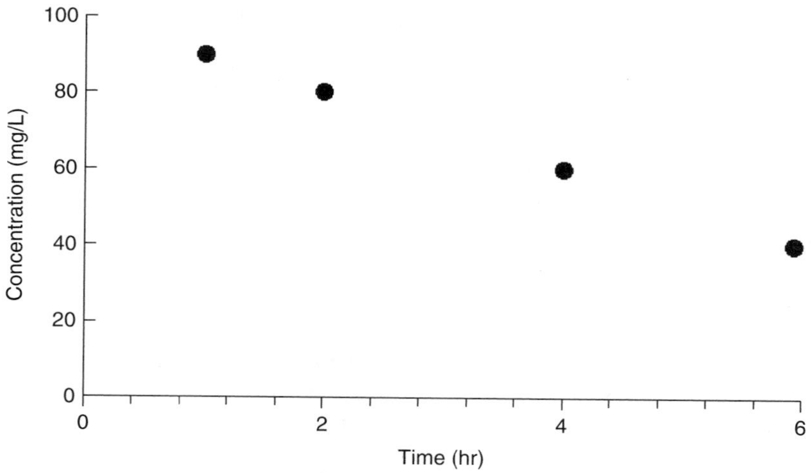

FIGURE 1.1 Plot of concentration vs. time data on rectangular coordinates.

TABLE 1.2
Comparison of the Stability of Drugs A and B in Intravenous Solution

	Concentration (mg/L)	
Time (hours)	**Drug A**	**Drug B**
0	100	75
1	85	70
2	60	65
4	30	55
6	0	45

When inspected, this data would reveal that drug concentration reduces with time. Alternatively, the result may be plotted on rectangular coordinates (Figure 1.1).

By convention, the independent variable is plotted on the x-axis, also known as the abscissa, while the dependent variable is plotted on the y-axis, also known as the ordinate. Both methods of representing the data yield the same conclusion, but the time required to arrive at that conclusion is less with graphical representation of data. Such representations also permit ready comparison of multiple data sets. Table 1.2 and Figure 1.2 summarize the data comparing the stability of two different drugs in an intravenous solution. From a quick examination of Figure 1.2 it is easy to conclude that drug A degrades more quickly in an intravenous solution that does drug B, but how quickly does it degrade? This can be determined by elucidation of the slope of the respective concentration vs. time curves, which can be described by the equation for a straight line:

$$y = mx + b$$

where m is the slope of the line, and b is the y intercept when $x = 0$. The slope can, therefore, be determined from any two points on the curve as

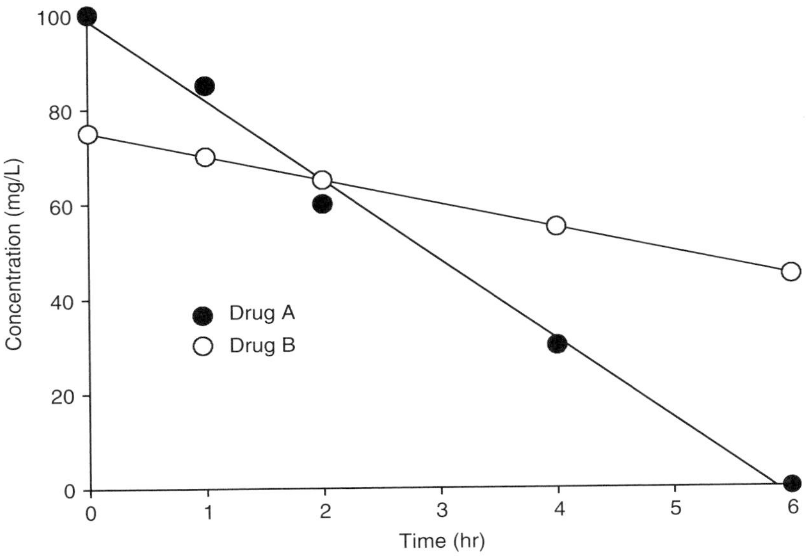

FIGURE 1.2 Stability of drug A and drug B in intravenous solution.

$$\text{slope} = \frac{\Delta y}{\Delta x}$$

$$\text{slope} = \frac{y_2 - y_1}{x_2 - x_1}$$

Thus, the slope for the decline in drug A would be

$$\text{slope} = \frac{60\ \text{mg} - 85\ \text{mg}}{2\ \text{hr} - 1\ \text{hr}} = -15\ \text{mg/hr}$$

The concentration of drug A declines at a rate of 15 mg/hr. In contrast, the slope for drug B is

$$\text{slope} = \frac{6\ 5\text{mg} - 70\ \text{mg}}{2\ \text{hr} - 1\ \text{hr}} = -5\ \text{mg/hr}$$

indicating that the concentration of drug B declines one third as rapidly as that of drug A. As stated previously, many processes of pharmaceutical interest are first order (rate of change of drug concentration is proportional to the remaining concentration as opposed to zero order decline in Figure 1.2; details in Chapter 8). Such processes are curvilinear when plotted on a standard rectangular plot but appear linear when plotted on a semi-log plot (Figure 1.3).

Plotting data on a semi-logarithmic graph obviates the need to convert each value to its corresponding logarithm prior to plotting. In order to determine the slope of a line from such a plot, the values must be converted to their logarithm. The equation for the slope of a line on a semi-log plot is given as

$$\text{slope} = \frac{\log y_2 - \log y_1}{x_2 - x_1}$$

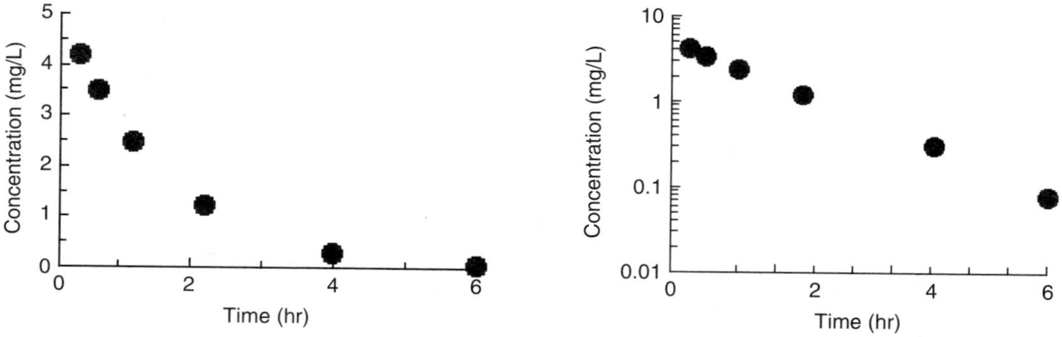

FIGURE 1.3 Plasma concentration vs. time data graphically represented on a rectangular plot (left panel) and a semi-log plot (right panel).

V. CALCULUS*

In the fall of 1665, probably well after he recovered from being knocked on his head with an apple, Isaac Newton began to consider the problem of planetary position. Describing the position of planets in the heavens was complicated by their nonstatic nature. Using Kepler's laws of planetary motion, Newton sought to prove the elliptical nature of their orbit around the sun. This challenge led Newton to develop an extremely important field in mathematics — calculus. The pharmaceutical sciences often deal with quantities that are not static; they change with time, temperature, humidity, pressure, or other factors. Examples include the decline in blood concentration of a drug with time, the decline in active content in a tablet with humidity, and the change in drug solubility with temperature. To quantify these changes necessitates the use of differential calculus. The general situation is one wherein a dependent variable, y, changes as a function of an independent variable, x. Expressed mathematically, y is a function of x; or

$$y = f(x)$$

Differential calculus provides the means by which the magnitude of change in y with a small change in x can be determined. The symbol Δ is used to denote change, such that Δy represents a change in y, while Δx represents a change in x. A vanishingly small change in x is denoted as dx, which is distinguished from a finite change in x (Δx). The symbol dx is not a product, but rather signifies that a very small amount of x is taken. The derivative of y with respect to x is given as dy/dx. The value of $\Delta y/\Delta x$ approaches dy/dx as Δy and Δx become smaller and smaller. The objective of differential calculus is to determine the value of the derivative. The meaning of this value can best be seen through an example.

Table 1.3 shows data for a square of metal with sides of length x and y for its area. As the metal is heated, it expands, resulting in an increase denoted as Δx. The area after heating is increased by an amount designated as Δy. Since $y = x^2$, it is apparent that

$$y + \Delta y = (x + \Delta x)^2$$

If the starting length of the square was 1 cm and the starting area 1 cm^2, heating results in the values of Δx shown in Table 1.3. It is apparent from this data that as Δx decreases, the value of

* It is assumed that students enrolled in a course in pharmaceutics have already completed one or more semesters of calculus. Therefore, this section is intended as a brief review, not a comprehensive introduction.

TABLE 1.3
Calculation of Change in Parameters in the Measurement of a Metal Square upon Heating

Original Length (x)	Original Area y or x^2	Increase in Length Δx	New Length $x + \Delta x$	New Area $y + \Delta y = (x + \Delta x)^2$	Increase in Area $\Delta y = (y + \Delta y) - y$	Ratio of Increase $\Delta y/\Delta x$
1	1	1.0	2.0	4	3	3.0
1	1	0.8	1.8	3.24	2.24	2.8
1	1	0.6	1.6	2.56	1.56	2.6
1	1	0.4	1.4	1.96	0.96	2.4
1	1	0.2	1.2	1.44	0.44	2.2
1	1	0.1	1.1	1.21	0.21	2.1
1	1	0.01	1.01	1.0201	0.0201	2.01
1	1	0.001	1.001	1.002001	0.002001	2.001

Source: Adapted from Daniels, F., *Mathematical Preparation for Physical Chemistry*, McGraw Hill, New York, 1956, 71.

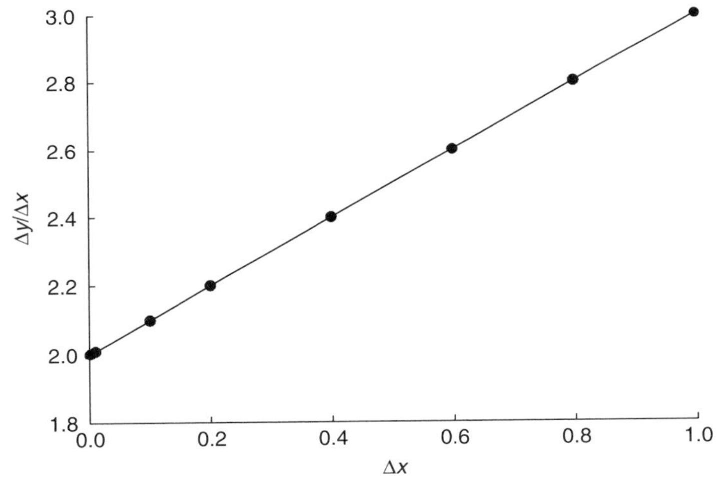

FIGURE 1.4 Plot of $\Delta y/\Delta x$ vs. Δx from the data in Table 3.

$\Delta y/\Delta x$ also decreases — but does not approach zero. As Δx decreases, $\Delta y/\Delta x$ approaches a limiting value that can be determined from the intercept of a plot of these values vs. Δx (Figure 1.4).

The limit that $\Delta y/\Delta x$ approaches as $\Delta x \to 0$ is called the *derivative* and is denoted as dy/dx. Note from the data in Table 1.3 and Figure 1.4 that the limiting value approaches 2, meaning that when $y = x^2$ and $x = 1$, $dy/dx = 2$. A similar study with an initial side length of 2 cm (i.e., $x = 2$) generates the data in Table 1.4. In this example, the limiting value is 4. Repeated experiments would reveal that the value $\Delta y/\Delta x$ approaches as $\Delta x \to 0$ is 2 times the value of x. Thus, when $y = x^2$, $dy/dx = 2x$. This same procedure can be done for a variety of functions yielding simplified rules for differentiation. Some of the most common rules are given in Table 1.5.

In differential calculus we denote changes in y as Δy and infinitely small changes as dy. It is intuitively obvious that the sum of all the parts into which y has been divided (Δy) should equal y. Mathematically,

$$\Sigma \Delta y = y$$

TABLE 1.4
Values for Repeat of Experiment in Table 1.3 but with Initial Length (*x*) of 2 cm

Δx	1	0.8	0.6	0.4	0.2	0.1	0.01	0.001
$\Delta y/\Delta x$	5	4.8	4.6	4.4	4.2	4.1	4.01	4.001

Source: Adapted from Daniels, F., *Mathematical Preparation for Physical Chemistry*, McGraw Hill, New York, 1956, 71.

TABLE 1.5
Common Rules for Differentiation and Integration

Function	Derivative
	Primary Rules
$y = x^n$	$dy/dx = nx^{n-1}$
$y = ku$	$dy/dx = k(du/dx)$
$y = u + v$	$dy/dx = du/dx + dv/dx$
$y = uv$	$dy/dx = u(dv/dx) + v(du/dx)$
$y = e^u$	$dy/dx = e^u(du/dx)$
$y = \ln u$	$dy/dx = (du/dx)/u$
	Secondary Rules
$y = k$	$dy/dx = 0$
$y = x$	$dy/dx = 1$
$y = kx$	$dy/dx = k$
$y = u^n$	$dy/dx = nu^{n-1}du/dx$
$y = e^x$	$dy/dx = e^x$
$y = \ln x$	$dy/dx = 1/x$

Note: Dependent variables *y*, *u*, and *v* are functions of *x*; *k* and *n* are constants.

$$\frac{dy}{dx} = x^n$$

$$\int dy = \int \alpha^n \; dx$$

$$y = \frac{1}{n+1}\left(x^{n+1}\right)$$

where Σ represents the process of summation. Likewise, the sum of the infinitely small parts (dy) should provide y. This process is referred to as *integration* and forms the basis of *integral calculus*. The symbol for this operation is a distorted S (representing summation) and is referred to as the integral sign, such that

$$\int dy = y$$

Differential calculus permits the determination of the relation between dy and dx when the relationship between y and x is known. In contrast, integral calculus allows one to determine the relation of y and x when the relationship between dy and dx is known. In this sense, integration is the reverse process of differentiation in the same manner that taking the square root of a number is the reverse of squaring that number. Therefore, the value for an integral can only be determined if the differential for the function has been found. For example, if

$$y = x^4/4$$

then upon differentiation

$$dy/dx = 4x^3/4 = x^3$$

Alternatively, if it is given that

$$\frac{dy}{dx} = x^3$$

or

$$dy = x^3 dx$$

or

$$\int dy = \int x^3 dx = \frac{x^4}{4}$$

Most commonly, one is interested in the integral between set limits. Integrals expressed with limits are referred to as *definite integrals*, while those with no specified limits are referred to as *indefinite integrals*. A definite integral is expressed and evaluated as

$$\int_1^3 x dx = \left[\frac{x^2}{2}\right]_1^3 = \frac{9}{2} - \frac{1}{2} = \frac{8}{2} = 4$$

Determination of a definite integral is very useful, as it allows the calculation of the area bounded by an irregular line; this is a procedure widely used in pharmacokinetic data analysis.

VI. STATISTICAL ANALYSIS

In the bottom of the ninth inning, with runners on first and third and the home team down by one run, the crowd groans as the shortstop comes out of the dugout and heads to home plate. Why? On the giant stadium screen is a picture of the upcoming batter *and* his batting statistics for the season. With runners in scoring position, he is batting an unimpressive 0.085. The batter's life would be less complicated if baseball fans hated statistics as much as most students. Any conversation about baseball is likely to be peppered with references to the myriad of statistics tracked in professional baseball. Why this infatuation with statistics among baseball fans? It is because statistics provide a means to present a player's performance that allows fans to compare, analyze, and predict how their favorite player will respond in a given game situation — especially in comparison to your colleague's favorite player.

In a similar manner, statistics are a valuable means by which pharmaceutical data can be compared, analyzed, and used to predict the behavior of drug products. Students in the pharmaceutical sciences must be familiar with simple statistical concepts and procedures, as well as the underlying basis for their use.

VII. METHODS TO SUMMARIZE DATA

When a drug product is produced, it is manufactured with a desired potency. However, variability in equipment used in the manufacturing process is such that each unit of the product made will not have the exact same potency. For the product to be useful clinically, it must not significantly deviate from the labeled potency, but how do we determine what constitutes *significant* deviation and whether or not a batch that has been manufactured is within the desired specifications? Answering such questions necessitates methods that allow us to summarize a data set derived with the specific intent of determining the potency of a manufactured batch of the product in question.

For example, suppose we want to determine the potency of a batch of 100 acetaminophen tablets that have been manufactured with an objective potency of 325 mg per tablet. After analyzing each tablet, the acetaminophen content is found to range between 302 and 341 mg, with most tablets exhibiting a content between 315 and 335 mg. Figure 1.5 illustrates the distribution of acetaminophen content in these 100 tablets. Representing the data as a frequency distribution plot is cumbersome and difficult to utilize for comparison and predictive purposes. For this reason, methods of reducing the data to a limited number of parameters are utilized. The most frequently used summary statistic is the *mean*. The mean provides a measure of the central tendency of the data. Since we have sampled the entire batch of tablets made, the mean value derived from this data would be referred to as the *population mean*. This can be calculated as

$$\text{Population mean} = \frac{\text{sum of all the values measured in the population}}{\text{size of population}}$$

Expressed for this specific example,

$$\text{Mean tablet content} = \frac{\text{sum of the content of all 100 tablets}}{\text{100 tablets}}$$

The symbol for the population mean is μ, ΣX signifies the sum of all X values, and N represents the size of the population.

$$\mu = \frac{\sum X}{N}$$

Utilizing this equation to analyze the data in Figure 1.5, the mean acetaminophen content of the tablets is determined to be 324 mg.

This mean value is referred to as the *arithmetic mean* and identifies the value each member of the sample would have to be in order for the values to all be identical and yet achieve the same sum total. Another measure of central tendency is the *geometric mean*. This measure of the central tendency of a group of data is useful when several quantities multiply together to produce a product. It identifies the value all members of the sample would have to be if they were identical and resulted in the same product. Central tendency is useful when changes are anticipated to occur in a relative manner. For example, drug is generally eliminated at a rate expressed as a fraction of the dose administered or the concentration in blood. Expressing the central tendency via the geometric mean is preferred when the data is highly skewed or expressed as ratios. It cannot be used when there are zero or negative values in the data set. The geometric mean, sometimes symbolized as *GM*, is defined as the *n*th root of the product of the observations. In symbolic form, for n observations X_1, X_2, X_3, ... X_n, the geometric mean is:

$$GM = \sqrt[n]{(X_1)(X_2)(X_3)\ldots(X_n)}$$

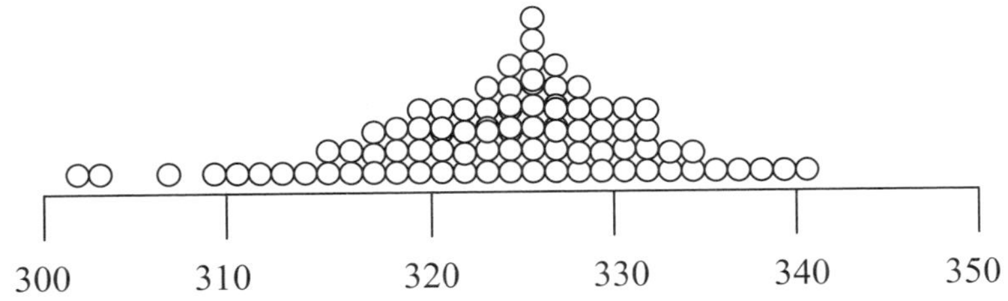

FIGURE 1.5 Distribution of acetaminophen content in 100 tablets. Each circle represents the content of one tablet.

Another way to express the central tendency is via the *median*. The median is the value that divides a data set in half. For a data set with an odd number of data points, the median is the middle value after the data has been arranged in either ascending or descending order. When the data set contains an even number of data values, the median is the average of the two middle values. What is the difference between the mean and the median? When the distribution of the data is symmetrical, the mean and median are identical. When the data is skewed, the mean and median can be quite different. The mean value is influenced by the actual values in the sample and can be altered significantly by one or two extreme values. The median is a robust measure that is no more influenced by extreme values than it is values near the median. For this reason, the median is the preferred measure when the distribution of the data is not symmetrical about the central tendency.

Another way to express the central tendency is via the *mode*. The mode is the value that occurs most frequently. It is mostly used for a large number of observations when the investigator wants to designate the "most popular" value.

The following tips help a researcher to decide which measure of central tendency is best with a given set of data:

- The *mean* is used for numerical data with symmetric distributions.
- The *median* is used for numerical data with skewed distributions.
- The *mode* is used mainly for bimodal distributions.

While the mean content of the tablets produced in the previous example is 324 mg, that information alone does not tell us anything about the variation in drug content observed within this batch of tablets. The simplest manner to express variation is provided by the *range* of observations made. Another way this can be expressed is by the difference between the highest and lowest tablet content, signified as R. For the distribution of data provided in Figure 1.5,

$$R = 341 - 302 = 39$$

This does not, however, provide a quantitative assessment of the variability in comparison to the central tendency (or mean) of the data. The magnitude by which each tablet differs from the mean tablet content can be determined and is expressed as the *deviation*. If the deviation of each tablet is determined, the *average deviation* can be calculated for the population. More commonly, variability is expressed in terms of the *standard deviation* (σ), which is determined as

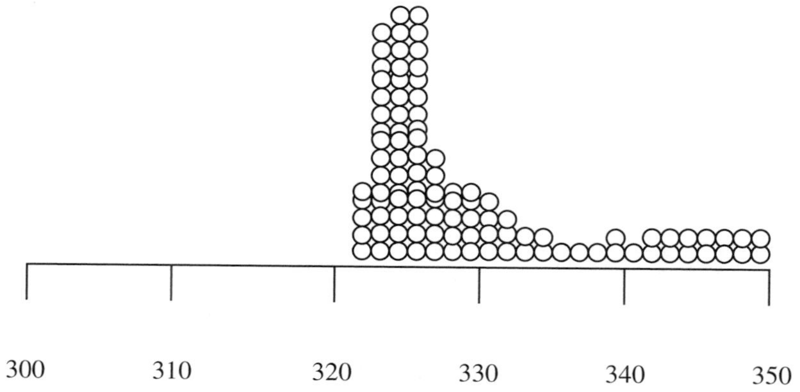

FIGURE 1.6 Distribution of acetaminophen content in 100 tablets of a different batch than that shown in Figure 1.5. Each circle represents the content of one tablet. Note that the data are not symmetrically distributed around the mean.

$$\sigma = \sqrt{\frac{\sum (X - \mu)^2}{N}}$$

Thus, the standard deviation in acetaminophen content among the tablets produced is 7 mg. This parameter provides important information about the distribution of the data. Roughly 68% of the observed values will fall within one standard deviation of the mean, while 95% will fall within two standard deviations. That means 68% of the values for the data in Figure 1.5 fall between 317 and 331 mg, while 95% will fall between 310 and 338 mg.

This characterization of the data based upon the mean and standard deviation assumes that the data exhibits *normal distribution*. Normal distribution of data occurs when the data exhibits symmetry about the mean. In other words, the values above the mean are essentially equal in number and degree of deviation to those below the mean. While the data in Figure 1.5 exhibit a normal distribution, that shown in Figure 1.6 does not. However, the mean and standard deviation is essentially the same for the data in Figures 1.5 and 1.6. Therefore, these summary statistics do not provide an accurate description of the distribution of the data. It is important to recognize that many of the comparative statistical tests that have been developed assume the data is normally distributed. If the data does not meet these criteria, different tests must be utilized to compare two or more groups (described later in this chapter).

In our example of tablet potency, we have measured the entire population of tablets; that is, we have assayed the drug content of every tablet made. This, obviously, would present a real dilemma in normal practice. If we sampled every tablet that was made, we would have no tablets for administration to patients. Therefore, if we want to know the validity of a manufactured batch of tablets, we will actually take a limited sample of the manufactured batch, with the expectation that our sample will be representative of the entire batch. By this method, however, we cannot actually determine the population mean or standard deviation. Instead, we will estimate these parameters from the sample data. The *sample mean* can be determined as

$$\text{Sample mean} = \frac{\text{sum of the values in the sample}}{\text{number of values in the sample}}$$

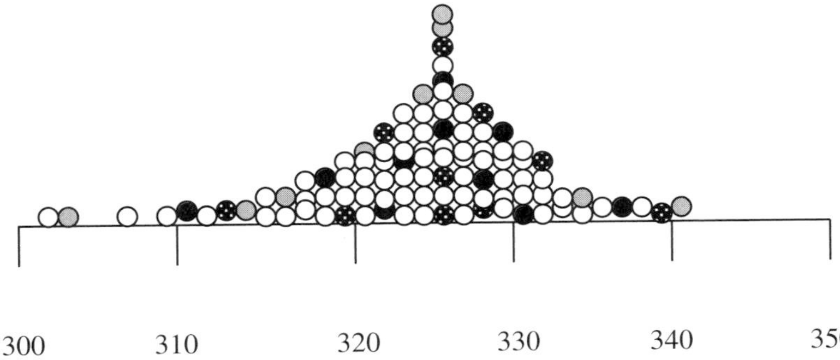

FIGURE 1.7 Distribution of acetaminophen content in 100 tablets. Each circle represents the content of one tablet. The three different fills for the circles represent the individual values drawn from three separate samples of the population. Unfilled circles represent tablets not sampled.

Expressed mathematically,

$$\bar{X} = \frac{\sum X}{n}$$

where $\bar{X}$ is the mean, and n is the number of observations.

In like manner, the *sample standard deviation* (SD) can be determined as

$$\text{SD} = \sqrt{\frac{\sum \left(X - \bar{X}\right)^2}{n-1}}$$

The sum of the squared deviations is divided by $n - 1$, rather than n, to compensate for the fact that the sample deviation tends to underestimate the population standard deviation.

In determining the standard deviation and mean of a sample from the population, one must recognize that they will not be the same as the population mean and standard deviation. A question that often arises is how accurate is the estimate of the population mean derived from the sample mean?

Consider the example of the content of acetaminophen tablets. Figure 1.7 shows this same data and values that would be obtained if three separate samples of 10 tablets each were obtained. The sample mean and standard deviation from each of these samples is provided in Table 1.6. It will be noted that the mean value of two of the samples was identical, but none of the three samples had a mean value identical to the population mean. The standard deviation for two of the samples was identical and was identical to the population standard deviation. The error in the estimate of the mean value is expressed as the *standard error of the mean* (SEM) and can be calculated as

$$\text{SEM} = \frac{\text{SD}}{\sqrt{n}}$$

Authors of journal articles frequently describe their data using the mean and standard error of the mean. While this is appealing to investigators, because the standard error of the mean is always smaller than the standard deviation, it is misleading. The standard deviation provides an estimate of the variation of values in the population from which the sample was drawn, while the standard error of the mean quantifies the accuracy of the estimate of the value of the mean. Statistical evaluation of the data for Sample 3 in Table 1.6 and Figure 1.7 illustrates these points.

TABLE 1.6
Mean and Standard Deviation Values for Three Random Samples Drawn from the Data Shown in Figure 1.5

	Sample Mean	Sample Standard Deviation
Sample 1	323	11
Sample 2	325	7
Sample 3	325	7

TABLE 1.7
Data from Sample 3 in Figure 7

Content (mg)	Deviation	$(\text{Deviation})^2$
312	−13	169
319	−6	36
322	−3	9
325	0	0
325	0	0
327	2	4
327	2	4
332	7	49
337	12	144
325	0	0
Σ 3251		**415**

The raw data for Sample 3 are shown in Column 1 of Table 1.7. The deviation of each tablet from the mean acetaminophen content is provided in Column 2, while the squares of the deviations are provided in column three. From this data, the standard deviation of the measurement can be determined for this sample.

The mean value for Sample 3 is given as

$$\bar{X} = \frac{\sum X}{n} = \frac{3251\text{ mg}}{10} = 325.1\text{ mg}$$

Based on the measured values of tablet content, the mean expressed in the appropriate number of significant figures is 325 mg. The sample standard deviation is given as

$$\text{SD} = \sqrt{\frac{\sum (X - \bar{X})^2}{n-1}} = \sqrt{\frac{415}{9}} = \sqrt{46.11} = 6.79\text{ mg}$$

Expressed as the appropriate number of significant figures, the standard deviation = 7 mg. As approximately 95% of all the values in the population fall within two standard deviations of the mean, this indicates that it would be a rare tablet that exhibited an acetaminophen content of less than 311 mg or more than 339 mg. In contrast, the standard error of the mean is given as

$$\text{SEM} = \frac{\text{SD}}{\sqrt{n}} = \frac{7\text{ mg}}{\sqrt{9}} = 2.33\text{ mg}$$

or 2 mg. If the data are reported as mean and standard error of the mean, many readers will conclude that the range of 95% of the tablets is 321 to 329 mg. This obviously results in a very different conclusion regarding the magnitude of variability in the tablet content. Though this results from a misapplication of the reported summary statistic, the inappropriate reporting of standard error of the mean by many authors perpetuates this misapplication. The appropriate use of the standard error of the mean is in the determination of the significance of the difference in mean values between two or more groups.

While the standard deviation is a useful way of describing the variation in a data set, it is often more useful to describe the *relative* variation rather than the absolute variation. This is especially true when comparing data sets. The most convenient method of describing the relative variation is through the use of the *coefficient of variation* (CV), which may be calculated as

$$\text{CV} = \frac{\text{SD}}{\bar{X}}$$

The coefficient of variation is often expressed as a percentage. A coefficient of variation of 0.05 indicates that the standard deviation is 5% of the value of the mean. Thus, a tablet batch that exhibits a mean content of 300 mg and a standard deviation of 15 mg exhibits the same relative variation as a batch of tablets with a mean content of 100 mg and a standard deviation of 5 mg (5%). The coefficient of variation is often used as a means of assessing the reproducibility of an analytical method. For example, Table 1.8 shows the results of an evaluation of an assay for acetaminophen across a range of concentrations. Each concentration was measured six times and the mean, standard deviation, and coefficient of variation determined. The data reveal that as the concentration in the sample decreases, the *absolute* variability (standard deviation) decreases. However, this may lead to erroneous conclusions since the mean value is also decreasing. When the relative variability is expressed as the coefficient of variation, it becomes clear that it *increases* as the concentration of drug decreases, indicating that assay performance is better at high concentrations than at low concentrations. This provides some measure of confidence in the validity of data derived across a range of concentrations.

The methods of summarizing data that have been described are useful when a continuum of data is available and meaningful. To return to the example of baseball, assume a pitcher throws four pitches to the first batter he faces. He throws two curveballs and two sliders. The curveballs all cross six inches to the left of home plate, while the two sliders cross six inches to the right of

TABLE 1.8
Assay Results of Measuring Various Concentrations of Standard Solutions of Acetaminophen

	Measured Content		
Actual Content (mg/L)	**Mean**	**Standard Deviation**	**Coefficient of Variation (%)**
1000	998	9	0.9
500	496	5	1.0
100	95	2	2.1
50	48	2	4.2
10	9	1	11.1

home plate. On average, the pitches crossed home plate exactly down the center, but in such a circumstance, average location is meaningless. Since each pitch missed the plate, the batter takes first base on balls. Similarly, there are situations in pharmaceutical manufacturing and clinical experimentation where these methods of summarizing data have little meaning.

PRECISION, ACCURACY, AND BIAS

Measurements made in the pharmaceutical sciences are inherently variable. This necessitates an understanding of the meaning of *precision* and *accuracy* as applied to measurements. *Precision* is a measure of the reproducibility of an event. If the measurements are close in magnitude they will exhibit a small standard deviation or coefficient of variation and will be considered precise. *Accuracy* reflects how close the measured value is to the true value, where the true value is the measure that would be observed in the absence of any error (e.g., the actual tablet content compared to the assayed result). *Bias* occurs when a systematic deviation from the true value occurs. These concepts are best understood in the context of an example.

Consider a pitcher throwing pitches to a batter. The pitcher's objective is to get all pitches within the batter's strike zone. Figure 1.8a illustrates the case where a pitcher throws four inaccurate pitches but is precise in his placement. In other words, all pitches miss the strike zone but are located at essentially the same place. These pitches represent a bias in their location because they systematically deviate to the low outside corner of the strike zone. Figure 1.8b represents a scenario where the four pitches all miss the strike zone and exhibit a biased location but are imprecise in their locations. Figure 1.8c illustrates a precise and accurate series of pitches — all four located within the strike zone. Figure 1.8d represents a series of inaccurate pitches that exhibit no systemic deviation and demonstrate an imprecise location on the part of the pitcher. For an analytical or manufacturing method to be useful, it must be both precise and accurate.

VIII. CORRELATION BETWEEN VARIABLES

Will knowing a patient's age allow me to predict the dose of a drug needed to obtain a desired concentration in the patient's blood? Can we predict the shelf-life of a solid dosage form based upon the temperature at which it will be stored? These are the types of questions that commonly arise in the biomedical sciences. In essence, what is being asked is whether or not one can quantitatively predict the value of one variable based upon the value of another variable. Such predictive capacity only exists when there is a strong correlation between the two variables.

Consider the situation where one wants to determine the concentration of dextromethorphan hydrobromide (a cough suppressant) in a series of over-the-counter cough preparations. A series of solutions of known dextromethorphan hydrobromide content are prepared, and the absorbance of each solution is determined at 254 nm. The results are tabulated in Table 1.9. From this tabulation, it is readily seen that as the concentration of dextromethorphan hydrobromide increases, the absorbance increases. This suggests that the absorbance of an unknown sample may be used to determine the dextromethorphan content. Figure 1.9 illustrates the relationship between these two variables.

If the absorbance of an unknown sample were found to be 1.233, from the data in Table 1.9 and Figure 1.9 we could conclude that the concentration of drug in that sample was approximately 5 mg/L. But what if the absorbance was 1.500? To determine the concentration of a sample with this absorbance would require some means of interpolating between the absorbance determined in the 10 and 5 mg/L concentration standard solutions. Figure 1.10 illustrates this same data with a solid line representing the best-fit line for the relationship between concentration and absorbance. The concentration of an unknown sample displaying an absorbance of 1.500 can be determined by extrapolating from 1.500 on the y-axis to the best-fit line, then from the point of intersection to the x-axis (dashed line). This indicates that the concentration of the unknown sample is 6 mg/L.

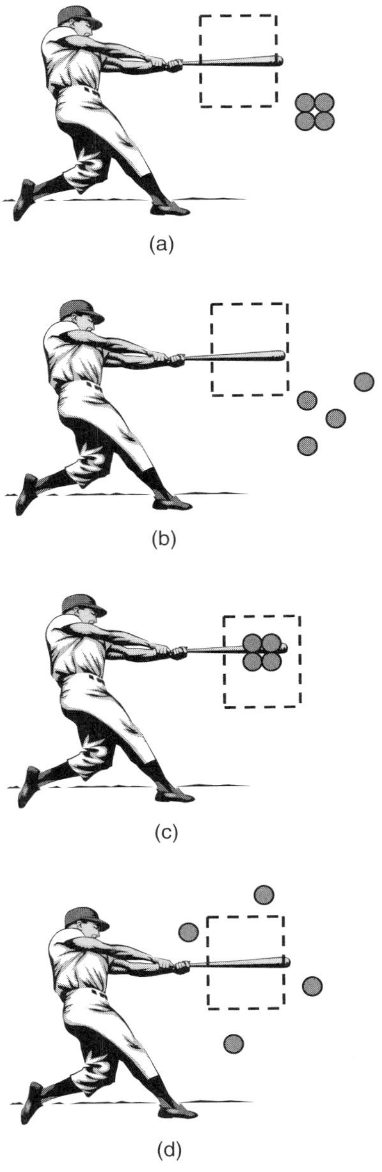

FIGURE 1.8 (a) Illustration of precise but inaccurate and biased results. The dashed box represents the strike zone. The gray circles represent the location of the four pitches. All four pitches were thrown in essentially the same location (they were precise) but missed the strike zone (they were inaccurate). Since they all deviated from the strike zone in the same location, they represent biased location. **(b) Illustration of imprecise, inaccurate, and biased results.** The four pitches were scattered (imprecise) and all missed the strike zone (inaccurate). The pitches were, however, all biased toward the lower outside corner of the strike zone. **(c) Illustration of precise, accurate, and unbiased results.** All four pitches are within the strike zone (accurate) and close to one another (precise). Since they do not deviate systematically from the desired location, they are unbiased. **(d) Illustration of imprecise and inaccurate but unbiased results.** All four pitches miss the strike zone (inaccurate) and are not located near one another (imprecise) but display no systematic deviation from the strike zone (unbiased).

How is the best-fit line determined? Consider the hypothetical relationship between x and y illustrated in Figure 1.11. Also illustrated in this figure are possible lines to describe the relationship between x and y. Line 2 seems to be the better choice, since it is closer to all the data points than any of the other lines presented. The further the line is from any specific point, the greater the variability between the data and the hypothetical best-fit line.

This indicates that the best-fit line is the line with the least deviation between the line and the observed data points. The deviations for Line 2 are represented in Figure 1.12. Since some deviations will be positive and others will be negative. Squaring the deviations will prevent them from offsetting one another. The procedure to determine the best-fit line is called *least squares regression analysis.*

The line that is derived from least squares regression analysis is termed the *regression line* of y on x. The equation for this line is given as

TABLE 1.9
Absorbance of Solutions of Dextromethorphan Hydrobromide of Known Content

Dextromethorphan Hydrobromide Concentration (mg/l)	Absorbance at 254 nm
10.0	2.399
5.0	1.233
2.5	0.684
1.0	0.247
0.5	0.132
0.2	0.048

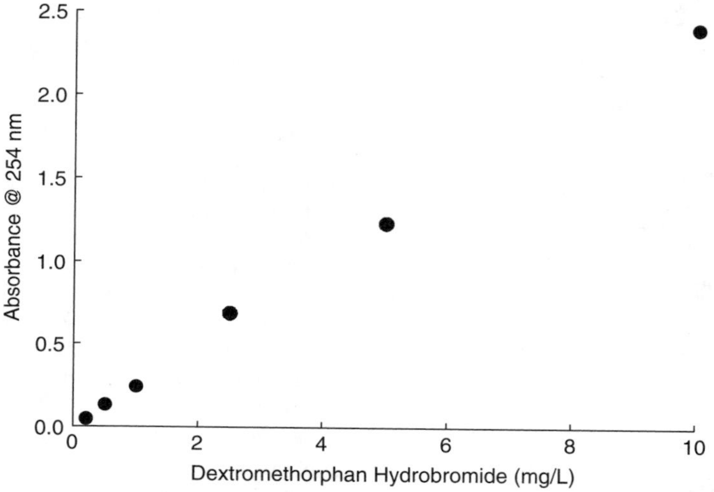

$$\hat{y} = a + bx$$

where y signifies the value of y on the regression for a given value of x. The intercept of this line, a, is given as

$$a = \frac{\left(\sum Y\right)\left(\sum X^2\right) - \left(\sum X\right)\left(\sum XY\right)}{n\left(\sum X^2\right) - \left(\sum X\right)^2}$$

where n is the points in the sample. Similarly, the slope is given as

$$b = \frac{n\left(\sum XY\right) - \left(\sum X\right)\left(\sum Y\right)}{n\left(\sum X^2\right) - \left(\sum X\right)^2}$$

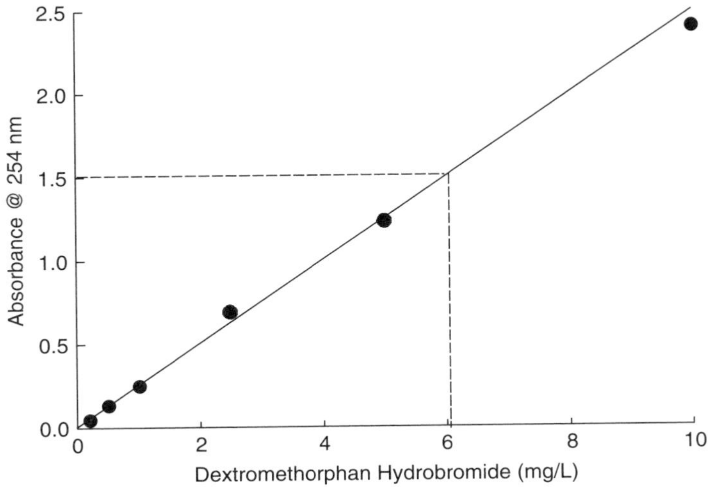

FIGURE 1.10 Replot of the data in Figure 1.9 together with the best-fit line (solid line) and extrapolation of the absorbance of an unknown sample to the standard concentration that would achieve that same absorbance value (dashed line).

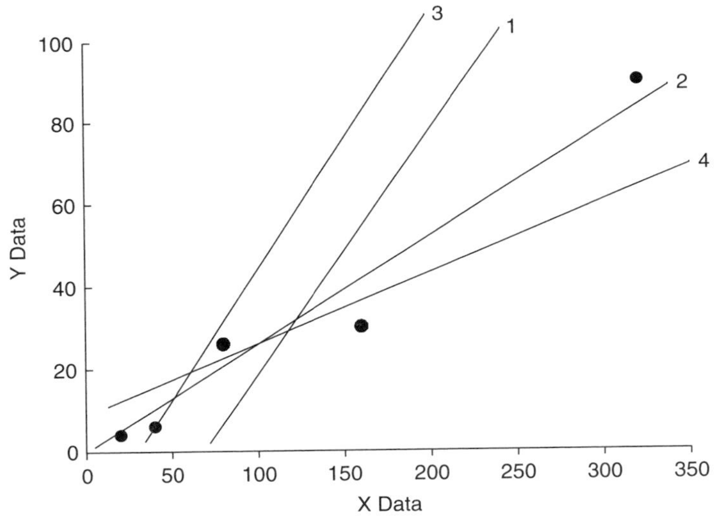

FIGURE 1.11 Hypothetical relationship between x and y data. Each line represents a possible best-fit line to express the relationship between x and y.

This procedure can be applied to the data in Table 1.9 to determine the regression line of absorbance on dextromethorphan hydrobromide content. A tabulation of the relevant parameters is given in Table 1.10.

These values can now be substituted into the equations for the slope and intercept of the regression line.

$$a = \frac{(4.743)(132.5) - (19.2)(32.18)}{6(132.5) - (19.2)^2} = \frac{628.448 - 617.86}{795 - 368.64} = \frac{10.588}{426.36} = 0.02 \text{ mg/L}$$

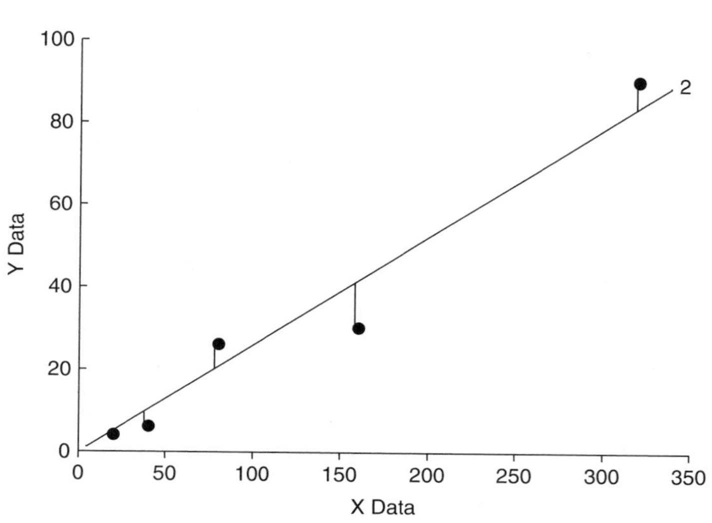

FIGURE 1.12 Deviations from the data and potential best-fit line. The lines from each data point to Line 2 represent the deviation for that location on the potential best-fit line.

TABLE 1.10
Parameter Values for Calculating the Regression Line for the Data in Table 1.9

Observed Concentration, X, in mg/l	Observed Absorbance, Y	X^2	XY
10	2.399	100	23.99
5	1.233	25	6.17
2.5	0.684	6.25	1.71
1.0	0.247	1.0	0.247
0.5	0.132	0.25	0.066
0.2	0.048	0.04	0.001
$\Sigma X = 19.2$	$\Sigma Y = 4.743$	$\Sigma X^2 = 132.5$	$\Sigma XY = 32.18$

Thus, the regression line is defined as

$$b = \frac{6(32.18) - (19.2)(4.743)}{6(132.5) - (19.2)^2} = \frac{193.08 - 91.066}{795 - 368.64} = \frac{102.014}{426.36} = 0.24 \text{ mg/l per absorbance unit}$$

$$= 0.02 \text{ mg/l} + (0.24 \text{ mg/l})x$$

Using this equation, if the absorbance of an unknown sample is determined, the concentration of that sample is readily calculated.

Linear regression analysis provides a useful means to determine how a dependent variable changes with respect to changes in an independent variable. As indicated, this can be extremely useful for predicting the value of a dependent variable based on a known value of the independent variable. There are numerous situations in which the goal is to determine the association of two variables, wherein neither can really be considered as the dependent or independent variable. In such circumstances, one is not necessarily seeking to identify a change in one variable as the *cause* of change in a second variable, but is merely seeking the strength of association. This can be quantified using the *correlation coefficient*, whose value lies between −1 and 1. This can be calculated as

$$r = \frac{\sum (X - \bar{X})(Y - \bar{Y})}{\sqrt{\sum (X - \bar{X})^2 \sum (Y - \bar{Y})^2}}$$

The value for r provides a measure of the strength of the association of the two variables. The value (+ or −) also indicates the type of relation between the variables. A positive value ($r > 0$) indicates that the two variables increase together, while a negative value ($r < 0$) indicates that one variable decreases while the other increases. The strongest association is observed when $r = \pm 1$. The further the correlation coefficient deviates from one, the weaker the association between the two variables. Numerous statistical tests are available to determine whether or not the strength of association between two variables is significant.

IX. COMPARISON OF TWO OR MORE GROUPS

One of the most important uses of statistics is its application in testing hypotheses. Such methods are commonly referred to as *tests of significance*. The objective of such tests is, in the face of the acknowledged variability of the measurements made, to determine if two or more groups are essentially identical. To place it in more practical terms, we use these tests to determine that a treatment group does not differ from a group that did not receive the treatment. This is referred to as the *hypothesis of no effect*, or the *null hypothesis*. It is important to recognize that all statistical tests of comparison of two or more groups are based on this hypothesis. If the result of the statistical test indicates that the null hypothesis should be rejected, then the alternative hypothesis should be accepted. This is stated in terms that the treatment did have an effect on the parameter being measured. For example, if two groups of patients with high blood pressure were studied, one of which received a new antihypertensive agent called BP-Lower and the other a placebo, the null hypothesis would be stated as "There is no difference in the blood pressure between subjects receiving BP-Lower and those receiving placebo." The alternative hypothesis would be stated as "Patients receiving BP-Lower have a different blood pressure than those receiving placebo."

In this example, the alternative hypothesis is stated in such a manner that the hypothesis does not anticipate the direction of change in blood pressure. This is, therefore, formulated as a two-sided alternative hypothesis. In some situations it is more appropriate to formulate a one-sided alternative hypothesis, such as "Patients receiving BP-Lower have a lower blood pressure than those receiving placebo."

Usually, a one-sided alternative is formulated when the effect can only occur in one direction.

A decision to accept or reject the null hypothesis cannot be made with absolute certainty. This is largely rooted in the fact that we are taking a sample of the population and evaluating results within that sample, not the population as a whole. Thus, we recognize in advance that such decisions are made with some risk of error. But how much error is acceptable? Worded slightly differently, what likelihood of being wrong are we willing to accept? This is referred to as the *alpha (α) error* (also called a type I error), which is the probability of erroneously stating that there is a difference between two or more groups when there is, in fact, no difference. Most commonly, this probability level is set as 5%.

The establishment of a level for the α error determines the magnitude of difference between the measured parameters that must be observed before the null hypothesis is rejected. Thus, setting the α error at 1% instead of 5% would require a larger difference between the groups in order to reject the null hypothesis. How does one determine if the magnitude of difference in the measured parameter reaches the level where it is appropriate to reject the null hypothesis within the predetermined level of α error? That is dependent upon the type of data analyzed, the number of groups studied, whether or not each group received each treatment, and several other considerations. It is beyond the scope of this chapter to present the basis and procedures for various statistical tests. This type of detail can be found in any one of a variety of biostatistics texts, many of which are at the introductory level (see Bibliography for example). What is essential to recognize is that each statistical test has been developed with certain underlying assumptions. Assessment of the validity of any statistical analysis must, therefore, begin with a determination of whether or not these assumptions are met by the data set being analyzed. For example, the *t-test* is the most commonly used statistical test used to compare experimental groups. This test assumes that data is interval in nature and drawn from a normally distributed population. There are numerous examples in the literature where this test has been applied to data sets that do not meet these criteria, resulting in suspect results.

Assuming the appropriate test has been utilized, tables can be consulted (often referred to as *critical value* tables) in any statistical text that will indicate if the difference between groups has reached the level where the null hypothesis is rejected. This is most commonly expressed in terms of a probability or p value. For example, if the α error is set as 5%, the critical value computed must be such that $p < 0.05$. When this level of difference is obtained, the groups are said to be statistically significantly different from one another. Again, it must be emphasized that such a conclusion does not express certainty, but rather a high probability that the observed difference between the groups is real.

While wrongly concluding that a difference exists between groups (i.e., a false positive) is the most common error considered in statistical analysis, it is not the only type of error with which one needs to be concerned. There is also the potential for wrongly concluding there is no difference between groups (i.e., a false negative). This is referred to as a *beta* (β) or *type II error*. This type of error most commonly arises when a small sample size has been studied. It is a common problem when small sample sizes are studied with parameter measurements that display substantial variability.

X. CONCLUSION

This chapter introduces the basic concepts of data representation, interpretation, and analysis with an introduction of basic statistical concepts. The field of pharmacy is becoming more and more complex and challenging every day. While pharmacists have to have a good understanding of the therapeutics and dispensing, a thorough understanding of the information presented in the label and patient package inserts is equally important for proper patient counseling and interaction with physicians. The authors hope that the concepts presented in this chapter along with the examples will help students comprehend the contents of the following chapters as well as other courses with ease.

TUTORIAL

1. Express the following sums or products to the appropriate number of significant figures.
   (a) 3.75 mg + 4.125 mg + 3.94 mg + 4.022 mg = ?
   (b) 5 tablets × 0.125 mg/tablet = ?
   (c) 16.1 ml + 15.0 ml + 22 ml = ?
   (d) 4 × 2.3 ml = ?

2. Write each of the following numbers in exponential form.
   (a) 0.125
   (b) 675
   (c) 10,200
   (d) 0.00000008
3. Write each of the following exponents in numerical form.
   (a) 5.2×10^3
   (b) 1.255×10^{-4}
   (c) 9.65×10^7
   (d) 3.2×10^{-1}
4. Find the logarithm of the following numbers.
   (a) 562
   (b) 3000
   (c) 0.0357
   (d) 4×10^5
5. Find the antilogarithm of the following numbers.
   (a) 1.1333
   (b) 2.1668
   (c) 0.0621

ANSWERS

1. (a) 15.84 mg, (b) 0.625 mg, (c) 53 ml, (d) 9.2 ml
2. (a) 1.25×10^{-1}, (b) 6.75×10^2, (c) 1.02×10^4, (d) 8×10^{-8}
3. (a) 5200, (b) 0.0001255, (c) 96,500,000, (d) 0.32
4. (a) 2.750, (b) 3.477, (c) –1.447, (d) 5.602
5. (a) 13.593, (b) 146.825, (c) 1.154

HOMEWORK

1. The content of a group of 10 capsules of glucosamine sulfate obtained from a local health food store was measured and found to contain 790 mg, 725 mg, 762 mg, 689 mg, 784 mg, 678 mg, 765 mg, 779 mg, 693 mg, and 700 mg of glucosamine sulfate. Calculate the mean, standard deviation, and coefficient of variation of this sample of glucosamine sulfate tablets.
2. The original quantity of a radioisotope was found to be 500 millicuries/ml. The amount of radioactivity found after 8 and 16 days was found to be 250 and 125 millicuries/ml,

respectively. Plot the radioactivity vs. time for this data and determine the rate constant of decay for this radioisotope.

3. A drug was administered as an intravenous bolus to a volunteer at a dose of 10 mg. Blood samples were taken at 3, 4, and 7 hours after administration. When drug concentration was measured in these samples, the results tabulated below were obtained.

Time (hours)	Concentration (mg/L)
3	30
4	20
7	6

From a plot of concentration vs. time, determine the initial concentration of drug immediately after the bolus dose was given (i.e., at time zero).

Answers to Homework

1.

Content (mg)	Deviation	$(\text{Deviation})^2$
790	53	2809
725	−12	144
762	25	625
689	−48	2304
784	47	2209
678	−59	3481
765	28	784
779	42	1764
693	−44	1936
700	−37	1369
Σ 7365		**17425**

$$\bar{X} = \frac{\sum X}{n} = \frac{7365 \text{ mg}}{10} = 736.5 \text{ mg}$$

Thus, appropriate number of significant figures 737 mg. The standard deviation is determined as

$$\text{SD} = \sqrt{\frac{\sum (X - \bar{X})^2}{n-1}} = \sqrt{\frac{17425}{9}} = \sqrt{1936} = 44 \text{ mg}$$

The coefficient of determination is determined as

$$\text{CV} = \frac{\text{SD}}{\bar{X}} = \frac{44 \text{ mg}}{737 \text{ mg}} = 0.06 \text{ or } 6\%$$

2. The first step is to determine whether the data should be plotted on a rectangular or semi-log plot. This will depend on whether the process being examined is zero-order or first-

order. If you do not recall which behavior radioactive decay displays, you can plot the data on both kinds of plots. Having done so, it will be observed that the data radioactivity declines linearly when plotted on a semi-log plot — indicating it is a first order process. The slope of the decline can be determined as:

$$\text{slope} = \frac{\log y_2 - \log y_1}{x_2 - x_1} = \frac{\log 250 - \log 500}{8 - 0} = -0.0376 \text{ day}^{-1}$$

A common mistake made when calculating the slope from a semi-log plot is to fail to take the log of each value in *y*. In addition, since this slope is derived from a $\log_{10}$ plot, to determine the actual rate constant of decay, the slope must be multiplied by 2.303 (recall that lnA = 2.303logA). Thus,

$$\text{Rate constant for decay} = 2.303 \times \text{slope} = 2.303(-0.0376 \text{ day}^{-1}) = -0.08665 \text{ day}^{-1}$$

3. 100 ng/ml
   This should be plotted on a semi-log scale, which will give a linear decline (i.e., first-order). The concentration is taken directly from the plot. Since the scale performs the log transform, no manipulation in this regard is needed.

CASES

In a mobile health clinic, blood systolic and diastolic blood pressures were checked in adult individuals between age 18 and 45. The results are tabulated below:

Patient #	**Age**	**Systolic BP (mm)**	**Diastolic BP (mm)**
1	22	115	78
2	34	124	82
3	43	122	84
4	24	128	86
5	36	131	85
6	39	134	88
7	31	129	89
8	40	198	98
9	19	118	81
10	45	222	153
11	44	132	92
12	38	121	80
13	30	131	84
14	20	128	82
15	44	91	61
16	33	128	87
17	35	134	86
18	18	112	79
19	21	121	84

Patient #	Age	Systolic BP (mm)	Diastolic BP (mm)
20	26	124	84
21	38	129	86
22	22	118	84
23	39	214	140
24	41	133	84
25	23	120	81

1. What is the *mean* age, systolic blood pressure (BP), and diastolic blood pressure of the population?
2. Calculate *standard deviation (SD)* of the age, systolic BP, and diastolic BP.
3. Calculate *geometric mean* (GM) of the age, systolic BP, and diastolic BP.
4. Calculate *median* of the age, systolic BP, and diastolic BP.
5. Calculate *mode* of the age, systolic BP, and diastolic BP.
6. Calculate percent coefficient of variation (%CV) of the age, systolic BP, and diastolic BP.
7. What other statistics may be relevant in representing the statistics of the data?
8. Is there any statistical *outlier*?

ANSWERS

	Age	Systolic BP (mm)	Diastolic BP (mm)
Mean	32.2	134.3	88.7
GM	30.9	131.6	87.3
SD	9.0	30.6	18.6
CV%	27.9	22.8	21.0
Median	34	128	84
Mode	22	128	84
Max	45	222	153
Min	18	91	61

The students are encouraged to calculate the figures using the formula given in the text. However, the above data was generated using Microsoft Excel spreadsheet.

(1–6) Please see the results above.

(7) Presenting max and min data gives the idea of range of the data set.

(8) There is NO simple rule to identify a statistical outlier in a given data set. However, the rule of thumb is that if data lies beyond ± 3SD from the arithmetic mean, it is considered an outlier. With that in mind, the maximum diastolic pressure (153 mm) is beyond ($88.7 + 3 \times 18.6 = 144.5$). The maximum systolic pressure (222 mm) barely missed the criteria.

REFERENCES

1. Bolton, S. Statistics, in Gennaro, A.R. (ed.), *Remington: The Science and Practice of Pharmacy,* Mack Publishing, Easton, PA, 1995.
2. Daniels, F. *Mathematical Preparation for Physical Chemistry,* McGraw-Hill, New York, 1956.
3. Glantz, S.A. *A Primer of Biostatistics*, 4th ed., McGraw-Hill, New York, 1997.
4. Zar, J.H. *Biostatistical Analysis*, Prentice Hall, Upper Saddle River, NJ, 1999.

2 Thermodynamics and States of Matter

Adam M. Persky, Laszlo Prokai, and Jeffrey A. Hughes

CONTENTS

0-4152-8863-0/05/$0.00+$1.50
© 2005 by CRC Press LLC

Learning Objectives

After finishing this chapter the student will have a through knowledge of:

- Forces that hold atoms and molecules together
- How these forces can potentially affect drug candidates and formulation
- Basics of radioactivity and use in pharmacy
- The utility of thermodynamics in dosage form design and stability

I. INTRODUCTION

It is important for the pharmacist to have a fundamental understanding of the states of matter in order to understand the various dosage forms. There are four states of matter — gas, liquid, solid, and plasma — and the pharmacist will encounter three of these (i.e., gas, liquid, solid) either through dispensing them or consulting on these pharmaceutical preparations. Examples of the three states of matter would include gas in metered dose inhalers or the gaseous state of inhaled anesthetics, liquids in syrups, and solid dosage forms of tablets and capsules. In this chapter we will review the forces that allow these states to exist, address the classes of bonds in molecules, and briefly review some aspects of thermodynamics.

Intermolecular forces are important not only pharmaceutically but also in our everyday life since they are the forces that cause some substances to be gas, some liquid, some solid, some soft like Vaseline® petrolatum, and some hard like wax. Why is O_2 a gas or H_2O a liquid or a sodium chloride a solid? The reason is differences in attractive forces between particles; particles of water have greater affinity for one another than particles of O_2 have for each other. These are some interesting questions we will address by learning about the forces that exist between particles.

II. STRUCTURE OF MATTER

The structure of matter and the bonds holding matter together are of paramount importance for pharmacists to understand since the drugs and dosage forms they will consult or dispense are composed of matter. No other member of the health care team has as detailed a chemical background as the pharmacist. The pharmacist is responsible for ensuring all the dosage forms are used to their fullest benefit. In this chapter we briefly review chemical background topics covered in prepharmacy curriculums and then continue into topics of more pharmaceutical importance. The reader is encouraged to review an introductory chemistry text for additional information if required.

A. Atoms

Atoms are the building components of molecules and of matter. Atoms have a very small, very dense, positively charged nucleus; the positive charge arises from **protons** surrounded by an equal number of negatively charged **electrons** resulting in a neutrally charged atom. Almost the entire weight of the atom arises from the nucleus.

Nomenclature describing subatomic elements includes the following terms: The **atomic number** of an element is given by the number of protons in the nucleus, which is the same as the number of surrounding electrons. The **atomic weight** of an atom is the sum of the neutrons and protons (e.g., for carbon-12, there are six protons and six neutrons). An **isotope** is two or more atoms with the same number of protons but different numbers of neutrons. For example, carbon has several isotopes such as carbon-12, carbon-13, and carbon-14 (Figure 2.1). In each case the number of protons is the same, but there are a different number of neutrons: carbon-12 has six

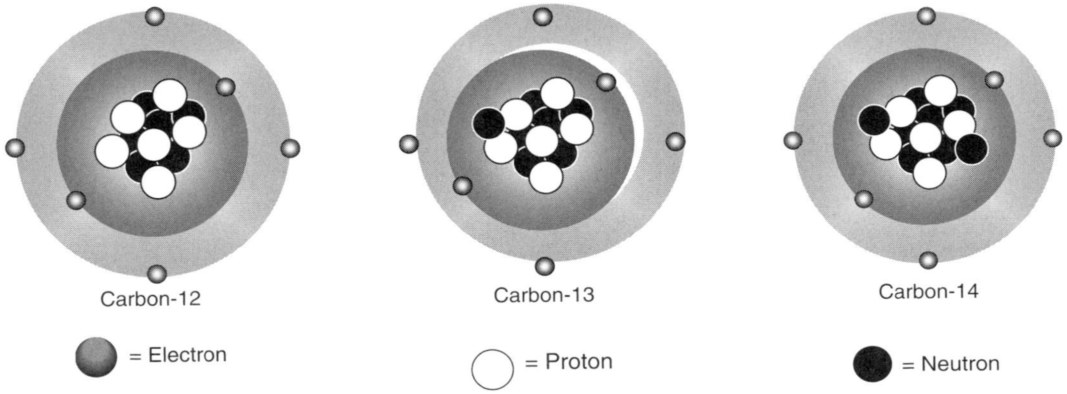

FIGURE 2.1 Isotopes of carbon.

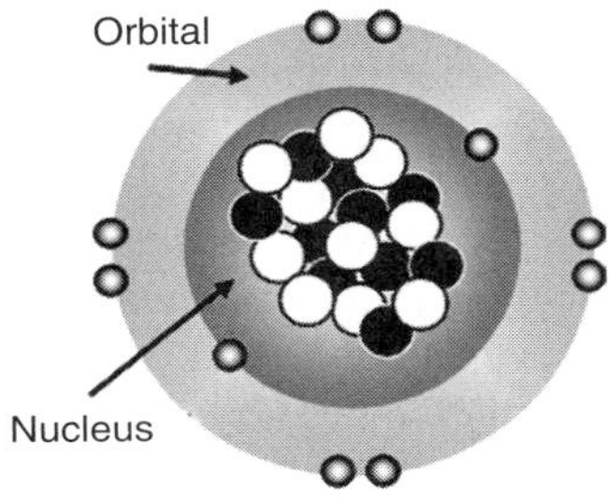

FIGURE 2.2 The rare gas neon and its filled outer shell of electrons.

neutrons while carbon-13 has seven. Isotopes of elements can be radioactive. Specific radioactive isotopes have nuclei that are unstable, leading to nuclear decay. Radioactive decay frequently produces **alpha** (α) particles, **beta** (β) particles, or **gamma** (γ) rays. Radioactive isotopes are used extensively in the practice of nuclear pharmacy and are discussed later in this chapter.

While the number of protons and neutrons governs many of the nuclear properties of atoms, it is the electrons that impact the chemical reactions. The atom's electrons are arranged in well-defined orbitals (electronegative shells, energy levels, electronic shells, bonding orbitals) around the nucleus, not just scattered randomly in space (Figure 2.2). The outermost orbital is the primary source of electrons for chemical bonding. The electrons are moving about 100,000 times faster than the speed of the nucleus (approximately at the speed of light). Perhaps one of the clearest examples of these orbitals is the rare, or noble, gases. The **rare gases**, neon, argon, krypton, xenon, and radon, each have eight electrons in their outer orbital. The unreactivity of the rare gases gives evidence that it is desirable to have a filled outer electron orbital.

B. Atomic Bonding

The electron arrangement of a full outer shell is considered to be ideal for maximum stability (e.g., the octet rule), as in the case of the extreme stability of noble gases. An atom that does not have eight electrons in the outermost orbital can attain this state by one of two methods. The first is theft of electrons from elements with lower electron affinity (i.e., ionic bonding) or by sharing electrons (i.e., covalent bonding). An element's affinity for electrons is often described as **electronegativity**. The periodic table gives a framework for predicting electron affinities; as one moves upward in a family of elements or to the right within a row of the periodic chart, the atom's affinity for electrons increases. This trend does not hold true for the rare gases since the octet rule is

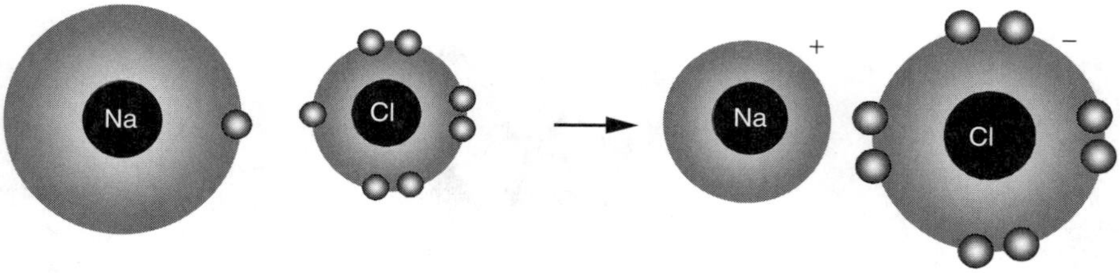

FIGURE 2.3 Example of the sodium chloride ionic bond.

satisfied. Thus, fluorine is the most electronegative (i.e., has the greatest electron affinity), with chlorine and oxygen being tied for the second most electronegative. The difference in electronegativities of the elements is one of the factors that determine if a theft of an electron can occur.

C. Ionic Bonding

An example of ionic bonding is the complete transfer of an electron from sodium (a low electronegative element) to chlorine (a high electron negative element) (Figure 2.3). The electron from sodium is no longer associated with the sodium nucleus. It requires considerably less energy to transfer only one electron to chlorine (Cl^-) than to transfer seven chlorine electrons in the other direction to sodium. The nature of the bond is two independent, oppositely charged spheres. When you have complete electron transfer you have an **ionic bond**, which is characteristic of salts. Note that the positive ion formed is smaller than the neutral counterpart. The shrinking of the positive ion is a result of loss of an electron from the outermost orbital while the remaining electrons are pulled in by the attractive forces of the positively charged nucleus. The negative ion of chlorine is much larger because of the addition of an extra electron in the chlorine orbital, and the increased repulsion between the electrons expands the ion. The net result of this exchange of electrons is the formation of sodium chloride that is held together by electrostatic interactions.

D. Covalent Bonding

When unequal sharing or theft on an electron is caused by large differences in electronegativity, an ionic or electrostatic bond develops. If the difference between the electronegativities of the two reacting elements is not large enough, no electron theft occurs. Rather than theft, the atoms involved will share electrons, resulting in the formation of a **covalent bond** (Figure 2.4).

A clear division between ionic and covalent bonds is not always clear. For example, as in the compound hydrogen chloride (HCl), which has a covalent bond, the electron is not equally shared between hydrogen and chlorine atoms because the chlorine atom, with a nucleus charge of +7, has a greater affinity for the electrons than the H atom, which has a nucleus charge of only +1. The result is that the H atom's electron now spends most of its time around the chlorine atom (Figure 2.5).

This pulling of electrons is also known as the **inductive effect**. Water molecules behave similarly (Figure 2.6).

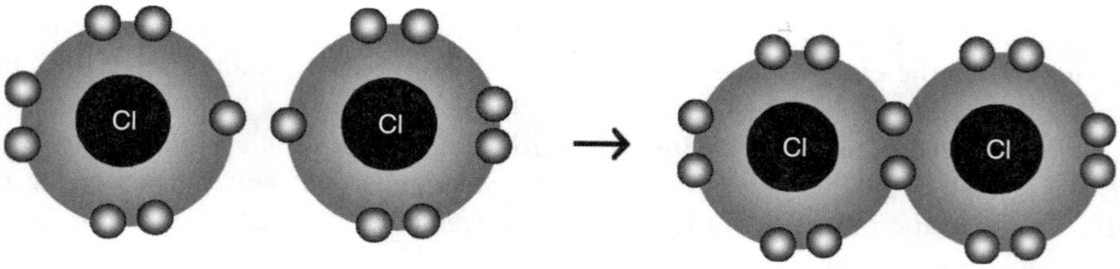

FIGURE 2.4 Electron sharing of chlorine.

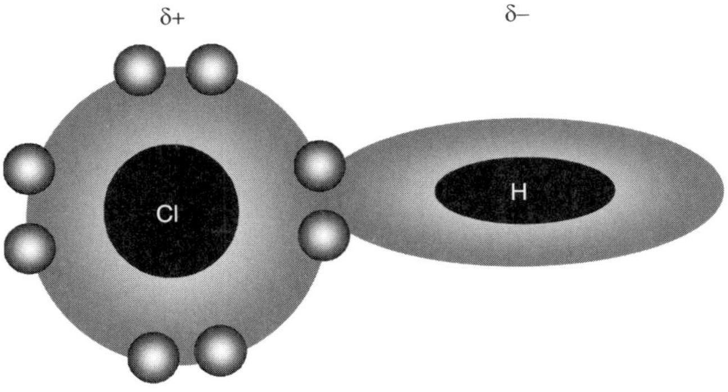

FIGURE 2.5 Electron sharing of HCl.

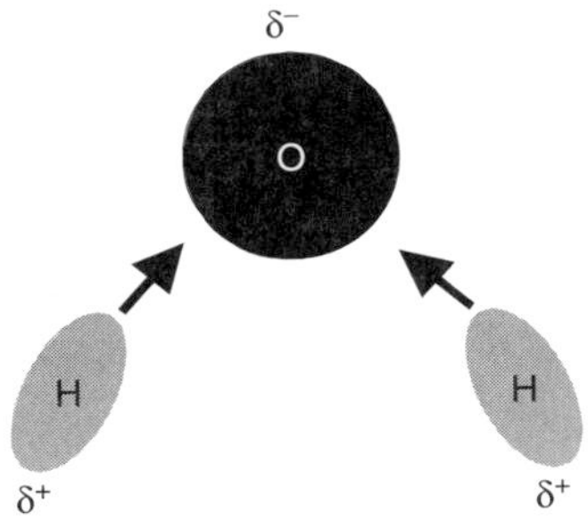

FIGURE 2.6 Polarity induction in water by electronegative oxygen.

E. Dipole Moments

Let us examine in greater detail this unequal sharing of electrons. Some highly electronegative elements are chlorine, bromine, iodine, oxygen, sulfur, fluorine, and nitrogen. Covalent bonds with these atoms often have charge separation within the bond. This separation is known as a **dipole**. Dipolar molecules can align themselves with respect to their dipole and provide rather strong **intermolecular** bonds (i.e., bonds between two molecules). An electric dipole consists of a positive ($+e$) and negative ($-e$) charge, separated by a distance, r. The dipole moment, μ, with the units of debye (or 10^{-18} electrostatic units • cm) is defined by the equation:

$$\vec{\mu} = e \cdot \vec{r}$$

The direction of the interaction is, by definition, from the negative charge toward the positive charge. If a molecule possesses separate regions of positive and negative charge, its dipole moment is permanent (e.g., permanent dipole moment), as seen in the molecules of water and HCl. Not all molecules have a permanent dipole moment. For example, when there is equal sharing of electrons, as in homonuclear diatomic molecules such as He_2 or N_2, there is no possibility for permanent asymmetry of charge distribution, so there is no permanent dipole moment. Even when a permanent charge separation between atoms exists, geometric consideration must be addressed. For example, Figure 2.7 shows three dichlorinated compounds. The position of electronegative chlorine on the benzene ring determines the resultant dipole vector (double arrow).

Even though a symmetric, linear triatomic molecule such as carbon dioxide has charge separation, the geometry of the molecule does not allow for a dipole moment (Figure 2.8).

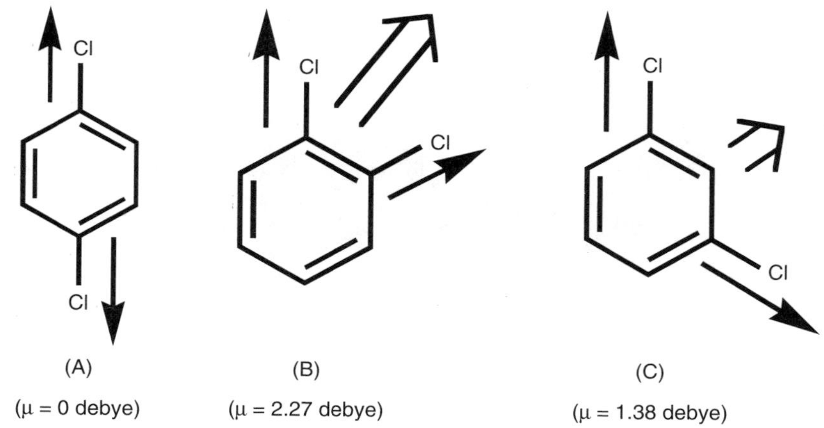

FIGURE 2.7 Dipole induced by position of an electronegative atom. Compound A has no net dipole moment since the two chlorines are directly opposite each other. Compound B has the greatest moment, and Compound C has an intermediate dipole moment.

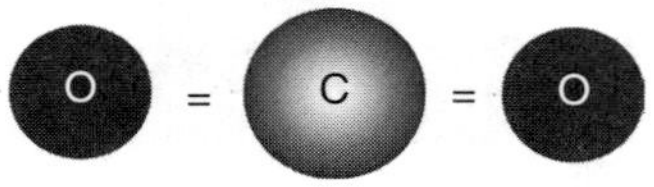

FIGURE 2.8 Carbon dioxide.

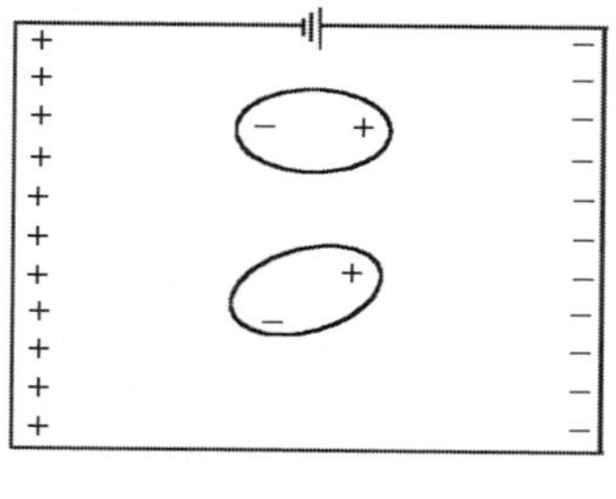

FIGURE 2.9 Alignment in an electric field.

A molecule without a permanent dipole (e.g., Cl_2) still has electrons in constant motion with respect to the positively charged nucleus. Therefore there is not always a precise center of the sphere of negative charge. Since the electrons are in constant motion around the nuclei, the average dipole moment of a molecule is zero when averaged over a long period of time. Thus, symmetric molecules only have instantaneous dipole moments. It can be reasoned that two molecules with instantaneous dipole moments will attract if they are antiparallel and will repel if they are parallel. The closer the molecules are, the greater the chance of the interaction.

There are practical issues for understanding dipole moments of molecules since this is one of the factors that determine how they will interact with other molecules. In some ways this is similar to the above discussion of how two atoms can interact. This effect also explains how molecules may align in an externally applied electric field (Figure 2.9). If a molecule containing a dipole is placed near an external positive charge, the molecule will orient the dipole so that its more negative portion is nearer to the external charge than its more positive region. This is called **orientation polarization**. Furthermore, the positive region of the molecule is repelled by the external charge so that there is an increase in the separation of the positive and negative regions of the molecule. If, however, the molecule is unpolarized (i.e., does not have a permanent dipole), it may become polarized in the electric field produced by the external charge because the positive and negative charges of the molecule will be distorted.

This type of polarization is known as **induced** or **distortion polarization**. The magnitude of the induced polarization of a molecule is given by a quantity called the distortion polarizabilty (α with units of 10^{-24} cm^3) defined by the relation:

$$\vec{\mu} = \alpha \cdot \vec{E}$$

in which μ is the dipole moment induced by the applied electric field, **E**. The distortion polarizability has the dimension of a volume, and its numerical value is roughly that of the average volume of the molecule, so large molecules are generally more polarizable than small molecules. Also, compounds that have unsaturated bonds (i.e., containing double or triple bonds) are more polarizable.

III. MOLECULAR BONDING FORCES

Atoms can interact to complete their outer shell by stealing an electron or by sharing an electron. These interactions of atoms can lead to dipole moments of the molecule. Now we will discuss the forces involved when molecules interact with each other. Table 2.1 shows the list energies involved in different types of bonds.

A. London Dispersion Forces

The weakest and yet most numerous interactive force is known as **London, van der Waal,** or dispersion force. For compounds that do not posses a permanent dipole moment, the question concerns how these molecules interact. The average electrostatic field of a nonpolar molecule is zero, but the molecule will have an instantaneous field, as explained above. The instantaneous field is easiest to envision in the hydrogen atom. A hydrogen atom consists of a single proton and one electron; at any instant, the electron and the positively charged proton constitute a dipole (Figure 2.10). This dipole continually changes direction as the electron whirls around the nucleus, and averaged over time, or at least over the period of one revolution, the net dipole moment is zero.

TABLE 2.1
Table of Intermolecular Forces and Their Bond Energies

Bond Type (synonym)	Bond Energy (kcal/mol)
Covalent	50–150 (210–630 kJ/mol)
Dipole-dipole (keesom force)	1–10 (4–42 kJ/mol)
Hydrogen bond	
OH••O	6 (25 kJ/mol)
CH••O	2–3 (8–13 kJ/mol)
OH••N	4–7 (17–30 kJ/mol)
NH••O	2–3 (8–13 kJ/mol)
FH••F	7 (30 kJ/mol)
Induced dipole (debye force)	1–10 (4–42 kJ/mol)
Ionic	100–200 (413–837 kJ/mol)
Van der Waals force (London, dispersion)	1–10 (4–42 kJ/mol)

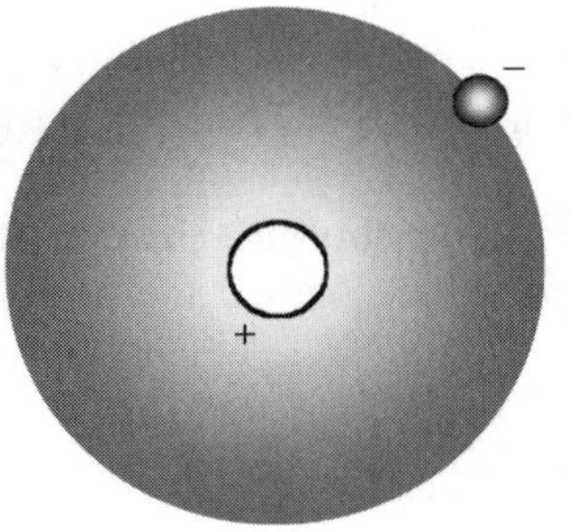

FIGURE 2.10 Polarity of a hydrogen molecule.

However, if two hydrogen atoms are present, at any instant one atom will sense the electric field of the other and cause a transient orientation of molecules. Thus, the more atoms in a compound, the greater this interaction. Also, the closer the atoms are, the greater the interaction. This is one reason high pressure allows some gases to liquefy.

1. Intermolecular Forces and Physicochemical Properties

Before discussing other intermolecular forces, lets look at some examples of how molecular force impacts the physiochemical status of the compound. The magnitude of the attraction between molecules of any substance determines whether that substance is a solid (strong attraction), liquid (medium attraction), or gas (weak attraction). For example, in a homologous series of molecules, the boiling point should increase with increasing molecular weight (Table 2.2). This is evident with simple alkanes (methane, ethane, propane, etc.), whose boiling point increases with carbon number. Boiling point (T_b) can be used to approximate intermolecular forces, and later we will see that the melting point (T_m) of a substance can also be used.

This section systematically examines the contributions of the various terms to interaction energy. We can do this by addressing T_b. T_b is an informative parameter because it indicates the strength of the cohesive forces (forces between the same molecules, in contrast to adhesive forces, in which the interactions are between different types molecules) between the molecules of a liquid. Hence the boiling point should increase as the interaction becomes stronger (e.g., as μ, α, or the van der Waals forces increase) as shown in Table 2.3. If a series of molecules contains a permanent dipole moment and the number of electrons is nearly the same, the dispersion energy should be the major factor influencing a measurable parameter such as T_b.

Comparisons of boiling points between isobutylene with dimethyl ether and ethylene oxide with trimethylamine demonstrate that London forces are a more important factor than the effect of permanent dipole on boiling point. That is, T_b for the first of each pair is higher than T_b for the second, because in each case the first has more electrons and hence a greater polarizability than the second, even though in each case the second has a greater dipole moment (Table 2.4).

TABLE 2.2
Carbon Number and Boiling Point

Number of Carbons	Alkane	Molecular Weight	T_b (K)
1	Methane	16.04	111.3
2	Ethane	30.07	184.4
3	Propane	44.06	230.8
6	Hexane	86.18	341.7
16	Hexadecane	226.44	553

TABLE 2.3
Influence of London Forces on Boiling Point

Compound	Molecular Weight	α (cm^3)	μ (Debye)	T_b (K)
Isobutane	58.12	8.36	0	263
Isobutylene	56.11	8.36	0.49	267
Trimethylamine	59.11	8.08	0.67	278

TABLE 2.4
Relationship between Structure, Boiling Point, and Adhesive Forces

	Propane	Dimethyl Ether	Ethylene Oxide
Structure	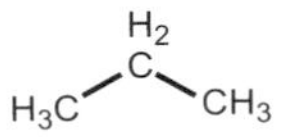	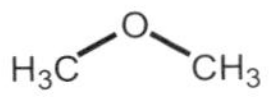	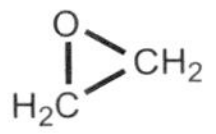
Molecular weight	44.1	46.07	44.05
$\alpha \times 10^{-24}$ (cm^3)	36.4	6.0	5.2
μ (Debye)	0	1.30	1.90
T_b (K)	231	248	284

B. Hydrogen Bonding

van der Waals forces are the most common forces responsible for intermolecular attraction. However, in molecules of biological importance another interaction, **hydrogen bonding** — also a weak interaction compared to covalent bond energy — plays a significant role in determining both intra- and intermolecular interactions and, in particular, the structure of macromolecules. Molecules containing hydroxyl and amino groups have boiling points much higher than expected simply from the values of their dipole moments.

The existence of a T_b higher than expected from the values of dipole implies that more energy must be put into the compound to maintain the gaseous state. This indicates that there must be an additional interaction between the molecules other than the van der Waals–London interaction. The attractive force we seek is that of the hydrogen bond. A proton interacting simultaneously with two atoms, which have strong negative charges, will form a hydrogen bond. Hydrogen bonds are typically symbolized by a dotted line (Figure 2.11).

Hydrogen bonds include electronegative atoms such as O and N as shown in Figure 2.11 and have a preferred direction similar to all permanent dipoles. That is, the interaction is strongest when the two electronegative atoms and the hydrogen atoms form a straight line. Other electronegative atoms such S, Cl, and F can also form hydrogen bonds. Experimentally it is found that the energy

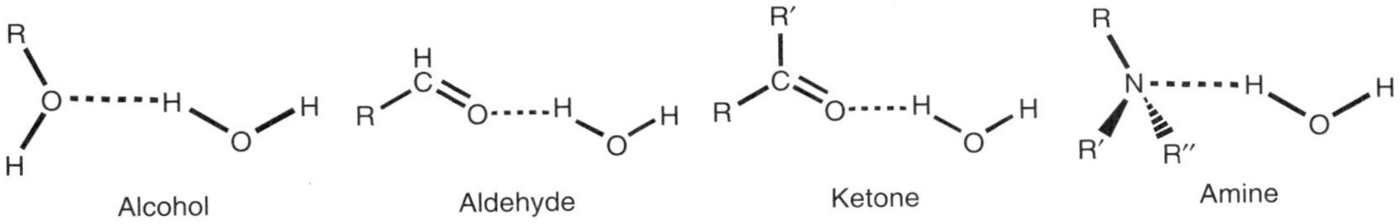

FIGURE 2.11 Types of hydrogen bonding.

of a hydrogen bond is in the range of 10 to 30 kJ per mol, which means that a hydrogen bond is stronger than a van der Waals attractive force and weaker than covalent bonds.

C. The Hydrophobic Interaction

The hydrophobic interaction is often considered a bond between molecules, but in actuality it is a lack of an attractive interaction. This interaction is actually not a bond at all. A **hydrophobic** (water-fearing) interaction refers to a clustering of nonpolar molecules in water. A benzene molecule cannot form hydrogen bonds with water molecules, but at the instant it is added to water, the ordered hydrogen-bonded water lattice is disrupted. As the water molecules attempt to reform hydrogen bonds, an ordered structure around the benzene molecule results. Interestingly, the degree of order of the water molecules around the benzene molecule is greater than that before the benzene molecule is added. As additional benzene molecules are added to the water, each will disrupt the water lattice and then cause excess ordering. In as much as the creation of order is thermodynamically unfavorable, an arrangement of two benzene molecules arranged to minimize the increases in order would be preferred. Since ordering happens at the surface of the benzene-water interface and since clustering of the two benzenes rescues the surface to volume ratio, clustering will occur. As the number of molecules increases, the cluster will simply form a second phase. This accounts for the observed phenomena that benzenes and water do not mix and oil forms on the surface of the water ($\Delta G_{AIR\text{-}BENZENE} < \Delta G_{BENZENE\text{-}WATER}$).

D. Dipole-Dipole Interaction between Molecules

Compounds with a dipole produce strong interactions. Water and alcohol are good examples. A strong dipole leads to high intermolecular forces. If you only consider molecular weight, oxygen (O_2; mol wt = 32) is a heavier molecule than water (H_2O; mol wt = 18), and you might expect oxygen to be more difficult to vaporize, but oxygen is a gas at room temperature, while water has a high boiling point. The molecules of water have very high attraction for one another, which arises from dipole interactions.

Dipole-dipole forces are only effective over very short distances. Hydrogen bonds are the strongest of the dipole-dipole interactions because the hydrogen atom is so small that the dipoles can approach very closely. Acetone has a large dipole moment (μ = 2.8), but the larger methyl groups prevent close approach of the dipoles, and therefore acetone has a much lower T_b than water (Figure 2.12).

Symmetrical molecules such as CCl_4 or benzene (Figure 2.13) do not have a dipole and thus cannot associate by means of dipole-dipole interactions.

In CCl_4, the interaction energy is due entirely to this type of dispersion force. Benzene, paraffin, and mineral oil are other compounds whose intermolecular forces can be attributed to these interactions. Since compounds with higher molecular weight have more electrons, they show greater dipole-induced forces than lower molecular weight compounds. Table 2.5 lists dipole moments for various drugs that can impact how the drug is formulated.

FIGURE 2.12 Dipole moments of acetone and water.

FIGURE 2.13 Carbon tetrachloride (left) and benzene (right).

TABLE 2.5
Dipole Moments of Selected Drugs

Drug	Dipole Moment (Debye)
Barbital	1.10
Phenobarbital	1.16
Ethanol	1.70
Water	1.84
Cholesterol	1.99
Methyltestosterone	2.07
Acetylsalicyclic Acid	2.07
Testosterone	4.17
Sulfanilamide	5.37
Phenacetin	5.67

Example 1

Molecules can have a variety of attractive and repulsion forces. Below is table of compounds that demonstrate the fact that one molecule can have multiple bonding forces.

	Dipole or H-bond	Van der Waal or hydrophobic	Ionic
Dopamine	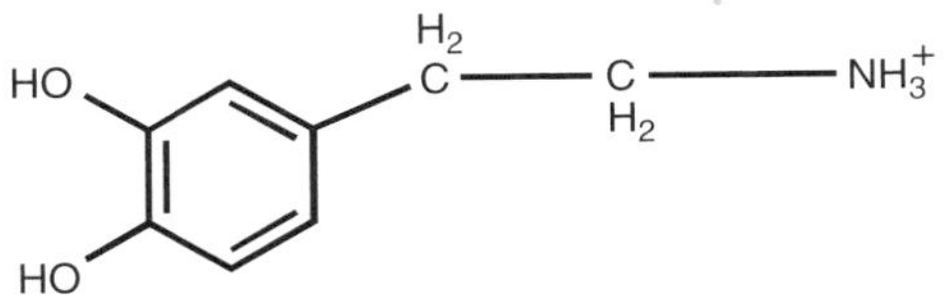		
γ-Aminobutyric Acid (GABA)	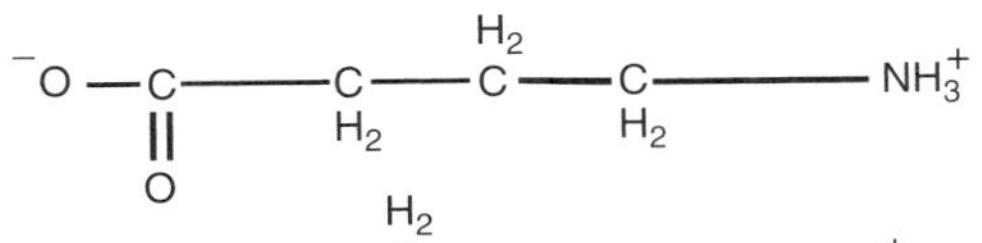		
Histamine	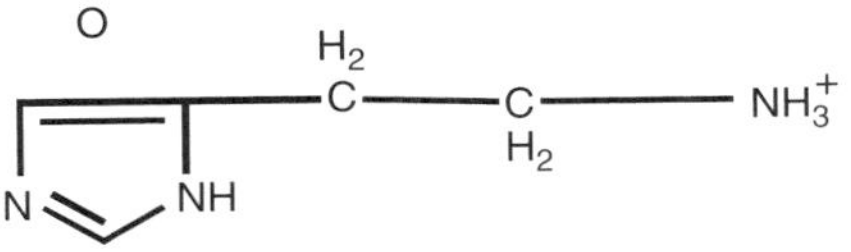		
Fluoxentine	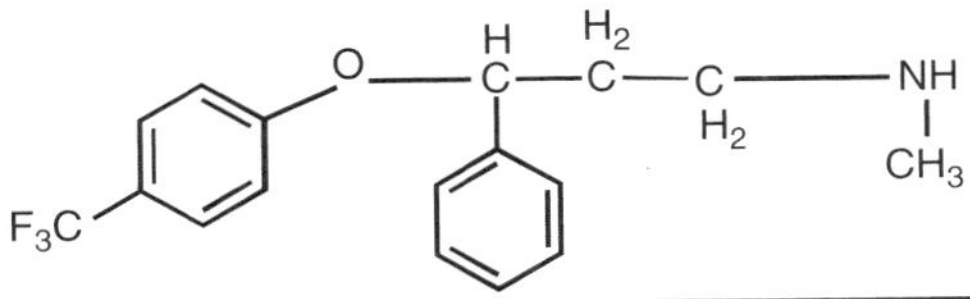		

IV. RADIOACTIVE DECAY

We have discussed why atoms like noble gases are stable and unreactive and why atoms or molecules react to enhance their stability by filling their outer orbital. This discussion has almost exclusively dealt with electrons. However, the nucleus of atoms can also cause instability in an atom and can lead to decay or breakdown until the nucleus reaches a stable state. This breakdown is often a result of the number of protons and neutrons within the nucleus. Radioactive decay is usually associated with the nucleus of the atom. The stability of the nucleus is determined by the neutron-to-proton (N/Z) ratio. For example, the isotopes of hydrogen either have no neutrons (normal hydrogen, called protium: ${}^{1}_{1}H$, one proton, one electron), one neutron (deuterium: ${}^{2}_{1}H$, one proton, one neutron, one electron), or two neutrons (tritium: ${}^{3}_{1}H$, one proton, two neutrons, one electron). Although isotopes of an element have almost the same chemical properties, the nuclear properties can be quite different, and therefore deuterium and tritium undergo decay to reach a more stable state. In general, atoms can decay by six different methods to reach a stable N/Z ratio. These six methods include spontaneous fission, α decay, β^- decay (negatron), β^+ decay (positron), electron capture, and isomeric transition. Regardless of the method of decay, in the overall process, energy, mass, and charge of the atom is conserved.

A. Spontaneous Fission

Spontaneous fission (splitting) is a process where a heavy nucleus degrades into two fragments and is followed by the emission of two to three neutrons. The probability of this process occurring is low but increases with the mass of the nuclei (i.e., heavier nuclei have a greater chance of spontaneous fission). Examples of atoms that undergo spontaneous fission are uranium (${}^{235}U$) and Californium (${}^{254}Cf$).

B. Alpha (α) Decay

An α particle is a helium ion containing two protons and two neutrons or nucleus of a He atom. Atoms that decay by α particle emission (e.g., radon, uranium) have their atomic number reduced by two and mass by four during this process, as in the case of uranium: alpha particles have high energy when emitted from the nucleus and can damage tissue. In some instances atoms that decay through alpha decay can be used to destroy unwanted tissue, as in cancer treatments.

$${}^{235}_{92}\mathrm{U} \rightarrow {}^{231}_{90}\mathrm{Th} + {}^{4}_{2}\mathrm{He}^{2+}$$

C. Beta (β^-) Decay

Neutron rich nuclei (e.g., have high N/Z ratios and decay by β^- particle and an antineutrino ($\bar{v}$) emission. In this process a neutron degrades into a proton and a β^- particle. For example,

$${}^{99}_{42}\mathrm{Mo} \rightarrow {}^{99m}_{43}\mathrm{Tc} + \beta^- + \bar{v}$$

A high-energy β^- particle can interact with high atomic number elements displacing orbital electrons leading to the production of x-rays, e.g., ${}^{32}P$, Pb.

D. Positron (β^+) Decay

Proton rich nuclei (e.g., ${}^{18}_{9}F$, ${}^{15}_{8}O$) have N/Z ratios that are lower than the stable nucleus and can decay by β^+ particle emission along with a neutrino (v). The daughter atom has an atomic number one less than the parent. Typically β^+ particles collide with electrons and are thus annihilated, producing two photons of gamma radiation, or gamma rays:

$$P \rightarrow \eta + \beta^+ + v$$

For example,

$$^{13}_{7}N \rightarrow ^{13}_{6}C + \beta^+ + v$$

Positron emitting isotopes are often used in nuclear medicine imaging studies to monitor function of particular organs or tissues.

E. Electron Capture

An alternative to β^+ decay is electron capture. An electron is captured from the electron shell, usually one close to the nucleus (e.g., K-shell), resulting in the transformation of a proton into a neutron and emitting a neutrino. The empty space left by the electron is filled by a higher level shell, which results in an x-ray. An example of electron decay is

$$^{67}_{31}Ga + e^- \rightarrow ^{67}_{30}Zn + v$$

F. Isomeric Transition

An isomeric state occurs when the nucleus of an atom is an excited state above the ground state. This excited state can last from picoseconds to years. Long-lived isomeric states are called metastable states (designated by "m") and are easily detected. For example, ^{99m}Tc is a metastable state of Tc and is used frequently in radiopharmacy. When a nucleus falls to ground state, the energy may be transmitted in the form of γ-rays or transferred to one of the atomic electrons, causing it to be emitted from the nucleus (internal conversion).

G. Application to Pharmacy

For most pharmaceutical applications of radioisotopes the isotope is attached to a drug through complexation or a covalent bond forming a radiopharmaceutical. In this composite molecule, the drug part dictates where the molecule goes in the body, and the radioisotope is used most often for monitoring the disposition of the radiopharmaceutical. The majority of radiopharmaceuticals utilize ^{99m}Tc as a diagnostic isotope. The decay of ^{99m}Tc produces an easily detectable gamma photon that produces little or no damage to the human body.

V. THERMODYNAMICS

A. Enthalpy, Entropy, and Free Energy

It is important to understand the energetics that determine whether a bond is energetically favorable; that is, is this bond worth forming? Important descriptors such as enthalpy, entropy, and free energy used for describing pharmaceutical systems are derived from thermodynamics. Thermodynamics (Greek *therme dunamikos*: movement, power of heat) deals with the quantitative relationships between heat and other forms (electrical, mechanical, and chemical) of energy. It is not possible to measure the absolute energy of a system, but it is possible to record changes in energy upon making a change in the system.

Thermodynamics is useful in allowing predictions to be made about the physical or chemical properties of a drug and other materials of pharmaceutical interest. These include:

- Whether it is possible that a particular physical or chemical process will occur, for example, whether ice will melt at a particular temperature, whether a drug will dissolve in a certain solution, the possibility that a compound will diffuse across a certain membrane, etc.
- To what extent a process will take place, for example, how much of a substance will dissolve in water, to what extent will a chemical reaction proceed, etc.
- Will heat or other forms of energy be absorbed or released during a particular process, and how much?

It is important to note that thermodynamics **cannot** make predictions about the rate or speed of a process, only whether it is possible for the process to occur.

Many pharmaceutically important processes are described thermodynamically, including:

- Solubility of a drug in a dosage form or in biological fluids
- Extent of ionization of an acidic or basic drug at various pH values
- Stability of a drug in various formulations
- Partitioning of a drug or excipient between immiscible phases
- Diffusion of a drug across biological membranes
- Binding of drugs to plasma proteins and to tissue components
- Interaction of drugs with enzymes
- The effect of temperature or pressure on a pharmaceutical process

B. Definitions in Thermodynamics

A **system** is the part of the universe being examined and may refer to a beaker, a cylinder of gas, a drug solution in an ampoule, or an experimental animal. **Surroundings** are the entire universe excluding the system. (A *boundary* separates the system from the surroundings). A system can be open (matter or energy can be exchanged with the surrounding), closed (energy but not matter can be exchanged with the surrounding), or isolated (neither energy nor matter can be exchanged with the surrounding).

Energy (E) is the condition of a system that gives it the capacity to do work. The classification of energy includes potential and kinetic.

Heat (Q) is a form of energy that is transferred from one system to another when a temperature difference is present. Heat and temperature are not the same.

Temperature (T) is a quality or an intensive property, meaning it is dependent only on the mean kinetic energy of the molecules in a substance, not on how many there are. (The temperature of a gram or a kilogram of a substance may be the same, but the amount of transferable heat they contain is different.)

Work (W) is mechanical energy transferred to the surroundings from the system or vice versa. In physical terms, work is given by the equation:

$$W = \text{force} \times \text{distance}$$

$$W = \text{mass} \times \text{acceleration} \times \text{distance}$$

$$W = \text{mass} \times \text{distance/time}^2 \times \text{distance}$$

$$W = \text{mass} \times \text{distance}^2 \times \text{time}^{-2}$$

So, dimensionally, work, $W = m \times l^2 \times t^{-2}$

where m = mass, d = distance, t = time

where m is mass of the object, l is the length it travels, and t is the time it takes to travel. The fundamental units of work and heat are the same, indicating that heat and work are interchangeable; by convention they have different units:

W (energy) is given in joules or ergs [10^7 erg (dyne·cm) = 1 joule (J) = 1 kg × m^2 × s^{-2}
Q (energy) is given in calories (1 cal = 4.184 J).

Internal energy (E) is the energy stored within the molecules of a system.
The state of the system is defined by various descriptors classified as:

- **extensive properties:** proportional to the number of molecules in the system
- **intensive properties:** not proportional to the number of molecules in the system.

Example 2

Extensive properties: mass (m), volume (V), heat content or enthalpy (H), Gibbs free energy (G), entropy (S), internal energy (E), heat capacity (C)

Intensive properties: temperature (T), refractive index, density (d), surface tension (γ).

Division of any external property of a substance by the number of molecules or moles of that material generates an intensive property. In general, changes of these properties (indicated by the greek *delta* [Δ] before the symbol of the property) are of interest.

Energy terms such as E, H, G, and S are termed **state functions**, because the energy change between an initial and final step is simply the difference of the energy values. This means the path from the initial to the final step is not important. By contrast, Q and W are not state functions and their quantity does depend on the path taken or the number of steps involved.

At equilibrium, certain systems have a simple equation that provides a relationship among the values of the properties. Such a relationship is referred to as an *equation of state*. For example, a system containing an ideal gas has the properties of pressure (P), volume (V), number of moles (n), and temperature (T, in Kelvin; 0°C = 273 K) related by the ideal gas law:

$$PV = nRT$$

where R is the gas law constant (1.98717 cal × K^{-1} × mol^{-1}, 8.3143 J × K^{-1} mol^{-1}, or 0.082 L × atm × K^{-1} × mol^{-1}, to be selected based on the other units in the equation).

Thermodynamics rests on specific conventions and definitions concerning various forms of energy and physical properties. In a strict sense, thermodynamics is the quantitative treatment of the equilibrium state of the various forms of energy and is based on three experimentally observable laws from which important descriptors (enthalpy, entropy, and free energy) applied in pharmaceutics can be derived. The foundation of much of what is known about thermodynamics is derived from gas expansion or compression studies. Apart from the theoretical significance of this area, such information is of considerable importance in the formulation and production of pharmaceutical aerosol products and gaseous anesthetics.

C. The First Law of Thermodynamics: Derivation of Enthalpy

Energy (mechanical, thermal, chemical, etc.) can be transformed from one form into another but cannot be lost, destroyed, or created. For the expansion or compression of ideal gases, where changes in internal energy (ΔE) can be expressed accurately from heat (Q) and work (W), the first law of thermodynamics is expressed as

$$\Delta E = Q - W$$

A useful *state function* can be introduced by considering the first law of thermodynamics: **enthalpy** or heat content ($H = E + PV$). It can be used to examine changes in heat in chemical reactions and certain physical processes. This subject area is referred to as **thermochemistry**.

In thermochemistry, heat of formation (ΔH_f°) is defined as the change of heat content (ΔH) in a reaction when 1.0 mol of a compound is formed from its elements in their standard states. Standard states are defined at 25°C and 1 atm. Elements at their standard states, by convention, have a ΔH_f° of zero. Since heat of formation values can be related to internal energy, they are useful in comparing stabilities of molecules. Such comparisons are limited to molecules with the same number and kind of atoms, that is, to isomers.

EXAMPLE 3

The *trans*-isomer of tamoxifen ($C_{26}H_{29}NO$) is marketed under the trade name Nolvadex as an antiestrogen. This agent is used against metastatic breast cancer and for adjuvant treatment of breast cancer. Its ΔH_f° is 49.94 kcal/mol. The *cis*-isomer of tamoxifen, which has lower biological activity, has a ΔH_f° of 49.83 kcal/mol. However, the *cis*-isomer is thermodynamically more stable than the *trans*-isomer.

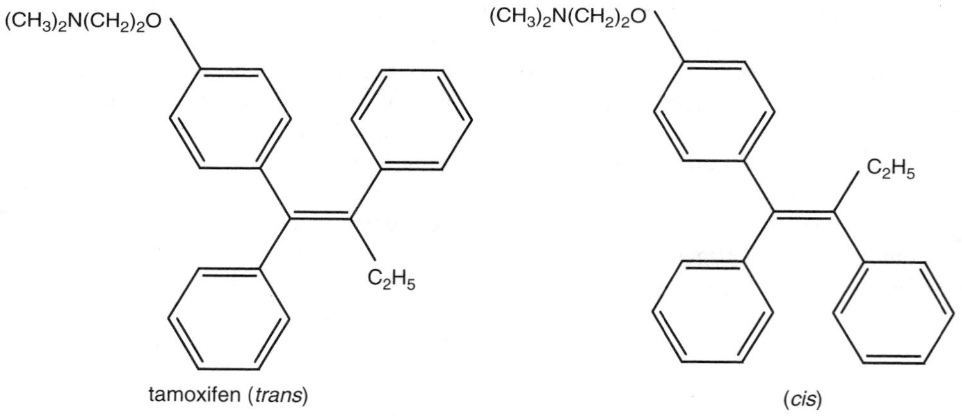

ΔH_f° values can be used to determine the heat of reaction (ΔH°) from the following relationship (law of Hess):

$$\Delta H_r^\circ = \Sigma\Delta H_{f\,(products)}^\circ - \Sigma\Delta H_{f\,(reactants)}^\circ$$

EXAMPLE 4

The combustion of methane:

$$CH_{4(g)} + 2\ O_{2(g)} \rightarrow CO_{2(g)} + H_2O_{(g)}$$

ΔH_f° for CH_4 = −17.8 kcal/mol, O_2 = 0, CO_2 = −94.1 kcal/mol, and for H_2O = −68.3 kcal/mol. Thus, $\Delta H^o_{combustion}$ = [(−94.1 kcal/mol) + (2 × (−68.3 kcal/mol))] − [(−17.8 kcal/mol) + (2 × 0 kcal/mol)] = −212.8 kcal/mol. This is, therefore, an exothermic reaction.

(Note: The (g) subscripts indicate gas phase.)

By definition, if ΔH_r° is positive, the process is endothermic (heat is absorbed); if ΔH_r° is negative, the process is exothermic (heat is evolved).

D. The Second Law of Thermodynamics: Definition of Entropy

This law introduces *entropy* (S), a state function that provides a quantitative description of randomness or disorder of the system and is fundamental for predicting the spontaneity of chemical reactions and physical changes. With the introduction of entropy, the second law of thermodynamics may be stated as follows: *For any spontaneous process in an isolated system, there is an increase in the value of entropy.*

The changes of entropy (ΔS) can often be calculated accurately, especially for ideal gases. When the temperature is held constant (isothermal process), where n is the number of moles in the system, V is volume, P is pressure, the indices refer to the properties in the starting (1) and

$$S = nR \bullet \ln\frac{V_2}{V_1} \quad \text{or} \quad S = nR \bullet \ln\frac{P_1}{P_2}$$

the final (2) state, and ln indicates the natural logarithm (see Chapter 1).

When the pressure is kept unchanged (isobaric process),

$$\Delta S = nC_P \bullet \ln\frac{T_2}{T_1}$$

where T is the absolute temperature, and C_p is the heat capacity at constant pressure (defined as the amount of heat needed to raise the temperature of a substance by 1 K).

E. The Third Law of Thermodynamics: Free Energy and Equilibria

The third law of thermodynamics states that the entropy of a pure, perfectly crystalline substance at absolute zero temperature (0 K) is defined as zero. Thus, an absolute value of the entropy (S) of a system in any state may be calculated.

A useful state function can be defined for a system's isothermal/isobaric processes: *Gibbs free energy* (G = H − TS). If ΔG < 0, the process is spontaneous; if ΔG > 0, the process is not spontaneous; if ΔG = 0, the system is in equilibrium. ΔG for ideal gases in an isothermal system is

$$\Delta G = nRT \bullet \ln\frac{P_2}{P_1} \quad \text{or} \quad \Delta G = nRT \bullet \ln\frac{V}{V_2}$$

The change in free energy of a solute associated with changes in concentration can also be examined using similar formalism:

$$\Delta G = nRT \bullet \ln\frac{a_2}{a_1}$$

where a is the activity (concentration for dilute solutions) at the initial and final concentrations. (The relationship between activity and concentration is $a = \gamma \times c$, where c is the concentration and γ is the activity coefficient; this equation considers that one may not deal with ideal solutions.)

EXAMPLE 5

The concentration of urea in plasma is 0.005 M, while its concentration in urine is 0.333 M. Calculate the free energy in transporting 0.10 mol of urea from plasma to urine. (The body temperature is about 310 K.)

$\Delta G = nRT\ln(a_2/a_1) = (0.1\ \text{mol})(1.987\ \text{cal}K^{-1}\text{mol}^{-1})(310\ \text{K})\ \ln(0.333\ \text{M}/0.005\text{M}) = 259$ cal Therefore, the process is not spontaneous.

The introduction of Gibbs free energy has significance in the thermodynamic treatment of chemical and phase equilibria. For an equilibrium process

$$\alpha A + \beta B \leftrightarrow \gamma C + \delta D$$

the equilibrium constant (K) is defined as

$$K = \frac{[C]^{\gamma} \bullet [D]^{\delta}}{[A]^{\alpha} \bullet [B]^{\beta}}$$

where the brackets indicate the concentration of the compounds, and

$$\Delta G = \Delta G^{\circ} + RT \bullet \ln K$$

where ΔG° is the change in standard free energy (ΔG° can be calculated by the analogy of the calculation of the reaction heat, ΔH°, by the law of Hess). In equilibrium, $\Delta G^{\circ} = 0$, so the temperature dependence of the equilibrium constant is also given by a thermodynamic relationship (van't Hoff equation):

$$\ln \frac{K_1}{K_2} = \frac{\Delta H^{\circ}}{R} \bullet \frac{(T_2 - T_1)}{T_2 \bullet T_1}$$

This equation allows one to estimate the heat of reaction (ΔH_r°) if the equilibrium constants at two temperatures are known or to determine the equilibrium constant at a particular temperature if ΔH° (standard heat of reaction) and one of the equilibrium constants are known.

F. HEAT OF VAPORIZATION AND HEAT OF SOLUTION

A relationship similar to the van't Hoff equation can be used to determine the vapor pressure at a particular temperature if the **heat of vaporization** (ΔH°_{v}) and one of the equilibrium vapor pressures (P) are known (Clausius-Clapeyron equation):

$$\ln \frac{P_2}{P_1} = \frac{\Delta H^{\circ}}{R} \bullet \frac{(T_2 - T_1)}{T_2 \bullet T_1}$$

Alternatively, ΔH°_v can be calculated for a liquid if its vapor pressures at two temperatures are known.

The phenomenon of evaporation and ΔH°_v can be explained by a closer look at the process. When a liquid is placed in a closed container at constant temperature, the molecules with the highest energy break away from the surface of the liquid and enter the gaseous state (evaporate), and some of the molecules subsequently return to the liquid state (condense). (Kinetic energy is not distributed evenly among molecules; some of the molecules have more energy than others and, hence, higher velocities.) When the rate of condensation equals the rate of evaporation at a definite temperature, the vapor becomes saturated and a dynamic equilibrium is established. The pressure of the saturated vapor above the liquid is known as the equilibrium *vapor pressure*. As the temperature of the liquid increases, more molecules approach the velocity necessary for escape and pass into the gaseous state. As a result, the vapor pressure increases with rising the temperature. If a liquid is placed in an open container and heated until the vapor pressure equals the atmospheric pressure, the vapor forms bubbles that rise rapidly through the liquid and escape into the gaseous state.

The temperature at which the vapor pressure of the liquid equals the external or atmospheric pressure is known as the **boiling point**. The pressure at sea level is 760 mmHg, and water boils at 100°C (212°F). At higher elevations, atmospheric pressure is lower and the boiling point decreases. The boiling point can be considered as the temperature at which the thermal agitation can overcome the attractive forces between the molecules of a liquid. ΔH°_v, like the boiling point of a compound and the vapor pressure at a particular temperature, indicates the magnitude of the attractive forces between molecules. The larger the ΔH°_v, the greater the attraction between molecules. ΔH°_v is absorbed when a liquid vaporizes. This heat is also liberated when a gas condenses. During evaporation and condensation, the temperature of the system does not change, while heat (Q) has to be supplied or removed to continue the process; therefore, ΔH°_v may be defined as a **latent heat**. In other processes such as warming up or cooling down a liquid, heat absorbed or released is accompanied by temperature change, and one may define it as a **sensible heat**.

For solubility, another similar equation can be employed,

$$\ln \frac{C_2}{C_1} = \frac{\Delta H^\circ}{R} \bullet \frac{(T_2 - T_1)}{T_2 \bullet T_1}$$

where C is the solubility (concentration of a saturated solution at the given absolute temperature), and ΔH_s is the differential heat of solution at saturation. This relationship explains in a quantitative fashion that the net energy change arising from solute-solute, solute-solvent, and solvent-solvent interactions causes heat to be either released or absorbed by the system when the solute transfers from the pure solid crystal to the solution. The magnitude and sign of ΔH_s determines the effect of temperature on solubility. Dissolution that absorbs heat ($\Delta H_s < 0$; e.g., potassium iodide, KI, dissolving in water) would be favored by an increase in temperature. However, if heat is given off upon dissolution of the solid ($\Delta H_s > 0$; e.g., calcium hydroxide, $Ca(OH)_2$, or methylcellulose dissolving in water), the material shows higher solubility at lower temperatures than at high temperatures.

GLOSSARY

Alpha (α) particles Two protons and two neutrons produced by radioactive decay. They are the same as the nucleus of a helium atom (two neutrons, two protons, no electrons). The mass of an alpha particle is about four times the mass of a single neutron or proton and has a positive charge of +2 (it has no electrons). The strong positive charge and its relatively slow speed (resulting from its large mass) limits the penetrating ability of the alpha particle.

Atomic number The number of protons in the nucleus.

Atomic weight The relative weight of an atom based on carbon's weight of 12.

Beta (β) particles Equivalent (mass and charge are the same) to an electron except for its source. Beta-emitting nuclides have too many neutrons. A neutron emits a beta particle and is then converted to a proton. Beta particles penetrate further than alpha particles of the same energy. Beta particles exist as two types, the negative electron (or negatron, e^-), and the positive electron (or positron, e^+) and are emitted by the nucleus during radioactive decay.

Boiling point The transition temperature of a liquid to a gas.

Covalent bond A bond in which there is equal sharing of electrons between atoms.

Dipole The charge separation within a molecule or atom that results in a separation of charge.

Electronegativity An element's affinity for electrons.

Electrons Negatively charged particles in an atom.

Energy Capacity to do work.

Enthalpy Heat content or sensible heat the sum of the energy and the product of pressure and volume ($H = E + PV$).

Extensive properties A property of a system that changes with the quantity of the material.

Gamma rays Emitted when the nucleus of a nuclide releases stored energy without releasing a particle. Gamma radiation is in the form of electromagnetic waves (or photons). Gamma rays are similar to x-rays but differ in their origin and energy. Gamma rays originate within the nucleus, and x-rays originate outside the nucleus. Gamma rays have a very high penetrating power because they have no charge or mass.

Heat (Q) A form of energy that is transferred from one body to another when a temperature difference is present.

Heat of vaporization (ΔH°_v) The amount of energy required to evaporate 1 mol of liquid under a constant pressure and temperature.

Hydrogen bonding Weak interaction between atoms formed through hydrogen and electrogenegative atoms.

Hydrophobic (water-fearing) A clustering of nonpolar molecules in water.

Inductive effect Origination of a charge on a molecule caused by pulling of electrons by another molecule or atom.

Intensive properties Properties of a substance that are independent of the quantity and shape of the substance.

Intermolecular bonds Bonds between two molecules.

Internal energy (E) The energy stored within the molecules of a system.

Ionic bond Complete electron transfer between two independent, neutral atoms that results in a bond formed between two oppositely charged atoms.

Isotope Two or more atoms with the same number of protons but different numbers of neutrons.

Latent heat The amount of heat given off or absorbed from 1 mol of a substance during a change in state.

London dispersion forces The interaction of two nonpolar atoms or molecules that arises because of fluctuating dipoles.

Protons Positively charged particles in the nucleus of an atom.

Radioactive Isotopes that have unstable nuclei leading to nuclear decay and emission of radiation.

Rare gases (also called noble gases) A chemically inert gas located in group 0 in the periodic table.

Sensible heat Heat that is absorbed or released. It is accompanied by a change in temperature but not a change in state.

State functions Energy terms such as enthalpy, heat, entropy, and Gibbs free energy.

Surrounding The entire universe excluding the system.

System The part of the universe that is being examined; may refer to a beaker, a cylinder of gas, a drug solution in an ampule, or an experimental animal.

Thermochemistry The study of heat changes in chemical reactions and certain physical processes.

van der Waals forces See London dispersion forces.

Work (W) Mechanical energy transferred to the surroundings from the system or vice versa.
x-ray Same as gamma rays except it originates from electron orbitals.

TUTORIAL

1. Which of following molecules have significant dipole moments?

(a)

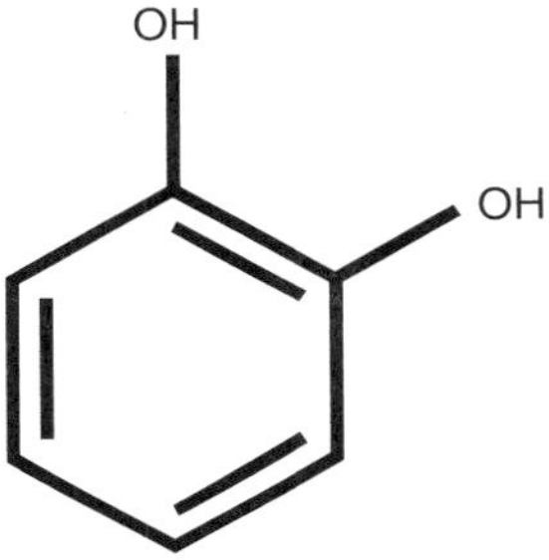

(b)

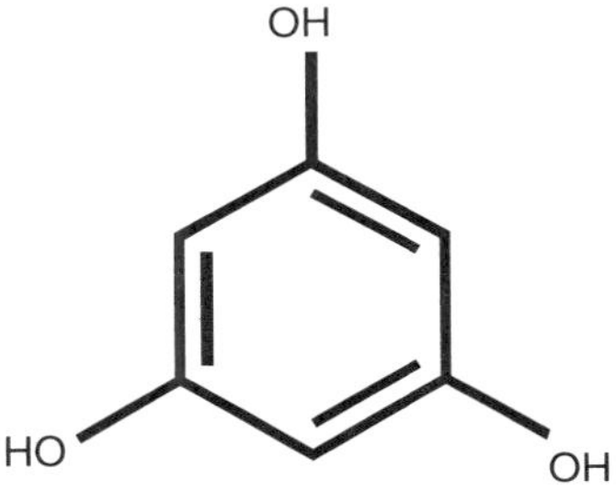

(c)

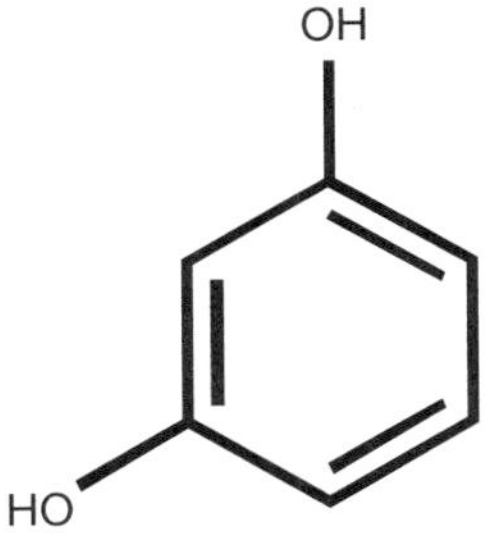

(d)

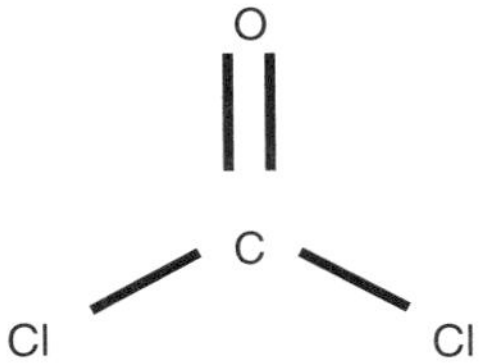

(e)

(f)

2. H_2O has a dipole moment. Can it be a linear molecule?
3. Explain the trends seen in this table

Number of Carbons	Alkane	State of Matter	Alcohols	State of Matter	Carboxylic Acids	State of Matter
1	Methane	Gas	Methyl alcohol	Liquid	Formic acid	Liquid
2	Ethane	Gas	Ethyl alcohol	Liquid	Acetic acid	Liquid
3	Propane	Gas	Propyl alcohol	Liquid	Propionic acid	Liquid
6	Hexane	Liquid	Hexyl alcohol	Liquid	Caproic acid	Liquid
16	Hexadecane	Liquid	Cetyl alcohol	Solid	Palmitic acid	Solid

ANSWERS

1. B, D
2. No. If the molecule was linear, the vector from the pull (attractive forces) of hydrogen from either side of oxygen would cancel each other out and result in no dipole.
3. Downward trends: as you increase carbon number you increase the number of van der Waals forces between molecules. Stronger intermolecular forces cause changes in states of matter. Not until 18 carbons do alkanes form solids. Side-to-Side Trends: Alkanes rely on van der Waal forces, which are relatively weak, to bind them together. Alcohols and acids can hydrogen-bond, forming stronger bonds than van der Waals forces form. This allows formation of liquid and solids when alkanes are gases and liquids at the same carbon number.

HOMEWORK

1. If fumaric acid reacts with water to give malic acid and if the standard (at 25°C) heats of formation for fumaric acid, malic acid, and water are −193.8, −264.2, and −68.3 kcal/mol, respectively, calculate the heat of reaction at 25°C.
2. Calculate the heat of reaction associated with the preparation of calcium hydroxide (1 mol of calcium oxide reacting with 1 mol of water to give 1 mol of calcium hydroxide), if the heat of formation of calcium hydroxide, water, and calcium oxide are −235.8, −68.3, and −151.9 kcal/mol, respectively.
3. If the concentration of HCl in the stomach is 0.145 M and that in the blood is 5.2×10^{-8} M, calculate the energy required for the transport of 2 mol of HCl from blood to the stomach at 37°C.
4. The active drug of a pharmaceutical formulation is known to decompose through a chemical process for which the following values have been obtained:

	50°C	**60°C**
ΔH (cal/mol)	1400	1625
ΔS (cal/mol K)	6.4	4.2

   Calculate the value of ΔG at each temperature.
5. At 45°C, at equilibrium, a protein denatures with a ΔH of 7 kcal/mol. Compute the entropy change.
6. The free energy of reaction of the dehydration of malate to fumarate at 25°C is 0.88 kcal/mol. What is the corresponding equilibrium constant, K?

Answers to Homework

1. $\Delta H = -2.1$ kcal/mol
2. $\Delta H = -15.6$ kcal/mol
3. $\Delta G = 18{,}283.2$ cal/mol = 76,497 Joules/mole
4. $\Delta G = -667$ cal/mol at 50°C and 226 cal/mol at 60°C.
5. $\Delta S = 22.0$ cal/mol K
6. $K = 0.226$

CASES

Theobroma oil (cacao butter) is a polymorphic natural fat and is often used in the preparation of suppositories. This excipient is an ideal diluent and vehicle since it is intended to melt when administered into the rectum or other body cavities. It consists mainly of a single glyceride and is intended to melt over a narrow temperature range (34 to 36°C). However, there are four common polymorphs of theobroma oil: the alpha (melting point 22°C), beta (melting point 34.5°C), beta prime (melting point 28°C), and gamma (melting point 18°C). The changes in polymorphic state of theobroma oil may impact its utility to patients.

1. What are polymorphs?
2. Which of the polymorphs is the most stable form?
3. What may happen if the pharmacist makes a suppository in the alpha polymorph form?
4. What should be the proper advice the pharmacist can give to the patient's parents?

Answers

1. Polymorphs are crystalline solids that are in different arrangements.
2. The polymorph with the highest melting point has the strongest bonds and thus is the most stable. For the case of theobroma oil, it would be the beta form.
3. The suppository would not be physically stable, as it might melt upon storage or melt in the patients hand when preparing for administration.
4. Cacao butter suppositories should be stored in the refrigerator to protect them against heating and phase conversion.

REFERENCES

Freifelder, D., *Principle of Physical Chemistry with Applications to the Biological Sciences*, 2nd ed., Boston, Jones and Bartlett, 1982.

Martin, A., *Physical Pharmacy*, 4th ed., Philadelphia, Lea and Febiger, 1993.

Saha, G.B., *Fundamentals of Nuclear Pharmacy*, 4th ed., New York, Springer, 1997.

3 Solubility

*Jeffrey A. Hughes, Adam M. Persky, Xin Guo, and Tapash K. Ghosh**

CONTENTS

* No official support or endorsement of this article by the FDA is intended or should be inferred.

0-4152-8863-0/05/$0.00+$1.50
© 2005 by CRC Press LLC

Learning Objectives

After finishing this chapter the student will have a thorough knowledge of:

- What a solution is
- How solutions and solution theory is applicable to pharmacy
- How to increase the solubility of a compound for formulation purposes

I. SOLUTION: AN INTRODUCTION

A **solution** is defined as a mixture of two or more components forming a homogenous molecular dispersion. A solution composed of only two substances is a binary solution, and the components making up the solution are the solvent and the solute. A solution can be a solid, liquid, or gas dissolved in a solid, a liquid, or a gas. For most pharmaceutical preparations of binary solutions, the solid is the solute and the liquid is the solvent. However, a solution for pharmaceutical use most often consists of a liquid containing one or more chemical substances. If two liquids are being mixed together and one of them is water, water is usually considered the solvent. In cases of mixing miscible (miscibility refers to the mutual solubility of the components in liquid–liquid systems) liquids, the terms of *solute* and *solvent* are usually not used. Solutions used in pharmacy consist of a wide range of solutes and solvents, as shown in Table 3.1.

In order to prepare a solution, a chemical substance must be soluble in a solvent. The **solubility** of a substance at a given temperature is defined as the concentration of the dissolved solute when it is in equilibrium with the undissolved solute. Intrinsic solubility is defined as the maximum concentration to which a solution can be prepared with a specific **solute** and **solvent** at a specific temperature. A solute is the dissolved agent, usually the less abundant component of a solution. The solvent, therefore, is the more abundant component of the solution, the component in which the solute is dissolved. When water is a component of the solution, it is usually considered the solvent. Solubility depends on many factors including chemical properties of the solute and the solvent, temperature, pressure, and pH. The concept of solubility is important to a pharmacist not only because it governs the preparation of solutions as a dosage form but also because a drug must be in solution before it can be absorbed by the body to exert biological activity.

TABLE 3.1
Examples of Pharmaceutical Solutions

Types of Solutions	Example
	Aqueous
Solution	Sodium fluoride oral solution, cimetidine HCl liquid
Syrup	Pseudoephedrine syrup
	Hydroalcoholic
Elixir	Digoxin elixir, diphenhydramine HCl elixir
Tinctures	Iodine tincture
Spirits	Camphor spirit
	Organic
Liniments	Calamine lotion
Collodion	Salicyclic acid collodion

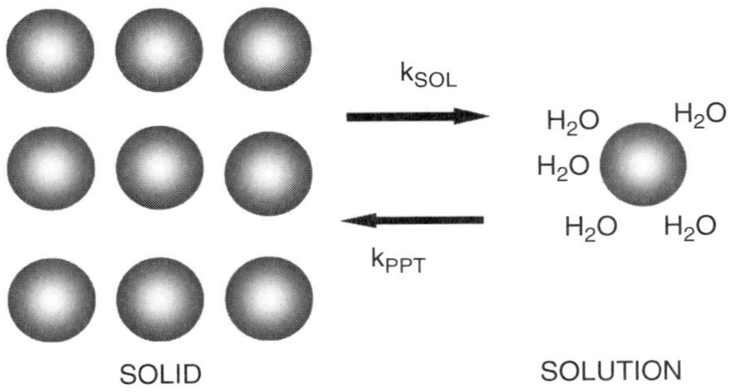

FIGURE 3.1 Definition of a saturated solution. At equilibrium the rate of a solute precipitating out of the solution is equal to the rate in which the solute enters the solution.

The dissolution of a solid solute into a liquid solvent starts with the contact of two phases, a liquid solvent phase and a solid solute phase. The process of dissolution includes two molecular events, the transfer of solute molecules from the solid phase to the liquid phase and the transfer of the dissolved solute molecules back to the solid phase from the liquid phase. The rate of transferring solute molecules from the solid phase to the liquid phase equals the activity of the solute molecules in the solid phase times a rate constant, k_{SOL} (Figure 3.1). The rate of transferring dissolved solute molecules from the liquid phase back to the solid phase equals the activity of the dissolved solute in the liquid phase times a rate constant of precipitation, k_{PPT} (Figure 3.1). In a **saturated solution** containing undissolved solid solute, the rate at which the molecules or ions leave the solid solute surface is equal to the rate at which the dissolved molecules return to the solid (Figure 3.1) and the system is in equilibrium. The solubility equilibrium constant of a substance is the ratio of the two rate constants at equilibrium in a given solution. Therefore, solubility is a thermodynamic property involving equilibrium, whereas the transfer of solute molecules into a solution (i.e., dissolution) is a kinetic process.

$$K_{SOLUBILITY} = \frac{k_{SOL}}{k_{PPT}}$$

An **unsaturated solution** contains the dissolved solute in a concentration lower than that of a saturated solution. A **supersaturated solution** contains a higher concentration of the dissolved solute than that at the dissolution equilibrium under the same conditions. Most pharmaceutical solutions are considered unsaturated. Since environmental conditions can impact solutions and their subsequent product activities, it is important for a pharmacist to control these conditions. For example, if a pharmaceutical solution is not capped, there may be evaporation of solvent from the solution, resulting in an alteration of the concentration of the solute and therefore a change in the dose a patient receives from the product.

II. EXPRESSIONS OF CONCENTRATION AND SOLUBILITY

The concentration of a drug or other substance in solution can be expressed in many different ways. Table 3.2 lists methods for expressing solution concentrations. Percentage is the most common method of expressing concentration in pharmacy practice. Percentage by weight (w/w) expresses the number of grams of solute in 100 grams of solution. Percentage by volume (v/v) expresses the number of milliliters of solute in 100 ml of solution. Percentage weight in volume (w/v) expresses the number of grams of solute in 100 ml of solution. In pharmacy, the term *per cent* traditionally is used without qualification. For mixtures of solids, per cent refers to a weight by weight percent;

TABLE 3.2
Commonly Used Concentration Expressions for Solutions

Concentration Expressions	Definition/Units
Percent weight	Gram of solute/100 g of solution
Percent volume	Milliliters of solute/100 ml of solution
Percent weight/volume	Gram of solute/100 ml of solution
Normality	Gram equivalent weights of solute/liter of solution
Molarity	Moles of solute/liter of solution
Molality	Moles of solute/1000 g of solvent
Mole fraction	Ratio of the moles of one component (e.g., the solute) of a solution to the total moles of all components (solute and solvent)
Milliequivalents per liter	Miligram equivalent weights (equivalent weight expressed in mg, mEq) of solute/liter of solution
Osmolarity	Osmoles[a] of solute/liter of solution
Osmolality	Osmoles[a] of solute/1000 g of solvent

[a] One osmole represents one mole of osmotically active entities, regardless of their nature.

for mixtures of liquids, a volume in volume percent is used. Unqualified use of the term *per cent* for solids or gases in liquids implies the use of weight in volume percent.

Other terms to describe concentrations include (a) **molarity**, which is the number of moles of solute per liter of solution, (b) **molality**, which is the number of moles of solute per kilogram of solvent, (c) **osmolality**, which is number of moles of particles (molecules or ions) dissolved in one kilogram of solvent, and (d) **mole fraction**, which is the ratio of moles of one component divided by total moles of the solution (i.e., solute + solvent).

An older method that is losing favor in current medical practice is the use of **milliequivalents (mEq)**. Milliequivalents is an expression of amount, which is analogous to millimoles. When using milliequivalents to express concentration, the term **milliequivalent per liter** is used to specify the volume of the solution in which the solute is dissolved. A concentration in terms of milliequilvalents per liter can be calculated by the weight concentration of the solute in milligrams per liter divided by the equivalent weight of the solute (g/Eq or mg/mEq), where the equivalent weight is defined as the formula/molecular weight of a compound divided by the net valence for the cation or anion.

Example 1

For 100 mg each of NaCl, $CaCl_2$, and $FeCl_3$ in 1 l of water, determine the concentration in, percent (w/v), molarity, milliequivalents (mEq), and milliosmolarity (mOsm).

Since this is a diluted solution, the volume of the final solution is approximately equal to the volume of the initial solvent (1 l). Therefore,

Percent (gram/volume):

NaCl: 100 mg NaCl in 1 l = 0.1 g/1000 ml = 0.01 g/100 ml = 0.01%
$CaCl_2$: 100 mg $CaCl_2$ in 1 l = 0.1/1000 ml = 0.01 g/100 ml = 0.01%
$FeCl_3$: 100 mg $FeCl_3$ in 1 l = 0.1/1000 ml = 0.01 g/100 ml = 0.01%

Molarity:

NaCl: NaCl has a molecular weight of 58.5 g/mol. So 0.1 g/58.5 g/mol/1 l = 0.0017 M = 1.7 mM.
$CaCl_2$: $CaCl_2$ has a molecular weight of 111 g/mol. So 0.1 g/111 g/mol/1 l = 0.0009 M = 0.9 mM.
$FeCl_3$: $FeCl_3$ has a molecular weight of 162.2 g/mol. So 0.1 g/ 162.2 g/mol/1 l = 0.00062 M = 0.62 mM.

Milliequivalent conversion:

NaCl: NaCl has a molecular weight of 58.5 g/mol and has a valence of 1 (Na = 1, Cl = −1) for an equivalent weight of 58.5 g or 1 mEq = 58.5 mg. So for 100 mg NaCl, 100 mg/58.5 mg/mEq = 1.7 mEq in 1 l.
$CaCl_2$: $CaCl_2$ has a molecular weight of 111 g/mol and has a valence of 2 (Ca = 2, Cl = −1 since there are two ions −1*2 = total valance of −2) for an equivalent weight of 55.5 g or 1 mEq = 55.5 mg. So for 100 mg $CaCl_2$, 100 mg/55.5 mg/mEq = 1.8 mEq in 1 l.
$FeCl_3$: $FeCl_3$ has a molecular weight of 162.2 g/mol and has a valence of 3 (Fe = +3, Cl = −1 since there are three ions −1*3 ions = total valance of −3) for an equivalent weight of 54.1 g or 1 mEq = 54.1 mg. So for 100 mg $FeCl_3$, 100 mg/54.1 mg/mEq = 1.85 mEq in 1 l.

Milliosmole conversion (if we assume complete ion dissociation in water):

When no ion interactions take place, then milliosmolality (mosm/Kg) = *i* x mM, where *i* is the number of ions formed and mm is millimolar concentration.

NaCl: Each NaCl molecule forms two particles, one Na^+ and one Cl^-, so a 1 M solution would be a 2 osmolar solution. For our example, (1.7 mM * 2) = 3.4 mOsm/l.
$CaCl_2$: Each $CaCl_2$ molecule forms three particles, one Ca^{2+} and two Cl^-, so 0.9 mM x 3 = 2.7 mOsm/l.
$FeCl_3$: Each $FeCl_3$ molecule forms four particles, one Fe^{3+}, and three Cl^-, so 0.62 mM × 4 = 24.8 mOsm/l.

(**Note:** Students should practice conversions into other units mentioned in Table 3.2.)

Less quantitative terms can also be used to describe solubility. These terms, as defined by the United States Pharmacopoeia (USP), are listed in Table 3.3 and are based on the parts of solvent needed to dissolve one part of the solute. However, as in pharmaceutical dispensing, aqueous solubility is of primary interest, the solvent generally refers to water, and solubility in general refers to water or aqueous solubility at room temperature.

TABLE 3.3
Terms of Solubility (USP)

Term	Parts of Solvent Required to Dissolve One Part of Solute
Very soluble	Less than 1
Freely soluble	1–10
Soluble	10–30
Sparingly soluble	30–100
Slightly soluble	100–1,000
Very slightly soluble	1000–10,000
Practically insoluble or insoluble	Greater than or equal to 10,000

III. THE ENERGETICS OF DISSOLUTION

A. Ideal Solutions vs. Real Solutions

The thermodynamic theories of solution and solubility are based on **ideal solutions**. In an ideal solution, the forces that hold the solute molecules together, the forces that hold the solvent molecules together, and the forces between solute and solvent molecules are equal in magnitude. This means that no net heat is released or absorbed during the dissolution process and that the spatial distribution of solute and solvent molecules is random and homogeneous. The solubility in this situation depends on temperature, the melting point of the solid, and the molar heat of fusion (ΔH_f), defined as the amount of heat absorbed when a solid melts. The following equation quantitatively describes the solubility of a solid in an ideal solution:

$$\log X_2^i = -\frac{\Delta H_f}{2.303 \bullet R} \bullet \frac{T_0 - T}{T_0 \bullet T}$$

or rewritten as

$$\log X_2^i = -\frac{\Delta H_f}{2.303 \bullet R} \bullet \frac{1}{T} + Y$$

In these equations, X_2^i is the ideal solubility (as a mole fraction), R is the gas constant, T_0 is the melting point of the solid solute in absolute degrees, T is the temperature of the solution (absolute in Kelvin), and Y is a constant. This equation can be used to calculate the molar heat of fusion (ΔH_f) by plotting the logarithm of solubility against the reciprocal of the absolute temperature. This results in a slope of $-\Delta H_f /2.303\ R$.

Unfortunately, not many solutions behave ideally, but rather are referred to as **nonideal solutions,** or real solutions, where the interactions between solute and solute, solute and solvent, and solvent and solvent molecules differ. In a nonideal solution, mixing of solute and solvent can generate heat or absorb heat from the surroundings. When talking about nonideal solutions, it is also customary to talk about the activity of a solute in solution. Activity (a_2), or the effective concentration, of a solute is defined as the concentration of the solute multiplied by its activity coefficient (γ_2). When concentration is substituted by activity in the ideal solution equation, the resulting equation is

$$\log X_2 = -\frac{\Delta H_f}{2.303 \bullet R} \bullet \frac{T_0 - T}{T \bullet T_0} + \log \gamma_2$$

where T_0 is the melting point of the solute. As the activity coefficient approaches unity ($\gamma_2 \rightarrow 1$), the solution becomes more ideal. For example, as a solution becomes more dilute, the activity increases and the solution becomes ideal. The log of the activity coefficient ($\log \gamma_2$) is a term that considers the work of solubilization, volume of solute, and the total volume of solvent. The work of solubilzation includes multiple steps of molecular interactions during the dissolution process, which will be described in the next section. Molecules are held together by various types of bonds (e.g., London forces, hydrogen bonds, dipole–dipole, etc.). These forces are intricately involved in solubility because solvent–solvent (B–B), solute–solute (A–A), and solvent–solute (B–A) interactions govern solubility.

An additional term used in solubility theory is the **solubility parameter**. The solubility parameter (δ) is a measure of cohesive forces between like molecules and is described by the following two equations:

$$\log \gamma_2 = \left(w_{22} + w_{11} - 2w_{12}\right) \bullet \frac{V_2 \bullet \Phi_1^2}{2.303RT}$$

and

$$\delta = \left(\frac{\Delta H_v - RT}{V_1}\right)^{\frac{1}{2}}$$

where ΔH_v is heat of vaporization of the solute, V_2 is the volume per mole of the solute as a liquid, T is temperature (in degrees Kelvin), R is the gas constant, V_1 is the molar volume of solution at a given temperature, and $\Phi_1{}^2$ is the volume fraction of solvent. The terms w_{11}, w_{22}, and w_{12} are described in the following section.

B. The Process of Dissolution

The process of **dissolution** involves the breaking of interionic or intermolecular bonds in the solute, the separation of the molecules of the solvent to provide a hole in the solvent for the solute molecule, and the interaction between the solvent molecules and the solute molecule or solute ion. The forces that govern these interactions are presented in Chapter 7. In order to dissolve a crystal of sugar in water, the individual sugar molecules must be freed from the crystal lattice. If a molecule of sugar is going to leave the crystal, it must first be freed from the attractive forces of its neighboring sugar molecules. At the same time, a hole must be made in the water to allow the freed sugar molecules to fit in. Finally, the sugar molecule enters the hole in the water (Figure 3.2).

The actual solubility of a substance represents the overall effect of the various factors involved in the transport of a solute particle from the solid phase to the solution phase. The driving force in dissolution is mainly the interaction of the solvent molecules with solute molecules or the solute ions. For example, although there is strong ionic interaction between sodium and chloride ions in a sodium crystal, it is compensated by strong ion–dipole bonding between water and the sodium and chloride ions, resulting in the dissolution of the solute crystal. The net result of the molecular interactions as the solute dissolves is manifested energetically as the **heat of solution** (ΔH_{sol}). The relationship of the heat of solution to energies of the stepwise molecular interactions is described in the following paragraphs. The end result is the equation

$$\text{work} = w_{22} + w_{11} - 2w_{12}$$

where w_{11} is the work to liberate a solute particle (step 1 of Figure 3.2), w_{22} is the work to vacate solvent particles (step 2 of Figure 3.2), and w_{12} is the interaction between the solute and solvent particles (step 3 of Figure 3.2). The interaction term w_{12} is multiplied by two because the surface area for solute–solvent interactions is twofold as the surface area where the solute–solute bonding or the solvent–solvent bonding is broken. The log γ_2 is a combination of the work involved in solubilization, the volume of the solute, and the volume occupied in the solvent. When the latter two factors are combined with the work of solubilzation, we get the following, known as the Hildebrand equation:

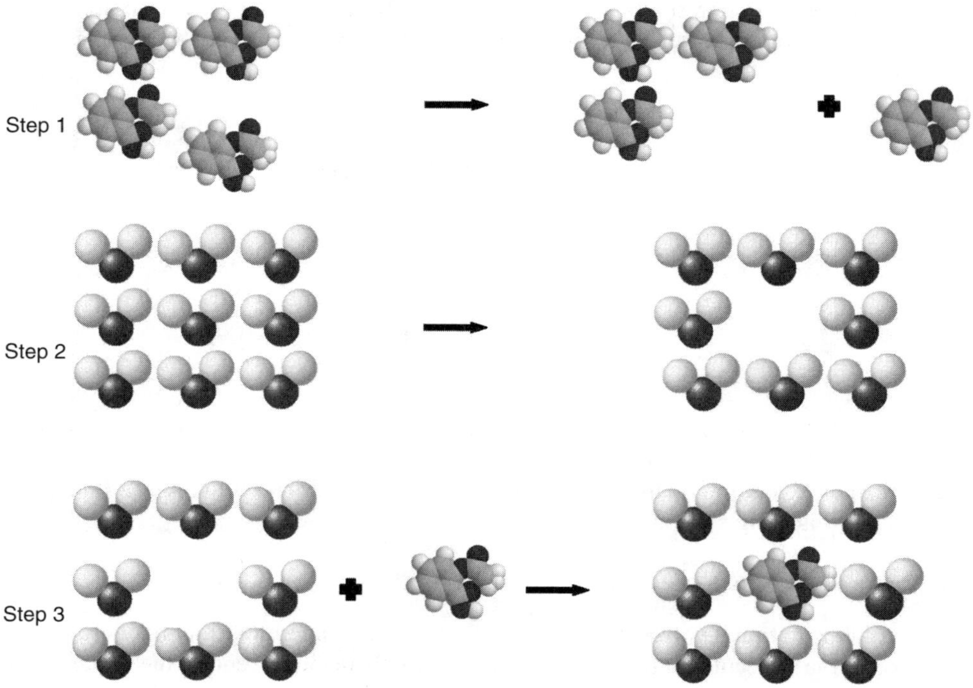

FIGURE 3.2 Steps of the dissolution of a solid solute. Step 1: one molecule of the solute breaks away from the bulk solid. Step 2: a hole opens in the solvent. Step 3: the freed solid molecule is intergrated into the hole in the solvent.

$$\log \gamma_2 = \left(w_{22} + w_{11} - 2w_{12}\right) \bullet \frac{V_2 \bullet \Phi_1^2}{2.303RT}$$

where V_2 is the molar volume of solute, and $\Phi_1{}^2$ is the volume fraction of solvent. For dilute solutions, $\Phi_1{}^2 = 1$.

Step 1: Separation of the Solute Particles

The initial step in dissolution is the work required to separate solute particles and is dependent on intermolecular forces of the solute. Assume in the dissolution process that a solute molecule is removed from the solute phase by breaking its bonding with the neighboring solute molecules and leaving a void. The energy needed for the removal of each solute molecule is $2w_{22}$, in which the subscript $_{22}$ refers to the interaction between solute molecules. Upon removal of the molecule from the solute phase, the void it has created closes, and half of the energy is regained. The total gain in potential energy, or the net work needed for the process, is thus w_{22}, which is half of the total interaction energy (half of $2w_{22}$).

Let us look at an important pharmaceutical example. Two structurally similar compounds, morphine and codeine, are shown in Figure 3.3. Each morphine molecule has two hydroxy groups that can form two hydrogen bonds with hydroxy groups from a neighboring morphine molecule in its crystal, resulting in a higher melting point (T_m = 250°C). Codeine only has one hydroxyl group and less intermolecular hydrogen bonding, resulting in a lower melting point (T_m = 154 to 158°C). Consequently codeine has a higher solubility than morphine because it is easier to free a codeine molecule from its crystal lattice. If hydroxy groups (or other hydrogen bonding moieties) of the same molecule are in close proximity and appropriate topology to each other, they can have

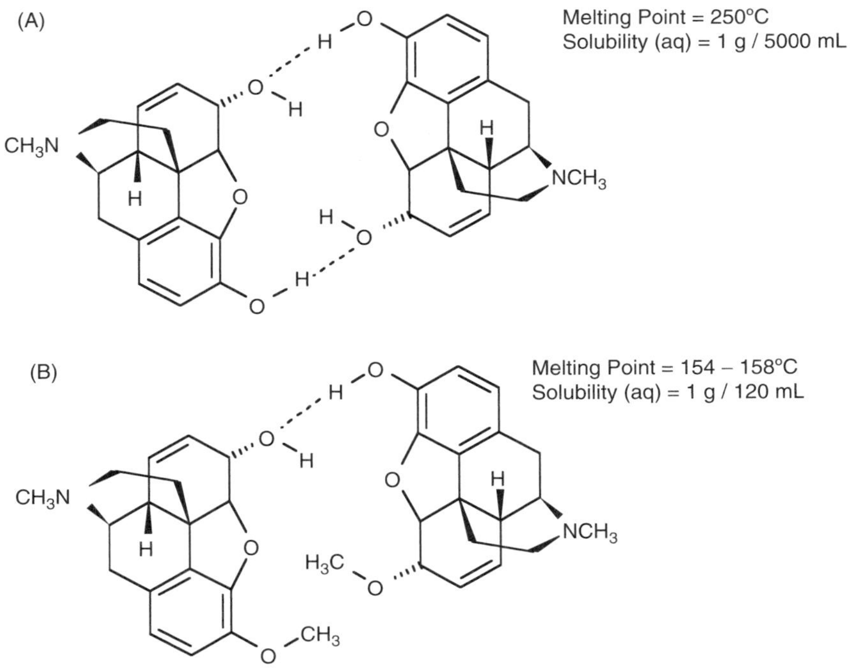

FIGURE 3.3 Intermolecular forces for morphine (A) and codeine (B).

intramolecular interactions. That is, one hydroxy group of a molecule interacts with another hydrogen-bond–acceptor/donor group from the same molecule. The intramolecular interactions tend to lower intermolecular interactions and hence lower the melting temperature.

Step 2: Separation of the Solvent Molecules

In the next step of the dissolution process, a void large enough to accommodate a solute molecule is formed in the solution phase. In this step, the work required to pull apart solvent molecules is described by w_{11}, which is equal to half the total interaction (half of $2w_{11}$) between a pair of solvent molecules. The intermolecular forces of the solvent are very important. If we look at the structure of water (Figure 3.4), we see it is highly ordered owing to hydrogen bonding of the highly polar molecule. Owing to its polar nature and its ability to hydrogen bond, water is a good solvent for polar molecules and has a large **dielectric constant.**

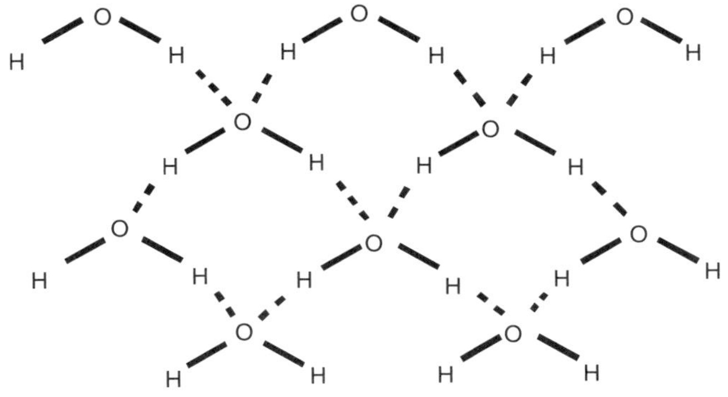

FIGURE 3.4 Structure of water. Dashed lines are hydrogen bonds.

The dielectric constant is a measure of the influence by a medium on the energy needed to separate two oppositely charged bodies. A vacuum is arbitrarily given a dielectric constant of 1. If you put two oppositely charged bodies into any medium, the medium tends to make it easier for the two oppositely charged bodies to separate. The energy required to separate two oppositely charged bodies is inversely proportional to the dielectric constant of the medium.

In water, which has a dielectric constant of 80.4, it requires only about 1/81 (0.0123 times) as much energy to separate two charged bodies as in vacuum, meaning that it is much easier to separate charged molecules in water than in vacuum. Since dielectric constant is a ratio of two amounts of energy, it has no units. The dielectric constant is also a measure of the degree of polarization in both an induced and a permanent dipole. Dipole moment is a function of the charge and the distance between the charges (see Chapter 2). Solvents that can form a hydrogen bonding network (such as water and alcohols) possess high dipole moments and high dielectric constants because of the long-chain structure of the associated pseudo-molecules.

Nonpolar solvents such as benzene do not have a sufficiently high dielectric constant to separate polar molecules such as the ions of NaCl. These solvents can only dissolve molecules held together by very weak intermolecular forces (induced dipole-induced dipole, or London forces), such as naphthalene. Because the solute–solute, solute–solvent, and solvent–solvent interactions are very weak, these types of nonpolar solutions behave close to ideal solutions.

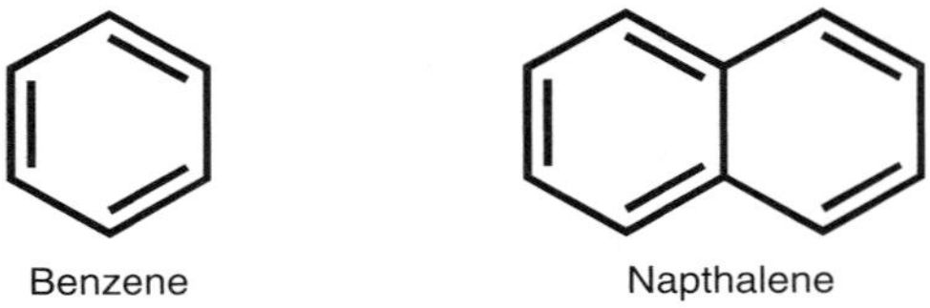

FIGURE 3.5 Structure of benzene and napthlene.

Water, however, cannot dissolve naphthalene because the bonding between water and naphthalene is much less than that between water molecules. The classification of solvents on the basis of polarity is often referred to as the rule of like dissolves like. In other words, if you want to dissolve a highly polar or ionic compound, you should use a solvent that is also highly polar or has a high dielectric constant. If you want to dissolve a compound that is nonpolar, you should use a solvent that is relatively nonpolar, or in other words, has a low dielectric constant. Dielectric constants of some commonly used liquids at 20°C are given in Table 3.4.

TABLE 3.4
Dielectric Constants of Some Liquids at 20°C

Substance	Dielectric Constant
Carbon tetrachloride	2.24
Benzene	2.28
Ethyl ether	4.34
Chloroform	4.8
Ethyl acetate	6.4
Phenol	9.7
Acetone	21.4
Ethanol	25.7
Methanol	33.7
Water	80.4
N-methylformamide	190

Step 3: Interactions between Solvent and Solute

Last, the solute molecule is placed in the void in the solution phase. The void in the solvent, created in Step 2, is now filled, and the net work or total interaction energy attributed to this final step is $-2w_{12}$.

If the sizes of the solute and solvent molecules are similar, the net energy for the entire dissolution process may be expressed by the following equation:

$$\text{work} = w_{22} + w_{11} - 2w_{12}$$

If this process is to proceed without having to add energy, work should be a negative value.

Dipole–dipole interactions are responsible for the dissolution of many pharmaceutical agents. The solubility of the low-molecular-weight organic acids, alcohols, amides, amines, esters, ketones, and sugars in polar solvents is a result of dipole–dipole interactions. Since water is one of the most commonly used solvents, hydrogen bond formation is often observed.

Alcohols dissolve in water predominantly because of hydrogen bonding. When the alkyl chain of a monohydric alcohol exceeds five carbons, the polarity is reduced and the longer alkyl chain alcohols are no longer water soluble unless the hydrocarbon is branched. Alcohols (e.g., ethanol) are capable of dissolving moderately polar solutes that are not soluble in water. Pharmacists often use ethanol as a co-solvent with water to increase the solubility of some hydrophobic molecules (e.g., in the production of elixirs).

In a given molecule, as the number of hydroxyl groups (or other polar groups) increases, its solubility in water often increases. Polyhydric alcohols such as glycerin, mannitol, and sorbitol are polar and highly water soluble. Phenols dissolve in water, glycerin, and alcohol. As the ratio of hydroxy groups to the number of carbon atoms increases, the water solubility increases. The solubility of resorcinol in water is about 15 times greater than phenol in water (see Example 2).

When strong dipole–dipole interactions are involved in dissolution, the solute–solvent interaction (A–B) may exceed the sum of the solvent–solvent interaction (B–B) and the solute–solute interaction (A–A), resulting in excess energy liberated as heat. Such dissolution is known as an exothermic process. Since the heat of solution is defined as the heat absorbed per mole dissolved, such a system has a negative heat of solution.

Usually a large negative heat of solution (i.e., exothermic reaction) means a higher solubility of the solute than a mixing with a smaller negative heat of solution. In certain dissolution processes, the solute–solute and solvent–solvent interaction may exceed the energy from the solute–solvent interaction. To complete the dissolution process, thermal energy is absorbed from the environment. This type of dissolution is known as endothermic, and the heat of solution is positive.

EXAMPLE 2

The addition of hydroxy groups (–OH) to a molecule can change the solubility in a polar solvent by increasing hydrogen bonding.

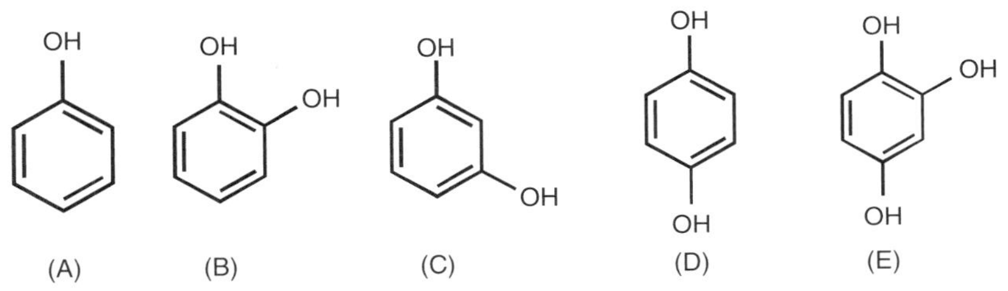

Compound A (phenol) has one –OH group and has a solubility of 1 g in 15 ml of water. **Compounds B** (pyrocatechol), **C** (rescorinol), and **D** (hydroquinone) all have two –OH groups, and their respective solubilities are 1 g in 2.3 ml, 1 g in 0.9 ml, and 1 g in 4 ml of water. **B**, **C**, and **D** are all more soluble than **A**. **D** is the least soluble among **B**, **C**, and **D** because it is symmetric, allowing tighter intermolecular packing than **B** and **C**.

Compound E (1, 2, 4 benzenetriol) has three –OH groups and is freely soluble in water.

IV. ELECTROLYTES AND SOLUBILITY PRODUCT

Ion–dipole interactions are the forces primarily responsible for the dissolution of electrolytes in polar solvents. The attraction of an electrical center of a dipole to an oppositely charged ion releases energy. The released energy is used to break the ion–ion bonds in the solid. Nonpolar and weakly polar solvents do not have sufficiently strong interactions with the electrolytes to dissolve them.

The traits of a good solvent for electrolytes include a high dipole moment, a small molecular size, and a high dielectric constant. The small molecular size allows a dipole to closely align with the electrolyte ions, and a high dielectric constant decreases the electrical interactions between the solvated ions. Water has all of these qualities.

In aqueous solutions, most ions are surrounded or hydrated by as many water molecules as the size of the ion allows. Anions are generally less highly hydrated than cations. The extent of interaction between ions and dipoles depends on the size and the charge of the ions. The effects of these two factors can be illustrated by a comparison of the solubility of aluminum chloride (1 g in 0.9 ml water) and sodium chloride (1 g in 2.8 ml water). The aluminum ion has a smaller radius than the sodium ion and would therefore be less hydrated, which affects the solubility. Interestingly, the quantity of water in a crystal affects the heat of solution. In a crystal hydrate, the ions are largely hydrated, and consequently the ion–dipole interaction energy is considerably less than that of the anhydrous solute during the dissolution process. The hydrates, therefore, usually have higher heat of solution and lower water solubility.

A. Soluble Electrolytes

As with molecular solids, it takes energy to separate ions from their crystal lattice such as sodium chloride. However, the ions can hydrate with water to give off heat. In sodium chloride, the energy to separate the ions from the crystal lattice is provided by the heat of hydration. In calcium hydroxide, the heat of hydration exceeds the heat needed to separate the ions and heat is released during dissolution. In sulfuric acid, the heat of hydration is also great. In potassium iodide, the heat required to separate the ions is greater than that supplied by hydration, and there is a cooling effect during the dissolution.

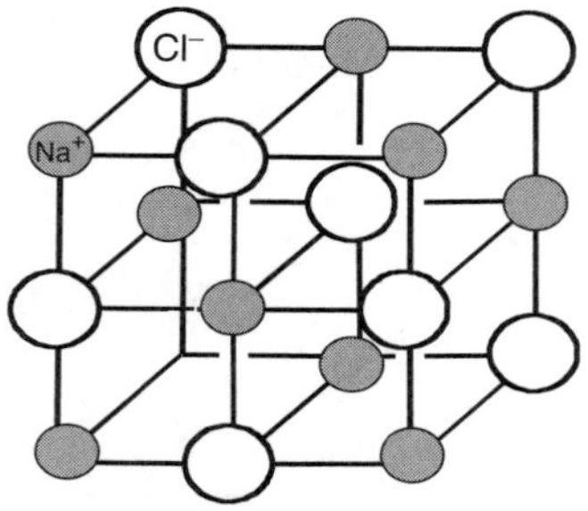

FIGURE 3.6 Crystal structure of sodium chloride.

B. Slightly Soluble Electrolytes (K_{sp})

The solubility of a slightly soluble electrolyte may be expressed as a solubility product. If the solute phase is composed of two or more definite ionic species, the solubility may be expressed as the product of the concentrations of the ions in a saturated solution, each concentration raised to the power corresponding to the number of each species of the ion produced. For the reaction

$$A_aB_{b(SOLID)} \leftrightarrow aA^+ + bB^-$$

the equilibrium constant (K) equals

$$K_{sp} = [A^+]^a[B^-]^b$$

Since the concentration of the solid is a constant in a solid crystal lattice, we can combine it with the equilibrium constant K to obtain the solubility product (K_{sp})

$$K_{sp} = [A^+]^a[B^-]^b$$

The solubility product can be used to determine the maximum concentration of an ion in a given solution at equilibrium with the solid solute phase. *This equation does not apply to freely soluble salts such as sodium chloride.*

Example 3

The solubility of calcium phosphate tribasic (formula = $Ca_3(PO_4)_2$) is 0.114 μM. Calculate the solubility product for calcium phosphate. Assume 100% dissociation.

$$Ca_3\left(PO_4\right)_{2(SOLID)} \Leftrightarrow 3Ca^{2+} + 2PO_4^{\ 3-}$$

$$K_{sp} = \left[3Ca^{2+}\right]^3\left[2PO_4^{\ 3-}\right]^2$$

$$K_{sp} = [3 \times 0.114 \times 10^{-6}]^3\ [2 \times 0.114 \times 10^{-6}]^2$$

$$= 2.08 \times 10^{-33}$$

V. FACTORS INFLUENCING SOLUBILITY

Dissolution is a process involving numerous interactions between solute–solute, solute–solvent, and solvent–solvent molecules. Therefore, it is often difficult to predict the solubility of solutes, especially those with complex molecular structures. The discussions in this section are general rules with important exceptions. However, it is safe to broadly group the factors influencing the solubility into two categories: the chemical properties of the solute and the solvent and the conditions of the solution such as the temperature and the pH.

A. Chemical Properties of the Solute and the Solvent

The chemical properties of the solute and the solvent molecules are the major factors that influence solubility. The most important rule in this regard is that solutes dissolve better in solvents of similar polarity, often referred to as the rule of like dissolves like. The rule of like dissolves like conforms with the thermodynamic theories of solutions discussed in Section III in that if two chemicals of very different polarity are mixed, a solution, that is, a molecular dispersion, cannot be formed because the energy gained from the solute–solvent interactions will in general not be sufficient to overcome the strong dipole–dipole or ionic interactions of the more polar chemical. Therefore, to dissolve a highly polar or ionic compound, one should use a solvent that is also highly polar or has a high dielectric constant. On the contrary, to dissolve a compound that is nonpolar, one should use a solvent that is relatively nonpolar, or in other words, has a low dielectric constant. That is why naphthalene has low solubility in water, and glucose has low solubility in benzene. Since most of the pharmaceutical solutions use water as the solvent, we will discuss in more detail the water solubility of inorganic and organic compounds.

1. Solubility of Inorganic Compounds in Water

Inorganic molecules in general have high polarity and hence are more soluble in water than in less polar organic solvents. Aqueous solutions of inorganic compounds have many important uses in pharmaceutical products, such as buffering the pH or monitoring the osmotic pressure of the product. When preparing a solution of multiple salts, one needs to take extra caution that none of the cations or anions in the mixture can form an insoluble salt. Such a mismatch will immediately cause the precipitation of the preparation.

Most inorganic compounds form crystals by strong ionic interactions, and the dissolution of an ionic compound relies on the strong ion–dipole interactions between the ions of the crystal and the water molecules to overcome such solute–solute attraction forces. If both the cation and the anion of an ionic compound are monovalent, the solute–solute attraction forces are easily overcome, and these compounds are generally water soluble. Examples include NaCl, KOH, LiBr, KI, and NH_4NO_3. The exceptions are silver and mercury salts of acetate, hydroxide, and halides (chloride, bromide, and iodide). If only one of the two ions in an ionic compound is monovalent, the solute–solute interactions are usually overcome, and the compounds are generally water soluble. Examples include $BaCl_2$, Na_2SO_4, and Na_3PO_4. Exceptions are hydroxides, which are poorly soluble in water, except for alkali metal cations and Li_2CO_3. When both the cation and the anion are multivalent, the solute–solute interaction may be too great to overcome, and the compounds generally have poor water solubility. Examples are $CaSO_4$, $BaSO_4$, $BiPO_4$, $ZnSO_4$, and $FeSO_4$ are exceptions to that rule. Hydroxides, oxides, and sulfides of compounds other than alkali metal cations are generally water insoluble. Ammonium and quaternary ammonium salts are water soluble, and acid salt usually has an improved solubility in water compared to its insoluble original salt.

2. Solubility of Organic Compounds

Virtually all drugs on the market are organic compounds. Compared with the inorganic molecules, most of the organic compounds are less polar and hence dissolve better in organic solvents than in water. However, from a pharmaceutical point of view, sufficient solubility is not the only factor in solvent selection. Other characteristics such as low toxicity, compatibility with other ingredients, stability, clarity, odor, color, and economy are also important in solvent selection. For most pharmaceutical applications, water is the preferred solvent because it meets more of the above criteria than other solvents.

Because of the high polarity of water, introduction of polar functional groups in organic compounds generally increases their water solubility. Examples of the polar functional groups include hydroxy groups, amines, ammoniums, aldehydes, and carboxylates. As discussed in Example 2, the introduction of the polar groups increases the dipole–dipole interaction between the organic molecule and the water molecules, which is favorable for the dissolution process. Introduction of hydrophobic functional groups, however, usually decreases the water solubility. The hydrophobic functional groups include the methyl, methylene, ethoxy, and halide (Cl^-, Br^-, I^-) groups.

However, owing to the complex structure of organic molecules, it is often difficult to predict their solubility from only the composition of the functional groups because the strength of solute–solute and solute–solvent interactions are also sensitive to the shape of the solute molecule as well as the three-dimensional orientation of its functional groups. For example, the solubility of n-butanol ($CH_3CH_2CH_2CH_2OH$) is 8.3 g in 100 ml water. Its branched analogue, 2-methyl-propanol [$(CH_3)_2CHCH_2OH$] carries the same functional groups and molecular weight but has a higher solubility of 9.6 g in 100 ml water. The more branched analogue, t-butanol [$(CH_3)_3COH$], is freely mixable with water. The more branched hydrophobic hydrocarbon chain leads to a higher water solubility of organic molecules compared with their straight-chain counterparts. In Example 2, Compound D is the least water soluble among B, C, and D because of its symmetrical orientation of the two hydroxy groups, which is favorable for strong hydrogen bonding between the solute molecules. In Figure 3.3, codeine has a more hydrophobic methoxy group than the hydroxy group in morphine. However, because the orientation of the two hydroxy groups in morphine is ideal for solute–solute hydrogen bonding, the methyl group in codeine decreases the solute–solute hydrogen bonding more than it decreases the solute–water hydrogen bonding, leading to a more favorable dissolution process and hence a higher water solubility for codeine.

To summarize, it may be inferred that (a) molecules with branched chains are more soluble in water than the corresponding straight chain compounds, (b) molecules with polar functional groups are usually more soluble in water, (c) water solubility decreases with increases in molecular weight, and (d) solubility in water or in other organic solvent increases with increase in structural similarity between solute and solvent molecules (the same concept of like dissolves like).

B. Solute Crystal Structure

A solid substance may exist in crystals of different shapes or habits (Figure 3.7). In a specific habit of crystal, the angles between the faces of the crystal are always constant. A crystal is made up of atoms, ions, or molecules in a regular geometric arrangement or lattice constantly repeated in three dimensions. This repeating pattern is known as the **unit cell** of a crystal.

One of pharmacists' common tools for altering apparent solubility is called comminution, or reduction of the size of the solid solutes by grinding or crushing their crystals. This process partially compensates for the energy needed to break the interactions between the solute molecules and improves both the apparent solubility and the dissolution rate. While this technique is effective, there is a limit to how much the total solubility can be increased.

The capacity for a substance to crystallize into more than one crystalline form is called **polymorphism**, and the different crystalline forms are called **polymorphs**. It is possible for all solids to have polymorphs. The color, hardness, solubility, melting point, and other properties of a solid compound depend on its polymorphic crystal form. If the change from one polymorph to another is reversible, the process is **enantiotropic**. However, if the transition from a metastable to a stable polymorph is uni-directional, the system is **monotropic**. Since polymorphs vary in melting point, which is related to solubility, they may have different solubilities. If a wrong polymorph is chosen during the formulation process, the metastable polymorph (i.e., the thermodynamically unstable form) can convert to a more stable polymorph, resulting in changes in solubility and consequently alterations in the bioavailability.

Relating back to step 1 of dissolution (solute–solute interaction) in Section III.B, the spatial arrangement of a solute molecule in a crystal affects its movement from the solid phase into the

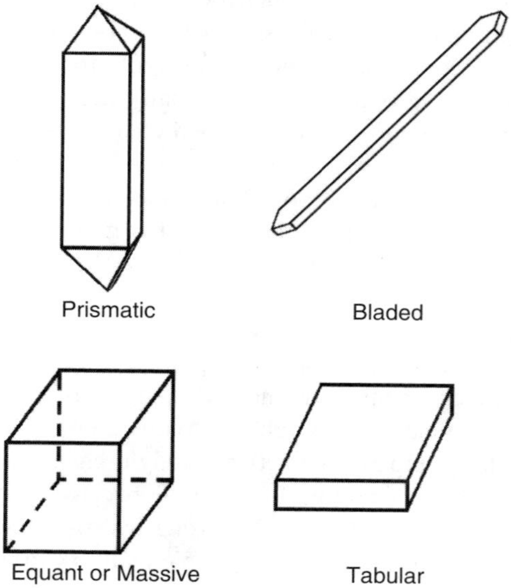

FIGURE 3.7 Examples of crystal habits.

solution phase. As a general rule, crystals of unsymmetrically packed molecules are more soluble than those of highly symmetrically packed molecules. Therefore, amorphous solids with irregularly packed molecules are usually more soluble than their crystalline counterparts.

C. Temperature

1. The General Trend of Solid Solutes in Liquid Solvents

If a substance is to be dissolved, the intermolecular forces between the solute molecules must be overcome. As the separation of solute molecules requires a certain amount of energy, which can be provided in the form of heat when the temperature is raised, the solubility of a solid in a solution depends on temperature, melting point of the solid, and molar heat of solution as outlined in Section III.A. With the increase of temperature, the solubility of solids in liquid solvents usually increases as predicted by the following equation:

$$\log \frac{S_2}{S_1} = \frac{\Delta H \left(T_2 - T_1\right)}{2.303 R.T_2.T_1}$$

where S_1 and S_2 are saturation solubilities of the solute in the given solvent at temperatures (in Kelvin) T_1 and T_2, respectively, ΔH is the molar heat of solution, and R is the universal gas constant. Though this equation applies to ideal solutions only, it may at least approximate solubility in nonideal conditions. Examination of the equation reveals that raising the temperature of a solution (from T_1 to T_2) will increase the solubility from S_1 to S_2 only if ΔH is positive (i.e., the dissolution process is endothermic) as in the case of dissolution of KNO_3, where there is a substantial absorption of heat during the dissolution process. On the contrary, if ΔH is negative (i.e., the dissolution process is exothermic), an increase in temperature will result in a decrease in solubility; for example, the water solubility of $Ca(OH)_2$ decreases with increasing temperature.

EXAMPLE 4

The molal solubility of $Ba(OH)_2.8H_2O$ in water at 20°C is 0.227 m. If the heat of solution of $Ba(OH)_2.8H_2O$ is 6719 cal/mole, predict the molal solubility of $Ba(OH)_2.8H_2O$ at 30°C.

Using the above equation,

$$S_{2(30°C)} = 0.332 \text{ molal}$$

(The student should keep in mind that S_2 and S_1 in the above equation should ideally be expressed as mole fraction solubility. However, as they appear as the fraction S_2/S_1, they can be expressed in any unit as long as they both are expressed in the same units.)

However, care needs to be taken when using heat to increase solubility because heat may destroy a drug or cause other changes in the solution. For example, if a sucrose solution is heated in the presence of an acid, sucrose will break down into glucose and fructose.

2. Solubility of Gas in Liquid

A special case of the effect of temperature on solubility is the dissolution of gas molecules into liquid solvents. Since the transfer of solute molecules from the gas phase into the liquid solvent phase significantly decreases their freedom of movement, the dissolution process involves a significant decrease of entropy ($\Delta S < 0$). As the temperature increases, the entropy term ($-T\Delta S$) of the free energy of dissolution ($\Delta G = \Delta H - T\Delta S$) increases, making it less favorable to dissolve gas molecules into the liquid solvent, meaning that as the temperature increases, the solubility of gas in liquid decreases.

D. pH

pH is one of the most important factors in the formulation process. The solubility and the stability of a drug in a formulation are often dependent on the pH of its solution.

1. Ionizable Substances

Weak acids and weak bases, which have ionizable groups (e.g., carboxylate groups in weak acids [HA] and amine groups in weak bases), dissolve in aqueous phase both in ionized and unionized forms, and the apparent solubility is the sum of the concentrations of the ionized form and the un-ionized form in a saturated solution.

$$S = S_0 + S_i$$

where S is the apparent total solubility of the compound; S_o is the solubility, or the concentration, of the un-ionized form of the compound in its saturated solution; and S_i is the concentration of the ionized form of the compound in its saturated solution.

In an aqueous solution, the ionized solute molecules have stronger ion–dipole interactions with the water molecules and therefore are more soluble than the un-ionized solute form. With the change of pH, the ratio of the ionized form and the un-ionized form changes, resulting in the change of the apparent total solubility (Figure 3.8). For a weak acid, the pH of the solution is related to the pK_a and the concentrations of the ionized and un-ionized forms of the solute by the Henderson–Hasselbalch equation:

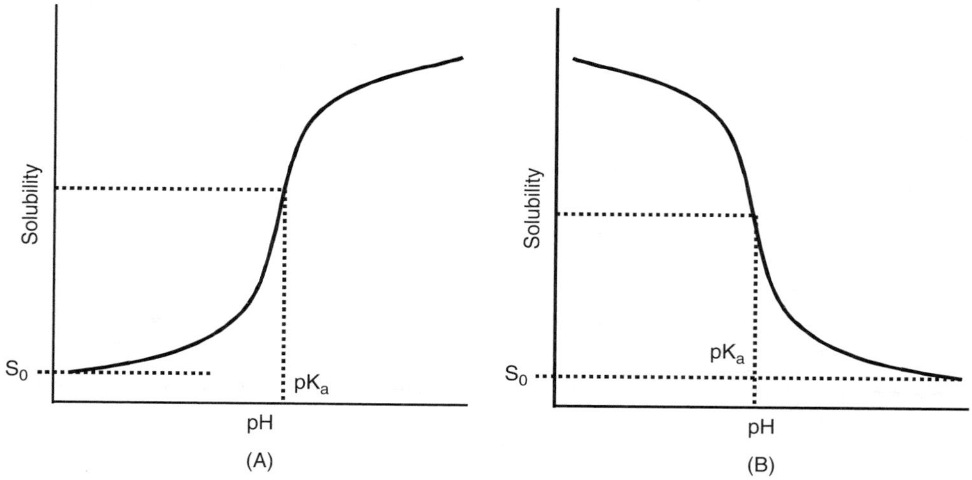

FIGURE 3.8 The effect of pH on a weak acid (A) and a weak base (B).

$$\mathrm{pH} = \mathrm{pK}_a + \log\left(\frac{[A^-]}{[\mathrm{HA}]}\right)$$

where the pK_a is the pH at which half of the compound is ionized and half is un-ionized (i.e., 50% A^- and 50% HA). For example, carboxylic acid groups (–COOH) have a pK_a of about pH 4, at which point 50% are COOH and 50% are COO^-. If the pH is above 4, then more COOH is converted to COO^-. The negatively charged carboxylate groups interact strongly with the partial positive charges of the hydrogens of water molecules. In a saturated solution, the un-ionized form of the weak acid ([HA]) is equal to S_o, and the ionized form of the weak acid ($[A^-]$) is equal to S_i, or $S - S_0$ ($S = S_0 + S_i$). Therefore, we can substitute [HA] and $[A^-]$ in the Henderson–Hasselbalch equation with S_0 and $S - S_0$ followed by rearrangements to obtain the following equation:

$$\mathrm{pH}_p = \mathrm{pK}_a + \log\left(\frac{S - S_0}{S_0}\right)$$

where pH_p is the pH below which the drug precipitates from solution as the undissociated acid, S is the total solubility, and S_0 is the intrinsic solubility of the undissociated acid. The dissociated or ionized form of a drug is much more soluble than the undissociated form. However, the dissociated form of the drug still has a maximum solubility by itself. For example, carboxylic acids have a pK_a of 4. If you were to take a methylpredinsolone hemisuccinate (solubility < 1 mg/ml) and add a base (e.g., NaOH), its carboxylic acid group will be deprotonated and ionized, thus increasing the solubility to about 200 mg/ml.

Similar rearrangements of the Henderson–Hasselbalch equation for a weak base give the following equation:

$$\mathrm{pH}_p = \mathrm{pK}_w - \mathrm{pK}_b + \log\left(\frac{S_0}{S - S_0}\right)$$

where pK_w is the dissociation constant of water, pK_b is the pK_b of the base, and pH_p is the pH above which the free base will precipitate out of solution.

EXAMPLE 5

Below what pH will homatropine (mol wt = 275) precipitate out from an aqueous solution of 2% homatropine hydrobromide (mol wt = 356)? Given water solubilities of homatropine and homatropine hydrobromide are 1 g/500 ml and 1 g/6 ml, respectively. The pKa of homatropine hydrobromide is 9.65.

S_0 = 1 g/500 mL = (1 g/500 mL) × (1000 mL/L) × (1 M/mol wt of free base) = 0.0072 M
S = 2% w/v = 2 g/100 mL = (2 g/100 mL) . (1000 mL/L) × (1 M/mol wt of salt) = 0.056 M

From the equation $pH_p = pK_w - pK_b + \log\left(\frac{S_0}{S - S_0}\right)$

$pH_p = 8.82$

2. Nonionizable Substances

pH has little effect on the solubility of nonionizable compounds. The solubility of these compounds can usually be improved by lowering the dielectric constant of the solvent with a co-solvent (such as ethanol in elixirs) rather than changing the pH of the solvent.

VI. SOLUBILITY ENHANCEMENT

Solubility is essential to a drug's success. Most drugs are organic and will not go into an aqueous solution easily. There are several different ways to enhance solubility, and the method of choice depends on the nature of the solute, the required degree of solubilization, and other formulation restrictions such as drug stability and compatibility with therapeutic use. This section discusses the pharmaceutical applications of the factors that effect solubility and additional techniques used to manipulate solubility.

For most pharmaceutical applications, water is the preferred solvent because it meets more of the above criteria than other solvents. However, for practical reasons, in many pharmaceutical preparations, an auxiliary solvent also known as a **co-solvent** is employed to impart desirable characteristics to the preparations. Such characteristics may include enhancement of solubility of the ingredients to impart physical and chemical stability to the overall preparation. The addition of a co-solvent can increase the solubility of hydrophobic molecules by reducing the dielectric constant of the solvent. Some of the solvents being used in pharmaceutical preparations include ethyl alcohol, glycerin, sorbitol, propylene glycol, and polyethylene glycols. Students should keep in mind that under no circumstances are methyl alcohol and ethylene glycol permitted in any pharmaceutical preparation because of their systemic toxicity and that a number of fixed oils such as corn oil, cottonseed oil, sesame oil, and peanut oil are used, particularly in the preparation of oleaginous injections. However, the use of peanut oil is now being discouraged because of the prevalence of allergy to peanuts. One problem with the use of co-solvents is the precipitation of the drug upon dilution during the administration of the drug solution into the body, resulting in pain or tissue damage.

Another method of enhancing solubility is to alter the structure of the drug molecule; this is one basis for the use of prodrugs. A **prodrug** is a drug that is therapeutically inactive when administered but becomes activated in the body by either chemical or enzymatic processing. The addition of polar groups such as carboxylic acids, ketones, and amines can increase aqueous solubility by increasing the hydrogen bonding and other molecular interactions between the drug

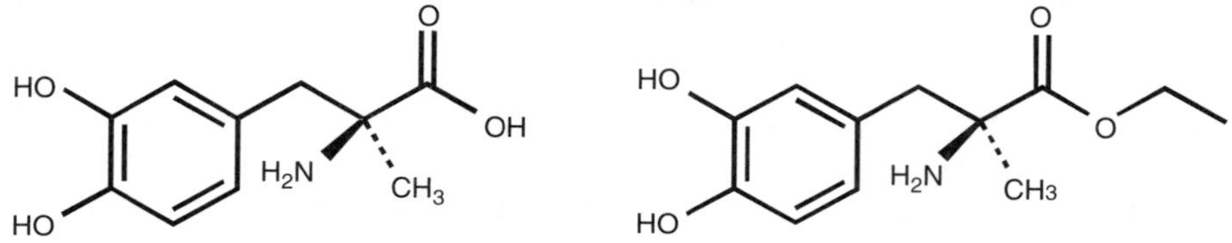

FIGURE 3.9 Structures of methyldopa (left) and methyldopate (right), the prodrug of methyldopa.

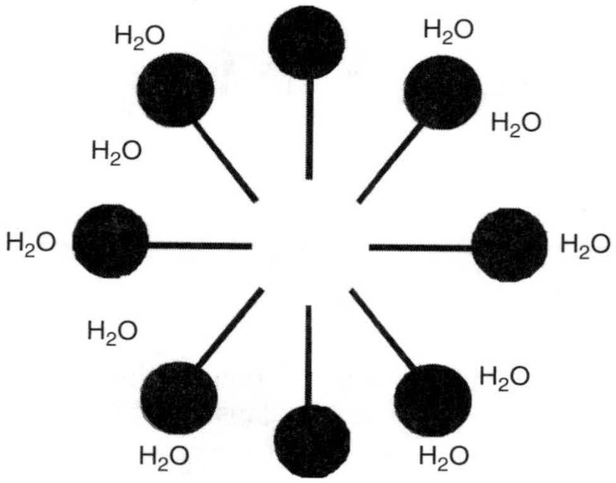

FIGURE 3.10 Formation of micelle in water.

and water. Another possible structure modification is to reduce interactions between solute molecules. One example of the use of prodrug design to enhance solubility is the conversion of methyldopa (solubility ~10 mg/ml) to methyldopate (solubility 10 to 300 mg/ml depending on pH) (Figure 3.9). Conversion of the carboxylic acid group to its ethyl ester in methyldopa decreases solute–solute hydrogen bonding. Therefore, this prodrug conversion reduces the melting point and increases the solubility. This is somewhat counter-intuitive because it is another exception to the trend that the addition of a hydrophobic group (ethyl in this case) usually reduces solubility in water. In the body, the ester is cleaved to yield the active drug, methyldopa.

Surfactants can also be used to enhance solubility. A surfactant, or surface-active agent, is amphipathic, meaning its molecular structure has a polar end (the circular head in Figure 3.10) and a nonpolar end (the tail). When a surfactant is placed in water it forms micelles (Figure 3.10) at concentrations above its critical micelle concentration. A nonpolar drug can partition into the hydrophobic core of the micelle while the polar ends of the surfactant molecules on the surface of the micelle solubilizes the complex in aqueous media.

The final method to be discussed is complexation of drug molecules using complexing agents. There are many types of complexing agents, and a partial list can be found in Table 3.5. Complexation relies on the relatively weak forces (London forces, hydrogen bonding, and hydrophobic interactions) between the drug and the complexing agent. As the concentration of the complexing agent is increased, so is the solubility, up to a maximum point. In some cases when the complexing agent is too concentrated, the complex can precipitate out of solution as more complexing agent is added.

VII. CONCLUSION

Solubility and dissolution is one of the most important concepts for a pharmacist to understand and to apply because for a drug to have a pharmacological effect, it must be able to interact with its target, and most often this interaction occurs in a solution phase. In order for most drugs to be absorbed by the gastrointestinal tract or pass other epithelial barriers, the drug must be in solution.

TABLE 3.5
List of Complexing Agents

Type	Example
Inorganic (ligands, coordination)	Hg, I_2 hexamminecobalt III chloride
Organic	Caffeine, phenols, ethers, ketones
Chelation	Ethylenediaminetetraacetic acid (EDTA)
Inclusion	Choleic acid, cyclodextrins
Molecular complexes	Polymers (polyethylene glycols, methyl and carboxy methyl cellulose, polyvinyl-pyrrolidone, etc.)

One question that might be asked then is why not make all dosage forms as solutions? The answer to this question encompasses a number of other concerns, such as patient convenience and stability of the resultant dosage form. Stability will be covered in a later chapter. In the preparation of a dosage form, a pharmacist has several tools at his disposal to aid in the dissolution of drug substances. The pharmacist's first tool for altering the apparent solubility is often comminution, that is, particle size reduction, and this process also impacts on the dissolution rate. While this technique is effective, there is a limit to how much the total solubility can be increased. Another common method is changing the pH, which generates the ionized form of a drug, which is more soluble in water. Different salts of the same drug can also be used to generate crystals with different lattice energies and hence different solubilities. If these methods do not work, then more advanced methods, such as micellization or complexation, may be used.

GLOSSARY

Dielectric constant The ability of a medium to diminish the strength of electrostatic interactions; also used as a measure of polarity. The higher the value, the more polar the substance.

Dissolution The process of a solid going into solution.

Enantiotropic transition A reversible change from one crystal from to another crystal form.

Habit The shape of a crystal.

Heat of solution The net change of enthalpy that results from solute–solvent interactions as the solute dissolves.

Ideal solution A solution in which the heat of solution is equal to the heat of fusion and no net heat is released or absorbed during the formation of the solution.

Equivalent weight The chemical equivalent weight of a molecule or atom of an element; equals to the molecular or atomic weight divided by valence.

Molarity Number of moles of solute per liter of solution.

Molality Number of moles of solute per kilogram of solvent.

Mole fraction Ratio of moles of one component to the total moles of a mixture.

Monotropic transition Uni-directional transition of a metastable polymorph to a stable polymorph.

Nonideal solution A solution that does not behave as an ideal solution over a wide range of temperature and pressure.

Osmolality Number of moles of particles (molecules or ions) dissolved in one kilogram of solvent.

Osmolarity

Polymorphism The existence of different crystal habits of the same compound.

Prodrug A drug that is therapeutically inactive when administered but becomes activated in the body by either chemical or enzymatic processing.

Saturated solution A solution that is in equilibrium between dissolved solute and undissolved solute at a given temperature and pressure.

Solubility The concentration of a solute in its saturated solution at a given temperature and pressure.
Solubility parameter (δ) A measure of the cohesive forces between similar molecules.
Solute An agent dissolved or to be dissolved in a solution.
Solution A homogenous molecular dispersion of a mixture of two or more components.
Solvent The agent that dissolves the solute into a solution.
Supersaturated solution A solution that contains a higher concentration of dissolved solute than a saturated solution would at a given temperature and pressure.
Unit cell A repeating pattern of regular geometric arrangement or three-dimensional lattice found in crystals.
Unsaturated solution A solution containing the dissolved solute in a concentration lower than that of a saturated solution for the same temperature and pressure.

TUTORIAL

1. The solubility of oxandrolone in ethanol is 1 g in 57 ml. What is the concentration of 57 ml of this solution in weight/volume, weight/weight, and molarity?

   Molecular weight of oxandrolone = 306.45 g/mol

   Molecular weight of ethanol = 46.07 g/mol

   Density of ethanol = 0.810 g/ml

2. The maximum solubility of benzene in water is 0.02 M. However, phenol can make a 0.7 M solution. Why is phenol more soluble in water than benzene? How would toluene's solubility compare to benzene and phenol?
3. The solubility of a drug in water is determined by:
   (a) Charge of the molecule
   (b) Lipophilicity
   (c) Log P (octanol:water partition coefficient)
   (d) Water temperature
   (e) All the above

ANSWERS

1. Weight per volume

$$X = \frac{1\ \text{g}}{57\ \text{ml}} = 0.0175\ \text{g/ml or } 17.5\ \text{mg/ml}$$

Weight per weight

Weight of 57 ml of ethanol

$$X = \text{density} * \text{volume} = 0.810\ \text{g/ml} * 57\ \text{ml} = 46.2\ \text{g}$$

$$Y = \frac{1\ \text{g}}{46.2\ \text{g}} = 0.0217\ \text{g/g}$$

Molarity

$$X = \text{grams} * \frac{\text{mol}}{\text{gram}} = 1\ \text{g} * \frac{1\ \text{mol}}{306.45\ \text{g}} = 0.00326\ \text{mol}$$

$$Y = 57\ \text{ml} * \frac{1\ \text{L}}{1000\ \text{ml}} = 0.057\ \text{L}$$

$$\text{molarity} = \frac{\text{mol}}{\text{liter}} = \frac{0.00326\ \text{mol}}{0.057\ \text{L}} = 0.0572\ \text{M or } 57.2\ \text{mM}$$

2.

Benzene

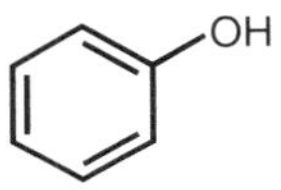

Phenol

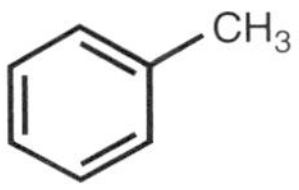

Toluene (solubility ~7 μM)

Phenol is more soluble because of the H-bonding between phenol –OH and water. The solubility of toluene in water is smaller than benzene and phenol because toluene has an additional nonpolar $–CH_3$ group than benzene.

3. E

HOMEWORK

1. The intrinsic solubility of a drug, sulfathiazole, is ~0.002 M and its pKa is 7.12. What are the lowest pHs at which a 1% and 5% solution of Na sulfathiazole (mol wt = 304) are completely soluble?
2. Silver iodide is a strong electrolyte and completely dissociates in aqueous solution at 25°C. If its aqueous solubility is 1.22×10^{-8} mole/liter, calculate its solubility product.
3. Below what pH will phenobaribital (mol wt = 232) begin to precipitate from a 1 g/100 ml sodium phenobarbital (mol wt = 254) solution (pKa for phenobarbital = 7.4; intrinsic water solubility of phenobarbital = 1 g/1000 ml)?
4. The solubility of boric acid in an aqueous solvent using 25% by volume of sorbitol as co-solvent is 2.08 molal at 35°C. Predict the molal solubility of boric acid in that mixed solvent at room temperature (25°C). (Given heat of solution of boric acid in this mixed solvent is 3470 cal/mol.)

5. Above what pH will physostigmine precipitate from a 1.25% w/v solution of the salt, physostigmine salicylate (pKa = 7.88)?

	Physostigmine	Physostigmine salicylate
Molecular Weight	275	413
Water Solubility	1 g/1000 ml	1 g/75 ml

ANSWERS TO HOMEWORK

1. 8.31 and 9.03
2. 1.5×10^{-16}
3. pH = 8.3
4. $S_{25°C}$ = 1.72 molal
5. pH = 6.92

CASES

CASE I

Simple syrup USP is made by dissolving 850 g of sucrose (mol wt 180 g/mol) in a sufficient quantity of water to make 1 l of solution. The process can be accelerated by moderate heating and stirring. Simple syrup is often used as a vehicle for the preparation of other liquid dosage forms.

1. What is the molarity of simple syrup USP?
2. Is the process of solution formation an endothermic or exothermic process?
3. Usually when making simple syrup it is suggested to use low heat instead of high heat. Can you hypothesize a reason?

ANSWERS

1. 4.7 M
2. Endothermic
3. Heating could lead to chemical reactions. In the case of sucrose, the molecule may decompose in acidic solutions.

CASE II

Elixirs are hydroalcoholic solutions (commonly containing ethanol) that pharmacists use in the preparation of particular oral dosage forms of certain drugs. Consider an elixir made with drug A (1600 mg), ethanol (15 ml), simple syrup (USP, 40 ml), and water (40 ml). The pharmacist prepares the formulation by adding the drug to the ethanol, followed by the water and finally the simple syrup. This product is then dispensed to the patient.

1. What is the percent w/v for drug A for this prescription?
2. What is the purpose of adding ethanol to this product?
3. If the pharmacist added the drug to the simple syrup USP first instead of the ethanol, what impact would this have on the formulation?
4. When dispensing this prescription, the pharmacist gives the following advice: "Please replace the cap tightly after each use." What is the rationale behind this statement?

ANSWERS:

1. 1.7%
2. The addition of ethanol to water lowers the dielectric constant of the solution, creating a solution that can better solubilize lipophilic compounds.
3. Simple syrup has a high viscosity (very thick), which would increase the time it takes for the drug to go into solution. It is always a good idea to use the less viscous liquid first in making oral dosage forms.
4. There are several reasons for making this recommendation. One reason relevant to the solubility is that ethanol might evaporate from the product. If this occurs, the dielectric constant will increase, which could lead to the precipitation of the drug.

CASE III

Suppose you are working for a pharmaceutical company and you find that a drug has three polymorphs with different solubilities as seen below. Which polymorph should be used in product development of an oral solid dosage form?

Polymorph	Water Solubility (mg/100 ml)	Melting Point (°C)
I	30	200
II	60	185
III	15	225

ANSWER

The best choice would probably be polymorph III from stability point of view. Since it has the lowest solubility, it probably has the higher melting point and is the most stable crystal lattice. Therefore this polymorph is probably the most thermodynamically stable and does not convert to other forms. However, because of the low solubility, it may encounter low bioavailability problems upon oral administration.

CASE IV

You are assessing the analgesic behavior of two compounds shown below. You know their melting points and aqueous solubilities. Complete the table to associate the two drugs with the solubility and melting point.

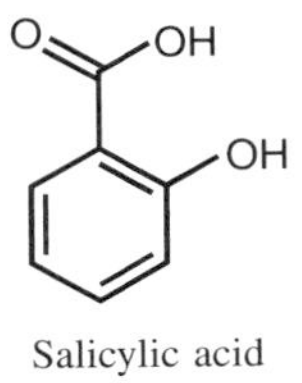

Salicylic acid

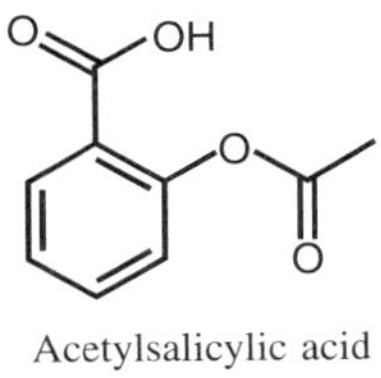

Acetylsalicylic acid

Solubility	Melting Point (°C)	Drug
1 g in 460 ml	157–159	
1 g in 300 ml	135	

Answer

Solubility	Melting Point (°C)	Drug
1 g in 460 ml	157–159	Salicyclic acid
1 g in 300 ml	135	Acetylsalicyclic acid

CASE V

In your pharmaceutics laboratory, you were asked to prepare one liter of 5.0% aqueous solution of sodium sulfadiazine for ophthalmic use. You prepared it and left home for the day. The following week, when you checked your preparation, you observed a precipitate at the bottom of the bottle.

1. What should you check first to identify the probable cause of the precipitation?
2. How will you confirm your hypothesis?
3. How will you fix it?
4. What other precaution should be taken to prepare an ophthalmic solution?

Answers

1. You should first check the pH of the solution, and calculate the pH below which sulfadiazine will precipitate out.
2. When you checked the pH of the solution, it was 9.0. Given the following information, first calculate the pH below which sulfadiazine will precipitate out.

Molecular weight of salt = 272 and of free acid = 250

Aqueous solubility of sulfadiazine = 1 g/13,000 ml of water

$$\text{pK}_\text{a} = 6.48$$

From the equation below,

$$\text{pH}_p = \text{pK}_a + \log\left(\frac{S - S_0}{S_0}\right)$$

Therefore you need to first calculate S and S_0.

S = 5% w/v = 5 g/100 ml = (5 g/100 ml) . (1000 ml/l) . (1 M/mol wt of salt) = 0.18 M

S_0 = 1 g/13,000 ml = (1 g/13,000 ml) . (1000 ml/l) . (1 M/mol wt of free acid) = 0.00031 M

$$S - S_0 = 0.18$$

$$\text{pH}_p = 6.48 + \log(0.18/.00031) = 9.24$$

3. Therefore, below pH 9.24 sulfadiazine is supposed to precipitate out. As you checked the pH to be 9.0, therefore it confirms the physical phenomena. What might have happened is that the pH was all right initially but shifted to a lower pH upon standing. Therefore when you prepare the solution again, you should adjust the pH of the solution above 9.24 with a suitable buffer system so that this pH shift will not occur again.
4. The ophthalmic solution also should be sterile.

CASE VI

When a student was asked to measure the boiling point of water, she found that some little bubbles appeared in a water sample from a pond when it was heated from room temperature to about 90°C. The bubbles kept coming out until the temperature rose to about 100°C, at which point bigger bubbles came out of the water at a faster pace. What would be her best way of measuring the boiling point of this sample?

Answers

Since the water sample was from a pond, it must have been exposed to air for some time, resulting in a water solution of gas molecules from the atmosphere. As the temperature of the water increased, the solubility of the gas in water decreased, and the gas molecules escaped water to form little bubbles. The true boiling process did not occur until water was heated to about 100°C, when bigger bubbles of vaporized water came out. Therefore, the best reaction to the phenomenon is to let the water boil for a little while to get rid of the excess dissolved gas and to reach a stable temperature, which will then be recorded as the boiling point of water.

REFERENCES

1. Ansel, H.C., Allen, L.V., and Popovich, N.G., *Pharmaceutical Dosage Forms and Drug Delivery Systems,* 7th ed., Lippincott Williams and Wilkins, Philadelphia, 1999.
2. Gennaro, A.R., *Remington: The Science and Practice of Pharmacy,* 20th ed., Lippincott Williams & Wilkins, Philadelphia, 2000.
3. Martin, A., Bustamante, P., and Chun A.H.C., *Physical Pharmacy: Physcial Chemical Priniciples in the Pharmaceutical Sciences,* 4th ed., Lea & Febiger, Philadelphia, 1993.

4 Physicochemical Factors Affecting Biological Activity

Hemant Alur, Blisse Jain, Surajit Dey, Thomas P. Johnston, Bhaskara R. Jasti, and Ashim K. Mitra

CONTENTS

0-4152-8863-0/05/$0.00+$1.50
© 2005 by CRC Press LLC

Learning Objectives

After finishing this chapter the student will have thorough knowledge of:

- Acid-base chemistry
- The methods used to determine ionization of weak acids and bases
- The characteristics of buffer solutions
- Partitioning and extraction of drugs
- Colligative properties
- How to calculate molecular weights from colligative properties
- How to calculate milliequivalents and osmolar strength of a solution
- How to adjust the tonicity of a pharmaceutical formulation

I. INTRODUCTION

Drugs are organic compounds, and as a result, their activity, their solubility in plasma, and their distribution to various tissues are all dependent on their physicochemical properties. Even the interaction of a drug with a receptor or an enzyme is dependent on characteristics of a drug molecule, such as ionization, polarity, and electronegativity. If we are to understand drug action, we must also understand the physicochemical parameters that make this action possible. The following sections explain the acid-base and physicochemical properties, which determine drug action.

II. ACID-BASE CHEMISTRY

A. Theories of Acids and Bases

According to the *Bronsted–Lowry theory,* an acid is a substance, charged or uncharged, that is capable of donating a proton, and a base is a substance, charged or uncharged, that is capable of accepting a proton.

Thus, in the reaction between CH_3COOH and water, CH_3COOH is the acid and water is the base.

$$\underset{\text{acid 1}}{CH_3COOH} + \underset{\text{base 2}}{H_2O} \leftrightarrow \underset{\text{acid 2}}{H_3O^+} + \underset{\text{base 1}}{CH_3COO^-}$$

Acid 1 and base 1 stand for acid–base pair, or conjugate pair, as do base 2 and acid 2. The strength of an acid or a base, to give up or take on protons, varies with the solvent. Solvents may be classified as protophillic, protogenic, amphiprotic, and aprotic. A protophillic, or basic, solvent is capable of accepting protons from the solute. Solvents such as acetone, ether, and liquid ammonia are examples of this class of solvents. A protogenic solvent donates proton and acids. Formic acid, acetic acid, and sulfuric acid represent this class of solvents. Amphiprotic solvents act as both proton acceptors and proton donors. Water and the alcohols are examples of this class of solvents. Aprotic solvents neither accept nor donate protons, that is, they are neutral. This class includes the hydrocarbons.

According to *Lewis electronic theory,* an acid is a molecule or ion that accepts an electron pair to form a covalent bond. A base is a molecule that donates an electron pair. An example of Lewis acid–base reaction is as follows:

$$H^+ + :NH_3 \leftrightarrow NH_4^+$$

B. Acid–Base Equilibria

Most chemical reactions proceed in both a forward and reverse direction if products of the reaction are not removed as they form. Therefore, equilibrium may be defined as a balance between the rates of forward and reverse reactions. Hence the ionization or protolysis of CH_3COOH can be written as:

$$CH_3COOH + H_2O \leftrightarrow H_3O^+ + CH_3COO^-$$

The velocity of forward reaction R_f, according to the law of mass action, is proportional to the concentration of the reactants:

$$R_f = k_1 \times [CH_3COOH] \times [H_2O]$$

Similarly, the velocity of the reverse reaction R_r can be written as

$$R_r = k_2 \times [H_3O^+] \times [CH_3COO^-]$$

k_1 and k_2 are proportionality constants known as specific reaction rates for forward and reverse reactions, respectively, and the brackets indicate concentrations.

1. Ionization of Weak Acids

Consider the following general equation for an uncharged weak acid, H:B:

$$H:B + H_2O \leftrightarrow H_3O^+ + B^-$$

At equilibrium

$$R_f = R_r \text{ or}$$

$$k_1 \times [HB] \times [H_2O] = k_2 \times [H_3O^+] \times [B^-]$$

Solving for the ratio k_1/k_2 yields

$$k = \frac{k_1}{k_2} = \frac{[H_3O^+][B^-]}{[HB][H_2O]}$$

In dilute solutions of weak acid, water is in sufficient excess to be regarded as constant. It can thus be combined with k_1/k_2 to obtain a new constant, K_a, the ionization constant or dissociation constant of weak acid.

$$K_a = \frac{[H_3O^+][B^-]}{[HB]}$$

An example of such an acid is acetic acid. Another example is m-methylphenol, which on treatment with an aqueous base, ionizes to the phenolate form as shown in Figure 4.1.

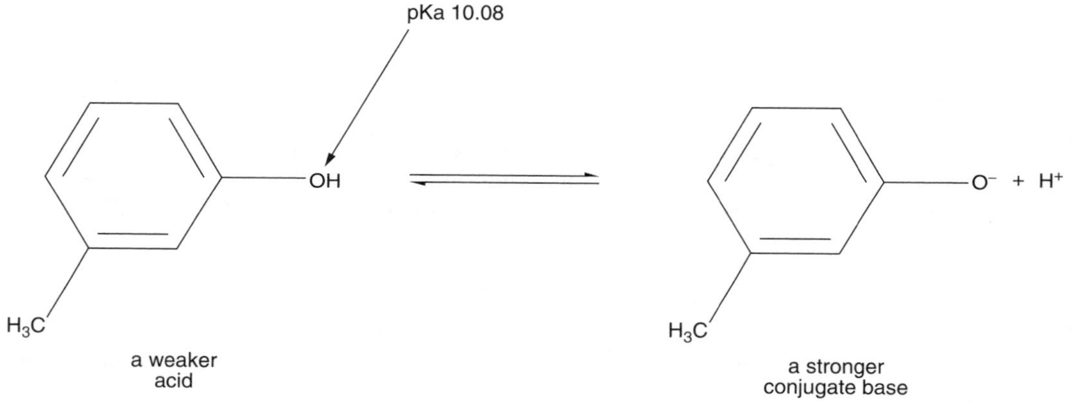

FIGURE 4.1 Ionization of m-methylphenol.

Similarly, in general, for charged acids, BH^+, the reaction is written as

$$BH^+ + H_2O \leftrightarrow H_3O^+ + B$$

and the acidity constant is

$$K_a = \frac{[H_3O^+][B]}{[BH^+]}.$$

Acetic acid has a pK_a of 4.75. What species would exist predominantly in solution at neutral pH?

Answer: Acetic acid, with a pK_a of 4.75, is a relatively stronger acid, and the conjugate base is a weaker base. This indicates that the compound would favor the B^- form at neutral pH, and this is reflected by the lower pK_a value.

---

2. Ionization of Weak Bases

Consider the following general equation for an uncharged weak base, B:

$$B: + H_2O \leftrightarrow OH^- + BH^+$$

At equilibrium

$$R_f = R_r$$

and the basicity constant for bases can be written as

$$K_b = \frac{[OH^-][BH^+]}{[B]}$$

An example of such a base is methylamine, the conjugate acid of which is cationic:

$$\underset{\text{a neutral base}}{H_3C{-}NH_2} + H^+ \rightleftharpoons \underset{\text{a cationic conjugate acid}}{H_3C{-}\overset{+}{N}H_3}$$

Similarly, in general, for charged bases, B^-

$$B^- + H_2O \leftrightarrow OH^- + HB$$

and the basicity constant is

$$K_b = \frac{[OH^-][HB]}{[B^-]}$$

---

m-methylphenol has a pK_a of 10.08. What species would exist predominantly in solution in water?

Answer: *m*-methylphenol, with a pK_a of 10.08, dissociates in water to form the corresponding phenolate. In this example, the BH form is a relatively weaker acid, and the conjugate base is a relatively stronger base. This is reflected in the high pK_a value and indicates that the equilibrium would lie to the left at neutral pH.

---

3. Ionization of Water

Water ionizes slightly to yield hydrogen and hydroxyl ions, in a manner similar to the dissociation of weak acids and bases. This autoprotolytic reaction can be written as

$$H_2O + H_2O \leftrightarrow H_3O^+ + OH^-$$

The equilibrium expression would be

$$k = \frac{[H_3O^+][OH^-]}{[H_2O]^2}$$

Since the molecular water is in great excess relative to the hydrogen and hydroxyl ion concentrations, $[H_2O]^2$ is considered as a constant and is combined with k to give K_w, known as the dissociation constant, or the ion product of water:

$$K_w = k \times [H_2O]^2$$

The value of the ion product of water is approximately 1×10^{-14} at 25°C and depends greatly on temperature. From the above two equations a common expression for the ionization of water at 25°C can be written:

$$[H_3O^+][OH^-] = K_w = \sqrt{1 \times 10^{-14}}$$

In pure water, the hydrogen and hydroxyl ion concentrations are equal:

$$\therefore [H_3O^+] = [OH^-] = \sqrt{1 \times 10^{-14}} = 1 \times 10^{-7}$$

C. Relationship between K_a and K_b

The relationship between the dissociation constant of a weak acid, HB, and that of its conjugate base, B^-, or between BH^+ and B, can be obtained as follows:

$$K_a K_b = \frac{[H_3O^+][B^-]}{[HB]} \times \frac{[OH^-][HB]]}{[B^-]} = [H_3O^+][OH^-] = K_w$$

and

$$K_b = \frac{K_w}{K_a} \text{ or } K_a = \frac{K_w}{K_b}$$

D. Physicochemical Properties of Drugs

Most conventional drug substances have the following properties:

- They have a molecular weight range of 100 to 850.
- They are either weak acids or weak bases.

Owing to varying pHs of the regions of the gastrointestinal tract, these weak acids and bases may be ionized or un-ionized. As such, using a weak acid such as aspirin (ASA) as an example, the following reaction scheme is applicable along with the ionization constant, K_a.

$$ASA_{(aq)} \leftrightarrow H^+_{(aq)} + ASA^-_{(aq)}$$

where ASA^- is the conjugate base.

Recall that K_a is defined as

$$K_a = \frac{\left[H^+\right]\left[ASA^-\right]}{\left[ASA\right]}$$

where K_a is defined as the ionization constant or the acid dissociation constant. Recall also that the pK_a is defined as

$$pK_a = -\log K_a = \log\left[1/K_a\right]$$

Thus, the stronger the acid:

- The larger the value of K_a
- The greater the degree of dissociation
- The smaller the value of the pK_a

Conversely, for the conjugate base, ASA^-, we can write an equilibrium expression:

$$ASA^-_{(aq)} + H_2O \leftrightarrow ASA_{(aq)} + OH^-_{(aq)}$$

where ASA is the conjugate acid.

Recall that K_b is defined as the base dissociation constant and is equal to

$$K_b = \frac{[ASA][OH^-]}{[ASA^-]}$$

$K_a K_b = K_w$ (which can be shown by multiplying the K_a and K_b expressions above), where K_w is the ionization constant of water at 25°C and is equal to

$$1 \times 10^{-14},\ K_b = \frac{K_w}{K_a} \text{ and } pK_a + pK_b = 14$$

That means that for the same pair of conjugate acid and base, K_a and K_b are inversely proportional. Typically, it is the value of pK_a that is listed in tables of drugs that are weak bases.

According to the pH partition hypothesis, the un-ionized form of a weakly basic or a weakly acidic drug partitions into the lipid bilayer of the membrane. Thus, it is important to be familiar with an expression that relates the percent or fraction of the drug in the ionized and the un-ionized state because of the regional differences in the pH of the gastrointestinal tract. This equation is called the *Henderson–Hasselbalch equation*.

E. The Henderson–Hasselbalch Equation

For a drug that is a weak acid,

$$\text{pH} = \text{p}K_a + \log\frac{[\textit{ionized}]}{[\textit{unionized}]}, \text{ where } [\textit{ionized}] = [\text{salt}] \text{ and } [\textit{unionized}] = [\text{acid}]$$

whereas for a drug that is a **weak base**, we have the expression

$$\text{pH} = \text{p}K_a + \log\frac{[\textit{unionized}]}{[\textit{ionized}]}, \text{ where } [\text{unionized}] = [\textit{base}] \text{ and } [\textit{ionized}] = [\text{salt}]$$

However, an expression that always applies whether we are dealing with a drug that is a weak acid or a weak base is the following:

$$\text{pH} = \text{p}K_a + \log\frac{[\textit{unprotonated}]}{[\textit{protonated}]}$$

where, for a drug that is a **weak base**, the unprotonated species is the base and the protonated species is the salt, and for a **weak acid**, the unprotonated species is the salt and the protonated species is the acid.

Using a weak acid as an example, its pK_a will determine what fraction of the drug exists in the un-ionized and ionized forms at a particular pH. For example, at a pH equal to the pK_a of the weakly acidic drug, 50% of the drug would exist in the ionized form and 50% of the drug would exist in the un-ionized form. Depending on the pH of the solution microenvironment, the amount of an acidic or basic drug, which exists in both the un-ionized and ionized states, can be calculated. Again, using a weak acid as an example, let us assume that we are in the stomach, where the pH might be from 1 to 3 (e.g., 2.1). Then, if we have an acidic drug with a pK_a of 3.5 such as aspirin, we can predict not only how much of the drug is in the ionized and un-ionized forms, but also whether absorption would be favored.

ASA has a pK_a of 3.5. At a gastric pH of 2.1, what percentages of the drug would exist in the ionized and un-ionized forms?

Answer:

$$2.1 = 3.5 + \log\frac{[\textit{ionized}]}{[\textit{unionized}]} \text{ or } 0.03981 = \log\frac{[\textit{ionized}]}{[\textit{unionized}]}$$

$$\text{or } [\text{ionized}] = 0.03981\ [\text{unionized}]$$

Also, we know that

$$[\text{ionized}] + [\text{unionized}] = 1$$

$$0.03981\ [\text{unionized}] + [\text{unionized}] = 1$$

$$[\text{unionized}] = \boxed{\mathbf{96.2\%}} \text{ and } [\text{ionized}] = \boxed{\mathbf{3.8\%}}$$

Would absorption be expected for this drug in the stomach at a pH of 2.1 in view of the pH partition hypothesis?

Answer: Yes, because the drug exists primarily in the un-ionized form.

When calculating percent ionized values, it is necessary to determine whether the B: or H:B form is the ionized species. This can be readily determined from the acid–base equation, where by convention the H:B form is on the left and the B: form is on the right.

F. General Principles

With a **weak acid**, as the pH of the surrounding solution environment is lowered, the fraction of the drug in the un-ionized state increases. Similarly, with a **weak base**, as the pH of the surrounding solution environment is raised, the fraction of the drug in the un-ionized state increases.

Schematically, another method used to obtain a quick estimate of whether a drug that is a weak acid or weak base exists in the un-ionized or ionized state as a function of the pH is the following:

	Weak Acid	Weak Base
If the pH is less than pK_a	Un-ionized	Ionized
If the pH is greater than pK_a	Ionized	Un-ionized

G. pH of Precipitation

Most drugs are salts of weak acids or bases. These salts are usually water soluble, whereas most of their un-ionized acid or base forms are practically insoluble. For compounds that are insoluble or only sparingly soluble and form acidic or basic solutions, the pH of the solution can significantly affect their solubility. Thus, if a drug solution of a salt of a weak base is made alkaline, free base might precipitate. Similarly, a free acid might precipitate if a salt solution of a weak acid is acidified. For example, consider the salt of a weak acid such as penicillin G potassium in solution. A saturated solution of this slightly soluble free acid form (represented as HPG) exists in equilibrium as

$$HPG_{(solid)} \leftrightarrow HPG_{(sol)} \text{ and}$$

$$HPG_{(sol)} + H_2O \leftrightarrow H_3O^+ + PG^-$$

where PG^- represents the soluble ionized form of penicillin G.

How do we know if the addition of an acid or base to this solution will cause precipitation or not?

This depends on:

- The solubility of the un-ionized acid
- The pH of the solution in which the drug is dissolved
- The dissociation constant of the acid

The pH of precipitation of a weakly acidic or basic drug can be found with a modified Henderson–Hasselbalch equation. Recall that the general form of the equation is

$$\mathrm{pH} = \mathrm{p}K_a + \log\frac{[unprotonated]}{[protonated]}$$

where, for a weak acid, the unprotonated species is the salt or the conjugate base and the protonated species is the free acid. Similarly, for a weak base, the unprotonated species is the free base and the protonated species is the salt or the conjugate acid.

In Figure 4.2, two H:B forms are shown that have the same pK_a. Phenols are weak acids and amines are weak bases; however, each compound has a B: and H:B form. m-Methylphenol is a **molecular acid**, meaning that it has a neutral charge. In this form, it is poorly water soluble. If you treat this compound with an aqueous base, it ionizes to the phenolate form. Since this form is anionic, it is now water soluble. The second compound, N-methylpiperidine, is a **cationic acid** (i.e., it is positively charged), and in this form it is water soluble. Treatment of this compound with an aqueous base produces a **molecular conjugate base**, which is no longer water soluble.

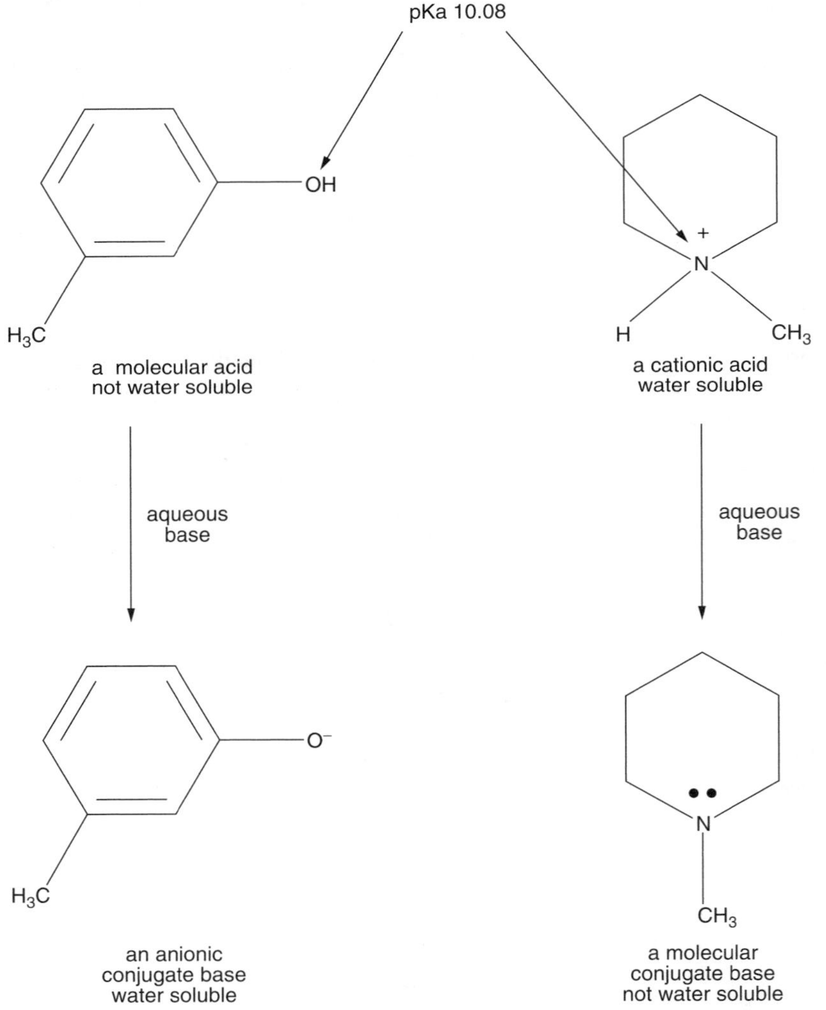

FIGURE 4.2 Deprotonation of a weak acid and a weak base having the same pKa.

H. Salt of a Weak Acid

Let us calculate the exact pH of precipitation for a salt solution of a weak acid. *It should be remembered that it is the un-ionized drug molecule that precipitates out of solution.* At the point of precipitation, the concentration of the protonated species represents the concentration of the free acid (un-ionized form), which will then be equal to the solubility of the free acid and is represented by S_0. (S_0 expressed in mass/volume is actually a concentration term.) At the same time, the concentration of the unprotonated form (conjugate base) will then be the difference between the initial amount of conjugate base added and the amount of conjugate base that is converted to the free acid as we lower the pH.

Now, the initial amount of conjugate base added equals the initial amount of salt added, which equals C_s. The amount converted to the free acid equals the solubility of the free acid, which equals S_0. Thus, the concentration of the unprotonated species is $C_s - S_0$, and the Hendersen–Hasselbalch equation to measure the pH of precipitation pH_p of this salt now becomes

$$pH = pK_a + \log\frac{C_s - S_0}{S_0}$$

From the literature, we can get values for the pK_a and solubility of the free or undissociated acid. The value for the concentration of salt will be known or can be determined, and thus, we can measure the pH of precipitation of a given salt of a weak electrolyte.

I. Salt of a Weak Base

Similarly, the salt of a weak base such as magnesium hydroxide (milk of magnesia) is only sparingly soluble and dissociates in water to give

$$Mg(OH)_{2(SOLID)} \leftrightarrow Mg^{2+}{}_{(aq)} + 2OH^{-}{}_{(aq)}$$

From the Henderson–Hasselbalch equation, an expression for its pH_p is

$$pH = pK_w - pK_b + \log\frac{S_0}{C_s - S_0}$$

Calculate the pH of precipitation of a 1% solution of sodium benzoate in water at 25°C. (The solubility of benzoic acid in water at 25°C is 3.4 g/l; the mol wt of benzoic acid is 122.12; the mol wt of sodium benzoate is 144.11, and the pK_a of benzoic acid is 4.2.)

Answer: Before calculating the pH_p of the sodium benzoate solution be careful to convert the concentrations to molar terms. The reason for this is that because of the differences in molecular weight, the amount of active drug given by the free acid form is different from the amount of active drug available from the salt.

Thus, the concentration of the salt added initially is

1 g/100 ml = 10 g/l = 10/144.11 = 0.069 mol/l

The solubility of benzoic acid is

$$3.4 \text{ g/l} = 3.4 / 122.12 = 0.028 \text{ mol/l}$$

$$\therefore \text{pH}_p = 4.2 + \log \frac{(0.069 - 0.028)}{0.028} = 4.36$$

Thus, the given 1% solution of benzoic acid will precipitate when the pH is lowered to **4.36**.

J. Solubility of Proteins and pH

The solubility of proteins and amino acids depends on, among other things, the pH of the solution. Protein itself can be either positively or negatively charged overall owing to the terminal amine (–NH2) and carboxyl (–COOH) groups and the groups on the side chain. It is positively charged at a low pH and negatively charged at a high pH. The intermediate pH at which a protein molecule has a net charge of zero is called the isoelectric point, I_c. In general, proteins have a net charge at a higher or lower pH and are thus more soluble. As a result, protein is the least soluble when the pH of the solution is at its isoelectric point.

When the pH of milk is lowered, precipitation or coagulation of milk protein (casein) occurs as its isoelectric point has been achieved. This is one of the common examples of protein isolation caused by changes in pH.

TUTORIAL

1. Methamphetamine, with a pK_a of 9.87, is dissolved in a solution at pH 7.87. Calculate the percent of drug ionized.
2. Diethylbarbituric has an un-ionized H:B form and an ionized B^- form. It has a pK_a of 8.0 and is dissolved in a fluid of pH 7. What would be the percent of drug in ionized form?
3. For a weak acid with a pK_a of 6.0, show how you would calculate the ratio of acid to salt at pH 5.
4. Suppose you have just added 100 ml of a solution containing 0.5 mol of acetic acid per liter to 400 ml of 0.5 M NaOH. What is the final pH? (The pK_a of acetic acid is 4.7.)
5. A weak acid, HA, has a pK_a of 5.0. If 1.0 mol of this acid and 0.1 mol of NaOH were dissolved in one liter of water, what would the final pH be?
6. Phosphoric acid (H_3PO_4) has three dissociable protons, with the pK_as shown below. Which form of phosphoric acid predominates in a solution at pH 4? Explain your answer.

Acid	pK_a
$H_3PO^-{}_4$	2.14
$H_2PO^{2-}{}_4$	6.86
$HPO^{3-}{}_4$	12.4

ANSWERS

1. Methamphetamine, with a pK_a of 9.87, is dissolved in a solution at pH 7.87 and ionizes as follows:

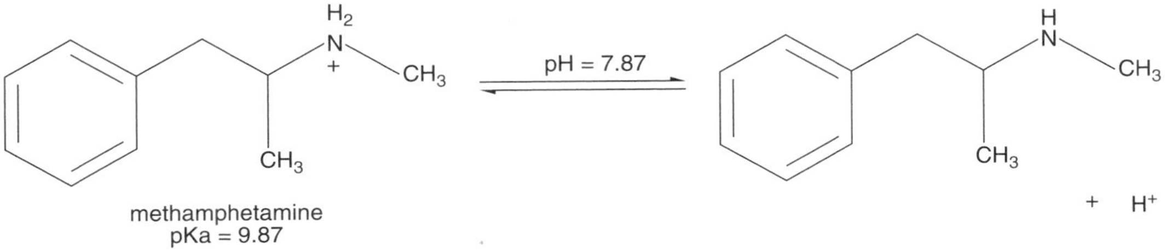

Thus, for methamphetamine at pH 7.87,

$$7.87 = 9.87 + \log\frac{[B]}{[BH^+]}$$

$$-2 = \log\frac{[B]}{[BH^+]}$$

$$\frac{[B]}{[BH^+]} = \frac{1}{100}$$

In this example $[BH^+]$ is ionized.

$$\%\text{ionized} = \frac{100}{101} = 99.01\%$$

2. Diethylbarbituric has an un-ionized H:B form and an ionized B^- form. It has a pK_a of 8.0 and is dissolved in a fluid of pH 7.

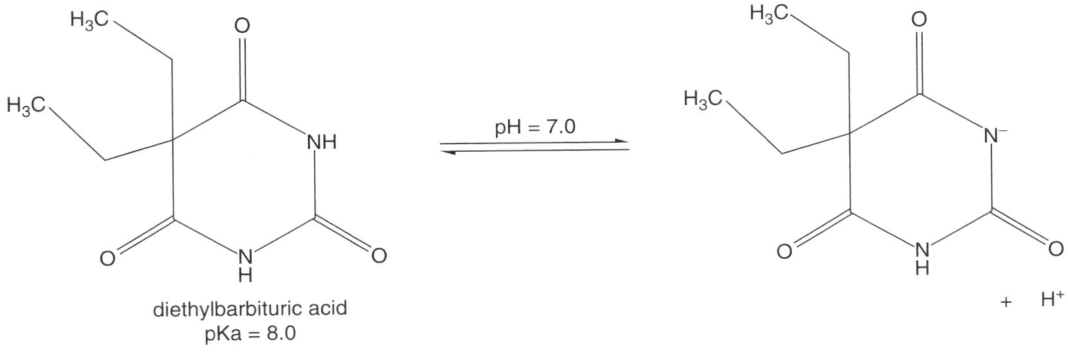

For diethylbarbituric acid at pH 7.0:

$$7.0 = 8.0 + \log \frac{[B^-]}{[H{:}B]}$$

$$-1 = \log \frac{[B^-]}{[H{:}B]}$$

$$\frac{[B^-]}{[H{:}B]} = \frac{1}{10}$$

In this example $[B^-]$ is ionized.

$$\%\text{ionized} = \frac{1}{11} = 9.09\%$$

3. For a weak acid with a pK_a of 6.0, at pH 5, we know that

$$\text{pH} = \text{p}K_a + \log \frac{[\text{conjugate base}]}{[\text{acid}]}, \text{ so p}K_a - \text{pH} = -\log \frac{[\text{conjugate base}]}{[\text{acid}]}$$

$$= \log \frac{[\text{acid}]}{[\text{conjugate base}]}$$

$$6.0 - 5.0 = \log \frac{[\text{acid}]}{[\text{conjugate base}]};\ \frac{[\text{acid}]}{[\text{conjugate base}]} = \text{antilog } 1 = 10$$

4. Addition of 200 mmol of NaOH (400 ml of 0.5 M) to 50 mmol of acetic acid (100 ml of 0.5 mM) completely titrates the acid so that it can no longer act as a buffer and leaves 150 mmol of NaOH dissolved in 500 ml and gives an $[OH^-]$ of 0.3 M. Given $[OH^-]$, $[H^+]$ can be calculated from the water constant:

$$[H^+][OH^-] = 10^{-14} \text{ or } [H^+] = 10^{-14}\ \text{M}^2/0.3\ \text{M}$$

pH is, by definition, log $(1/[H^+])$

$$\text{pH} = \log\left(0.3\ \text{M}/10^{-14}\ \text{M}^2\right) = 12.48$$

5. Combining 1 mol of weak acid with 0.1 mol of NaOH yields 0.9 mol of weak acid and 0.1 mol of salt.

$$\text{pH} = \text{p}K_a + \log\ [\text{conjugate base}]/[\text{acid}] = 5.0 + \log\left(0.1/0.9\right) = 4.05$$

6. At pH 4, the first dissociable proton (pK_a = 2.14) has been titrated completely, and the second (pK_a = 6.86) has just started to be titrated. The dominant form at pH 4 is therefore $H_2PO_4^-$, the form with one dissociated proton.

III. BUFFER SOLUTIONS

The pH of a solution is the negative logarithm of the molar hydrogen ion concentration. From a pharmaceutical perspective, it is often important to control the pH of a solution to minimize drug degradation, to improve patient comfort and compliance, and to improve the efficacy of delivery. A **buffer solution** is able to resist a significant change in pH when a limited concentration of acid or base is added to it either directly or as a result of some other chemical reaction. Buffers play an important role in cellular processes because they maintain the pH at an optimal level for biological processes.

A. How Do We Define a Buffer?

A buffer solution is a system, particularly an aqueous system, that has the ability to resist a change in its pH on the addition of limited amounts of an acid or a base or on dilution with a solvent.

1. Buffer Components

Since acids react with bases (and not with other acids), while bases react with acids (and not with other bases), buffered solutions characteristically have two components: a *weak acid* and its *conjugate base* (soluble salt of the acid), for example, acetic acid and sodium acetate. Thus, any extra H_3O^+ will be neutralized by CH_3COO^- in the buffer

$$H_3O^+ + CH_3COO^- \leftrightarrow CH_3COOH + H_2O$$

and any extra OH^- that is added will be neutralized by the acid

$$CH_3COOH + OH^- \leftrightarrow CH_3COO^- + H_2O$$

Similarly, a basic buffer has a weak base and a soluble salt such as ammonium hydroxide and ammonium chloride.

The salt and the acid/base should be present in comparable concentrations, meaning that the two concentrations should not differ by more than one order of magnitude. The most important characteristic of a buffer solution is its **pH**, which can be calculated using the Henderson–Hasselbalch equation, and its **buffer capacity**, which can be calculated from the Van Slyke equation.

B. How Buffers Work (Common Ion Effect) and Buffer pH

1. Weak Acid Buffers

Buffer systems are a special application of the common ion effect. For example, consider the equilibrium involved in the dissociation of acetic acid:

$$CH_3COOH + H_2O \leftrightarrow CH_3COO^- + H_3O^+$$

The acid dissociation constant for the weak acid is given by the equation

$$K_a = \frac{\left[H_3O^+\right]\left[Ac^-\right]}{\left[HAc\right]} \tag{4.1}$$

Now, if a solution of sodium acetate is added to this solution, as the concentration of acetate ions is increased, the equilibrium will shift in the direction of the reactants to maintain the K_a constant (in accordance with the Le Chatelier's principle). Thus, the ionization of acetic acid is repressed upon the addition of the common acetate ion, and the pH of the final solution decreases. This is called the common ion effect. From Equation 4.1,

$$\left[\mathrm{H_3O^+}\right] = \frac{K_a\left[\mathrm{HAc}\right]}{\left[\mathrm{Ac^-}\right]} \tag{4.2}$$

As the acid dissociates only slightly, we can make two assumptions:

$$[\mathrm{HAc}]_{\text{equilibrium}} = [\mathrm{HAc}]_{\text{from the initial acid concentration}} = [\text{acid}]$$

$$[\mathrm{Ac^-}]_{\text{equilibrium}} = [\mathrm{Ac^-}]_{\text{from the initial salt concentration}} = [\text{salt}]$$

Thus, the expression for the hydronium ion concentration now becomes

$$\left[\mathrm{H_3O^+}\right] = \frac{K_a[\text{acid}]}{[\text{salt}]}$$

which, when expressed in the logarithmic form with the signs reversed, gives the *Henderson–Hasselbalch* equation, or the *buffer equation*, as applied to a buffer system of a weak acid and its salt. This is written as

$$\mathrm{pH} = \mathrm{p}K_a + \log\frac{[\text{salt}]}{[\text{acid}]} \tag{4.3}$$

From Equation 4.3 we can see that two factors govern the pH in a buffered solution: (a) the $\mathrm{p}K_a$ of the weak acid, and (b) the ratio of the initial molar concentrations of the acid and its salt. So, if we prepare a solution where

$$[\text{salt}] = [\text{acid}] \text{ then } \log\frac{[\text{salt}]}{[\text{acid}]} = \log 1 = 0$$

Therefore the pH of the solution will be equal to the $\mathrm{p}K_a$ of the weak acid.

In what concentration ratio should sodium acetate and acetic acid be added to make a solution with a pH of 4.0?

Answer: We know that

$$\mathrm{pH} = \mathrm{p}K_a + \log\frac{[\text{salt}]}{[\text{acid}]} \quad \text{or,} \quad \mathrm{pH} - \mathrm{p}K_a = \log\frac{[\text{salt}]}{[\text{acid}]}$$

$$4 - 4.76 = \log\frac{[\text{salt}]}{[\text{acid}]} \quad \text{or} \quad -0.76 = \log\frac{[\text{salt}]}{[\text{acid}]} \quad \text{or} \quad 0.76 = \log\frac{[\text{acid}]}{[\text{salt}]}$$

Thus,

$$\log\frac{[\text{acid}]}{[\text{salt}]} = \text{antilog}0.76 = \boxed{\mathbf{5.76}}$$

Hence, a molar ratio of acid to salt of 5.76/1 is required to make a pH 4 solution of acetic acid and sodium acetate.

---

2. Weak Base Buffers

Similarly, it can be shown that the ionization of a weak base such as ammonium hydroxide is repressed in the presence of a common ion (e.g., addition of ammonium chloride), and the following expression can be obtained for a buffer of a weak base and its salt:

$$\left[\text{OH}^-\right] = \frac{K_b[\text{base}]}{[\text{salt}]}$$

Thus, the following expression can be obtained,

$$\text{pH} = \text{p}K_w - \text{p}K_b + \log\frac{[\text{base}]}{[\text{salt}]} \tag{4.4}$$

using the fact that

$$K_w = \left[\text{H}_3\text{O}^+\right]\left[\text{OH}^-\right]$$

It should be remembered that weak bases and their salts are not commonly used, as they depend on K_w, which in turn varies with temperature changes. Also, these buffers are unstable and volatile, which might affect the ionic strength and hence their buffer action, as will be shown in the next section.

It should be noted at this point that buffered solutions do change in pH after the addition of an acid or a base, but the amount of change is much less than what would occur in a nonbuffered solution. The amount of change depends on the strength of the buffer and the ratio of salt and acid (or salt and base).

C. Factors Affecting the pH of Buffer Solutions

1. Ionic Strength

The equilibrium constant for a weak acid is written in terms of concentration of different species, assuming that the solutions are dilute. For a more exact estimation of the constant, we need to consider activities instead of concentrations, and thus Equation 4.1 becomes

$$K_a = \frac{a_{\text{H}_3\text{O}^+}a_{\text{Ac}^-}}{a_{\text{HAc}}} \tag{4.5}$$

Substituting for the activity of the species in terms of activity coefficients and molar concentrations, we get

$$K_a = \frac{\left(\gamma_{H_3O^+} c_{H_3O^+}\right) \times \left(\gamma_{Ac^-} c_{Ac^-}\right)}{\gamma_{HAc} a_{HAc}} \tag{4.6}$$

where a represents activity, γ represents the activity coefficient, and c represents the concentration of different species. On dropping the activity coefficient of HAc (which can be considered as one) and writing the above equation in the negative log form, we get a modified Equation 4.3, which accounts for the activities of different species as well.

$$\mathrm{pH} = \mathrm{p}K_a + \log\frac{[\text{salt}]}{[\text{acid}]} + \log \gamma_{Ac^-} \tag{4.7}$$

The activity coefficient can then be written in terms of ionic strength, μ, from the *Debye–Huckel* expression for an aqueous solution of a univalent ion at 25°C having an ionic strength not greater than about 0.1 or 0.2. We thus get a final expression for measuring the pH of buffer solutions of weak acids and their salts, also accounting for the ionic strength of the solution:

$$\mathrm{pH} = \mathrm{p}K_a + \log\frac{[\text{salt}]}{[\text{acid}]} - \frac{0.5\sqrt{\mu}}{1+\sqrt{\mu}} \tag{4.8}$$

From Equation 4.8 we can see that the addition of neutral salts to a buffer solution can change its pH.

2. Dilution

What happens if a buffer solution is diluted? It will obviously lead to a change in its ionic strength. The addition of small amounts of water might not change the pH of the solution but can cause small positive (increase) and negative (decrease) deviations because of its effect on the ionic strength. Also, water can act as both a weak base and a weak acid, which might be another reason for the deviations.

3. Temperature

The pH of basic buffers is affected more by temperature than is the pH of acidic buffers (recall that the pH of basic buffers is expressed in terms of K_w, which is affected more by a change in temperature). Kolthoff and Tekelenburg[1] determined the change in the pH of a buffer solution with a change in temperature and expressed it in terms of the *temperature coefficient*.

D. Buffer Capacity and Van Slyke's Equation

pH is one important property of buffer solutions. Now we will take a look at the other important characteristic of buffers: the buffer capacity. No buffer has an unlimited capacity; that is, buffers can only absorb so much acid or base before they are destroyed. For example, if enough strong acid were added to neutralize all of the buffer's basic component, then additional strong acid will make the pH drop rapidly.

The capacity of a buffer is the amount of acid or base it can handle before the pH of the solution changes drastically. It is the magnitude of resistance offered by a buffer to pH changes. Van Slyke[2] was the first person to introduce the concept of buffer capacity, and he defined it as the ratio of strong base or acid added to the buffer to the small change in pH brought about by this addition. Thus, if buffer capacity is represented by β, then it can be represented approximately as

$$\beta = \frac{\Delta\beta}{\Delta \text{pH}} \quad (4.9)$$

where $\Delta\beta$ is the small amount of base or acid added (in gram equivalents/liter) to the buffer solution, which produces a small change in pH (ΔpH). A more exact representation of buffer capacity is given by Van Slyke's equation:

$$\beta = 2.3C\frac{K_a\left[\text{H}_3\text{O}^+\right]}{\left(K_a + \left[\text{H}_3\text{O}^+\right]\right)^2} \quad (4.10)$$

where C represents the sum of molar concentrations of the salt and acid (or base), that is, the total buffer concentration.

E. Factors Effecting Buffer Capacity

1. Buffer Concentration

The buffer's pH is a function of its pK_a and the ratio of concentrations of salt and acid (or salt and base), but the buffer's capacity depends upon *actual* concentrations of the buffer components. This can also be seen from Equation 4.8. When C, or the total buffer concentration, is increased, buffer capacity also increases. This can be expected because as we increase the concentrations of the salt and acid (in a weak acid buffer), the alkaline reserve and the acid reserve of the buffer, respectively, increase, and thus, the buffer can withstand higher increments of acid or base without significantly changing its pH.

2. Maximum Buffer Capacity

A buffer shows its maximum capacity when its pH is equal to its pK_a or when $[H_2O^+]$ is equal to K_a. Substituting this in Van Slyke's equation we get

$$\beta_{\text{maximum}} = 2.303C\frac{\left[\text{H}_3\text{O}^+\right]^2}{\left(2\left[\text{H}_3\text{O}^+\right]\right)^2} \quad (4.11)$$

or

$$\beta_{\text{maximum}} = 0.576C \quad (4.12)$$

Calculate the maximum buffer capacity of a boric acid buffer with a total concentration of 0.03 mol/l.

Answer: The maximum buffer capacity of that buffer would be

$$\beta_{\max imum} = 0.576C$$

$$C = 0.03 \text{ mol/l}$$

$$0.576 \times 0.03 = \boxed{\mathbf{0.017}}$$

IV. PARTITIONING AND EXTRACTION OF DRUGS

Partitioning is the ability of a compound to distribute in two immiscible systems. It is of paramount importance, as many pharmaceutical processes such as absorption from the gastrointestinal tract after oral administration, diffusion across skin and other epithelia, distribution following entry into systemic circulation, extraction and isolation of pure drugs after synthetic manufacturing or from crude plant sources, formulation of a stable dosage form (emulsion, suspensions, etc.), and assay of plasma concentrations are all based on partitioning phenomenon.

All drugs traverse one or more biological membranes from the time they are administered until they are eliminated. Biological membranes are made of protein and lipid materials, but they act primarily as lipid barriers. Passive diffusion is the predominant mechanism by which many drugs are transported, and thus the lipophilic nature of the molecules is important.

The ability of drug to permeate across biological membranes has traditionally been evaluated using its partitioning in octanol (representing lipid materials) and water systems. Occasionally, other organic solvents such as chloroform, ether, and hexane have been used as lipid vehicles to evaluate drug-partitioning behavior. When a drug is placed in an immiscible system of octanol and water, the drug is distributed into each solvent and eventually reaches equilibrium. The ratio of drug concentration in each phase is termed its *distribution coefficient* or *partition coefficient,* and can be expressed as

$$\text{Distribution or partition coefficient (K)} = \frac{\text{concentration in octanol, } C_o}{\text{concentration in water, } C_w} \tag{4.13}$$

The concentration in aqueous phase is estimated by an analytical assay, and concentration in octanol or other organic phases is deduced by subtracting the aqueous amount from the total amount placed in the solvents.

A. Importance of Partitioning in the Field of Pharmacy

1. *Transport/permeation of drugs*: Transport of drug molecules from dosage forms to target sites often involves several partitioning processes. For example, in oral dosage form, followed by dissolution in the gastrointestinal (GI) tract, the drug has to first partition between GI fluid and the inner GI membrane, followed by a second partitioning between the outer GI membrane and systemic circulation before it will reach the target organ. Similarly, the drug has to partition between its vehicle and skin before the efficacy of a drug molecule can be achieved following topical application on skin.
2. *Formulation of drugs:* Partitioning may be a factor in the stability of finished dosage forms where both oil and water phases exist, for example, in emulsion systems. Antimicrobial preservatives must be present in the aqueous phase of an emulsion to prevent spoilage. If the preservative chosen partitions into the oil, then additional preservatives may be needed to maintain an adequate water phase concentration. (This problem becomes more complicated if the preservative is a weak acid, HA, whose partitioning is pH dependent.)
3. *Industrial processes:* Many manufacturing processes involve extracting desirable (or undesirable) substances from a liquid phase with another immiscible liquid. Such processes depend upon partitioning and permit one to obtain purer drugs and chemicals.
4. *Analytical procedures:* Often these procedures involve an extraction or partitioning step to obtain the desired substance in a system where it can be analyzed. HPLC (high performance liquid chromatography) is the most widely used procedure for assaying drug concentrations in biological fluids and is fundamentally dependent upon a partitioning step.

Example: 0.15 grams of compound X in 100 ml of octanol is shaken with 10 ml of water. After equilibrium was achieved, the water layer contained 0.067 grams of compound X. What is the partition coefficient (K) of compound X?

$$\text{Concentration in water} = \frac{0.067 \text{ gm}}{10 \text{ ml}} = 0.0067 \text{ gm/ml}$$

$$\text{Concentration in octanol} = \frac{(0.15 - 0.067) \text{ gm}}{100 \text{ ml}} = 0.083 \text{ gm/ml}$$

So,

$$\text{K of compound X} = \frac{0.083 \text{ gm/ml}}{0.0067 \text{ gm/ml}} = 0.124$$

Note: If both solvents are used in equal volume, then K can be expressed as a ratio of the amount in each phase instead of concentration.

Example: 500 ml of a 0.2 M aqueous solution of Drug A (mol wt = 225) is extracted with 100 ml of ether. If the partition coefficient, K, = 35, how many grams of drug solution remain in water after extraction?

$$K = 35 = \frac{C_o}{C_w} = \frac{\left[(\text{Amt})_o / V_o\right]}{\left[(\text{Amt})_w / V_w\right]} = \frac{\left[(\text{Amt})_o / 100 \text{ ml}\right]}{\left[(\text{Amt})_w / 500 \text{ ml}\right]}$$

However, the **amount** in water after extraction $(\text{Amt})_w$ = (the amount in water before extraction, Amt_w^{orig}) – (the amount in oil after extraction, Amt_o).

Now,

$$\text{Amt}_w^{orig} = (0.2 \text{ mol/l})(0.5 \text{ l})(225 \text{ g/mol}) = 22.5 \text{ g}$$

$$\therefore K = \frac{(\text{Amt})_o / 100 \text{ ml}}{\left[(\text{Amt})_w^{orig} - (\text{Amt})_o\right] / 500 \text{ ml}}$$

$$\frac{(\text{Amt})_o / 100 \text{ ml}}{\left[22.5 \text{ gm} - (\text{Amt})_o\right] / 500 \text{ ml}} \quad \text{or} \quad 35 = \frac{(\text{Amt})_o}{\left[22.5 \text{ gm} - (\text{Amt})_o\right]} \times \frac{500 \text{ ml}}{100 \text{ ml}}$$

Solving for $(\text{Amt})_o$, we obtain

$$(\text{Amt})_o = \frac{787.5}{40} = 19.6875 \text{ gm}$$

$$\therefore (\text{Amt})_w = 22.5 \text{ gm} - 19.6875 \text{ gm} = 2.8125 \text{ gm}$$

B. Partitioning as a Function of pH

Equation 4.13 holds true if a compound is not an electrolyte, but the majority of pharmaceuticals are either weakly acidic or weakly basic. Thus, it is important to understand their partitioning behavior. Weak electrolytes exist in un-ionized form in octanol or lipid phase, whereas in aqueous phase, both the un-ionized and ionized forms exist in equilibrium. Thus, partition coefficient K calculated as described above represents the apparent partition coefficient, and Equation 4.13 can be rewritten as

$$K_{app} = \frac{[\mathrm{HA}]_o}{[\mathrm{HA}]_w + [\mathrm{A}^-]_w} \tag{4.14}$$

However, by definition, the true partition coefficients for such a weak electrolyte [HA] is the ratio of un-ionized concentrations only in each phase as shown below

$$K = \frac{[\mathrm{HA}]_o}{[\mathrm{HA}]_w} \tag{4.15}$$

However, $[\mathrm{HA}]_w$ in water ionizes to $[\mathrm{H}^+][\mathrm{A}^-]$ as defined by its ionization constant K_a as described earlier in this chapter:

$$K_a = \frac{[\mathrm{H}^+][\mathrm{A}^-]_w}{[\mathrm{HA}]_w} \text{ or } [\mathrm{A}^-]_w = \frac{K_a[\mathrm{HA}]_w}{[\mathrm{H}^+]_w} \tag{4.16}$$

Thus, K_{app} can be rewritten substituting for $[\mathrm{A}^-]_w$ from Equation 4.15 as

$$K_{app} = \frac{[\mathrm{HA}]_o}{[\mathrm{HA}]_w + \dfrac{K_a[\mathrm{HA}]_w}{[\mathrm{H}^+]_w}} \tag{4.17}$$

Taking $[\mathrm{HA}]_w$ as a commonality in the denominator, Equation 4.16 can be rearranged as

$$K_{app} = \frac{[\mathrm{HA}]_o}{[\mathrm{HA}]_w\left\{1 + \dfrac{K_a}{[\mathrm{H}^+]_w}\right\}} \text{ or } K_{app} = \frac{[\mathrm{HA}]_o}{[\mathrm{HA}]_w\left\{\dfrac{[\mathrm{H}^+]_w + K_a}{[\mathrm{H}^+]_w}\right\}} \tag{4.18}$$

Equation 17 can be further rearranged to get

$$K_{app} = \frac{[\mathrm{HA}]_o}{[\mathrm{HA}]_w} x \frac{[\mathrm{H}^+]_w}{[\mathrm{H}^+]_w + K_a} \tag{4.19}$$

Substituting Equation 4.15 yields

$$K_{app} = K^* \frac{[H^+]_w}{K_a + [H^+]_w} \tag{4.20}$$

Equation 4.20 indicates that the apparent partition coefficient is a function of H^+ or pH, whereas the true partition coefficient is independent of pH.

This function of $[H^+]$ can be examined in the following equilibrium:

$$[HA] \overset{K_a}{\longleftrightarrow} [A^-] + [H^+]$$

where the total concentration of the acid species present is $[HA]_T$, which is given by

$$[HA]_T = [A^-] + [HA]$$

and the fraction of the total acid present as either [HA] or $[A^-]$ can be calculated using the equilibrium expression

$$K_a = \frac{[H^+][A^-]}{[HA]}$$

as follows:

$$f_{[HA]} = \frac{[HA]}{[HA]_T} = \frac{[HA]}{[HA]+[A^-]} = \frac{[H^+]}{[H^+]+K_a} \tag{4.21}$$

$$f_{[A-]} = \frac{[A^-]}{[HA]_T} = \frac{[A^-]}{[HA]+[A^-]} = \frac{K_a}{[H^+]+K_a} \tag{4.22}$$

and finally, by substitution from Equation 4.20,

$$K_{app} = K^* f_{[HA]} \tag{4.23}$$

Therefore, K_{app}, the apparent partition coefficient, will vary with pH in the same manner that [HA] varies with pH since K_p is a true or constant value.

It should also be noted that

$$f_{[HA]} = \frac{[HA]}{[HA]+[A^-]} \text{ and } f_{[A^-]} = \frac{[A^-]}{[HA]+[A^-]}$$

$$f_{[HA]} + f_{[A^-]} = \frac{[HA]}{[HA]+[A^-]} + \frac{[A^-]}{[HA]+[A^-]} = \frac{[HA]+[A^-]}{[HA]+[A^-]} = 1$$

Now, let us see how $f_{[HA]}$ and $f_{[A^-]}$ vary with pH, using benzoic acid (pK_a = 4.20, $K_a = 6.31 \times 10^{-5}$) as an example over the pH range 2.0 to 6.5.

pH	[H+]	$f_{[HA]}$	$f_{[A^-]}$	$f_{[HA]} + f_{[A^-]}$
2	1.00E-02	0.9937	0.0063	1
2.5	3.16E-03	0.9804	0.0196	1
3	1.00E-03	0.9406	0.0594	1
3.25	5.62E-04	0.8991	0.1009	1
3.5	3.16E-04	0.8336	0.1664	1
3.75	1.78E-04	0.7383	0.2617	1
4	1.00E-04	0.6131	0.3869	1
4.2	**6.31E-05**	0.5000	0.5000	1
4.5	3.16E-05	0.3337	0.6663	1
4.75	1.78E-05	0.2200	0.7800	1
5	1.00E-05	0.1368	0.8632	1
5.5	3.16E-06	0.0477	0.9523	1
6	1.00E-06	0.0156	0.9844	1
6.5	3.16E-07	0.0050	0.9950	1

The above data are plotted below:

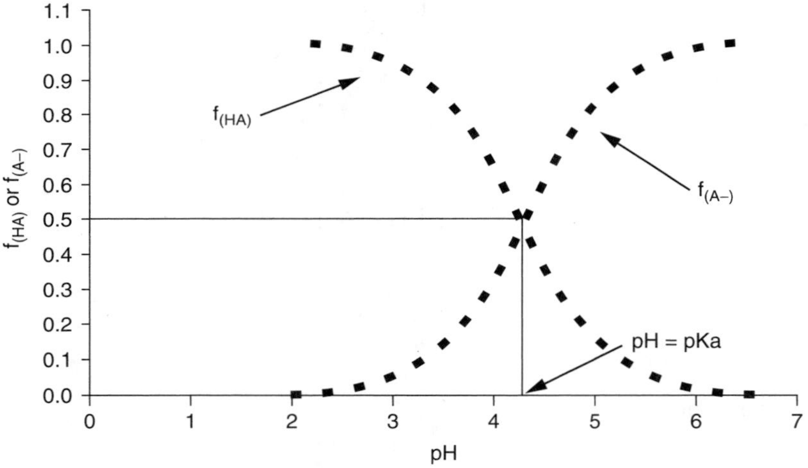

A plot showing the influence of pH on the ratio of ionized and unionized benzoic acid.

Since $f_{[HA]}$ exhibits a sigmoidal dependence on pH, and $K_{app} = K^* f_{[HA]}$, it is expected that K_{app} would also exhibit a sigmoidal dependence on pH since K is a constant and the product $K^* f_{[HA]}$ should take the shape of $f_{[HA]}$. This is illustrated below for the partitioning of benzoic acid (pK_a = 4.20) between peanut oil and water, for which K = 5.33.

pH	K	$f_{[HA]}$	$K_{app} = K^* f_{[HA]}$
2	5.33	0.9937	5.296
3	5.33	0.9406	5.01
3.5	5.33	0.8337	4.44
4	5.33	0.6131	3.27
4.5	5.33	0.3339	1.78
5	5.33	0.1368	0.73
6	5.33	0.0156	0.08

The above data are plotted below:

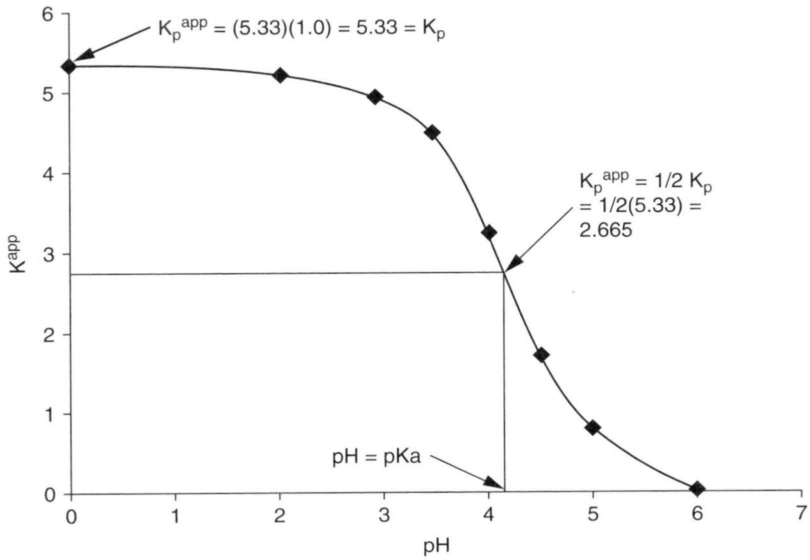

A plot showing the influence on the partitioning of benzoic acid.

The following conclusions can be made from the above data and curves:

- Both, the $f_{[HA]}$ and the $f_{[A^-]}$ curves are sigmoidal in their dependency on pH.
- At $f_{[HA]} = f_{[A^-]} = 0.5$, the pH = p$K_a$ of the acid involved in the plot. (If the pK_a differs, the plot shifts along the pH-axis.)
- $f_{[HA]} + f_{[A^-]} = 1$ at all values of pH.
- $f_{[HA]}$ varies from 1 to 0 and since $K_{app} = K^* f_{[HA]} = (5.33)(f_{[HA]})$, K_{app} varies from 5.33 to 0.
- When pH = pk_a, $K_{app} = 0.5(K) = 0.5(5.33) = 2.665$.

In the case of partitioning of a weak base, B, between oil and water, it can be shown that the pH dependency of f_B will be the same as that for $f_{[HA]}$ shown earlier. Therefore, K_{app} will also be sigmoidal because it is directly proportional to f_B (fraction of the weak base). Some examples of other natural processes (besides partitioning) that depend on either HA (or A^-) or B (or BH^+) and also show a sigmoidal dependency on pH include chemical reactions that involve only the HA (or A^-) or the B (or BH^+) form of the substance as the reactive species and partitioning processes such as absorption of drugs through the gut wall or through the cornea, where the un-ionized drug form (HA rather than A^- and B rather than BH^+) is preferentially absorbed.

In the context of the above discussion, Van Slyke's equation and the maximum buffer capacity can be derived alternatively as follows:

Recall that the Van Slyke equation is

$$\beta = (2.3)(C)\frac{[H^+]K_a}{(K_a + [H^+])^2} = 2.3(C)\frac{[H^+]}{K_a + [H^+]} \bullet \frac{K_a}{K_a + [H^+]}$$

or

$$\beta = 2.3(C) f_{[HA]} \cdot f_{[A^-]}$$

From this equation, one can expect that β assumes a bell-shaped dependency on pH. For example, if $f_{[HA]}$, $f_{[A^-]}$ and the product $f_{[HA]} \cdot f_{[A^-]}$ are calculated for acetic acid ($pK_a = 4.76$, $K_a = 1.7 \times 10^{-5}$), the data obtained will be as shown below:

pH	$f_{[HA]}$	$f_{[A^-]}$	$f_{[HA]} \cdot f_{[A^-]}$
2	0.998303	0.0017	0.0017
3	0.983284	0.0167	0.0164
3.5	0.948949	0.0511	0.0484
4	0.854701	0.1453	0.1242
4.5	0.650206	0.3498	0.2274
4.76	0.508671	0.4913	0.2499
5	0.37037	0.6296	0.2332
6	0.055556	0.9444	0.0525
6.5	0.018249	0.9818	0.0179

The above data are plotted in Figure 4.5:

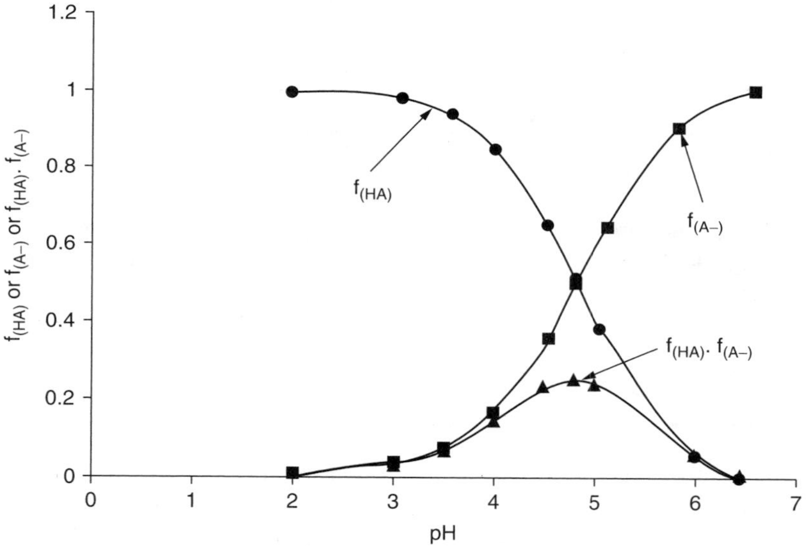

A plot showing the influence of pH on the fraction ionized and unionized.

The following conclusions can be made from the data and curves:

- The function $f_{[HA]} \cdot f_{[A^-]}$ is bell shaped and reaches a maximum when $f_{[HA]} = f_{[A^-]} = 0.5$, i.e., $f_{[HA]} \cdot f_{[A^-]} = (0.5)\,(0.5) = 0.25$ at maximum.
- When $pH = pK_a$, $f_{[HA]} \cdot f_{[A^-]}$ as well as β is at maximum and as $\beta - 2.3(C)\, f_{[HA]} \cdot f_{[A^-]}$, $\beta_{max} = 0.576$ C.

C. Single vs. Multiple Extractions

Extraction is a process in which partitioning phenomena are encountered in drug development. Whether in the isolation of pure compounds from crude mixtures or the measurement of analytes from plasma samples during pharmacokinetic studies, extraction plays an important role. The efficiency with which one solvent extracts a compound from another immiscible solvent is dependent upon the distribution phenomenon and is related by the following equation:

$$K = \frac{\text{Conc. of solute in extracting solvent}}{\text{Conc. in original solvent}} \tag{4.24}$$

or

$$K = \frac{(W - W_1)/V_2}{(W_1)/V_1}$$

where W and W_1 are the weight of solute added and the weight of solute remaining in the original solvent, and V_1, and V_2 are the volumes of original and extracting solvents, respectively.

Example: If 30 ml (V_1) of water containing 15 mg (w) of salicylate is extracted with 30 ml (V_2) of chloroform, 2.59 mg (W_1) of salicylate remains in the water layer. Assuming water and chloroform are immiscible, what is the distribution or partition coefficient of salicylate in the chloroform-water solvent system?

$$K = \frac{(W - W_1)/V_2}{(W_1)/V_1} = \frac{(15 - 2.59)/30}{(2.59)/30} = 4.8$$

The extraction efficiency can be improved by using the same amount of extraction solvent but dividing into multiple portions of equal volumes, what is also known as multiple extraction. This is because each extraction is a separate equilibrium process. The general equation describing the amount extracted W_n, after n extractions is given by

$$W_n = W\left\{\frac{V_1}{V_1 + KV_2}\right\}^n$$

Such a relationship assumes complete immiscibility of the two solvents involved. The presence of other solutes or salts may also affect the outcome. The following example should make clear the increased efficiency of multiple extractions.

Example: From the above example, how much amount of the drug would be left in water if salicylate is extracted sequentially with two portions of 15 ml of chloroform each?

From the example, K = 4.8. The amount of salicylate left in water after first extraction would be

$$4.8 = \frac{(\text{Amt}_{cl})/V_{Cl}}{(\text{Amt}_{w})/V_{w}} = \frac{(15 - x)/15}{(x)/30} \text{ and}$$

$$x = 4.41 \text{ mg}$$

The amount of salicylate left in water after second extraction would be

$$4.8 = \frac{(\text{Amt}_{cl})/V_{Cl}}{(\text{Amt}_{w})/V_{w}} = \frac{(4.41 - x)/15}{(x)/30} \text{ and}$$

$$x = 1.30 \text{ mg}$$

Thus, a total of 13.7 mg (15 – 1.3 mg) of salicylate was extracted from water after two sequential extractions of 15 ml each when compared to a single extraction using 30 ml that extracted only 12.41 mg. Hence multiple sequential extractions using small portions of extraction solvent is more efficient than single extraction.

Alternatively, the above problem could also be solved in a single step by using the multiple extraction formula given above.

As $n = 2$, W_2 (amount remaining in water after two extractions) can be calculated as

$$W_2 = 15\left\{\frac{30}{30 + (4.8)(15)}\right\}^2 = 1.30 \text{ mg}$$

V. COLLIGATIVE PROPERTIES

Solutions, especially liquid solutions, generally have markedly different properties than either the pure solvent or the solute. For example, a solution of sugar in water is neither crystalline like sugar nor tasteless like water. Some of the properties unique to solutions depend only on the number of dissolved particles and not their identity. Such properties are called colligative properties. **Colligative properties** can only be applied to solutions. They depend on the number of solute particles (un-ionized and ionized species) in the solution rather than the weight concentration and nature of solute.

The colligative properties are:

- Lowering of the vapor pressure of solvent by solute
- Elevation of the boiling point of solvent by solute
- Depression of freezing point of solvent by solute
- Establishment of osmotic pressure of solvent by solute

The characteristics of colligative properties are:

- Colligative properties are related and a change in any one of them will be reflected in the corresponding changes in the others.
- A given colligative property of equimolar solutions of electrolytes and nonelectrolytes will not be identical.
- Colligative properties of different solutes in a solution are additive.

A. Lowering of Vapor Pressure

Raoult's law states that the vapor pressure, p_1, of a solvent over a dilute solution equals the vapor pressure of the pure solvent, $p_1°$, times the mole fraction of the solvent in solution, x_1. If the solute is nonvolatile, then the vapor pressure of the solvent equals the total pressure of the solution.

Raoult's law equation may be mathematically written as

$$p = p_1^0 x_1 \tag{4.25}$$

Since the total mole fraction must equal unity,

$$x_1 + x_2 = 1$$

in which x_1 is the mole fraction of the solvent, and x_2 is the mole fraction of the solute.

Equation (4.25) can be written as

$$p = p_1^0 \left(1 - x_2\right) \tag{26}$$

$$p_1^0 - p = p_1^0 x_2 \tag{4.27}$$

$$\frac{p_1^0 - p}{p_1^0} = \frac{\Delta p}{p_1^0} = x_2 = \frac{n_2}{n_1 + n_2} \tag{4.28}$$

In dilute solution, mole fraction $n_2/(n_1 + n_2)$ can often be replaced by n_2/n_1 because solvent is in such large excess that $n_1 \equiv n_1 + n_2$. Thus, Equation 4.28 can be written as

$$\frac{\Delta p}{p_1^0} = x_2 \cong \frac{n_2}{n_1} \tag{4.29}$$

In Equation 4.29 Δp is the lowering of vapor pressure. The lowering of vapor pressure is dependent on the solute mole fraction, which is given by the ratio of the moles of solute divided by the total moles of solute and solvent.

$\Delta p/p_1^0$ is the relative vapor pressure lowering, which only depends on the solute mole fraction. Thus, relative lowering of vapor pressure is a colligative property.

The vapor pressure of a solution can be determined directly by means of a manometer where the vapor pressure lowering is obtained by subtracting the vapor pressure of the solution from the vapor pressure of pure solvent. A vapor pressure osmometer is also used for measurement of vapor pressure lowering. For precise determination of vapor pressures, the *isopiestic* (Greek: *isopiestic* means equal pressure) method is used (Figure 4.3).

At 35°C, the vapor pressure of water is 43.4 mmHg. What is the vapor pressure of a 1.00 molal solution of NaCl?

Answer: First, we need to figure out the mole fraction of water in the solution, then apply the above equation for the vapor pressure lowering. 1.00 molal means 1 mol solute in 1 kg water.

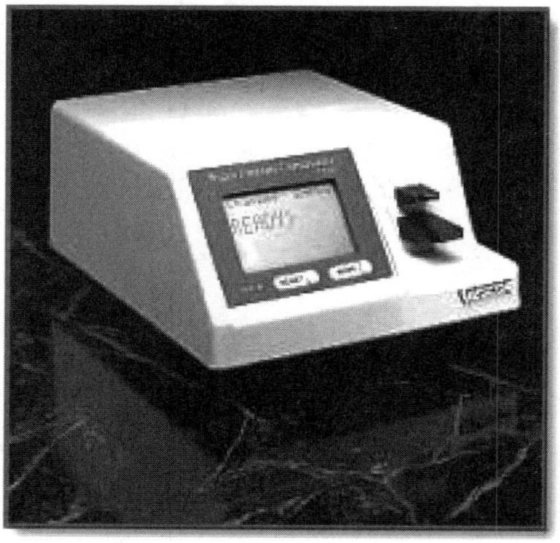

FIGURE 4.3 A modern vapor pressure osmometer.

1 kg water = 1000 g water. Water has a molecular weight of 18.01 g/mol, so 1000 kg water/18.01 g/mol = 55.5 mol of water.

We have to be careful about the number of moles of solute. NaCl ionizes into Na^+ and Cl^- ions in water, so we have 2 mol of ions in the solution, not 1.

We have 2 mol of ions and 55.5 mol of water in 1 l of solution, so we have a total of 57.5 mol of solution. The mole fraction of water is thus

$$X_{\text{water}} = \frac{55.5\text{moles(water)}}{57.5\text{moles(total)}} = \boxed{\mathbf{0.965}}$$

B. Boiling Point Elevation

Normal boiling point is defined as the temperature at which the vapor pressure of the liquid becomes equal to the external atmospheric pressure (760 mm Hg). The boiling point of a solution containing a nonvolatile solute would be higher than the pure solvent because the solute would lower the vapor pressure of the solvent. The increase in boiling point can be written as $T - T_0 = \Delta T_b$. The ratio of the elevation of boiling point, ΔT_b, to the lowering of vapor pressure, Δp, is approximately a constant at 100°C. It can be written as

$$\frac{\Delta T_b}{\Delta p} = k \text{ or } \Delta T_b = k * \Delta p \tag{4.30}$$

As p^0 is a constant, Equation 4.30 can be written as

$$\Delta T_b \propto \frac{\Delta p}{p^0}$$

Now we know $\Delta p/p^0$, the mole fraction of the solute; therefore, Equation 4.30 can be rewritten as

$$\Delta T_b = k * x_2 \tag{4.31}$$

Equation 4.31 shows that boiling point elevation depends only on x_2, the mole fraction of the solute, and hence it is a colligative property.

In dilute solutions,

$$x_2 = \frac{\Delta p}{p_1^0} \cong \frac{n_2}{n_1}$$

$$\frac{n_2}{n_1} = \frac{w_2/M_2}{w_2/M_1}$$

in which w_2 and w_1 are the weights in grams of the solute and the solvent, respectively, and M_1 and M_2 are the molecular weights of the solute and the solvent, respectively. If relative vapor pressure lowering is expressed in molal concentrations, then w_1 equals 1000 g.

Then

$$n_2/n_1 = \left(w_2/M_2\right)/\left(1000/M_1\right) \tag{4.32}$$

The moles of solute, m, can be written as w_2/M_2, and Equation 4.32 can be rearranged as

$$x_2 = \frac{n_2}{n_1} = \frac{M_1}{1000}m$$

Substituting the value of x_2 in Equation 4.31, we get

$$\Delta T_b = \frac{kM_1}{1000}m \tag{4.33}$$

Since k and M_1 are both constant for a given solvent system, they can be replaced by another constant, K_b. Equation 4.33 can be rewritten as

$$\Delta T_b = K_b m \tag{4.34}$$

ΔT_b is the boiling point elevation and K_b is the ebullioscopic constant, or the molal elevation constant. It can be defined as *the ratio of the boiling point elevation to the molal concentration for a solution that is ideal.* K_b has a characteristic value for each solvent as shown in Table 4.1. An instrument called a Cottrell boiling point apparatus measures boiling point elevation.

TABLE 4.1
Ebullioscopic and Cryoscopic Constants for Commonly Used Solvents

Solvent	Freezing Point (°C)	Boiling Point (°C)	K_b (Ebullioscopic Constant)	K_f (Cyroscopic Constant)
Water	0.0	100.0	0.51	1.86
Ethyl alcohol	−114.5	78.4	1.22	3.00
Chloroform	−63.5	61.2	3.54	4.96
Benzene	5.5	80.1	2.53	5.12
Acetone	−94.82	56.0	1.71	2.40

The normal boiling point of benzene is 80.1°C. It has a boiling point elevation constant of 2.53°C/m. If we make up a 0.500 molal solution of Br_2 in benzene, what is the boiling point of the mixture?

Answer:

$$\Delta T_b = K_b m$$

$$\Delta T_b = 2.53^{\circ}\text{C}/m * 0.500 \text{ molal}$$

$$\Delta T_b = 1.27^{\circ}\text{C}$$

Since the boiling point is normally 80.1°C, the boiling point of the mixture is 80.1 + 1.27 = **81.4°C**.

C. Freezing Point Depresssion

The normal freezing point of a compound is defined as the temperature at which the solid and liquid phases coexist under an external pressure of 1 atm. In general, solutions have a lower freezing point than the pure solvent. If a solute is dissolved in a liquid, the vapor pressure of the liquid solvent is lowered below the solid solvent. Then the temperature will drop to reestablish equilibrium between the liquid and the solute. Thus, the freezing point of the solution is always lower than that of the pure solvent. The depression of the freezing point of a solution with respect to the pure solvent is analogous to boiling point elevation. Both pure solid and pure liquid phases of the solvent can coexist with their vapor, and therefore, both phases have a nonzero vapor pressure. In solution, solvent coexists with its vapor, **so, at the freezing point of a solution, the solvent in the solution and the solvent in the solid (which is composed only of solvent) must coexist**. For the solid phase, we can consider only the pure solid solvent vapor pressure curve, while in solution, we must consider the solution vapor pressure curve. Thus, freezing point corresponds to the point where the solution vapor pressure curve intersects that of the pure solid solvent. We can compare this curve to that of the pure liquid solvent as shown in Figure 4.4.

Figure 4.5 shows the normal freezing point for water (solvent) as a function of molality in several solutions containing sucrose (a nonvolatile solute). Note that the normal freezing point of water decreases as the concentration of sucrose increases.

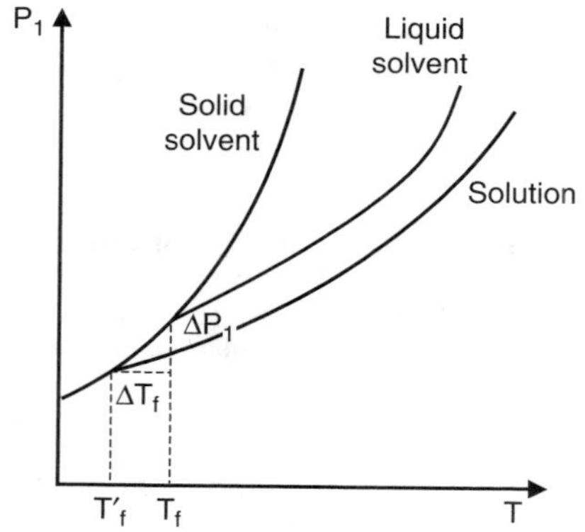

FIGURE 4.4 Vapor pressure curves of solid, liquid, and solution.

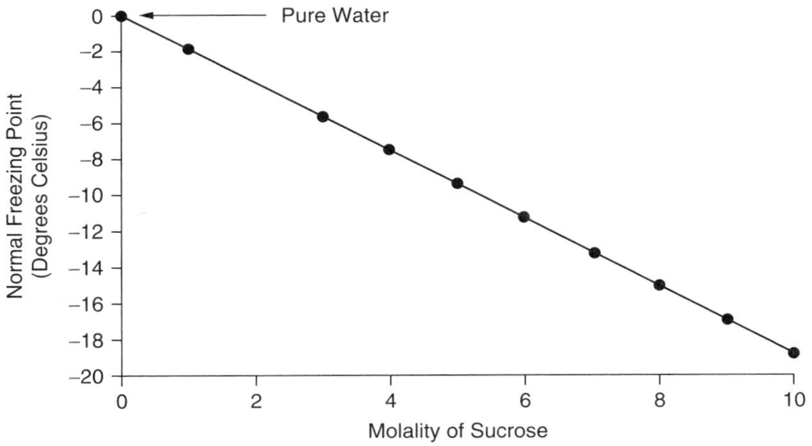

FIGURE 4.5 Normal freezing point for water (solvent) as a function of molality in several solutions containing sucrose.

As the concentration of the solute in the solution increases, the freezing point depression also increases proportionally. Thus, the freezing point depression is proportional to the molal concentration of the solute. The equation for this is

$$\Delta T_f = K_f m \tag{4.35}$$

Since m can be expressed as a function of the molecular weight of the solute and the solvent,

$$m = \frac{1000 w_2}{w_1 M_2}$$

Equation 4.31 can be expressed as

$$\Delta T_f = K_f \frac{1000 w_2}{w_1 M_2} \tag{4.36}$$

ΔT_f is the freezing point depression and K_f is the cryoscopic constant, or molal depression constant. Since freezing point depression is a function of the moles of the solute, it is a colligative property. Several methods exist for the experimental determination of freezing point depression, but perhaps the most commonly used is the Beckmann freezing point apparatus (Figure 4.6).

Freezing point depression is a solution property that is very useful to avoid being inconvenienced by cold weather. Sprinkling salt on icy roads clears up the ice, because the salt lowers the freezing point of the ice. *If you think about the equation, you should also be able to figure out why calcium chloride ($CaCl_2$) is used to de-ice roads rather than sodium chloride.* Calcium chloride has a Van't Hoff factor of 3, as opposed to sodium chloride's Van't Hoff factor of 2. Thus, one mole of calcium chloride is 1.5 times as efficient at lowering the freezing point as sodium chloride. Antifreeze in a car engine is another example of an application of freezing point depression. A mixture of ethylene glycol and water in the radiator protects the engine from both freezing and boiling. The concentration of ethylene glycol required to lower the freezing point to −10°C is 5.38m.

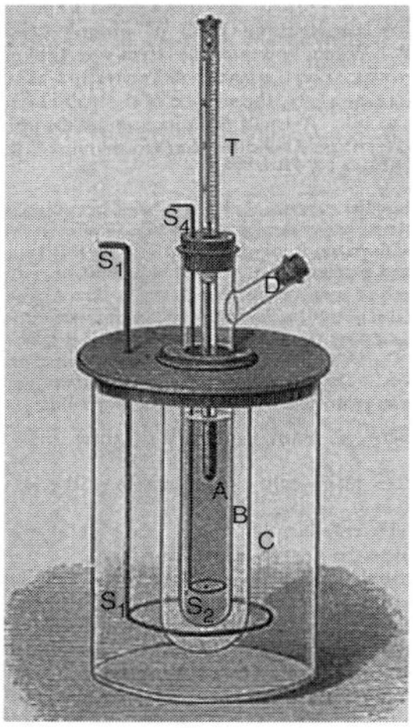

FIGURE 4.6 Beckmann freezing point apparatus.

---

1. What is the freezing point of a solution containing 1.8 g of glucose and 1 l of water? (In this dilute solution, K_f is 1.86.)

   Answer:

   $$\frac{1000w_2}{w_1 M_2} \text{ or } = 1.86 \times = 0.0186°\text{C}$$

   Therefore the freezing point depression of the aqueous solution is **-0.0186°C**.

2. The freezing point of a solution, which contains 1.00 g of an unknown compound, A, dissolved in 10.0 g of benzene, is 2.07°C. The freezing point of pure benzene is 5.48°C. The molal freezing point depression constant of benzene is 5.12°C/molal. What is the molecular weight of the unknown compound?

   Answer:

   Step 1: Calculate the freezing point depression of the solution

   $$\Delta T_f = (\text{freezing point of pure solvent}) - (\text{freezing point of solution})$$

   $$(5.48°\text{C}) - (2.07°\text{C}) = 3.41°\text{C}$$

   Step 2: Calculate the molal concentration of the solution using the freezing point depression

   $$\Delta T_f = (K_f)(m)$$

   $$m = (3.41°\text{C})/(5.12°\text{C/molal})$$

   $$m = 0.666 \text{ molal}$$

Step 3: Calculate the molecular weight of the unknown using the molal concentration.

$$m = 0.666 \text{ molal} = 0.666 \text{ mol A/kg benzene}$$

$$\text{moles A} = (0.66 \text{ mol A/kg benzene})(0.100 \text{ kg benzene}) = 0.66 \times 10^{-3} \text{ mol A}$$

$$\text{molecular weight of A} = (1.00 \text{ g A})/(6.66 \times 10^{-3} \text{ mol A})$$

$$\text{molecular weight of A} = \mathbf{150\ g/mol}$$

D. Osmotic Pressure

Osmosis (Greek for push) is defined as the spontaneous net movement of water across a semipermeable membrane from a region of higher water concentration to one of lower water concentration. A semipermeable membrane is a membrane that selectively allows the transport of solvent molecules but does not allow solute molecules to pass through. The concept of **water concentration** can be explained as follows.

Addition of a solute to water will lower the concentration of water in the solution relative to the concentration of water in pure water, for example, the concentration of water in pure water is 55.5 M. However, the concentration of water in a 1 M glucose solution is 54.5 M (55.5 – 1). In other words, the solute (glucose) will dilute or displace water molecules and lower the water concentration. *Concentration of water in a solution depends upon the number of solute particles in the solution, not on their chemical composition.*

Osmotic pressure is the colligative property that we are concerned with when we wish to adjust the tonicity of a solution. *The osmotic pressure can also be defined as the pressure required to offset the movement of solvent through a semipermeable membrane from a dilute aqueous solution to a more concentrated one.* To generalize, osmotic pressure is a measure of a fluid's ability to hold or attract water when it is separated from another aqueous fluid by a semipermeable membrane. Osmotic pressure is a colligative property, as it is proportional to the number of particles in solution. The osmotic pressure of a solution is directly proportional to the solute concentration and **inversely** proportional to the water concentration of that solution. The osmotic pressure of a nonelectrolyte depends on the concentration of the solute dissolved in the solvent, whereas for an electrolyte that can dissociate into ions, the osmotic pressure depends on both the concentration and the degree to which the electrolyte dissociates in solution. This is because the osmotic pressure of an electrolyte is greater than a nonelectrolyte because the former potentially yields a greater number of particles upon dissociation.

Thermodynamically speaking, the chemical potential of a solvent molecule in solution is less than that in pure solvent. The solvent molecule therefore passes spontaneously into the solution until the chemical potential of solvent and solution are equal. Osmotic pressure is defined as the pressure required to prevent osmosis in solutions.

In 1886, Van't Hoff found out that there exists a relationship between osmotic pressure, concentration, and temperature and suggested a relationship that was similar to the ideal gas law equation. The equation can be written as

$$\pi V = nRT \tag{4.37}$$

in which π is the osmotic pressure in atmospheres, V is the volume of solution in liters, n is the number of moles of solute, R is the universal gas constant equal to 0.082 l atm mole^{-1}K^{-1}, and T is the absolute temperature in Kelvin. Equation 4.37 can also be expressed as

$$\pi = \frac{n}{V}RT = mRT \tag{4.38}$$

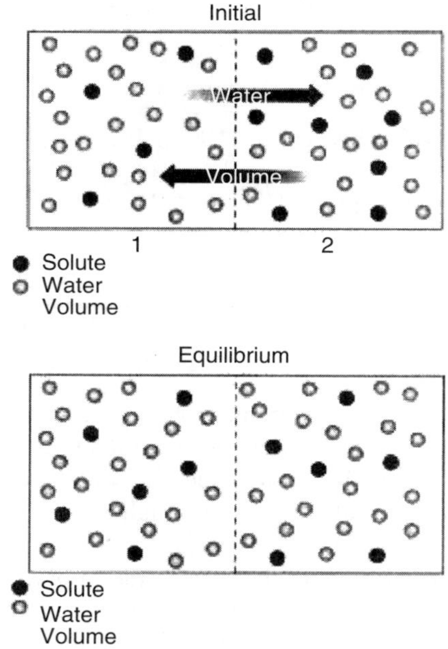

FIGURE 4.7 Osmosis occurring as a result of differences in concentration between two solutions.

in which *m* is the molarity of the solution (concentration of the solute in moles per liter). Osmotic pressure can be measured by osmometer.

In Figure 4.7 two solutions of different concentrations (dilute and concentrated) are separated by a semipermeable membrane. Solvent molecules pass from the dilute solution (compartment 1) to the concentrated one (compartment 2) as a result of osmosis, and after a few hours, the levels in the two compartments become equal. This is essentially the situation occurring across plasma (cell) membranes, which are highly elastic.

E. Concept and Calculation of Osmolar Strength

The total solute concentration is known as its **osmolarity**. One osmole equals 1 mol of solute particles. (1 Osm = 1000 mOsm). Thus, 1 M glucose solution has a concentration of 1 Osm/l. In pharmacy, the unit commonly used to measure osmotic concentration is a milliosmole, which is represented as mOsmol. Considering a nonelectrolyte as an example, 1 millimole (mmol) (1 formula weight in milligrams) of dextrose represents 1 mOsmol. However, with electrolytes, the relationship is quite different. There are two issues to consider. The first is how many particles will exist when the electrolyte dissociates; the second is to what degree does the electrolyte in question dissociate. If, for simplicity, it is assumed that an electrolye is 100% dissociated in water and forms an ideal solution, each ion is an osmotically active particle. Then, 1 mmol of NaCl in solution dissociates into 2 mOsmol of total particles. Using NaCl and $CaCl_2$ as examples and assuming complete dissociation, then 1 mmol of NaCl and $CaCl_2$ give 2 mOsmol (one Na^+ and one Cl^- ion) and 3 mOsmol (one Ca^{2+} and two Cl^- ions) of total particles, respectively. (*Note:* The osmolarity of plasma, provided largely by Na^+ and Cl^- ions, is about 0.3 Osm/l [300 mOsm/l]).

Occasionally, one needs to know the milliosmolar value of a single ion of an electrolyte. This is obtained by dividing the ion's concentration in milligrams per liter by its atomic weight. A convenient rule to follow is that the milliosmolar value of the whole electrolyte in solution is equal to the sum of the milliosmolar values for the separate ions that comprise the electrolyte.

The United States Pharmacopeia indicates that the ideal osmolar concentration may be calculated according to the following equation:

$$\text{mOsmol/liter} = \frac{\text{wt. of substance (g/l)}}{\text{molecular weight (g)}} \times \text{number of species} \times 1000$$

This equation is the osmolar concentration in milliosmoles per liter for ideal solutions. These can be interpreted as meaning for solutions that are within the physiologic range (normal serum osmolality is 275 to 295 mOsmol/kg) or for more dilute solutions. Deviation from ideality and from the above equation occurs when the solution is highly concentrated with respect to an electrolyte. This nonideality results because of particle-to-particle interaction within the solute typically cause the value of the calculated osmolar concentration to decrease from the value one would obtain for an ideal (dilute) solution. Many equations in pharmaceutics and in chemistry have these limitations, that is, deviation from ideality at higher solute concentrations. As an example, the measured osmolarity of a 0.9% sodium chloride (normal saline) solution is approximately 286 mOsmol/l, as compared with the value one would obtain using the above equation. Since a 0.9% sodium chloride solution contains 9 g of sodium chloride per liter and its molecular weight is 58.5 g/mol, the ideal value for the osmolar concentration would be

$$\text{mOsmol/liter} = \frac{9\text{g/l}}{58.5\text{g}} \times 2 \times 1000 = 308 \text{ mOsmol/l}$$

Clinical pharmacists should appreciate the difference in the values of osmolar concentration that a dilute vs. concentrated parenteral solution may yield, since some pharmaceutical manufacturers label electrolyte solutions with the ideal or stochiometric osmolarities calculated with the above equation, whereas others label the product with the experimentally determined value. A distinction should also be made between the terms osmolarity and osmolality. Osmolarity is the milliosmoles of solute per liter of solution, whereas osmolality is the milliosmoles of solute per kilogram of solvent.

It is important to know the osmolar strength of parenteral solutions. In fact, the United States Pharmacopeia requires that intravenous fluids, nutrients, electrolytes, and the osmotic diuretic mannitol are required to state the osmolar concentration. This information allows for determination of whether the given solution is hypo-osmotic, iso-osmotic, or hyper-osmotic in relation to biological fluids.

Two grams of glucose, molecular weight 180, is dissolved in 250 ml of solution at 25°C. What is the osmotic pressure of the solution?

Answer:

$$\text{Osmotic pressure, } \pi V = nRT$$

$$= \frac{nRT}{V}$$

$$\text{Moles of glucose} = \frac{2.0}{180} = 0.111$$

$$\partial = \frac{0.0111 \times 0.082 \times 298}{0.25} = 1.075 \text{ atm}$$

$$\text{Osmotic pressure of solution} = \boxed{\mathbf{1.075\ atm}}$$

Three types of intravenous solution exits:

- *Isotonic solution:* normal saline; used to replace body fluids
- *Hypertonic solution:* more concentrated than plasma; used to draw water from tissue
- *Hypotonic solution:* less concentrated than plasma; used to rehydrate tissue

Solutions that have the same osmotic pressure are termed *isosmotic*. Many pharmaceutical solutions intended for mixing with body fluids are designed to have the same osmotic pressure and are termed *isotonic*. Solutions of lower osmotic pressure than that of a body fluid are termed *hypotonic*, whereas those having a higher osmotic pressure are *hypertonic*.

VI. PREPARATION OF ISOTONIC SOLUTIONS

There are two common methods for adjusting the tonicity of a pharmaceutical formulation: the NaCl equivalent method and the freezing point depression method.

A. Sodium Chloride Equivalent Method to Adjust Tonicity

With electrolytes, the calculations involved in preparing an isotonic solution are more complicated than with solutions containing non-ionizable solutes. This is because the osmotic pressure depends more on the number than the kind of particles and the osmotic pressure increases with the degree of dissociation of the electrolyte. Conversely, the degree of dissociation and the net osmotic pressure developed are inversely related. That is, as the dissociation increases, the quantity of the substance required to produce any given osmotic pressure decreases. Using sodium chloride as an example and assuming that NaCl (mol wt = 58.5 gm/mol) is approximately 80% dissociated, then each 100 molecules of NaCl yield 180 particles, or 1.8 times as many particles as are yielded by 100 molecules of a nonelectrolyte. This dissociation factor, commonly designated by the letter *i*, must be included in the proportion when we seek to determine the strength of an isotonic solution of sodium chloride. Thus,

$$\frac{1.86°\text{C} \times 1.8}{58.5\ \text{g}} = \frac{0.52°\text{C}}{x\ \text{gm}}$$

$$x = 9.09\ \text{gm}$$

and approximately 9.09 g of NaCl in 1000 g of water should make a solution isotonic with blood or lacriminal fluid. A 0.9% (w/v) NaCl solution is considered isotonic with body fluids.

Isotonic solutions may then be calculated using the formula below:

$$\frac{\text{Molecular weight} \times 0.52°\text{C}}{1.86 \times \text{dissociation}(\text{i})} = \text{gm of solute per 1000 gm of water}$$

Unfortunately, the value of *i* has not been experimentally determined for many medicinal salts. However, weak solutions of most of these medicinal substances (salts of weak acids and bases) approximate the dissociation of NaCl in water. If the number of ions produced by the dissociation of a medicinal salt is known, we may use the following values of *i* as an approximation.

Nonelectrolytes and substances that dissociate slightly	1.0
Substances that dissociate into two ions	1.8
Substances that dissociate into three ions	2.6
Substances that dissociate into four ions	3.4
Substances that dissociate into five ions	4.2

A problem arises when an isotonic pharmaceutical solution that contains one or more medicinal salts has to be prepared or its tonicity adjusted. Because the isotonicity of a pharmaceutical solution is typically adjusted with sodium chloride, it has to be determined how much NaCl is represented by the other medicinal substance(s). This is called the sodium chloride equivalent method of adjusting the tonicity of a pharmaceutical solution.

Most pharmaceutical textbooks list tabulated sodium chloride equivalent values, or E-values, for many drug substances (Table 4.2). The amount of sodium chloride represented by that substance is determined by multiplying the number of grams of a substance in a particular pharmaceutical solution by its sodium chloride equivalent.

The procedure used to calculate isotonic solutions using the sodium chloride equivalent method may be outlined as follows:

Step 1: Calculate the amount (in grams) of sodium chloride represented by the ingredient(s) in the pharmaceutical solution to be made isotonic. Multiply the amount of each substance by its sodium chloride equivalent (consult table).

Step 2: Calculate the amount of sodium chloride alone that would be contained in an isotonic solution of the volume required, that is, the amount of sodium chloride in a 0.9% solution of the specified volume. (Such a solution would contain 0.009 g/ml of NaCl.)

Step 3: Subtract the amount of sodium chloride represented by the ingredient(s) in step 1 from the amount of NaCl alone that would be represented in the specific volume of an isotonic solution from step 2. The answer represents the amount (in grams) of NaCl to be added to make the solution isotonic.

Step 4: If an agent other than NaCl, such as boric acid, dextrose, sodium, or potassium nitrate is to be used to make the solution isotonic, divide the mount of NaCl from step 3 by the sodium chloride equivalent of the other substance.

What amount of NaCl is required to adjust the tonicity of a 1-l bag containing a 1% w/v morphine sulfate (5 H_2O) (mol wt = 759; E-value, or NaCl equivalent value, = 0.11) so that it is an isotonic solution?

Answer: Since 1% is equal to 1 g/100 ml, then 1 l of a 1% (w/v) solution of morphine sulfate would contain 10 g of the drug.

Step 1: 0.11 × 10 g = 1.1 g of NaCl represented by the morphine sulfate
Step 2: 1000 ml × 0.009 g/ml = 9 g of NaCl in 1000 ml of an isotonic NaCl solution
Step 3: 9 g (from step 2) – 1.1 g (from step 1) = 7.9 g of NaCl required to make the 1-l solution isotonic

---

How many grams of NaCl should be used in compounding the following prescription?

Rx: Pilocarpine nitrate:	0.30 g
Sodium chloride:	q.s.
Purified water, ad:	30 ml
Make isoton. sol.	

Recall that a 0.90% NaCl solution is isotonic, corresponding to 9 g in 1000 ml or 0.270 g in 30 m l. This amount must be reduced by the contribution of the pilocarpine nitrate, whose NaCl equivalent is 0.22. This amount is 0.30 g * 0.22 = 0.066 g. Thus, NaCl to be added (q.s.) is 0.270 g – 0.066 g = 0.204 g.

TABLE 4.2
Sodium Chloride Equivalents (E-Values)

Substance	MolWt	Ions	*i*-factor	E-Value
Antazoline phosphate	363	2	1.8	0.16
Antipyrine	188	1	1	0.17
Atropine sulfate H_2O	695	3	2.6	0.12
Benoxinate HCl	345	2	1.8	0.17
Benzalkonium chloride	360	2	1.8	0.16
Benzyl alcohol	108	1	1	0.30
Boric acid	61.8	1	1	0.52
Chloramphenicol	323	1	1	0.10
Chlorobutanol	177	1	1	0.24
Chlortetracycline HCl	515	2	1.8	0.11
Cocaine HCl	340	2	1.8	0.16
Cromolyn sodium	512	2	1.8	0.11
Cyclopentolate HCl	328	2	1.8	0.18
Demecarium bromide	717	3	2.6	0.12
Dextrose (anhydrous)	180	1	1	0.18
Dextrose H_2O	198	1	1	0.16
Dipivefrin HCl	388	2	1.8	0.15
Ephedrine HCl	202	2	1.8	0.29
Ephedrine sulfate	429	3	2.6	0.23
Epinephrine bitartrate	333	2	1.8	0.18
Epinephryl borate	209	1	1	0.16
Eucatropine HCl	328	2	1.8	0.18
Fluorescein sodium	376	3	2.6	0.31
Glycerin	92	1	1	0.34
Homatropine hydrobromide	356	2	1.8	0.17
Idoxuridine	354	1	1	0.09
Lidocaine HCl	289	2	1.8	0.22
Mannitol	182	1	1	0.18
Morphine sulfate 5 H_2O	759	3	2.6	0.11
Naphazoline HCl	247	2	1.8	0.27
Oxymetazoline HCl	297	2	1.8	0.20
OxytetracyclineHCl	497	2	1.8	0.12
Phenacaine HCl	353	2	1.8	0.20
Phenobarbital sodium	254	2	1.8	0.24
Phenylephrine HCl	204	2	1.8	0.32
Physostigmine salicylate	413	2	1.8	0.16
Physostigmine sulfate	649	3	2.6	0.13
Pilocarpine HCl	245	2	1.8	0.24
Pilocarpine nitrate	271	2	1.8	0.23
Potassium biphosphate	136	2	1.8	0.43
Potassium chloride	74.5	2	1.8	0.76
Potassium iodide	166	2	1.8	0.34
Potassium nitrate	101	2	1.8	0.58
Potassium penicillin G	372	2	1.8	0.18
Procaine HCl	273	2	1.8	0.21
Proparacaine HCl	331	2	1.8	0.18
Scopolamine HBr 3 H_2O	438	2	1.8	0.12
Silver nitrate	170	2	1.8	0.33
Sodium bicarbonate	84	2	1.8	0.65

TABLE 4.2 (continued)
Sodium Chloride Equivalents (E-Values)

Substance	MolWt	Ions	*i*-factor	E-Value
Sodium borate 10 H_2O	381	5	4.2	0.42
Sodium carbonate	106	3	2.6	0.80
Sodium carbonate H_2O	124	3	2.6	0.68
Sodium chloride	58.5	2	1.8	1.00
Sodium citrate 2 H_2O	294	4	3.4	0.38
Sodium iodide	150	2	1.8	0.39
Sodium lactate	112	2	1.8	0.52
Sodium phosphate, dibasic, anhydrous	142	3	2.6	0.53
Sodium phosphate, dibasic 7 H_2O	268	3	2.6	0.29
Sodium phosphate, monobasic, anhydrous	120	2	1.8	0.49
Sodium phosphate, monobasic H_2O	138	2	1.8	0.42
Tetracaine HCl	301	2	1.8	0.18
Tetracycline HCl	481	2	1.8	0.12
Tetrahydrozoline HCl	237	2	1.8	0.25
Timolol maleate	432	2	1.8	0.14
Tobramycin	468	1	1	0.07
Tropicamide	284	1	1	0.11
Urea	60	1	1	0.59
Zinc chloride	136	3	2.6	0.62
Zinc sulfate 7 H_2O	288	2	1.4	0.15

B. Freezing Point Method to Adjust Tonicity

It is generally accepted that blood serum and lacrimal fluid in the eye freeze at –0.52°C. Utilizing freezing point depression for preparation of a solution containing a nonelectrolyte is less complicated than for one containing an electrolyte.

When one gram molecular weight of any nonelectrolyte, that is, a substance with negligible dissociation such as boric acid, is dissolved in 1000 g of water, the freezing point is approximately –1.86°C. By simple proportion, the weight of any nonelectrolyte that should be dissolved in each 1000 gm of water can be calculated if the solution is to be isotonic with body fluids.

For example, boric acid has a molecular weight of 61.8 g/mol and hence, 61.8 g in 1000 gm of water should produce a freezing point of –1.86°C. Therefore,

$$\frac{1.86°\text{C}}{61.8\text{ g}} = \frac{0.52°\text{C}}{x\text{ gm}}$$

$$x = 17.3\text{ gm}$$

In summary, 17.3 g of boric acid in 1000 g of water, with a weight-in-volume strength of approximately 1.73%, should yield a solution isotonic with lacrimal fluid or blood serum.

How many milligrams each of NaCl and chloramphenicol are required to prepare 500 ml of a 1% solution of chloramphenicol isotonic with blood?

Answer: To make the solution isotonic with blood, the freezing point will have to be lowered to –0.52°C. From the table below, a 1% solution of chloramphenicol has a freezing point lowering of 0.06°C. Thus, a sufficient amount of NaCl must be added to lower the freezing point an additional 0.46°C (0.52°C – 0.06°C). From the table below, we can also note that a 1% solution of NaCl lowers the freezing point by 0.58°C. Therefore by proportion,

$$\frac{1\%(\text{NaCl})}{0.58°\text{C}} = \frac{x\%(\text{NaCl})}{0.46°\text{C}}$$

where $x = 0.793\%$ (the concentration of NaCl needed to lower the freezing point by 0.46°C).

Thus, to make 500 ml of the chloramphenicol solution,

500 ml of 1% = 5 gm = 5000 mg of chloramphenicol, and

500 ml of 0.793% = 3.965 gm = **3965 mg of NaCl**

If more than one medicinal or pharmaceutical compound is to be included in the pharmaceutical solution to be made isotonic, the sum of the freezing points is subtracted from the required value in determining the additional lowering of the freezing point required of the agent used to provide isotonicity.

Table 4.3, albeit not exhaustive, provides the freezing point depression values, designated as $\Delta T_{f(1\%)}$, for a variety of medicinal and pharmaceutical substances.

C. Molecular Weight Determination from Colligative Properties

The four colligative properties that have been discussed can be used to calculate the molecular weights of nonelectrolytes that are present as solutes in solution.

TABLE 4.3
Freezing Point Depression for Various Agents

Compound	Freezing Point Depression; 1% Solutions ($\Delta T_{f,1\%}$) in °C
Atropine sulfate	0.07
Boric acid	0.29
Butacaine sulfate	0.12
Chloramphenicol	0.06
Chlorobutanol	0.14
Dextrose	0.09
Dibucaine HCl	0.08
Ephedrine sulfate	0.13
Epinephrine bitartrate	0.10
Ethylmorphine HCl	0.09
Glycerin	0.20
Homatropine HBr	0.11
Lincomycin	0.09
Morphine sulfate	0.08
Naphazoline HCl	0.16
Physostigmine salicylate	0.09
Pilocarpine nitrate	0.14
Sodium bisulfite	0.36
Sodium chloride	0.58
Sulfacetamide sodium	0.14
Zinc sulfate	0.09

1. From Lowering of Vapor Pressure

From Equation 4.28 we know

$$\frac{p_1^0 - p}{p_1^0} = \frac{\Delta p}{p_1^0} = x_2 = \frac{n_2}{n_1 + n_2}$$

which reduces to Equation 4.29 if n_1 is very large compared to n_2.

$$\frac{\Delta p}{p_1^0} = x_2 \cong \frac{n_2}{n_1} = \frac{w_2/M_2}{w_1/M_1}$$

This equation can be rearranged to give the molecular weight of the solute, M_2.

$$M_2 = \frac{w_2 m_1 p_1^0}{w_1 \Delta p} \tag{39}$$

2. From Boiling Point Elevation

From Equation 4.34 we know that

$$\Delta T_b = K_b m$$

Since m can be expressed as

$$m = \frac{w_2/M_2}{w_1} \times 1000 = \frac{1000 w_2}{w_1 M_2}$$

this equation can be rearranged to give the molecular weight of the nonvolatile solute, M_2.

$$\Delta T_b = K_b \frac{1000 w_2}{w_1 M_2}$$

This equation can be rearranged to give the molecular weight of the nonvolatile solute, M_2.

$$M_2 = K_b \frac{1000 w_2}{w_1 \Delta T_b} \tag{4.40}$$

3. From Freezing Point Depression

From Equation 4.35 we know that

$$\Delta T_f = K_f m$$

Now since m can be expressed as

$$m = \frac{w_2/M_2}{w_1} \times 1000 = \frac{1000 w_2}{w_1 M_2}$$

This equation can be rearranged to give the molecular weight of the nonvolatile solute, M_2.

$$\Delta T_f = K_f \frac{1000 w_2}{w_1 M_2}$$

This equation can be rearranged to give the molecular weight of the nonvolatile solute, M_2.

$$M_2 = K_f \frac{1000 w_2}{w_1 \Delta T_f} \tag{4.41}$$

4. From Osmotic Pressure

From Equation 4.39 we know that

$$\partial = \frac{nRT}{V}$$

or

$$\pi = mRT = \frac{cRT}{M_2}$$

where M_2 is the molecular weight of the nonvolatile solute.

This equation can be rearranged to give

$$M_2 = \frac{cRT}{\pi} \tag{4.42}$$

12.1 grams of a new drug was dissolved in water to make up a 750-ml solution. At 25°C the osmotic pressure was 0.6 atm. What is the molecular weight of the solute?

Answer:

$$\partial = \frac{nRT}{V}$$

$$\pi = mRT = \frac{cRT}{M_2} \quad \text{c is in g/liter}$$

$$\pi = \frac{12.1 \times 0.0821 \times 298}{M_2}$$

$$M_2 = \frac{296.04}{0.6} = 493.4 \text{ g/mol}$$

Therefore, the molecular weight of the solute = $\boxed{\mathbf{493.4\ g / mol}}$

VII. CONCEPT AND CALCULATION OF MILLIEQUIVALENTS (MEQ)

Many systems of our body such as plasma, interstitial fluid, and intracellular fluid consist of solutions of electrolytes and nonelectrolytes. The electrolytes dissolve in body fluids to produce varying kinds and amounts of cations and anions, which help maintain the osmotic balance of the body, preserve the acid-base balance in the body, and carry electrical current required for nerve conduction. Some important cations in the body include *calcium*, *sodium*, *magnesium*, and *potassium*; common anions include *chloride*, *bicarbonate*, *phosphate*, and *sulfate*. *A chemical balance of these positively and negatively charged particles is necessary for maintenance of good health.* In pathological states, when this electrolyte balance is disturbed, it becomes necessary to administer salts to restore electroneutrality.

A. How Do We Know How Much to Give?

The normal physiological plasma content of Na^+ ions, for example, is 142 mEq/l, and that of chloride ions is 103 mEq/l.

B. What Is a MilliEquivalent?

The valence of a compound gives an idea about its reactivity or its combining power. The **equivalent (Eq)** expresses the chemical activity, or combining power, of a substance relative to the activity of 1 mg of hydrogen. An equivalent is an amount of material that will provide an Avogadro's number of electrically charged particles such as OH^-, H^+, or electrons. The formula weight has also been modified to give an idea about a compound's reactivity to give the *equivalent weight*, which is formula weight divided by the valence of that compound. Thus, for atoms

$$\text{Equivalent weight} = \frac{\text{atomic weight}}{\text{valence}}$$

Thus, 1-g atom of hydrogen consisting of 6.02×10^{23} atoms of hydrogen combines with 6.02×10^{23} atoms of chlorine and half this many atoms of oxygen. The atomic weight of chlorine is 35.5 and that of oxygen is 16. So, 1 g of hydrogen combines with 35.5 g of chlorine and 16/2, or 8, g of oxygen. In other words, the equivalent weight of chlorine is 35.5 and that of oxygen is 8 because one equivalent of chlorine, or 35.5 g, and one equivalent of oxygen, or 8 g, combine with 1 g of hydrogen.

This example shows that the equivalent weight of chlorine is the same as its atomic weight but that of oxygen is half of its atomic weight, meaning that the atomic weight of chlorine contains one equivalent, while the atomic weight of oxygen contains two equivalents.

This concept is also applicable to molecules. The equation for the equivalent weight of molecules is

$$\text{Equivalent weight}\ (\text{g/Eq}) = \frac{\text{atomic weight}\ (\text{g/mole})}{(\text{Equivalents/mole})}$$

Thus, the equivalent weight of KCl is 74.5 g/Eq, which is the same as its molecular weight because the combining power or the valence of both the ions is 1 for KCl.

What would be the equivalent weight of $CaCl_2$? The equivalent weight of $CaCl_2$ is half of its molecular weight The valence of Ca ions is 2, so its equivalent weight would be 40/2, or 20 g/Eq, and although the valence of chloride ions is unity, two atoms are present totaling a weight of 35.5 × 2 = 71 g, while its equivalent weight is just half of that, or 35.5 g/Eq. So the equivalent weight of $CaCl_2$ is 20 + 35.5 = 55.5 g/Eq, which is half the molecular weight of $CaCl_2$, which is 111 g.

Now to answer the question of what a milliequivalent is. In the field of medicine, amounts of electrolytes are usually described in terms of milliequivalents. This is because electrolytes are present in small amounts, and it is more convenient to express them as milliequivalents than equivalents, where a milliequivalent is 1/1000 of an equivalent.The plasma concentrations of some electrolytes expressed in terms of milliequivalents are as follows:

Electrolyte	Na^+	K^+	Cl^-	HCO_3^-	HPO_4^{2-}
Plasma concentration (mEq/l)	142	5	103	27	2

Thus, 1 M CH_3COOH = 1 N CH_3COOH (1 Eq of H^+ per mole of acid). 1 M H_2SO_4 = 2 N H_2SO_4 (2 Eq of H^+ per mole of acid).

Example: From the table given above, we can see that the plasma concentration of Cl^- ions is 103 mEq/l. Suppose you want to make 500 ml of an aqueous solution of KCl, which has the same concentration of chloride ions. How many milligrams of KCl would you add? The equivalent weight of KCl is the same as its molecular weight, which is 74.5 g/Eq.

We know that

$$\text{Equivalent weight}\left(\text{g/Eq}\right)=\frac{\text{weight}\left(\text{g/L}\right)}{\left(\text{Eq/L}\right)}=\frac{\text{weight}\left(\text{mg/L}\right)}{\left(\text{mEq/L}\right)}$$

So, 74.5 g/Eq = weight (in mg/l) 103 mEq/l or, weight (milligrams/liter) = 74.5 * 103 = 7673.5 mg/l.

Thus, the amount of KCl required to make 1 l of an aqueous solution with 103 mEq/l of Cl^- ions = **7673.5 mg**.

Therefore, the amount of KCl required to make 500 ml of such a solution = 7673.5/2 = 3836.75 mg or approximately **3.84 g**.

Calculate the milliequivalent weight of aluminium sulfate.

Answer: Aluminium sulfate, $Al_2(SO_4)_3$, has two aluminium and three sulfate ions. Thus, the molecular weight or the formula weight of $Al_2(SO_4)_3$ is

Al (2 atoms) = 2 × 27 = 54

S (3 atoms) = 3 × 32 = 96

O (12 atoms) = 12 × 16 = 192

Molecular weight = 54 + 96 + 192 = 342

Total positive or total negative valence of $Al_2(SO_4)_3$

$$Al = +3 \times 2 = \boxed{+6}$$

$$SO_4 = -2 \times 3 = \boxed{-6}$$

Hence, the equivalent weight of $Al_2(SO_4)_3$ would be

$$342\ g/6 = 57\ g/Eq$$

The milliequivalent weight of $Al_2(SO_4)_3$ would be

$$342\ mg/6 = \boxed{\mathbf{57\ mg\ /\ mEq}}$$

Thus, one mole of $Al_2(SO_4)_3$ has six equivalents of $Al_2(SO_4)_3$. It also has the same number of equivalents of aluminium and sulfate ions.

VIII. CONCLUSION

Solutions applied to delicate membranes of our body should be prepared carefully so that they will not cause significant discomfort. All the physico-chemical properties described in this chapter are important in this respect. A pharmacist should have a thorough understanding of buffers, solubility, partition coefficient, and colligative properties in order not only to dispense appropriate extemporaneous preparations in hospital and retail pharmacies but also to work as a formulation scientist in the pahermaceutical industry.

HOMEWORK

1. Explain the effects of pH on the solubility of a weak acid (base).
2. What is a buffer? How are buffers used in pharmaceutics?
3. What is buffer capacity?
4. Define an acid.
5. What is a hydronium ion?
6. Define basicity of an acid.
7. How are acids classified on the basis of their strength?
8. How are acids classified on the basis of their basicity?
9. Explain why a weak acid remains a weak acid even if it is concentrated.
10. Explain why a strong acid remains a strong acid even if it is dilute.
11. What are the applications of the pH value?

12. What are acid–base indicators?
13. What is the pH of a solution that is 0.1 M in NH_3 and 0.5 M in NH_4Cl? (K_b for NH_3 = 1.8×10^{-5}):
    (a) 4.0
    (b) 7.2
    (c) 8.6
    (d) 10.0
    (e) 11.1

14. Which of the following solutions behave as buffers?
    1. 100 ml of 0.2 M HOAc + 50 ml of the strong base 0.4 M $Ba(OH)_2$
    2. 100 ml of 0.1 M HCl + 100 ml of 0.1 M NaOH
    3. 100 ml of 0.2 M NH_3 + 50 ml of 0.1 M NH_4Cl
    4. 100 ml of 0.1 M F- + 100 ml of 0.2 M HF
       (a) 1, 2, and 4
       (b) 2, 3, and 4
       (c) 2 and 4
       (d) 3 and 4
       (e) 4 only

15. A buffer solution contains 0.11 mol of acetic acid and 0.15 mol of sodium acetate in 1.0 l. What is the pH of the buffer after the addition of 0.01 mol of HCl? (K_a for acetic acid = 1.8×10^{-5}):
    (a) 4.58
    (b) 4.74
    (c) 5.02
    (d) 4.81
    (e) None of the above

16. 0.64 g of adrenaline in 36.0 g of CCl_4 produces a boiling point elevation of 0.49°C. What is adrenaline's molecular weight? (Ans: M = 182.5 g/mol)
17. At 4°C, the vapor pressure of water is 6.1 mmHg. What is the vapor pressure of a 3.00 molal solution of NaCl? (Ans: 5.50 mm Hg)
18. Determine the boiling point and freezing point of a solution that contains 50.0 g of $CaCl_2$ (111 g/mol) in 200 g of water? ($CaCl_2$ dissociates as: $CaCl_2 \rightarrow Ca^{2+} + 2Cl^-$) Given: ΔT_b = 0.51°C/m, ΔT_f = 1.86°C/m. (Ans: T_b = 103.45°C, T_f = −12.57°C)
19. Calculate the relative vapor pressure lowering at 25°C for a solution containing 45.2 g of glucose in 1 l of water. The molecular weights of glucose and water are 180 g/mol and 18.02 g/mole, respectively. If the actual pressure at 25°C is 19mm Hg, what is the final pressure? (Ans: Δp = 0.0855 mm, P_1= 18.91 mm)
20. 102.7 g of sucrose was dissolved in 1000 g of water to make a solution. This solution gave a boiling point elevation of 0.221°C. Calculate the molal elevation constant (ebbulioscopic constant) for water. Molecular weight of sucrose is 342.3 g/mole. (Ans: K_b = 0.737°C.kg/mole)
21. The presence of impurities in a compound increases its boiling point while reducing the freezing point. Seawater consists of about 3.5% (by weight) dissolved solids, almost all of which is NaCl. Given that K_b equals 0.52°C m^{-1}, what would be the normal boiling point of seawater? (Ans: 100.64°C)

CASES

CASE I

Dextrose in water is commonly administered to provide calories for metabolic needs, hydrate the body, and spare protein catabolism. Keeping in mind that dextrose is a nondissociating solute, what concentration of dextrose in water (%[w/v]) would be isotonic with blood?

Hint: Isotonic dextrose solution has a freezing point of –0.52°C. We know that the molecular weight of dextrose (anhydrous) is 180 g. Since dextrose dissolves but does not dissociate in water, only one particle will result from one molecule (i = 1.00). One gram molecular weight of particles of any substance will lower the freezing point of 1000 ml of water to −1.86°C. The extent of freezing point depression is proportional to the number of particles of solute – in this case, dextrose. Thus, if 180 g of dextrose dissolved in 1000 ml will lower the freezing point to −1.86, how many grams will be required to lower the freezing point of 1000 m to –0.52 (in order to be isotonic)? Do not forget to convert grams/1000 ml to grams/100 ml, that is, %(w/v).

CASE II

The following prescription was given to a pharmacist to dispense:

R_x
Atropine sulfate 2%
Aqua. dist. q.s. ad. 30 ml
M.ft. isotonic solution

1. Where is the solution likely to be applied?
2. How does the active ingredient work?
3. If the solution is not made isotonic, what may happen?
4. How much NaCl is required to make it isotonic.

Answers:

1. Eye
2. In the eye, atropine blocks the effect of acetylcholine on the sphincter muscle of the iris and the accommodative muscle of the ciliary body. This results in dilation of the pupil (mydriasis) and paralysis of the muscles required to accommodate for close vision (cycloplegia). Thus, opthalmic solution of atropine sulfate is used to treat cycloplegic refraction or pupillary dilation in acute inflammatory conditions of the iris and uveal tract.
3. If atropine sulfate solution is administered by itself into the eye, a burning sensation would result because of irritation. Thus, the tonicity of preparation has to be adjusted before administration.

4. A. Determine the amount of NaCl to make 30 ml of an isotonic solution

$$\frac{0.9\text{ g}}{100\text{ ml}} = \frac{X}{30\text{ ml}}$$

$$X = 0.27\text{ g}$$

B. Calculate the contribution of atropine sulfate to the NaCl equivalent

$$30\text{ ml} \times 2\text{ g}/100\text{ ml} = 0.6\text{ g atropine sulfate}$$

$$\text{E of atropine sulfate} = 0.13$$

$$0.6\text{ g} \times 0.13 = 0.078\text{ g}$$

C. Determine the amount of NaCl to add to make the solution isotonic by subtracting (B) from (A)

$$0.27\text{ g} - 0.078\text{ g} = 0.192\text{ g or 192 mg}$$

Other substances may be used, in addition to or in place of NaCl, to render solutions isotonic. This is done by taking the process one step further and calculating the amount of the substance that is equivalent to the amount of NaCl calculated in step C.

For example, boric acid is often used to adjust isotonicity in ophthalmic solutions because of its buffering and anti-infective properties. If E for boric acid is 0.50, then the amount of boric acid needed to replace the NaCl in Step C can be calculated:

$$\frac{0.192\text{ g NaCl}}{X\text{ g boric acid}} = \frac{0.50\text{ g NaCl equiv.}}{1\text{ g boric acid}} \quad \text{or } X = 0.38\text{ g}$$

or $X = 0.38$ g. Or, more simply $0.192\text{ g} \div 0.50 = 0.38$ g.

Thus, 0.38 g, or 380 mg, of boric acid would be required to render the previous ophthalmic solution isotonic.

CASE III

If you are a practicing pharmacist working in a specialty pharmacy, how do you prepare the following isotonic buffered solution?

R_x

Procaine HCl 2%
Aqua. dist. q.s. ad 15 ml
M.Ft. Isotonic, buffered injection

Answer:

In some cases, the drug is dissolved in an appropriate volume of water (V-value) to make the solution isotonic, and the volume of the solution is made up by an isotonic buffer. The formulation requires 0.3 g of Procaine HCl. **V-value** tables can be found in standard references and are tabulated to tell how many milliliters of water, when added to 0.3 g of drug, will result in an isotonic solution. For Procaine HCl, 7 ml of water added to 0.3 g of drug will make an isotonic solution. Therefore

0.3 g Procaine HCl is dissolved in 7 ml of water and then a sufficiently buffered, isotonic vehicle of appropriate pH is added to make 15 ml. (*Note:* This is way to make an isotonic solution was not described in the text.)

The pH of an isotonic Procaine HCl solution is 5.6. Therefore, an isotonic buffer of approximately that pH would be used. One commonly used isotonic buffer is the Sorenson's modified phosphate buffer. The following table is available in the United States Pharmacopeia.

Sorensen's Phosphate Vehicle

0.0667 M NaH_2PO_4 (ml)	0.0667 M Na_2HPO_4 (ml)	Resulting pH	NaCl Required for Isotonicity (g/100 ml)
90	10	5.9	0.52
80	20	6.2	0.51
70	30	6.5	9.5
60	40	6.6	0.49
50	50	6.8	0.48
40	60	7.0	0.46
30	70	7.2	0.45
20	80	7.4	0.44
10	90	7.7	0.43
5	95	8.0	0.42

The closest Sorensen's buffer to pH 5.6 would be pH 5.9, so to complete the formulation, 8 ml of pH 5.9 Sorensen's buffer would be added to the 7 ml of Procaine HCl solution. The individual amounts of the Sorenson's buffer to add can be determined as follows:

$$90 \text{ ml}/100 \text{ ml} \times 8 \text{ ml} = 7.2 \text{ ml of } 0.0667 \text{ M } NaH_2PO_4$$

$$10 \text{ ml}/100 \text{ ml} \times 8 \text{ ml} = 0.8 \text{ ml of } 0.0667 \text{ M } Na_2HPO_4$$

$$0.52 \text{ g}/100 \text{ ml} \times 8 \text{ ml} = 0.04 \text{ g of NaCl}$$

Therefore, the compounding procedure would be to weigh 0.3 g Procaine HCl and 0.04 g NaCl, add 7 ml of H_2O, 7.2 ml of 0.0667 M NaH_2PO_4, and 0.8 ml of 0.0667 M Na_2HPO_4. Filter and sterilize the solution and package in a sterile final container.

The limitation to this approach is that the final formulation pH may be a little different than the desired pH.

CASE IV

Outline the strategy to prepare the following isotonic buffered solution using the buffer table shown above.

Ampicillin sodium: 30 mg/ml
Sodium chloride: q.s.
Make 15 ml of sterile, buffered, isotonic solution at pH 6.6

Answer:

A different method from the one used above is to use the Sorensen's buffer as the entire solvent of the formulation. In this situation, the sodium chloride equivalent of the active drug is subtracted from the "NaCl required for isotonicity" listed in the table. This method has the advantage that the pH of the final solution will be the pH of the selected Sorensen's buffer.

1. A ratio calculation will show that 0.45 g of ampicillin sodium is needed for this formulation.

$$\frac{30 \text{ mg}}{\text{ml}} \times 15 \text{ ml} = 450 \text{ mg} = 0.45 \text{ g}$$

2. The sodium chloride equivalent for ampicillin sodium is 0.16. Therefore, the drug will contribute osmotic pressure as if it was 0.072 g of sodium chloride.

$$0.45 \text{ g} \times 0.16 = 0.072 \text{ g}$$

3. To have a pH of 6.6, 9.0 ml of monobasic sodium phosphate solution and 6.0 ml of dibasic sodium phosphate solution are needed. Then to adjust the isotonicity, 0.0735 g of sodium chloride is needed, but the ampicillin sodium equivalent will account for 0.072 g, so an additional 0.0015 g of sodium chloride must be added.

$$0.49 \text{ g/100 ml} \times 15 \text{ ml} = 0.0735 \text{ g of NaCl}$$

$$0.0735 \text{ g} - 0.072 \text{ g} = 0.0015 \text{ g of NaCl}$$

Therefore, the compounding procedure would be to weigh 0.45 g of ampicillin sodium and 0.0015 g of sodium chloride. Add 9.0 ml of monobasic sodium phosphate solution and 6.0 ml of dibasic sodium phosphate solution. Filter and sterilize the solution and package in a sterile final container.

CASE V

Absorption of a drug molecule is influenced by its solubility, pH, and partition coefficient in a given liquid formulation. The absorption of isoproterenol HCl from an *in vitro* percutaneous absorption study is shown in the following table.

Formulation	Solubility (mg/ml)	Partition Coefficient	C × P	Absorption (mg/cm²)
	C	P		J
1	389	0.55	212.5	14.7
2	357	0.47	166.5	18.1
3	335	1.25	418.1	21.6
4	288	1.43	414.4	21.3
5	243	1.53	371.6	21.6
6	92	5.06	464.1	19.7
7	9	—		

Note: Items in double-lined boxes are not significantly different.

1. Based on the solubility alone, which formulation(s) results in maximum of isoproterenol HCl?
2. Which formulation(s) should provide the highest absorption based on partition coefficient alone?

3. Which formulation(s) provides the highest absorption based upon the product, C × P?
4. Which of the aforementioned parameters, C, P, or C × P correlates well with observed absorption listed in the last column?

Answers:

1. 1,2,3 (because they are not significantly different)
2. 6
3. 3 to 6 (because they are not significantly different)
4. C × P is best because it takes both solubility and partition coefficient into consideration.

CASE VI

As mentioned in the above case study, the drug pK_a and vehicle pH could influence drug absorption. Thus, the pH for optimum absorption can be predicted using Henderson–Hasselbalch equation.

$$pH = pK_a + \log\text{ (conjugate base/conjugate acid)}$$

Isoproterenol HCl consists of two ionizable groups and consequently has two pK_a:

$$pK_{a1} = 8.83 = 9 \text{ (amino group is weakly basic)}$$

$$pK_{a2} = 10.19 = 10 \text{ (phenolic group is weakly acidic)}$$

1. Write the chemical structure of isoproterenol HCl.
2. Circle the amino group and the phenolic groups.
3. Calculate the pHs at which the ratio of un-ionized to ionized forms of isopreternol would be 100 to 1, 10 to 1, and 1 to 1 considering both pK_a values.
4. Estimate the pH for optimum absorption.

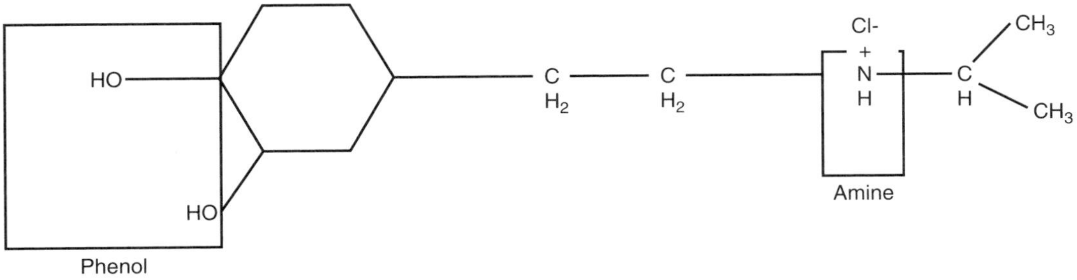

Answers:

1. and 2. See above
3. For amine:

$$\text{at } 100:1, pH = 9 + \log\frac{100}{1} = 11$$

$$\text{at } 10:1, pH = 9 + \log\frac{10}{1} = 10$$

$$\text{at } 1:1, pH = 9 + \log\frac{1}{1} = 9$$

For phenol:

$$\text{at } 100:1, \text{pH} = 10 + \log\frac{1}{100} = 8$$

$$\text{at } 10:1, \text{pH} = 10 + \log\frac{1}{10} = 9$$

$$\text{at } 1:1, \text{pH} = 10 + \log\frac{1}{1} = 10$$

4. Drugs are absorbed when they are predominantly in un-ionized states, that is 100:1 un-ionized to ionized ratio in this case. Since there are two ionizable groups in isoproterenol HCl, the answer is the average pH of the two (pH 11 for amino group and pH 8 for phenolic group) ionizable groups (11+ 8)/2 = 9.5.

REFERENCES

1. Kolthoff, I.M. and Takelenberg, F., in *Rec. Trav. Chim.,* 46, 33, 1925.
2. Van Slyke, D.D., in *J. Biol. Chem.,* 52, 525, 1922.
3. Cartensen, J.T., *Drug Stability: Principles and Practices*, 2nd ed. Marcel Dekker, New York, 1995.
4. Gennaro, A.R., *Remington: The Science and Practice of Pharmacy*, 20th ed., Lippincott Williams & Wilkins, Philadelphia, 2000.
5. Martin, A., *Physical Pharmacy*, 4th ed.Lea & Febiger, Philadelphia, 1993.
6. Zatz, J.L., *Pharmaceutical Calculations*, 3rd ed. John Wiley, New York, 1995.

5 Micromeritics and Rheology

Sunil S. Jambhekar

CONTENTS

0-4152-8863-0/05/$0.00+$1.50
© 2005 by CRC Press LLC

Learning Objectives

After finishing this chapter students will have thorough knowledge of:

- Ways to express particle size
- Methods to determine particle size
- Flow properties of powder
- Rheology of liquids and semisolids
- Methods to evaluate flow properties

I. MICROMERITICS

A. Introduction

The term *micromertics* was introduced by Dallavale in 1948 to describe the science of small particles. Dallavale found that the behavior and characteristics of small particles brought together information on particle size measurement, size distribution, and packing arrangements. Additionally, Dallavale discovered that the use of small particles ranged from the manufacturing of pharmaceutical dosage forms to atmospheric studies and many other fields.

In the field of pharmacy, micromertics has become an important area of study because it influences a large number of parameters in research, development, and manufacturing of dosage forms such as suspension to be reconstituted, tablet, and capsule. For instance, the amount of water needed to reconstitute a suspension of a fixed volume is influenced by the particle size of the powdered suspension. The content uniformity of a drug in a solid dosage form, particularly low dose and highly potent drugs, can also be affected by the particle size and size distribution of a drug. The particle size range and distribution of particles in a given product can influence its efficacy, stability, and safety. This is particularly true in the case of solid dosage and suspension dosage forms, where the particle size of poorly water soluble drug is known to influence dissolution behavior and, therefore, bioavailability. Since a drug dissolves more rapidly when its surface area is increased, many poorly soluble and slowly dissolving drugs are marketed in micronized or microcrystal forms. Smaller particles generally permit more rapid and complete absorption of drugs.

B. Particle Size and Size Distribution

Particle shape plays an important role in particle size determination. The simplest definition of particle size diameters is based on the assumption that particles are spherical. In reality, however,

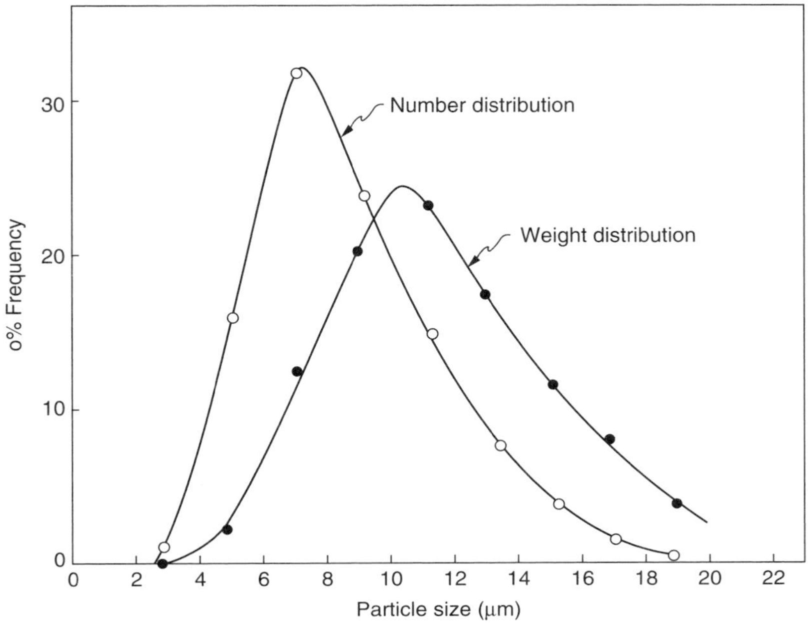

FIGURE 5.1 A frequency distribution plot.

particles possess different shapes, for example, rod, cubical, granular, etc. As the size of the particle increases, so does its tendency to have an irregular shape. This complicates the true particle size determination of a powder. Particle shape coefficients have therefore been derived for different geometries and various *equivalent diameters* have been developed to relate the size of such particles to that of a sphere with identical diameter, surface area, or volume.

Any collection of particles is usually polydisperse. It is therefore necessary to not only know the size of certain particles but also to know how many particles of the same size exist in a sample. Thus, we need an estimate of the size range and the number or weight traction of each particle size. This is the particle size of the sample.

1. Determination of Size Distribution

Since most pharmaceutical substances contain a range of sizes, the size distribution of the powder particles must also be determined to characterize the powder and determine the particle size. When the number or weight of particles within a certain size range is plotted against the size range or mean particle size, a frequency distribution curve is obtained. A typical example of such a curve is illustrated in Figure 5.1. The number distribution implies that the data were collected by employing counting techniques, and weight distribution representing the data were obtained by weighing technique.

The ideal resulting curve in Figure 5.2 is a normal or *Gaussian* distribution, with the standard deviation measuring the distribution around the mean. It is immediately apparent from frequency distribution curves (Figures 5.1 and 5.2) and data presented in Table 5.1 what particle size occurs most frequently within the sample. This is termed *mode*. In a perfectly symmetrical distribution, such as that represented in Figure 5.2, the mean, median, and mode values are the same. The standard deviation, as indicated in Figure 5.2, indicates the uniformity of particle size within the sample. In Figure 5.2, there is a 68% probability that a particle in the sample will be within ±1 standard deviation and a 95% chance that it will be within ±2 standard deviation.

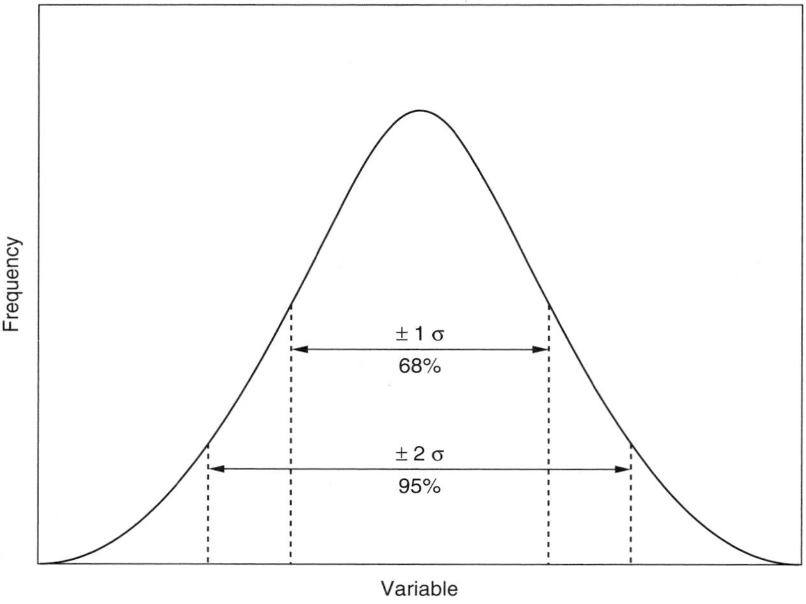

FIGURE 5.2 Typical plot of normal or Gaussian size frequency distribution curve.

TABLE 5.1
Determination of the Average Particle Size of Powder Measured by an Optical Microscope

Size Group (μ)	Midsize, *d* (μ)	Number of Particles in Each Size Group, *n*	*nd*
40–59.9	50	20	1,000
60–79.9	70	40	2,800
80–99.9	90	120	10,800
100–119.9	110	100	11,000
120–139.9	130	40	5,200
140–159.9	150	10	1,500
		$\Sigma n = 330$	$\Sigma nd = 32{,}300$

$$d_{av} = \frac{32{,}300}{330} = 98\ \mu$$

Most pharmaceutical powders, however, as a consequence of milling, tend to give skewed distributions. Such distributions can be normalized by plotting frequency vs. logarithm of particle size as illustrated in Figure 5.3 and Table 5.2. Such a size distribution is referred to as a log-normal distribution.

A log normal distribution has several properties of interest. When a logarithm of particle size is plotted against the cumulative percent frequency on a probability scale (Figure 5.4 and Table 5.1), a linear relationship can be observed. Such a linear plot has the advantage of permitting characterization of a log-normal distribution curve by means of two parameters: the slope of the line and the geometric mean diameter (d_g), which is equal to the median diameter.

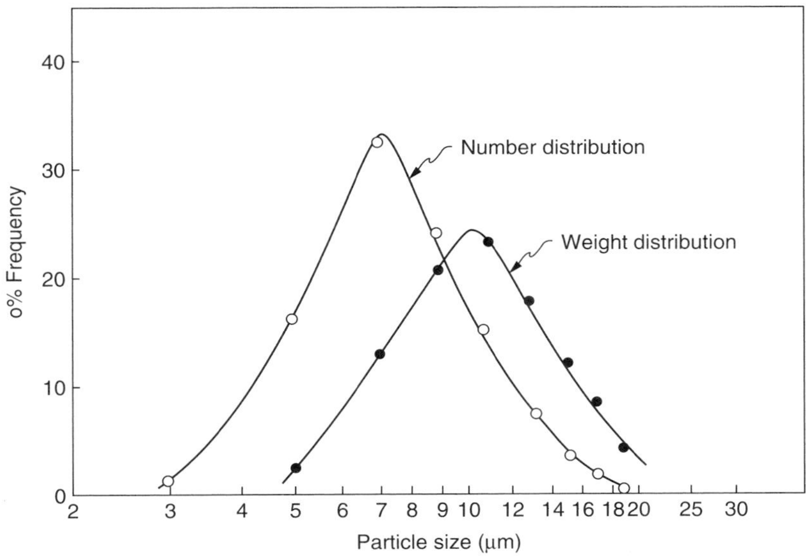

FIGURE 5.3 Frequency distribution plot of data showing log-normal relation.

TABLE 5.2
Size Weight Distribution of a Powder for Particle Size Measurement

Sieve Number (Passed/Retained)	Arithmetic Mean of Openings (mm)	Weight Retained (g)	Percent Retained	(Percent Retained) × (Mean of Opening)
10/20	1.420	0.81	0.84	1.1928
20/40	0.630	23.60	24.58	15.4854
40/60	0.335	52.30	54.49	18.2541
60/80	0.214	13.75	14.32	3.0645
80/100	0.163	4.72	4.92	0.8020
100/120	0.137	0.82	0.85	0.1164
		96.00	100.00	38.9152

$$d_{av} = \frac{\sum \text{(percent retained)} \times \text{(average size)}}{100} = \frac{38.91}{100} = 0.389 \text{ mm}$$

The slope of the line represents the standard deviation:

$$\text{slope} = \text{std. dev} = \frac{50\% \text{ size}}{16\% \text{ undersize}} \tag{5.1}$$

or

$$\text{std. dev} = \frac{84.13\% \text{ undersize}}{50\% \text{ size}} \tag{5.2}$$

Using the geometric mean diameter (d_g) and the standard deviation obtained from number and/or weight distribution, one can convert one type of diameter to another.

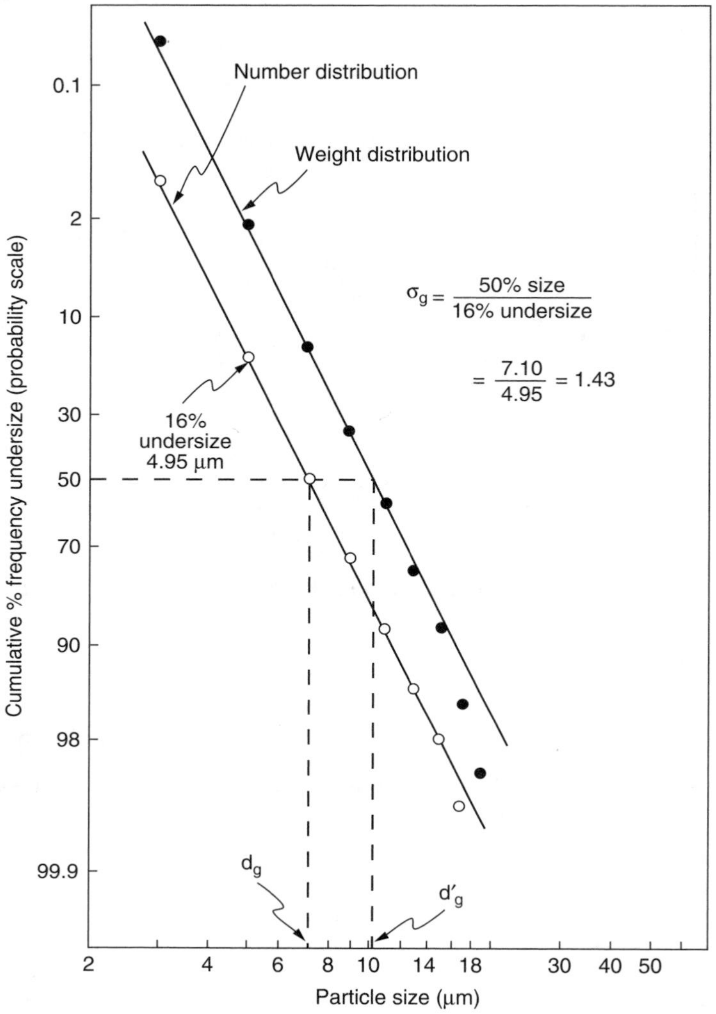

FIGURE 5.4 Typical log-probability plots of data.

2. Types of Diameter

Particle size (diameter) can be described by different expressions as outlined below:

a. *Mean Particle Diameter (d_{ave})*

The mean is the sum of all individual diameters divided by the total number of particles.

$$d_{ave} = \frac{\sum nd}{\sum n} \tag{5.3}$$

where d_{ave} is the diameter of the spherical particles, and n is the number of particles in the sample. Using Equation 5.3 and the values for $\sum nd$ and $\sum n$ presented in Table 5.1, one can determine the average diameter (d_{ave}) as follows:

$$d_{ave} = \frac{\sum nd}{\sum n} = \frac{32{,}300}{330} = 98\ \mu$$

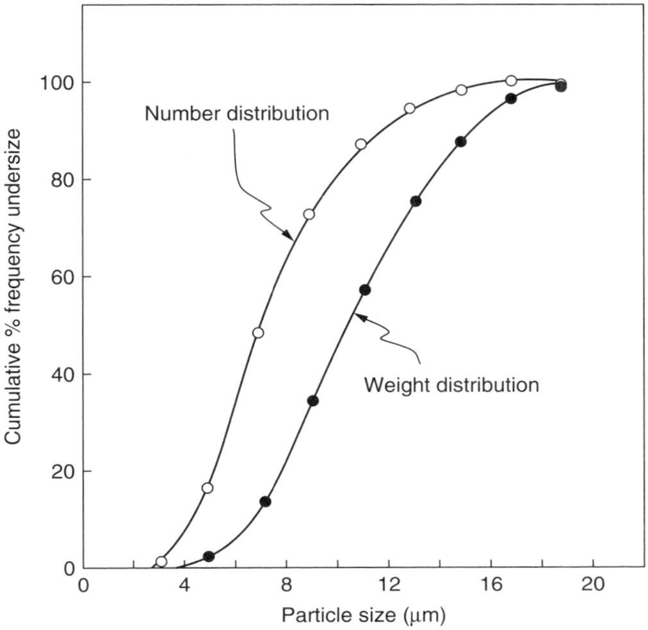

FIGURE 5.5 Cumulative frequency plot of data.

This diameter is sensitive to extreme values and thus is not often particularly useful. It may give underweight to smaller particles, and this average represents the size present in the greatest number.

b. Median Diameter (d_{med})

This is a diameter for which 50% of the particles are less than the stated size. It is influenced by outlying values and is preferable to the mean diameters as a single measure of particle size. The determination of the median diameter is possible following plotting of values of cumulative percent less than the stated size against the particle size (Figure 5.5 and Table 5.1).

c. Mode Diameter

Mode diameter represents the particle size occurring most frequently in a sample; it is used far less frequently than mean and median diameter. If the particle size distribution were symmetrical, the mode, median, and mean would be equal.

d. Mean Volume Surface Diameter (d_{vs})

This diameter is used to express powder particle sizes in terms of surface area per unit volume. A unit volume of monosize particles with a diameter, d_{vs}, will have a total surface area identical with the surface of a unit volume or actual sample having a mean surface value diameter. This diameter is used to express the size of particles in studies dealing with, for example, adsorption, coating, and surface degradation. This diameter can be computed by employing the following example.

$$d_{ave} = \frac{\sum nd^3}{\sum nd^2} \tag{5.4}$$

where the terms n and d are as defined previously.

3. Methods of Particle Size Determination

For a detailed discussion of the numerous methods of particle size analysis the reader should refer to the texts mentioned in the references. Microscopy, sieving, and sedimentation methods are discussed in the following section. None of the methods permit the true direct measurement of the particle size, though microscopy allows one to view the actual particle under the microscope. In other methods such as adsorption and air permeability, the surface area of the powder sample is computed from the particle size distribution, which, in turn, is useful to determine the particle size.

a. Microscopy

In this method the average diameter of a particulate system is obtained by measuring the particles at random along a given fixed line. As the particles lie in a random manner, their diameters are measured. To provide statistically sound data, a minimum of 200 particles should be measured.

The eyepiece of an optical micrometer has a cross hair, which is actuated by a calibrated micrometer drum. The cross hair is lined up with one edge of the particle, and the micrometer drum reading is recorded. The cross hair is moved to the opposite edge of the particle, and the micrometer drum reading is recorded again. The difference between the two readings is the measure of the particle size. The data obtained for 200 or more particles is recorded and tabulated, as shown in Table 5.1, to obtain the size distribution and the mean particle size. The average diameter is determined by using Equation 5.4. This method is useful for the particle size range of 0.2 to 100 μm. A disadvantage is that the diameter is obtained from the dimensions of a particle: length and breadth. Furthermore, the number of particles that must be counted to obtain a good estimation of the distribution makes the method slow and tedious.

b. Sieving

This method uses a series of standard sieves calibrated by the National Bureau of Standards. Sieves are made up of wire mesh with openings of known size. The term *mesh* is used to denote the number of openings per linear inch. The method is generally used to measure coarser particles; however, the method can be employed to measure particles as small as 44 μm.

In determining particle size by this method, a nest of sieves with the coarsest on top is placed on the shaker, and the powder sample of known weight is placed on the top of the sieve. The powder is classified as having passed through one sieve and being retained on the adjacent finer sieve. Particle diameter is considered as the size of the opening in the larger or finer sieve or as the size of the arithmetic or geometric mean of the opening of the two sieves. Whichever size is chosen, it should be stated and used throughout the study. The data presented in Table 5.2 and Figure 5.3 illustrate particle size measurement by this method.

For example, the diameter of particles that pass a 40-mesh sieve and are retained on a 60-mesh sieve (i.e., 40/60) may be defined in terms of the larger sieve, that is, 0.42 μm. The same particle might be expressed as the arithmetic mean of the opening of two sieves:

$$\frac{0.42+0.25}{2}=0.335\ \mu\text{m}$$

The size of the particles can also be expressed as the geometric average of the two sieve openings:

$$\left(0.42\times 0.25\right)^{1/2}=0.324\ \mu\text{m}$$

The weight of the powder retained on each sieve is weighed and, assuming log-normal distribution, the cumulative percent by weight of powder retained is plotted on a probability scale against

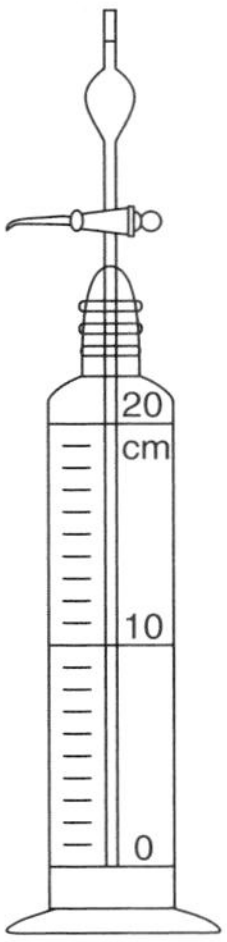

FIGURE 5.6 Andreasen apparatus for determining particle size by the sedimentation method.

the logarithm of the arithmetic mean size of the opening of each two successive screens, as illustrated in Figures 5.4 and 5.5. The geometric diameter (d_g) and the geometric standard deviation (δg) can be obtained from the straight line by using Equations 5.1 and 5.2.

Errors in computation may arise from a number of variables including sieve loading and duration and the intensity of agitation, which may produce attrition of the powder.

c. *Sedimentation*

For particles of subsieve size, which are often encountered in pharmacy, the sedimentation method as based on the Stoke's equation (Equation 5.5) is primarily used. The powder to be measured is suspended in a liquid in which material is completely insoluble. The suspension is placed in a calibrated pipette (Figure 5.6) from which the samples are withdrawn from fixed depth at various times. The samples are evaporated to dryness and the residues are weighed.

Each successive sample withdrawn has a particle size smaller than that corresponding to the settling velocity, because all larger particles will have fallen below the tip of the pipette. The effective, or Stoke, diameter is calculated by the Stoke's equation:

$$d_{st} = \sqrt{\frac{18\eta h}{(\rho_s - \rho_o) gt}} \qquad (5.5)$$

where (d_{st}) is the Stoke diameter, η is the viscosity of the suspending liquid in poises, h is the distance between the liquid surface and the tip of the pipette when the sample is drawn, $\rho_s - \rho_o$ is the difference in density between the particle and the suspending medium, g is the gravitational constant, and t is the time in seconds from the start of the measurement.

The percentages of the weight of the original samples are obtained from the weight of the dried samples. The data are plotted as cumulative percentages of weight against the particle size (Figure 5.4). Percentage by weight of any particle in the sample may be determined from such a plot.

In utilizing this method, one should realize that the diameter obtained is the relative size equivalent to that of a sphere with the same settling rate. The shape factor will affect the rate of settling; in general, the greater the irregularity and the deviation from the spherical shape, the slower will be the rate of setting. Furthermore, in using the Stokes equation (Equation 5.5), it is assumed that the particles to be measured are uniformly distributed and that they fall independently of each other. Since such assumptions are valid for dilute suspensions, it is a common practice to prepare suspensions that do not exceed 2% concentration to ensure that there is no interference among particles.

C. Properties of Powders

The preceding sections have been concerned mainly with size distribution, particle size, and methods of size measurement. The size distribution and the surface area of powders are the two fundamental properties of any collection of particles. Additionally, numerous derived properties are based upon these fundamental properties, and those of particular relevance to pharmacy are discussed in this section. These properties play an important role in the content uniformity of solid dosage forms, drug dissolution, and the manufacturing of dosage forms.

1. Densities of Powders

Density is universally defined as weight/volume. Determining the volume of particles containing microscopic cracks, internal pores, and capillary space, however, presents some difficulties. For convenience, three types of densities may be defined: true density, granule density, and bulk density.

a. *True Density* (ρ)

The true density of a solid is the mass per unit volume (grams per cubic centimeters), exclusive of all voids that are not fundamental parts of the molecular packing arrangement. True density is normally measured by helium pycnometer, where the volume occupied by the known mass of powder is determined by measuring the volume of the gas displaced by the powder. The true density is an intrinsic property of a powder.

b. *Bulk Density* (ρ_a)

Bulk density is defined as the mass of powder divided by its bulk volume. The bulk density is determined by using the standard procedure where 50 cm^3 of powder, which has been previously passed through a 20-mesh sieve, is carefully introduced into a 100 ml graduated cylinder, which is then dropped at 2-second intervals on the hard wood surface three times from the height of 1 inch.

The bulk density of powder depends primarily on the particle size distribution, the particle shape, and the tendency of particles to adhere to one another. The particles may pack in such a way as to leave large gaps between their surfaces, resulting in a light powder of low bulk density. However, smaller particles may sift between larger ones to form a heavy powder or one with higher bulk density.

The terms *light* and *heavy* as applied to powders mean low bulk density or large bulk volume and high bulk density or small volume, respectively. It should be noted that these terms have no relationship to true density.

c. *Granule Density* (ρ_g)

Granule density may be determined by a method similar to that for true density. Mercury is used because it fills the void spaces but fails to penetrate into the internal pores of the particles. The volume of the particles together with their interparticle spaces then yield the granule volume, and from the knowledge of the powder weight, the granule density is determined.

2. Porosity (ε)

In real particulate systems, porosity or void cannot be calculated, but it must be experimentally determined. Porosity is the percent void space: void × 100. Void, however, may be calculated from the true volume (V_{true}) and the bulk volume (V_{bulk}) of the powder material as follows:

$$\text{Void} = \frac{V_{bulk} - V_{true}}{V_{bulk}} \tag{5.6}$$

or

$$\text{Void} = 1 - \frac{V_{\text{true}}}{V_{\text{bulk}}} \tag{5.7}$$

The true volume refers to the volume occupied by the solid substance exclusive of space larger than the spacing between molecule, ions, or atoms in the crystal lattices of the solid. Bulk volume, however, is the total volume occupied by the powder or particles placed in a container; it includes true volume plus the volume of internal pores and the volume of the spaces between the particles.

Although the bulk volume may vary according to the treatment of the powder, a satisfactory measurement of bulk volume may be made by a standardized procedure. Consider, for example, a 50-ml sample of a powder passed through a 20-mesh sieve and placed in a 100-ml cylinder. The cylinder is dropped on a wood surface three times from a height of 1 inch at 2-second intervals. The volume in milliliters is the bulk volume.

The void or porosity (percent of void) can be determined from the bulk and true density of the powders as follows:

V_{bulk} = weight of the powder/bulk density (i.e., W/ρ_a)

V_{true} = weight of the powder /true density (i.e., W/ρ)

Substituting for V_{bulk} and V_{true} in Equation 5.7 yields

$$\varepsilon = \left(1 - \frac{W/\rho_a}{W/\rho}\right) \times 100 \tag{5.8}$$

Cancellation of the term W in Equation 5.8 yields

$$\varepsilon = \left(1 - \frac{\rho_a}{\rho}\right) \times 100 \tag{5.9}$$

where ρ_a and ρ are the bulk and true densities, respectively, of the powders.

Application of Equations 5.7 and 5.9 to determine the void and porosity of the powder is illustrated in the following example.

The true density of magnesium carbonate is 3 g/cm³. It is observed that 3.5 grams of light magnesium carbonate occupied 50 cm³ in a graduated cylinder. The data available, therefore, permit one to determine the void and the porosity (ε) of this powder.

The true volume (V_{true}) of the powder is 3.5 grams/3 g/cm^3

The true volume (V_{true}) of the powder is 1.1667 cm^3

The bulk volume (V_{bulk}) of the powder = 50 cm^3

$$\text{Void} = \frac{V_{\text{bulk}} - V_{\text{true}}}{V_{\text{bulk}}} = \frac{50 - 1.1667}{50} = 0.977$$

Porosity (ε) = void × 100 = 97.7%

The porosity of the powder may also be determined from the true and bulk densities of the powder.

$$\text{Void} = \; = 1 - \frac{W/\rho_a}{W/\rho} = 1 - \frac{\rho_a}{\rho}$$

The bulk density (ρ_a) of the powder is 3.5 grams/50 cm^3 or 0.07 grams/cm^3.

$$\text{Void} = 1 - \frac{0.07\ \text{g/cm}^3}{3.0\ \text{g/cm}^3}$$

Porosity (ε) = 1 – 0.0233 = 0.9767 = 97.67%

Porosity is a characteristic of a powder, and in real powders extreme values range from 10 to 90%, though no single material has such a great range. The porosity of a system of perfectly spherical particles is independent of the size; in real powders, however, the void is generally increased as the particle size is reduced because the particles are not of a single size and are not spherical.

3. Specific Surface

The specific surface is the area per unit volume (S_v) or per unit weight (S_w) and may be derived from the following equations:

$$S_v = \frac{\text{surface area of particles}}{\text{volume of particles}}$$

$$S_v = \frac{n\alpha_s d^2}{n\alpha_s d^3} = \frac{\alpha_s}{\alpha_v d} \tag{5.10}$$

where n is the number of particles. The surface area per unit weight is therefore

$$S_w = \frac{S_v}{\rho} \tag{5.11}$$

Substituting from Equation 5.10 into Equation 5.11 yields the general equation

$$S_w = \frac{\alpha_s}{\rho . d_{vs} \alpha_v} \tag{5.12}$$

where d_{vs} is the volume surface diameter; α_s/α_v is the ratio of surface areas and volume shape factors and for spherical or nearly spherical particles, and $\alpha_s/\alpha_v = 6.0$. Equation 5.12, therefore, simplifies into

$$S_w = \frac{6}{\rho . d_{vs}} \tag{5.13}$$

4. Particle Number

Another important property in particle technology is the number of particles (n) per unit weight. It is obtained as follows. Assuming that the particles are spherical,

$$\text{Volume of 1 particle} = \frac{\pi d_v^3 n}{6} \tag{5.14}$$

and, since mass (m) = volume times ρ,

$$\text{Weight of single particle} = \frac{\pi d_v^3 n \rho}{6} \tag{5.15}$$

The number of particles per gram is then obtained from the proportion

$$n = \frac{6}{\pi d_v^3 \rho} \tag{5.16}$$

From the mean volume diameter (d_v^3) of a spherical powder particle and its true density (ρ), one can determine the number of particles per unit of powder as follows.

The mean volume diameter of a spherical powder particle is 2.42×10^{-4} cm, and its true density is 3g/cm^3.

$$n = \frac{6}{3.14 \times \left(2.41 \times 10^{-4}\right) \times 3\ \text{g/cm}^3} = 4.55 \times 10^{10}$$

$$n = 4.55 \times 10^{10} \text{ number of particles/gram of powder}$$

It is clear from Equation 5.16 that, since the true density (ρ), π, and 6 are constant for a powder, the smaller the value of the mean volume diameter (i.e., the smaller the particle size), the greater will be the number of particles per unit weight of the powder.

D. Flow Properties of Powder

The frictional forces in a loose powder can be measured by the angle of repose, Ø, which permits characterization of the flow of the powder. This is the maximum angle possible between the surface of a pile of powder and the horizontal plane. The tangent of the angle of repose is equal to the coefficient of friction between particles:

$$\tan Ø = \mu \tag{5.17}$$

where Ø is the angle of repose and μ is coefficient of friction. From Equation 5.17, it is clear that the rougher and more irregular the surface of particles, the higher will be the angle of repose and the less will be the free flow of powders.

Flow of the powder or granules may present difficulties in the manufacture of tablet dosage form, which, in turn, affects the content uniformity of a drug in a dosage form. To improve the flow characteristics of granules, glidants are frequently added to the formulation. Examples of commercially used glidants are magnesium stearate, starch, and talc.

II. RHEOLOGY

A. Introduction

The term *rheology* refers to the science of flow properties of matter, and viscosity is the resistance offered when one part of the liquid flows past another. Therefore, principles of rheology play an important role in liquid, and semiliquid, and semisolid dosage forms such as solution, suspension, emulsion, cream, pastes, etc. The manufacturer of medicinal and cosmetic creams, pastes, and lotions must be able to pour the products with an acceptable consistency and smoothness for each batch. Rheology is also involved in the mixing and flow of materials, their packaging into containers, and their removal prior to use. The rheology of a particular product, which can range in consistency from fluid to semisolid, can affect the acceptability of the product to the patient, its physical stability, and even its bioavailability by affecting the rate of drug absorption from the gastrointestinal tract and dosage forms.

Materials are classified into two general types depending upon their flow properties: Newtonian and non-Newtonian.

1. Newtonian Systems

The force required for one layer of a liquid to slip past another layer with a given velocity depends directly on the *viscosity* of the liquid and on the areas of layers exposed to each other and inversely on the distance separating the two surfaces.

The difference in velocity (dv) between two planes of a liquid separated by an infinitesimal distance (dx) is the velocity gradient or a rate of shear (dv/dx). The force per unit area (F/A) needed to bring about this flow is known as the shearing stress.The higher the viscosity of a liquid, the greater is the shearing stress (i.e., force/unit area) needed to yield a certain rate of shear. Viscosity tends to prevent a fluid from flowing when subjected to an applied force. The tenacity with which a moving layer of fluid drags adjacent layers of fluid along with it determines its viscosity. In other words, viscosity describes the internal friction of a moving fluid. A fluid with high viscosity resists motion because its molecular makeup gives it a lot of internal friction. A fluid with low viscosity flows easily because its molecular makeup results in very little friction when it is in motion. The reciprocal of viscosity is called ***fluidity*** and is given the symbol ϕ, that is $\phi = 1/\eta$.

When the relationship between the shear stress and rate of shear is directly proportioned, it is described as a Newtonian flow (Figure 5.7).

$$F/A = \eta\, dv/dx \tag{5.18}$$

where F/A is the shearing stress (dyne.cm^{-2}); dv/dx is the rate of shear (sec^{-1}), and η is the coefficient of viscosity (dynes.sec.cm^{-2}), usually referred to simply as viscosity. Rearrangement of Equation 5.18 yields

$$\eta = \frac{F/A}{dv/dx} \tag{5.19}$$

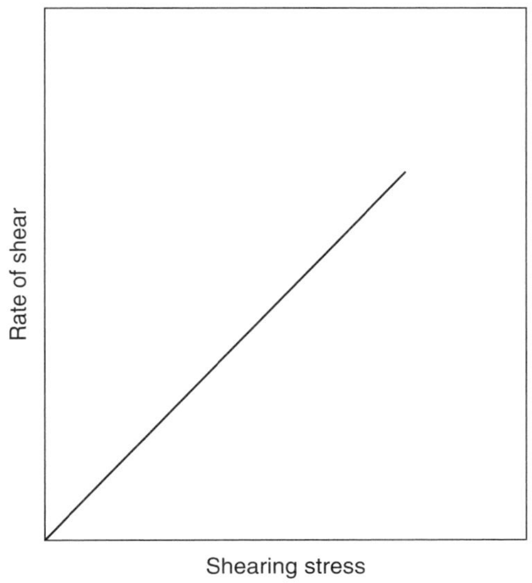

FIGURE 5.7 Representative flow curve for a liquid displaying a Newtonian flow.

$$\eta = \frac{\text{dynes.cm}^{-2}}{\text{sec}^{-1}}$$

$$\eta = \text{dynes.sec.cm}^{-2}$$

or

$$\eta = \frac{\text{dynes.sec}}{\text{cm}^2} = \frac{\text{g.cm/sec}^2\text{.sec}}{\text{cm}^2}$$

$$\eta = \text{g.cm}^{-1}\text{.sec}^{-1}$$

For convenience, the poise has been designated as the unit of viscosity. A liquid has a viscosity of one poise when a shearing stress of 1 dyne.cm^{-2} is needed to maintain a velocity of 1 cm.sec^{-1} between two parallel planes of liquid 1 cm apart. A more convenient unit employed in pharmacy is centipoises (1 centipoises = 0.01 poise).

Two types of viscosities are often encountered: **dynamic viscosity** and **kinematic viscosity**. So far, this discussion has been limited to dynamic viscosity. Kinematic viscosity is defined as the dynamic viscosity divided by the density of the fluid. Because density is an intrinsic property in itself, it can be argued that kinematic viscosity is not a precise measure of internal fluid friction. However, kinematic viscosity is the preferred unit when the shear stress and shear rate of the fluid are influenced by the density. Kinematic viscosity is given in units of centistokes, while the dynamic viscosity is given in centipoise; the conversion from dynamic to kinematic is given by dividing the dynamic viscosity by the fluid density in grams per cubic centimeter. Since some manufacturers specify viscosity in centipoise and others in centistokes, it is important to know the difference between the two and to be able to convert from one to the other.

Pharmacists deal more frequently with non-Newtonian material and should therefore have suitable methods for the study of these systems. Substances that fail to follow the Newtonian equation of flow (Equation 5.18) are non-Newtonian; liquids and solid heterogeneous dispersions such as colloidal solutions, emulsions, ointments, and liquid suspensions make up this class. Non-Newtonian systems are further classified as plastic, pseudoplastic, and dilatant materials.

2. Non-Newtonian Systems

a. *Plastic Flow*

A plastic material does not flow until a certain minimum shearing stress, the yield value, is applied. Therefore, the plastic flow curve does not pass through the origin (Figure 5.8) but rather intersects the shear stress axis at a particular point.

At stress below yield value the material acts like an elastic material. The yield value, therefore, becomes an important property of some dispersions. The slope of the rheogram (Figure 5.8) is termed the *mobility,* and reciprocal of the slope provides plastic viscosity. The coefficient of plastic viscosity, U, is defined by the following expression:

$$U = \frac{F - f}{G} \tag{5.20}$$

where f is the yield value (dyne.cm^{-2}), F is the force/unit area (dyne.cm^{-2}), and G is the rate of shear (sec^{-1}).

Figure 5.8 also indicates that at low rate of shear, the curve is nonlinear; however, at higher shearing rate the curve approaches linearity. The yield value of a plastic system is increased by the increased concentration of the dispersed phase.

Tomato catsup is a good example of this type fluid; its yield value often prevents it from pouring from the bottle until the bottle is shaken or struck, allowing the catsup to gush freely. Once the yield value is exceeded and flow begins, plastic fluids may display Newtonian, pseudoplastic, or dilatant flow characteristics.

b. *Pseudoplastic Flow*

This type of flow is generally exhibited by long chain or linear polymers, e.g., methylcellulose, sodium carboxymethylcellulose, and tragacanth. At rest, the linear polymers or macrometers are

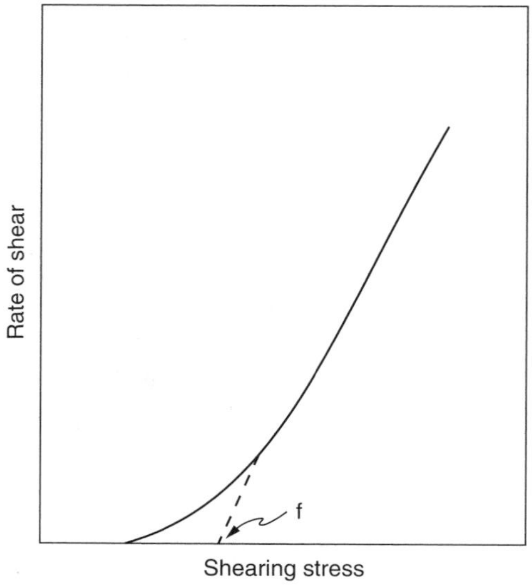

FIGURE 5.8 Representative flow curve for a liquid displaying a simple plastic flow.

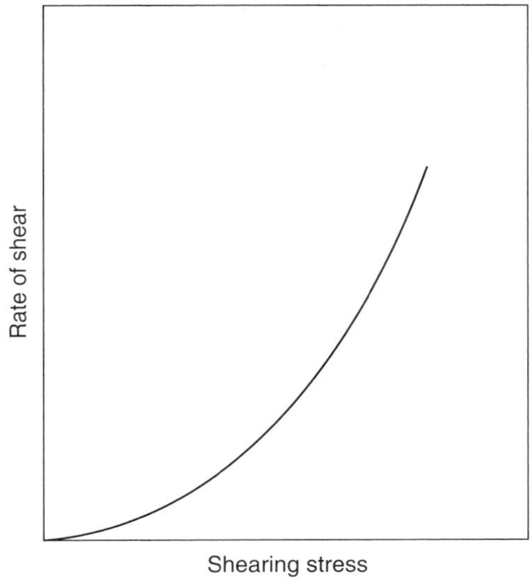

FIGURE 5.9 Representative flow curve for a liquid displaying a simple pseudoplastic flow.

dispersed at random in the dispersion medium. As shearing stress is applied (Figure 5.9), the macromolecules become aligned with the long axis parallel to the direction of flow. With this ordered alignment the molecules pass one another with less frictional resistance, and the viscosity is decreased. If the shearing stress is decreased, the orientation of the macromolecules becomes more random, and greater frictional resistance to the flow is reflected in an increased viscosity. Since only a molecular alignment is involved, there is no lag time.

This type of fluid displays decreasing viscosity with increasing shear rate, as shown in Figure 5.9. Probably the most common of the non-Newtonian fluids, pseudoplastics include paints, emulsions, and dispersions of many types. This type of flow behavior is sometimes called shear-thinning.

c. *Dilatant Flow*

This type of flow exhibits an increase in viscosity as the shearing stress is increased (Figure 5.10) and is exhibited by polyphasic systems, for example, concentrated suspensions and zinc oxide paste. The behavior of a dilatant curve is explained by postulating that at rest, a minimum volume of dispersion medium, which diminishes the resistance to flow, surrounds each of the particles. When shearing is applied, the particles are rearranged and the dispersion medium is displaced from around the particles, which produces increased friction between particles and increased resistance to flow. Dilatancy is also referred to as shear-thickening flow behavior.

3. Thixotropy and Rheopexy

Some fluids display a change in viscosity with time under conditions of constant shear rate. There are two categories to consider: thixotropy and rheopexy.

a. *Thixotropy*

As shown in Figure 5.11, a thixotropic fluid undergoes a decrease in viscosity with time, while it is subjected to constant shearing. Flow occurs when the yield value is exceeded, and the viscosity decreases as the rate of shear increases (Figure 5.11). If the rate of shear is immediately decreased, the down curve does not coincide with the up curve. If the shear is continued at a maximum rate for time t_1 and t_2, the viscosity falls and a decreased shear rate curve C and D, respectively, are obtained.

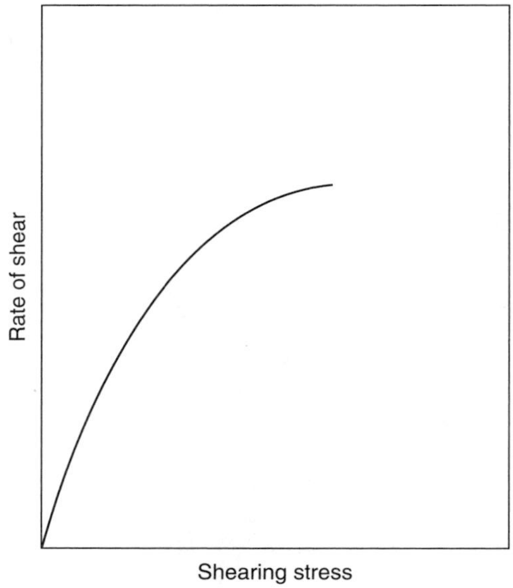

FIGURE 5.10 Representative flow curve for a liquid displaying a dilatant flow.

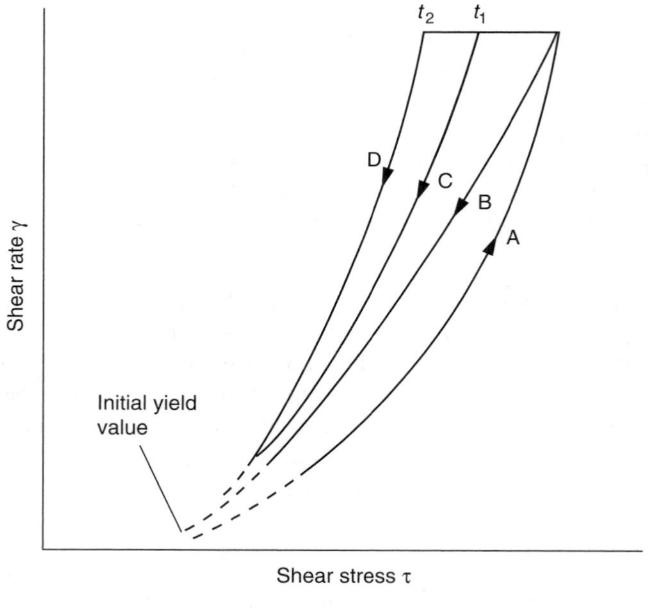

FIGURE 5.11 Flow curves for thixotropic dispersions.

The fall in the viscosity is caused by the breaking of the polar bonds between the particles aligned parallel to the streamline in liquid. When the system is completely rested, Brownian movement allows the starch to form gel again. The time required to form the gel structure is setting time, which is affected by electrolytes. Examples of gel include aluminum hydroxide gel and bentonite magma. Thixotropy is also frequently observed in materials such as greases, heavy printing inks, and paints.

b. *Rheopexy*

Rheopexy is essentially the opposite of thixotropy, in that the fluid's viscosity increases with time as it is sheared at a constant rate. Rheopectic fluids are rarely encountered.

B. Determination of Rheological Properties of Liquids and Semisolids

The successful determination and evaluation of the rhelogic properties of a particular system depends, to a large extent, on choosing the correct instrument and method. Selecting the method and instrument that operate at a single rate of shear is likely to provide inaccurate results, particularly with non-Newtonian systems. Furthermore, one can obtain a complete rheogram for a system only by employing instruments that operate at a variety of rates of shear. Therefore, the use of one shear rate instrument, even as a quality control, is wrong, if the liquids are non-Newtonian type.

Other rheological properties such as tackiness or stickiness and spreadability are difficult to measure by means of conventional apparatuses. Individual properties such as viscosity yield value and thixotropy and other properties that contribute to the total consistency of non-Newtonian liquids can be analyzed to some degree of satisfaction by using reliable apparatuses.

The many types of viscometers have been discussed in detail in the literature (2, 5, 7). For the purpose of this discussion two simple methods will be discussed for illustration.

1. Capillary Viscometer

The viscosity of a Newtonian liquid can be determined by measuring the time required for the liquid to pass between two marks as gravity causes it to flow through a capillary. Figure 5.12 shows an Ostwald viscometer, an adoption of an original Ostwald viscometer. The time of flow of the liquid under test is compared with the time required for a liquid of a known viscosity to pass between the two marks. If η_1 and η_2 are the viscosities of the unknown and reference standard liquids, respectively, and ρ_1 and ρ_2 are the densities of the unknown and reference standard liquids and t_1 and t_2 are the respective flow times, the absolute viscosity of the unknown liquid, η_1, can be determined by substituting the value in the following equation (Equation 5.21).

$$\frac{\eta_1}{\eta_2} = \frac{\rho_1 t_1}{\rho_2 t_2} \tag{5.21}$$

where η_1/η_2 is the relative viscosity of the test liquid.

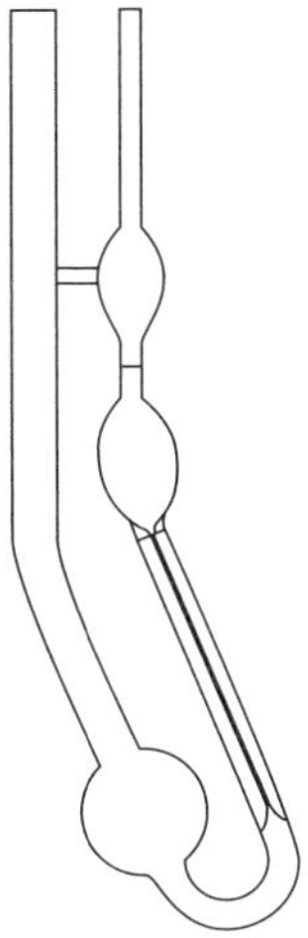

FIGURE 5.12 An Ostwald viscometer.

The following example illustrates the use of an Ostwald capillary viscometer and Equation 5.21 to determine the relative viscosity of an unknown solution.

At 25°C water has a density (ρ_1) of 0.997 g/cm^3 and viscosity (η) of 0.00895 poise. The time of flow of water in a capillary viscometer is 10 sec. A 65% sucrose solution requires 890 sec, and its density (ρ_2) is 1.313 g/cm^3. The viscosity of the 65% sucrose solution can be determined by employing Equation 5.21.

$$\frac{\eta_1}{\eta_2} = \frac{\rho_1 t_1}{\rho_2 t_2}$$

$$\frac{\eta_1}{0.0895} = \frac{1.313 \times 890}{0.997 \times 10}$$

$$\eta_2 = 1.04 \text{ poise (viscosity of 65\% sucrose solution)}$$

---

2. Falling Sphere Viscometer

In this type of viscometer, a glass or a steel ball rolls down an almost vertical glass tube containing the test liquid at a known constant temperature. The rate at which a ball of particular density and diameter falls is an inverse function of the viscosity of the liquid. The Hoeppeler viscometer, shown in Figure 5.13, is a commercial instrument that uses this principal. The test liquid and the ball are placed in the inner glass tube and allowed to reach temperature equilibrium with the water in the surrounding constant temperature jacket. The tube and the jacket are then inverted, which puts the ball at the top of the inner glass tube. The time it takes the ball to fall between two marks is

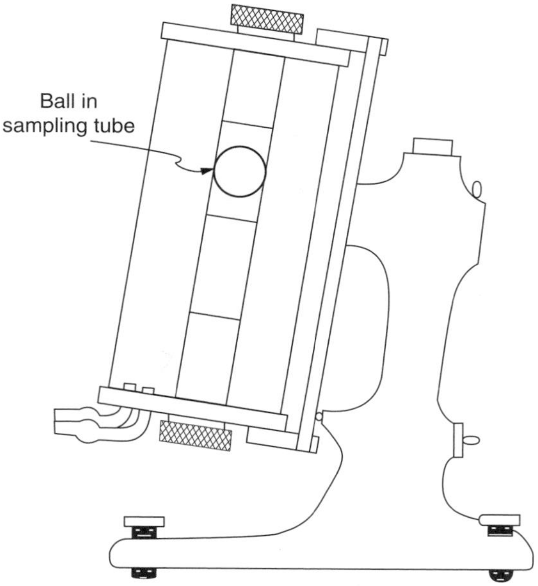

FIGURE 5.13 A Hoeppler falling ball viscometer.

accurately measured. The viscosity of the Newtonian liquid is then calculated from the following equation:

$$\eta = t\left(S_b - S_f\right)B \tag{5.22}$$

where t is the time in seconds it takes the ball to fall between the two points, and S_b and S_f are the specific gravities of the ball and the liquid being tested. B is the constant for a particular ball, and its value is supplied by the manufacturer. The instrument can be used over the viscosity range of 0.5 to 200,000 poises. For accurate results, it is recommended that the ball be selected such that the time value is not less than 30 seconds.

For descriptions of other methods used to measure the viscosity of non-Newtonian liquids, readers are referred to the textbooks listed in the references.

C. Factors Affecting Rheological Properties and Measurement of Viscosity of Liquids and Semisolids

1. Temperature

One of the most obvious factors that can effect the rheological behavior of a material is temperature. The viscosity of a fluid decreases with a decrease in density that occurs when the temperature increases often by about 2% per degree Celsius; it also increases with an increase in pressure. Some materials are quite sensitive to temperature, and a relatively small variation will result in a significant change in viscosity. Others are relatively insensitive. Consideration of the effect of temperature on viscosity is essential in the evaluation of materials that will be subjected to temperature variations in use or processing, such as emulsions, suspensions, and hot-melt adhesives.

2. Shear Rate

Non-Newtonian fluids are more common in the real world, making an appreciation of the effects of shear rate a necessity for anyone engaged in the practical application of rheological data. When a material is to be subjected to a variety of shear rates in processing or use, it is essential to know its viscosity at the projected shear rates. If these are not known, an estimate should be made. Viscosity measurements should then be made at shear rates as close as possible to the estimated values.

3. Measuring Conditions

The condition of a material when its viscosity is measured can have a considerable effect on the measurement results. It is therefore important to be aware of, and to control as much as possible, the environment of any sample being tested. First, the viscosity measurement techniques should be adhered to: variables such as viscometer model, spindle/speed combination, sample container size, absence or presence of the guard leg, sample temperature, sample preparation technique, etc. Second, less obvious factors that may affect viscosity must be considered. For example, the sample material may be sensitive to the ambient atmosphere, as is the case with blood and mucus. The third factor that can affect viscosity measurements is the homogeneity of the sample. It is usually desirable to have a homogeneous sample so that more consistent results may be obtained.

4. Time

The time elapsed under conditions of shear affects thixotropic and rheopectic (time-dependent) materials, but changes in the viscosity of many materials can occur over time even though the material is not being sheared. Aging phenomena must be considered when selecting and preparing samples for viscosity measurement.

5. Pressure

Pressure compresses fluids and thus, increases intermolecular resistance. Increases in pressure tend to increase viscosity. For example, the flow properties of a highly concentrated suspension (above 70 to 80% of particles by volume) where there is insufficient liquid to completely fill all the voids between the particles results in a three-phase mixture (i.e., solids, liquids, and usually air). Owing to the presence of air, the mixture is compressible, and the more it is compressed, the greater is the resistance to flow.

6. Composition and Additives

The composition of a material is a determining factor of its viscosity. When this composition is altered, either by changing the proportions of the component substances or by the addition of other materials, a change in viscosity is quite likely. For example, the addition of different polymers can change the viscosity of dispersed systems, and polymers of many types are used to control the rheological properties of dispersed systems as discussed below.

One of the major characteristics to study is the sample material's state of aggregation. If the clumps (flocs) occupy a large volume in the dispersion, viscosity will tend to be higher than if the floc volume was smaller. When flocs are aggregated in a dispersion, reaction of the aggregates to shear can result in shear-thinning (pseudoplastic) flow. At low shear rates, the aggregates may be deformed but remain essentially intact. As the shear rate is increased, the aggregates may be broken down into individual flocs, decreasing friction and therefore viscosity. If the bonds within the aggregates are extremely strong, the system may display a yield value described earlier. The magnitude of the yield value depends on the force required to break these bonds.

If a material's flocculated structure is destroyed with time as it is sheared, a time-dependent type of flow behavior will be observed. Since flocs begin to link together after destruction, the rate at which this occurs effects the time required for viscosity to attain previous levels. If the relinking rate is high, viscosity will be about the same as before. If the relinking rate is low, viscosity will be lower. This results in thixotropy.

The attraction between particles in a dispersed phase is largely dependent on the type of material present at the interface between the dispersed phase and the liquid phase. This in turn effects the rheological behavior of the system. Thus, the introduction of flocculating or deflocculating agents into a system is one method of controlling its rheology.

The stability of a dispersed phase is particularly critical when measuring the viscosity of a multiphase system. If the dispersed phase has a tendency to settle, producing a nonhomogeneous fluid, the rheological characteristics of the system will change. In most cases, this means the measured viscosity will decrease. Data acquired during such conditions will usually be erroneous, necessitating special precautions to ensure that the dispersed phase remains in suspension.

III. CONCLUSION

Knowledge of particle size and size distributions is useful to pharmacists, as they have direct impacts ranging from control of flow properties in manufacturing machines to ultimate disintegration, dissolution, and bioavailability. In addition, understanding rheology is equally useful for developing desirable as well as stable formulations. For topical products, appropriate viscosity is crucial in achieving a desirable feeling and consistency so that the product will be easy to apply, will feel good to the user, and will remain in contact with the affected area for the specified time. "Mouth feel" of oral liquid products is also often enhanced by formulating the product with appropriate viscosity. Applications of these important concepts have been discussed in detail in the respective pharmaceutical formulation development chapters.

HOMEWORK

1. After determining the particle size of 151 powder particles by the microscopic method, the $\sum$ nd was reported to be 4,037.5. Determine the average diameter of the powder particle.

$$d_{ave} = \frac{\sum nd}{\sum n} = \frac{4037.5}{151}$$

$$d_{ave} = 26.74\ \mu\text{m}$$

2. When 100 g of magnesium oxide powder is placed in a graduated cylinder, the powder occupies the volume of 82.67 cm^3. The true density of this powder is 3.63 g/cm^3. **Determine the porosity of this sample of magnesium oxide powder.**

$$\text{The bulk density } (\rho_a) \text{ of a powder} = \frac{100\text{ g}}{82.67\text{ cm}^3} = 1.209\text{ g/cm}^3$$

$$\%\text{ Porosity } (\varepsilon) = \frac{3.63 - 1.209}{3.63} = 66.69\%$$

3. A tablet contains 0.1 g of a drug, weighs about 0.3 g, and has a bulk volume of 0.0955 ml. **Determine the bulk density (ρ_a) of the tablet.**

$$\text{The bulk density } (\rho_a) \text{ of a tablet} = \frac{0.30\text{ g}}{0.0955\text{ ml}} = 3.141\text{ g/ml}$$

4. The true density (ρ) of the mixture of a drug and inactive ingredients of the tablet is 3.202 g/cm^3. **Determine the porosity (ε) of the tablet in Question 3.**

$$\%\text{ Porosity } (\varepsilon) = \frac{3.202 - 3.14}{3.202} = 1.93\%$$

5. **Determine the specific surface, S_w,** of powder particles of sulfamethoxyzole powder with the true density (ρ) of 1.5 g/cm^3 and average particle diameter (d_v) of 2 μm. It is assumed that the particles are spherical.

$$\text{Specific surface } (S_w) = \frac{6}{(\rho) \times (d_v)}$$

$$\text{Specific surface } (S_w) = \frac{6}{(1.5\text{ g/cm}^3) \times (2 \times 10^{-4}\text{ cm})}$$

$$\text{Specific surface } (S_w) = 2 \times 10^4\text{ cm}^2/\text{g}$$

6. The time required for water to flow through an Oswald's viscometer was 297.3 seconds. The identical volume of carbon disulfide flowed through the same viscometer in 85.1 seconds. The densities of water and carbon disulfide, at 20°C are reported to be 0.9982 and 1.2632 g/cm^3, respectively. The viscosity of water is 1.002 cps. **Determine the relative viscosity of carbon disulfide.**

$$\eta_{relative} = \frac{(\rho \times \text{time}) \text{ for sample}}{(\rho \times \text{time}) \text{ for reference}} \times \eta_{reference}$$

$$\eta_{relative} = \frac{(1.2632 \times 85.1) \text{ for sample}}{(0.9982 \times 297.3) \text{ for reference}} \times 1.002 \text{ cps}$$

$$\eta_{relative} = 0.363 \text{ cps at } 20°\text{C}$$

CASES

CASE I

When powdered suspensions such as amoxicillin (250 mg/5 ml) and Augmentin[R] (250 mg/5 ml) are reconstituted in a pharmacy by the addition of water, it is necessary to add different volumes of water to make the final volume of each suspension identical (150 ml). Explain why.

Answers:

1. Each antibiotic powder will occupy different volume in a container.
2. Each powdered formulation will have a different bulk density and, therefore, a different bulk volume.
3. The particle size and size distribution of each formulation may be different.

CASE II

After pouring the uncoated tablets from the original container onto the counting tray for dispensing to the patient, occasionally you may have noticed a tablet that splits open from the top into two parts. Explain why.

Answers:

1. The air pressure inside the tablet may cause it split.
2. The porosity (ε) or the percent void of the tablet may cause it to split open.
3. The porosity is a derived property of a powder, which is influenced by the bulk density and the true density of the powder material used for the preparation of the tablet and the compression pressure used in the manufacture of a tablet.

CASE III

Erythromycin suspension is very viscous and, therefore, difficult to pour from the original container into the dispensing bottles. It is also stored in a refrigerator. Prior to dispensing the desired volume of the suspension, the original container is shaken rigorously to make it easier to pour into the dispensing bottle. Explain why is it easier to pour after shaking.

Answers:

1. Suspensions like erythromycin are described a shear-thinning systems.
2. When it is shaken, it is subjected to higher forces, which, in turn, increase the rate of shear. The increase in the rate of shear lower the viscosity of the suspension and enable it to flow easily.

CASE IV

While working in a pharmacy or pharmaceutics laboratory, you may have observed that a powder such as magnesium carbonate or bismuth subcarbonate is available in two grades: heavy and light. Equal amounts of two grades of each powder will occupy different volumes. Explain why.

Answers:

1. The terms *light* and *heavy* refer to the difference in the bulk densities of the powders; the true density, however, is constant for each powder.
2. The bulk density (ρ_a) or the lightness of the powder is influenced by the particle size; the smaller the particle size, the larger will be the bulk density and, therefore, the larger will be the volume occupied by the identical weight of powders.
3. The particle size also affects the porosity (ε) of a powder owing to the differences in the bulk density and, therefore, the bulk volume of the identical weight powders.

REFERENCES

1. Dallavalle, J.M., *Micromeritics*, 2nd ed., Pitman, New York, 1948.
2. Florence, A. and Attwood, D., *Physicochemical Principles of Pharmacy,* 2nd ed., Chapman and Hall, New York, 1988.
3. Martin, A., Bustamante, P., Chun, A.H.C., *Physical Pharmacy: Physical Chemical Principles in the Pharmaceutical Sciences,* 4th ed., Lea & Febiger, Philadelphia, 1993.
4. Newman, A., Micromeritics in physical characterization of pharmaceutical solids, in Brittain, H., ed., *Drugs and the Pharmaceutical Sciences,* vol. 70, Marcel Dekker, New York, 1995.
5. Parrott, E., *Pharmaceutical Technology: Fundamental Pharmaceutics,* Burgess, Minneapolis, MN, 1970.
6. Randall, C., Particle size distribution in physical characterization of pharmaceutical solids, in Brittain, H., ed., *Drugs and the Pharmaceutical Sciences,* vol. 70, Marcel Dekker, New York, 1995.
7. Shotton, E. and Ridgway, K., *Physical Pharmaceutics,* Clarendon Press, Oxford, UK, 1974.

Module II

6 Principles and Applications of Surface Phenomena

*Laszlo Prokai, Vien Nguyen, Bhaskara R. Jasti, and Tapash K. Ghosh**

CONTENTS

* No official support or endorsement of this article by the FDA is intended or should be inferred.

0-4152-8863-0/05/$0.00+$1.50
© 2005 by CRC Press LLC

Learning Objectives

After finishing this chapter, the students will be able to understand the concepts about

- Surface and interfacial tension
- Different types of surfactants and their characteristics
- Electrical properties at the interface
- Dispersed systems and their applications
- Suspension and its theory and preparation
- Emulsion, the theory of emulsification, and different types of emulsifiers

I. INTRODUCTION

Surfaces and interfaces are a concept of fundamental importance in materials science and technology. All materials have a surface. When two or more materials are put together, interface forms. In composite materials, the interfaces play a crucial role in determining their mechanical, optical, and electronic properties.

The boundary between two homogenous phases is not to be regarded as a simple geometrical plane but rather as a film of a characteristic thickness: The material in this *surface phase* shows properties differing from those of the materials in the contiguous homogenous phases. Thus, a surface is an interface, which is defined as *the boundary of separation of two phases, where a phase is a mass of substance (solid, liquid, or gas) that possesses a well-defined boundary.* There are several types of interfaces at the boundaries between any two states of matter (gas, liquid, or solid), which can be gas–liquid, gas–solid, and liquid–solid, or at boundaries between two immiscible phases of the same state: liquid–liquid or solid–solid interfaces. However, there cannot be gas–gas interface, as two gaseous materials would mix together rather than form an interface (Table 6.1).

TABLE 6.1
Classification of Interfaces

Phases	Types and Examples of Interfaces
Gas–gas	No interface possible
Gas–liquid	Liquid surface (e.g., body of water exposed to atmosphere)
Gas–solid	Solid surface (e.g., table top)
Liquid–liquid	Liquid–liquid interface (e.g., emulsion
Liquid–solid	Liquid–solid interface (e.g., suspension)
Solid–solid	Solid–solid interface (e.g., powder particles in contact with each other)

II. SURFACE AND INTERFACIAL TENSIONS

A. Surface Tension

The molecules forming the interface behave differently from those in the bulk of each phase. The differences are manifested in specific phenomena such as capillary action, wetting, adsorption, etc. These phenomena can be best exemplified by considering a liquid surface (interaction between gas molecules and between gas and liquid are negligible compared to a much stronger interaction between liquid molecules). The particles in liquids are supposed to move with considerable speed in all directions; they collide frequently with one another, whereupon a change in direction occurs, without any loss or gain in energy. Each particle exerts forces of attraction, called cohesive forces, which are manifested only at a small distance from the surface of the particle. There are two types of intermolecular binding forces: *cohesion* (the attraction of like molecules) and *adhesion* (the attraction of unlike molecules). The molecules at the surface of a liquid have potential energies greater than those of similar molecules in the interior of the liquid. This is because attractive interactions of molecules at the surface with those in the interior of the liquid are greater than those with widely separated molecules in the gas phase. The net effect is that these molecules at the surface exert an inward force, which is defined as *surface tension* (Figure 6.1).

Surface tension describes the interface between liquids and vapor. In the bulk of the phase, the binding forces between the molecules are equivalent in every direction, while the molecules on the surface are subjected to intermolecular forces in the direction of the interface and toward the bulk

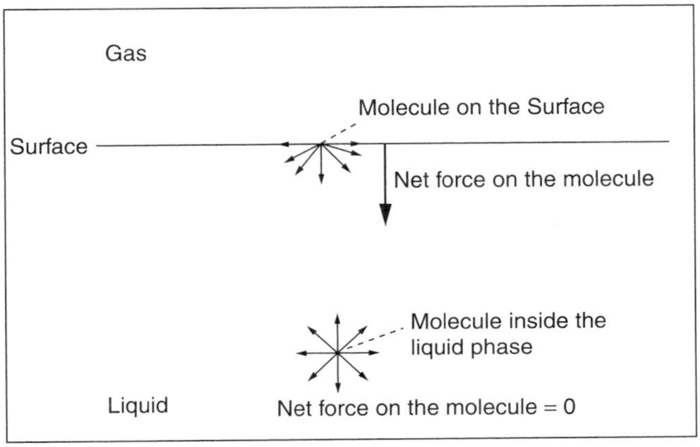

FIGURE 6.1 Attraction forces acting on molecules at the surface and in the bulk of the liquid.

TABLE 6.2
Surface Tension of Liquids at 20°C

Substance	Surface Tension (dyne/cm)
Water	72.8
Glycerin	63.4
Castor oil	39.0
Olive oil	35.8
Cottonseed oil	35.4
Liquid petrolatum	33.1
Oleic acid	32.5
Benzene	28.9
Chloroform	27.1
Carbon tetrachloride	26.7

of the phase. These forces tend to decrease the surface. If one increases the surface (i.e., brings molecules from the bulk of the phase to the surface), work (W) should be done against the forces in the surface. The surface free energy per unit area, or surface tension, is a measure of this work; it is the minimum amount of work required to bring the molecules to the surface from the interior to expand the surface by unit area:

$$W = \gamma \cdot \Delta A$$

where W is the work done (ergs), γ is the surface tension (dyne/cm), and ΔA is the increase in area (cm^2).

There are several methods used to determine the exact values of this important constant. One of the well-known methods is the *capillary rise method*. When we dip a capillary into a liquid, we observe that the liquid rises in the interior tube. The height of the column is proportional to the surface tension. If the radius of the capillary and the specific gravity of the liquid are known, the surface tension can be determined. Another method is the *tensiometer method*. It is an apparatus that is particularly suitable for rapid determinations of surface tension. A thin wire ring is placed on the surface of the liquid, and the force is determined that is necessary to lift the ring until it is completely detached from the surface. This force is proportional to the surface tension. Some representative values of surface tension are listed in Table 6.2.

B. Interfacial Tension

When a substance is adjacent to another substance, as is the case at the interface between two liquids, then there is interfacial tension. Some tension or excess energy is also present at the interfaces between two condensed matters. Figure 6.2 shows a schematic representation of intermolecular cohesive forces at the interface between two liquids, A and B.

When the surfaces of liquids A and B are separated from each other, they have their own surface tensions, γ_a and γ_b. When the two surfaces come into contact with each other to form an interface, an attractive force between molecules of A and B appears. This attractive force partially compensates for the excess energy of the molecules present at the surface of A and B and consequently reduces the surface tension of two liquids.

In term of work done with the newly created interface between liquids A and B,

$$W_a = \gamma_a + \gamma_b - \gamma_{ab}$$

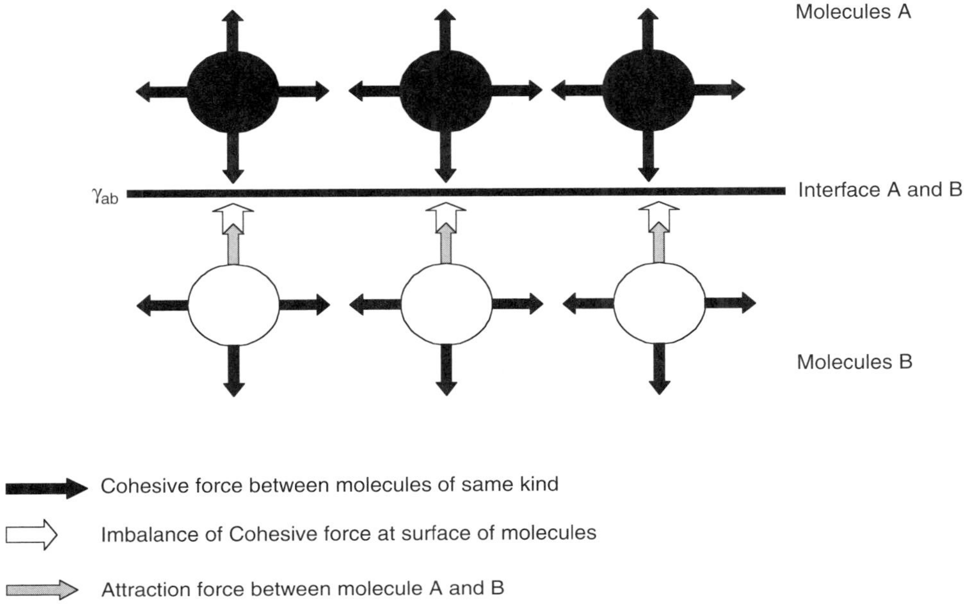

FIGURE 6.2 Schematic representation of intermolecular cohesive forces at an interface.

where W_a is the adhesion work done (ergs), γ_a and γ_b are the surface tension of liquid A and B, respectively (dynes/centimeters), and γ_{ab} is the interfacial tension of A and B (dynes/centimeters).

Table 6.3 contains a summary of some known interface tensions related to liquid–liquid interfaces where one phase, water, is kept the same, while the other phase is varied.

TABLE 6.3
Standard Interface Tensions between Water and Pure Liquids (at 20°C)

Substance	Interface Tension Against Water (dyne/cm)
Mercury	375
n-Hexane	51.1
n-Octane	50.8
Carbon disulphide	48.0
Carbon tetrachloride	45.1
Benzene	35.0
Chloroform	32.8
Nitrobenzene	26.0
Olive oil	22.9
Oleic acid	15.6
Ethyl ether	10.7
n-Octanol	8.5
Caprylic acid	8.2
Isobutanol	2.1
n-Butanol	1.6

III. ELECTRICAL PROPERTIES AT THE INTERFACE

A. General Aspect

The influence of electrical charges on a surface is very important to the surface's physical chemistry. For instance, aqueous suspensions are not only influenced by the interfacial properties, but also by attraction and repulsion caused by electrostatic charges.

1. Origin of Charge

When a particle is dispersed in a surrounding liquid environment, as is the case with colloidal systems and suspensions, the dispersed particles in liquid media may become charged. The development of charge may occur either by selective adsorption of a particular ionic species present in solution or from ionization of a polar group present on the particle. For example, when a particle is added to pure water, charge may be imparted on the particle owing to adsorption of hydronium (H_3O^+) or hydroxyl (OH^-) ion. However, if the particle is a carboxylic acid, it may acquire a negative charge because of ionization of –COOH groups at the surface of the particle, in which case the charge is a function of its ionization constant, pKa, and pH of the medium. Occasionally a difference in dielectric constant between the particle and its dispersion medium may impart a charge on the surface of the particle.

2. Electric Double Layer

When a solid particle comes in contact with an aqueous solution of an electrolyte, either cations or anions from the electrolyte solution may get preferentially adsorbed on the surface of the solid particle, imparting a net charge (positive or negative depending on the nature of preferential adsorption of ions) to the surface of the particle. Assume that owing to preferential adsorption of cations, the surface has a positive charge or potential. The rest of the cations and all the anions remain in the solution. Once the adsorption is complete, the cations on the surface will repel the cations in the solution, while the anions in solution are attracted to the positively charged surface by electric forces. Simultaneously, the thermal motion in the solution tends to produce an equal distribution of all the ions. As a consequence, excess anions approach the surface, while the remaining are distributed in decreasing amounts as one proceeds away from the charged surface. Eventually at a distance from the surface of the particle the concentration of anions and cations will be equal, resulting in electric neutrality. Even though pockets of unequal distribution of ions prevail, the system as a whole is electrically neutral.

This situation is illustrated in Figure 6.3, where X is the surface of the solid on which "potential determining ions" (in this case cations, but it could could be anions and depends on the particle) are adsorbed. Solvent molecules containing the counter ions are tightly bound to the solid surface; this is shown as plane Y in the figure. Z represents the plane from and beyond which electrically neutral region exists. Because of the strong attraction between the solvent molecules and counterions, the region between planes X and Y creates a very tightly bound practically immobile layer, and therefore plane Y becomes the shear plane for all practical purposes instead of the true surface X. In the region bounded by planes Y and Z, there is an excess of negative ions. The potential at Y is still positive, as there are fewer anions in the tightly bound layer than there are cations adsorbed onto the surface of the solid. Beyond plane Z, the distribution of ions is uniform and electric neutrality prevails.

Thus, at the interface, there are two layers of charge: a tightly bound first layer next to the particle (X to Y) and a more diffuse layer (from Y to Z) constituting an electrical double layer that extends from X to Z. It is possible that the total charge of the counterions in the region of X to Y may equal or exceed the positive ions present on the particle surface, resulting in a net neutral or

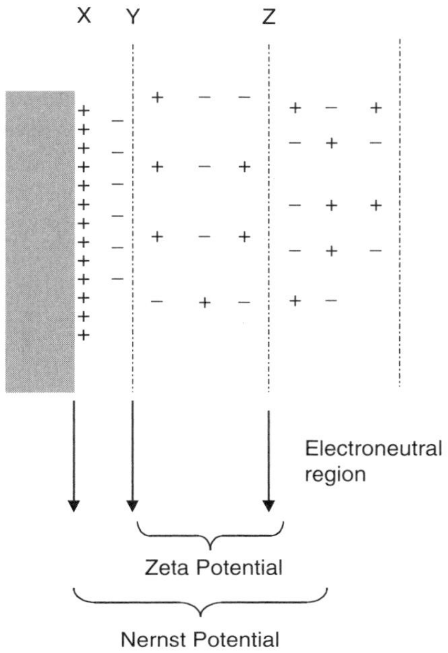

FIGURE 6.3 An artist's view of the electric double layer at the surface of separation between two phases.

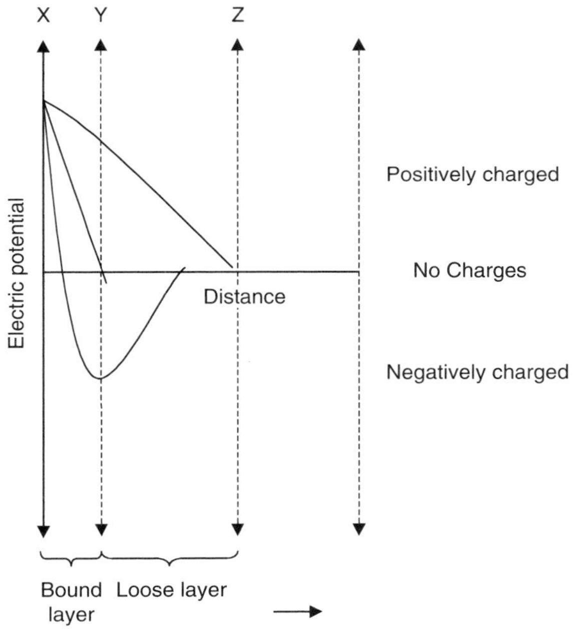

FIGURE 6.4 Potential gradients at a solid–liquid interface.

negative charge at Y as shown in Figure 6.4. One should keep in mind that if the potential-determining ion is negative, the counterions will be positively charged, and the above discussion will still apply.

3. Nernst and Zeta Potentials

Determination of the electric potential changes from the surface of the particles to various regions discussed in the above section is important in the formulation of dispersed systems such as suspensions, emulsions, and colloids. The potential difference between the actual particle surface X and the electron neutral region Z is the electrothermodynamic (Nernst) potential, E. The potential difference between the shear plane Y and electroneutral region Z is known as the electro kinetic, or Zeta potential. The potential difference drops rapidly adjacent to the particle surface and more gradually as the distance from the particle surface increases, as illustrated in Figure 6.4. This is because the counterions close to the surface act as a screen that reduces the electrostatic attraction between the charged surface and those counterions further away from the surface.

Zeta potential, rather than the Nernest potential, controls the degree of repulsion between adjacent, similarly charged, dispersed particles. Therefore, measurement of zeta potential is a useful tool in the stability of dispersion systems. For suspensions, when the attractive forces overcome the repulsive forces of the adjacent molecules, the solid particles come together, resulting in flocculation (an indication of instability). This occurs when the zeta potential falls below a threshold value. Zeta potential can be reduced either by increasing the concentration of electrolyte present in the system or by increasing the valency of the counterions.

IV. SURFACTANTS

A. Definition and Characteristics

Surface-active agents commonly known as surfactants are a group of substances that, when present at a low concentration in a system, adsorb onto the surfaces or interfaces of the system and alter to a marked degree the surface or interface free energies. They have two regions in their chemical structure, one hydrophilic and the other hydrophobic. The hydrophilic region can be an ion, polar, or water-soluble group. The hydrophobic group consists of a hydrocarbon chain and is also known as the lipophilic group. Consequently, surfactants are often defined as being amphiphilic in that they have an attraction to both aqueous and oil phases. When the surface-active agent is dissolved in a solvent, the presence of the hydrophobic group in the interior of the solvent may distort the solvent liquid structure, increasing the free energy of the system. In an aqueous solution of a surfactant, this distortion of the water by the hydrophobic group of the surfactant and the resulting increase in the free energy of the system when it is dissolved means that less work is needed to bring a surfactant molecule than a water molecule to the surface. The surfactant therefore concentrates at the surface. Since less work is now needed to bring molecules to the surface, the presence of the surfactant decreases the work needed to create surface tension.

However, the presence of the hydrophilic group prevents the surfactant from being expelled completely from the solvent as a separate phase, since that would require dehydration of the hydrophilic group. The amphiphilic nature of these molecules is such that part of the molecule will tend to be expelled from the bulk of the liquid in which they are dissolved, while the other part of the molecule will freely interact with the liquid. This leads to a tendency for the molecules to perform two functions. First, they will be forced to liquid interfaces; second, they will tend to align in an ordered fashion at those interfaces.

Surfactants are among the most versatile products used in the chemical industry, appearing in such diverse products as motor oils, pharmaceuticals, and detergents. Specifically, they are necessary (excipients) in heterogeneous dosage forms such as dispersed systems (i.e., suspensions and emulsions). Surfactants are the key material in processes involving adhesion, coating, mixing, and many other phenomena in the medical, pharmaceutical, and chemical industries. They are compounds that dramatically lower the surface tension of water. Surfaces and interfaces can be dramatically changed when surfactant molecules are applied to an area. They modify not only the interactions between two materials at the interface, but also the bulk properties of the composite.

B. Types of Surfactants

Surfactants exhibit a variety of interesting phenomena such as formations of micelles, vesicles, layer, and gels. They are classified by their hydrophilic groups. Generally, surfactants can be classified into four groups.

1. Anionic

These are surfactants in which the hydrophilic portion of the molecule carries a negative charge: R-COO^-, R-SO_4^-, or R-SO_3^- (where R represents the hydrocarbon-based chain). Anionic surfactants are electrolytes, and a surface-active ion is an anion when the surfactants dissociate in water. The anionic surfactants adsorb on various kinds of substrates and give them an anionic charge. This action of the anionic surfactant contributes to the strong detergency and high foaming power of the agent. Consequently, anionic surfactants are used most widely and extensively in detergents, shampoos, and body cleansers.

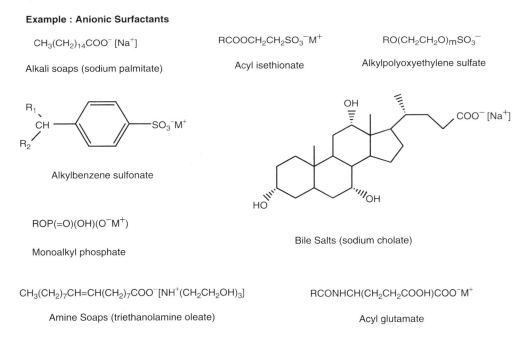

Example : Anionic Surfactants

$CH_3(CH_2)_{14}COO^-$ [Na^+]

Alkali soaps (sodium palmitate)

$RCOOCH_2CH_2SO_3^-M^+$

Acyl isethionate

$RO(CH_2CH_2O)_mSO_3^-$

Alkylpolyoxyethylene sulfate

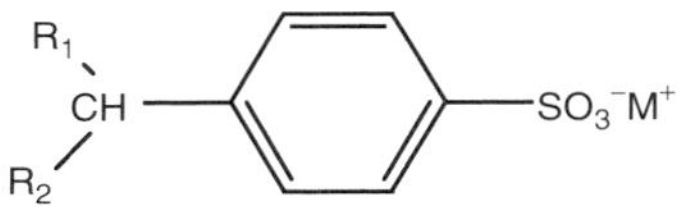

Alkylbenzene sulfonate

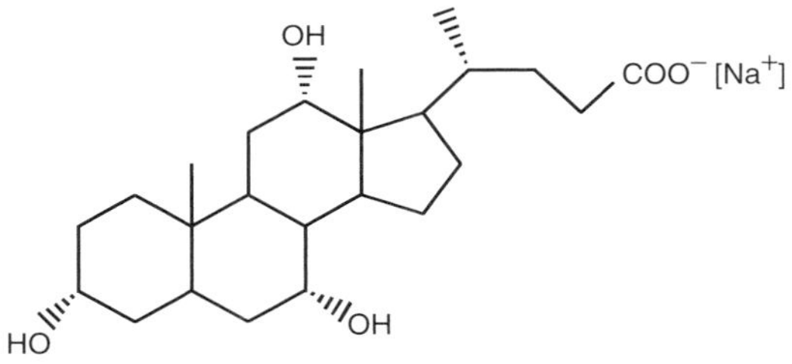

ROP(=O)(OH)(O^-M^+)

Monoalkyl phosphate

Bile Salts (sodium cholate)

$CH_3(CH_2)_7CH{=}CH(CH_2)_7COO^-[NH^+(CH_2CH_2OH)_3]$

Amine Soaps (triethanolamine oleate)

$RCONHCH(CH_2CH_2COOH)COO^-M^+$

Acyl glutamate

- *Soap (sodium salt of fatty acid):* Soap is the oldest known surfactant. It is used mostly as toiletry soap bars for body cleansers and sometimes for fabric detergent. Soap is a salt of weak acid (fatty acid) and strong base (sodium hydroxide) and shows an alkaline nature, with pH 10 in aqueous solutions.
- *Alkyl sulfate:* Sodium dodecyl sulfate is the most typical alkyl sulfate compound and is used extensively in scientific research. It is an essential component of shampoos and a foaming agent for toothpaste.
- *Alkylpolyoxyethylene sulfate:* This is an improved surfactant of alkyl sulfate since it has greater solution stability at low temperatures, more tolerance for hard water, and causes fewer skin irritations. This surfactant is one of the rare anionic agents that are soluble

in water in the presence of calcium or magnesium ions. However, disadvantage is the caking phenomenon that results from its hygroscopic nature.

- *Alkylbenzene sulfonate:* This is one of the most extensively used anionic surfactants as the main component of fabric, dishwashing, and industrial-use detergents. The agent is also applicable as an emulsifier (which will be discussed later on in this chapter) and a dispersing agent for agricultural chemicals. Some other types of anionic surfactants include monoalkyl phosphate, acyl isethionate, acyl glutamate, N-acyl sarcosinate, and alkyl succinate.

2. Cationic

These are surfactants in which the cation is the surface-active component. They are also electrolytes, with a positive charge. Most materials have negative charges in an aqueous media, and the cationic surfactant molecules adsorb by orienting their hydrophilic head group toward the surface of the materials. This characteristic of the cationic surfactant is fully utilized in several products such as fabric softeners and hair conditioners. Some cationic surfactants can be also used as a germicide because of their strong ability to adsorb to negatively charged materials in the living bodies of bacteria. All the cationic surfactants used at present are derivatives of alkylamines.

- *Alkyltrimethylammonium salt:* This surfactant is the most typical cationic surfactant. Tertiary amine, $RN(CH_3)_2$, is reacted with methylchloride or methylbromide to be converted to quaternary ammonium salt.
- *Dialkyldimethylammonium salt:* This is synthesized by the reaction of dialkylamine with methylchloride. It has two long hydrocarbon chains in one molecule and is well known as the first synthetic surfactant found to form vesicles.
- *Alkylbenzyldimethylammonium salt:* This cationic surfactant is popular as a germicide. It is often called invert soap because of its opposite charge from soap. The reaction of tertiary amine with benzylchloride is used to produce this surfactant.

Example : Cationic Surfactants

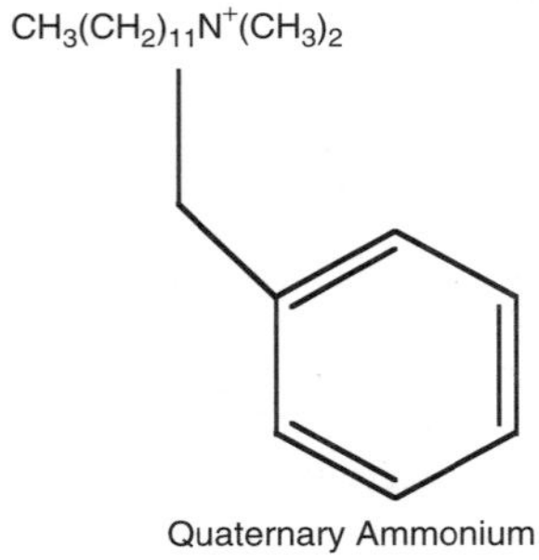

Quaternary Ammonium Salts

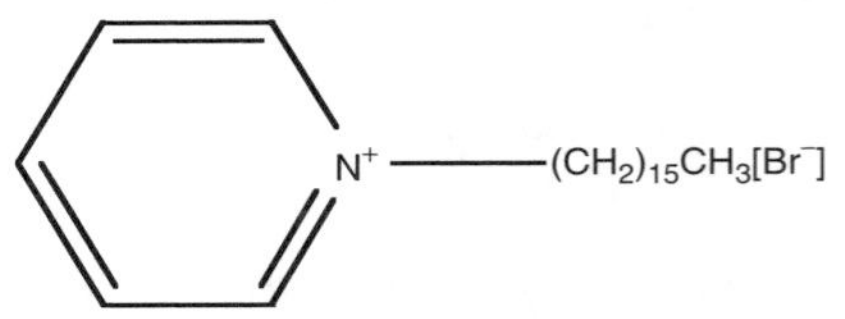

Pyridinium Salts (cetylpyridinium bromide)

$RNH_3^+X^-$

Alkylammonium Salts

$RN^+(CH_3)_3X^-$

Alkyltrimethylammonium salts

3. Amphoteric

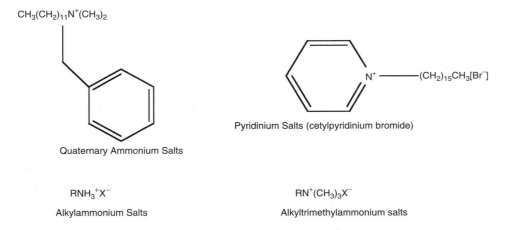

Amphoteric surfactants can behave either as anionic or cationic molecules, depending upon the pH. Thus, this surfactant can be in the anionic, nonionic, cationic, or zwitter-ionic state depending on the pH value of the solution it is in. Amphoteric surfactants are rarely used as a main product component. Their primary use is as an important co-surfactant that boosts the detergency and the foaming power of anionic surfactants.

- Alkyldimethylamine oxide: This compound is a typical booster for anionic surfactants such as alkylsulfate and alkylpolyoxyethylene sulfate, mainly used in dishwashing detergents. The amine oxide is synthesized from tertiary amine, by oxidation reaction with hydrogen peroxide.
- Alkylcarboxy betaine
- Alkylsulfobetaine
- Amide-amino acid type surfactants

Example : Ampholytic Surfactants

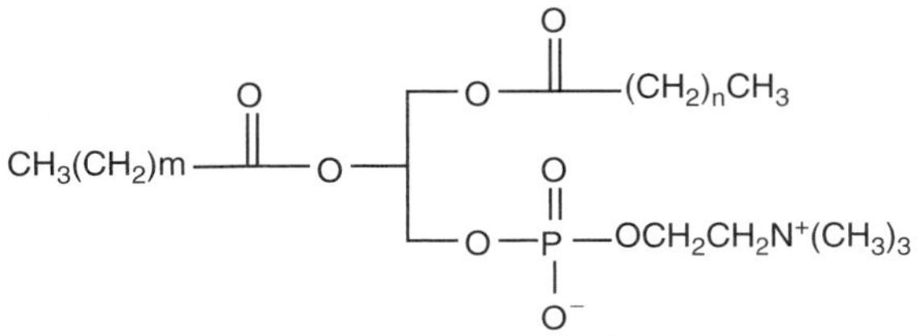

Phospholipids (lecithin)

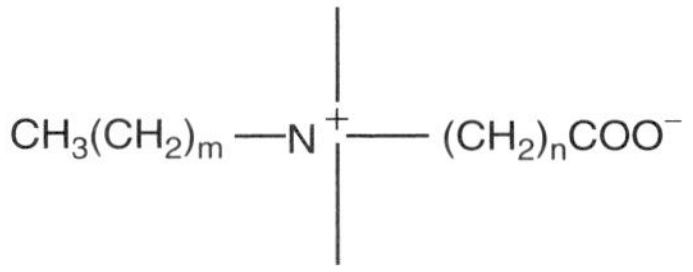

Ampholite Soaps (Betaine)

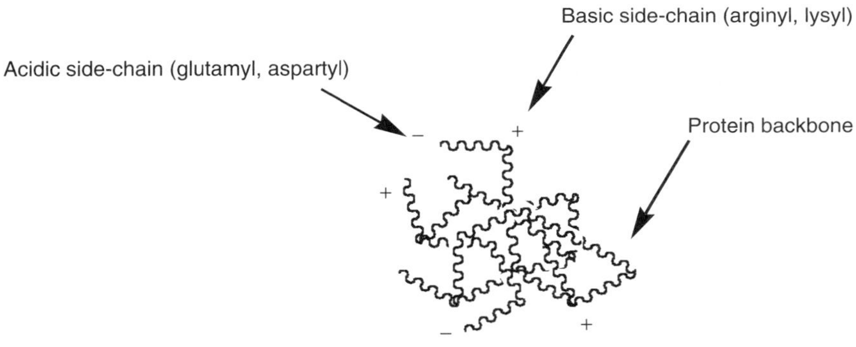

Proteins (gelatin, casein)

4. Nonionic

Nonionic surfactants have a water-soluble hydrophilic region. They are not electrolytes and have some nondissociative hydrophilic group. Polyoxyethylene groups, saccharides (sorbitan, sugar, glucose), and hydroxyl groups are examples of nonionic hydrophilic groups. Nonionic surfactants are mostly tolerant in aqueous solutions of added salts and hard water. Most common in this group are Spans and Tweens. Sorbitan fatty acid esters, such as sorbitan monopalmitate (Spans), are oil-soluble emulsifiers that promote water-in-oil emulsions. Polyethylene glycol sorbitan fatty acid esters (Tweens) are water-soluble emulsifiers that promote oil-in-water emulsions.

Example : Nonionic Surfactants

$CH_3(CH_2)_nOH$

Fatty alcohols
(n = 11: lauryl; n = 15: cetyl; n = 17: stearyl)

$HO(CH_2CH_2O)_n{-}(CH_2)_{11}CH_3$

Polyethyleneglycol (PEG) Ether
(PEG-200 lauryl ether)

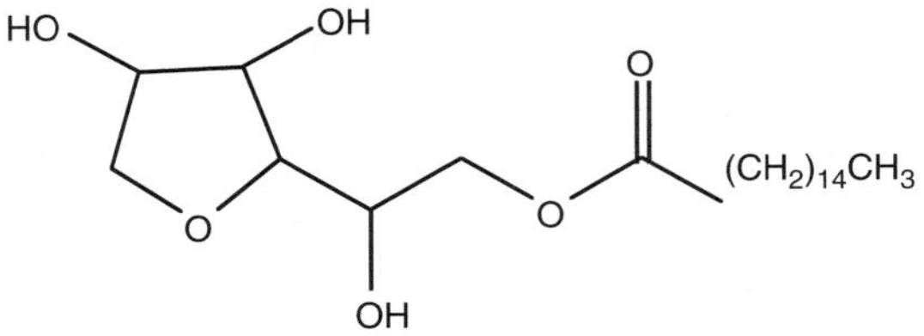

Spans: Sorbitan Ester of Fatty Acids
(sorbitan monopalmitate)

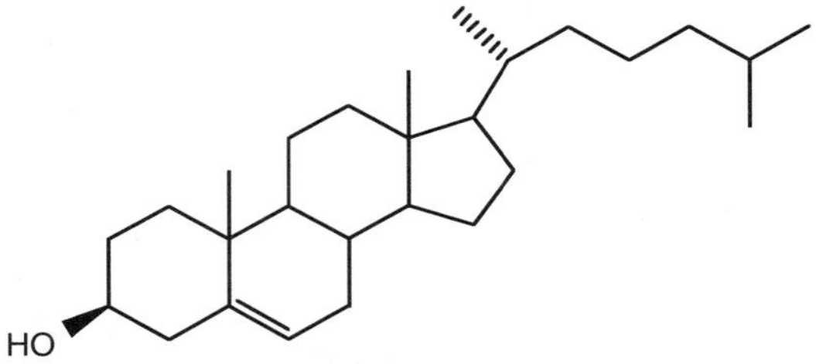

Cholesterol

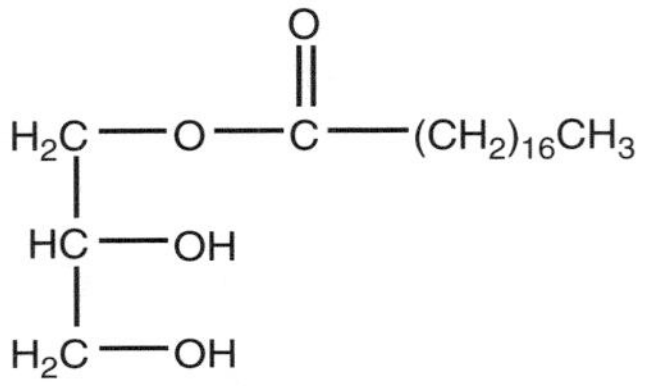

Partial Fatty Acid Esters of
Multivalent Alcohols
(glycerol monostearate)

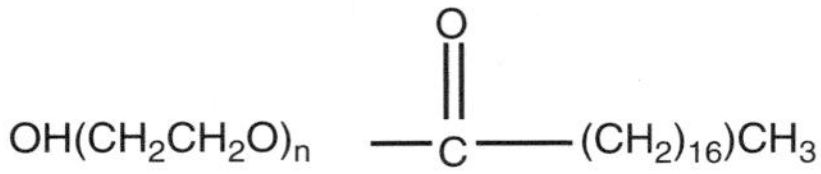

Cremophor: Polyethyleneglycol (PEG)
Fatty Acid Ester
(PEG-400 stearate)

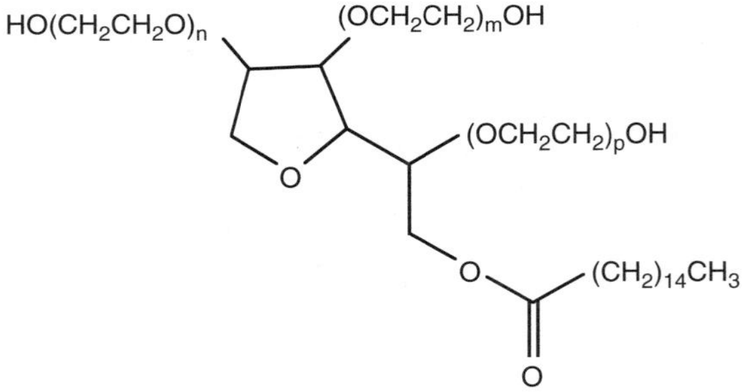

Polysorbates,Tweens: PEG-Sorbitan Fatty Acids Esters
(PEG-200-sorbitan monostearate, Polysorbate 60)

C. Wetting Agent

Wetting and dispersion are the two main phenomena resulting from the adsorption of a surfactant at solid–liquid interfaces. Wetting means that the contact angle between a liquid and a solid is zero or close to zero so that the liquid spreads easily over the solid. A primary function of surfactants is to act as wetting agents. A wetting agent encourages interfacial contact between a liquid and another phase. It can be added to allow spreading on a rough surface. It is important to note that all wetting agents are not surfactants, but all surfactants are wetting agents. Any molecule that lowers the surface tension of a liquid can act as a wetting agent by helping the liquid to spread on a surface. As surfactants accumulate at the interface, they form effective bridges between a liquid

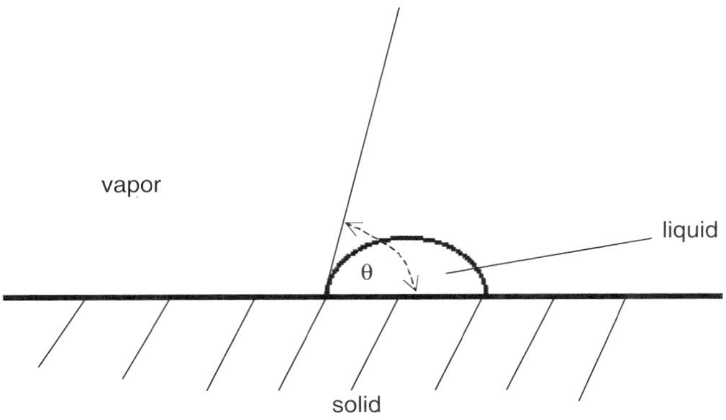

FIGURE 6.5 Contact angle.

and a solid. Some of the examples of wetting agents are alcohol and glycerin (hygroscopic liquids). Wetting usually is described by the contact angles that characterize how easily liquids spread on a given solid surface.

D. Contact Angle

The tendency for a liquid to spread is estimated from the magnitude of the contact angle (θ) formed, which is defined as the angle formed between the tangent drawn from the drop at the three-phase interface and the solid surface over which it spreads (Figure 6.5).

The contact angle of a liquid on a flat surface is determined by the chemical factor of materials of the liquid and solid. If a contact angle is measured on a perfectly smooth, and totally homogeneous, clean flat surface, with a pure liquid, then there should be only one value for. For instance, the contact angle for pure water on clean glass is zero. The contact angle between a solid and a liquid may be 0°, signifying complete wetting, or it may approach 180°, at which wetting is insignificant. However, there is one more important factor that determines the contact angle: the geometrical (microscopic roughness) factor of the solid surfaces. When the solid surface is rough, it can be shown that

$$\cos \theta_r = r(\gamma_a - \gamma_{ab})/\gamma_b$$

where θ_r is the contact angle on the rough surface, r is the roughness factor, γ_a is the surface tension of the solid, γ_b is the surface tension of the liquid, and γ_{ab} is the interfacial tension between the solid and the liquid. The smaller the contact angle is, the better the liquid spreads on the surface.

V. DISPERSED SYSTEMS

A substance is dispersed if it is more or less uniformly distributed throughout another substance, which is commonly called the dispersion medium. Therefore, dispersion is defined as a homogeneously mixed state of small particles in a continuous liquid medium. In liquid or semisolid dosage forms, there are preparations containing nondissolved drug distributed throughout a vehicle. In these preparations, the substance distributed is referred to as the dispersed phase, and the vehicle is termed the dispersing phase or medium. Together, they produce a dispersed system. Dispersed systems are often thermodynamically unstable, because any free energy associated with the large interfacial area between the dispersed phase and the continuous phase can be decreased by the aggregation or coalescence of the dispersed phase. If the particles of the dispersed phase are solid materials that are insoluble in the dispersion medium, the dispersed system is called suspension.

TABLE 6.4
Dispersed Systems

Types of Dispersion	Size of Particles
Coarse dispersions	10–50 μm
Fine dispersions	0.5–10 μm
Colloidal dispersions	1 nm–0.5 μm
Molecular dispersions	<1 nm

In the case of emulsions, the dispersed phase is a liquid that is neither soluble nor miscible with the liquid of the medium.

- ***Dispersed phase:*** the phase forming the particles
- ***Dispersion medium:*** the medium in which the particles are distributed
- ***Disperse system:*** the whole system

Dispersed Phase + Dispersion Medium = Dispersed System

The particles of the dispersed phase may vary widely in size from large particles visible to the naked eye down to particles of colloidal dimensions. For instance, if we pour some fine chalk powder into a beaker of water and agitate gently, the chalk powder will become uniformly distributed, or dispersed, in the water. This type of dispersion is called a coarse dispersion, because the dispersed particles are rather coarse, which means they are much larger than the particles of the dispersion medium. Coarse dispersions are unstable; if we discontinue agitation of the chalk dispersion, the powder will soon separate, or settle out, because the chalk powder is specifically heavier than the water. If instead of chalk powder, we choose finely powdered sugar, which will readily dissolve in the water, this type of dispersion is called a molecular dispersion. In molecular dispersions, the dispersed particles are roughly of the same size as the particles of the dispersion medium. Between these two types of dispersions there are other types: fine dispersion and colloidal dispersion, in which the particles are smaller than the coarsely dispersed particles but larger than the molecularly dispersed particles. Table 6.4 summarizes the size of the particles in the various types of dispersions.

The range of different particle sizes from highest to lowest shows an increased degree of dispersion. However, if we start from the bottom of the scale and build up larger particles from smaller ones, the process is called aggregation. Aggregation is the converse of dispersion, where we have to induce the cohesive forces to link the larger particles together. Cohesive forces and aggregations play an important role in emulsions, which will be discussed in Section VII in this chapter. Complete and uniform distribution of the dispersed phase in the medium is essential to the accurate administration of uniform doses. Largely because of their bigger size, dispersed particles in a coarse dispersion have a greater tendency to separate from the dispersing phase than particles of a fine dispersion. Colloid dispersions possess specific features worthy of further discussions.

A. Colloidal Dispersion

In a true solution, such as one of urea or sucrose, the particles of solute distributed in the solvent are believed to be of molecular size. However, a **suspension** or **emulsion** contains particles large enough to be visible to the naked eye, or at least in the microscope, distributed in a liquid medium.

Between these two extremes are the colloidal systems; that is, the essential characteristic of the colloidal state is the existence of particles that are larger than molecules but not large enough to be seen in the microscope. It is impossible to draw a distinct line between colloidal solutions and true solutions on one side, and between colloidal solutions and suspensions on the other; nevertheless colloidal solutions have certain properties that, as a general rule, place them in a category distinct from the other systems mentioned. The upper size limit of particles in the colloidal state is usually taken as the lower limit of microscopic visibility, that is, about 0.2 μm, whereas the lower limit is 3 mμ (50 Å).

B. Properties of Colloids

One property that distinguishes colloid systems from true solutions is that colloidal particles scatter light. If a beam of light, such as that from a flashlight, passes through a colloid, the light is reflected (scattered) by the colloidal particles, and the path of the light can therefore be observed. When a beam of light passes through a true solution (e.g., salt in water) there is so little scattering of the light that the path of the light cannot be seen, and the small amount of scattered light cannot be detected except by very sensitive instruments. The scattering of light by colloids, known as *the Tyndall effect*, was first explained by the British physicist John Tyndall. When an ultramicroscope is used to examine a colloid, the colloidal particles appear as tiny points of light in constant motion; this motion, called *Brownian movement*, helps keep the particles in suspension.

Adsorption is another characteristic of colloids, since the finely divided colloidal particles have a large surface area exposed. The presence of colloidal particles has little effect on the colligative properties (boiling point, freezing point, etc.) of a solution. The particles of a colloid selectively adsorb ions and acquire an electric charge. All of the particles of a given colloid take on the same charge (either positive or negative) and thus are repelled by one another. If an electric potential is applied to a colloid, the charged colloidal particles move toward the oppositely charged electrode; this migration is called *electrophoresis*. If the charge on the particles is neutralized, they may precipitate out of the suspension. A colloid may be precipitated by adding another colloid with oppositely charged particles; the particles are attracted to one another, coagulate, and precipitate out. For the most part, colloidal sols (solid) and gels used in pharmacy are aqueous preparations. The presence and magnitude of a charge on a colloidal particle is an important factor in the stability of colloidal systems. Stabilization is accomplished by two means: providing the dispersed particles with an electric charge and surrounding each particle with a protective solvent.

The addition of large amounts of hydrophilic colloid stabilizes the system. This concept is known as *protection*, and the added hydrophilic colloid is called a *protective colloid*. The protective property is expressed most frequently in terms of the *gold number*. The gold number is the minimum weight (measured in milligrams) of the protective colloid required to prevent a color change. Table 6.5 shows the gold numbers of some common protective colloids.

TABLE 6.5
Gold Numbers of Protective Colloids

Protective Colloids	Gold Number
Gelatin	0.005–0.01
Albumin	0.1
Acacia	0.1–0.2
Sodium Oleate	1–5
Tragacanth	2

VI. SUSPENSIONS

Suspensions are preparations that contain fine drug particles distributed uniformly throughout a vehicle. A pharmaceutical suspension is usually a coarse dispersion in which insoluble solid particles are dispersed in a liquid system. The particles usually have diameter greater than 0.1 µm.

A. Theory of Suspension

Work must be done to reduce a large solid material into small particles and disperse them in a continuous medium. This communition process results in the generation of surface free energy that makes the system thermodynamically unstable as the resultant small particles are highly energetic and tend to regroup in such a way as to decrease the total area and reduce the surface free energy. The particles in a liquid suspension therefore tend to form light, fluffy conglomerates known as floccules that are held together by weak van der Waals forces. Under certain conditions, the particle may adhere by stronger forces to form aggregates known as cakes.

The formation of any type of aggregate, either floccules or cakes, is perceived as a measure of the system's tendency to reach a more thermodynamically stable state. An increase in the work, W, or surface free energy, ΔG, brought about by dividing the solid into smaller particles and eventually increasing the total surface area, ΔA, is given by

$$\Delta G = \gamma_{SL} \bullet \Delta A$$

in which γ_{SL} is the interfacial tension between the liquid medium and the solid particles.

In order to approach a stable state, the system tends to reduce the surface free energy that may be achieved either by a reduction of interfacial tension or by a reduction of the interfacial area. Interfacial tension can be reduced by the addition of a surfactant. Reduction of interfacial area, on the other hand, is achieved through formation of floccules or cakes.

B. Purposes of Suspension

There are several reasons for preparing a suspension. First, some drugs are unstable in solution phase but stable in suspension. In this case, the suspension can assure the stability while delivering the drugs. Second, for many patients, the liquid formulation is preferred over the solid form of the same drug since it is easier to swallow. Also, it would be easy to control the range of doses for a liquid dosage form than a solid dosage form such as tablets. This is a big advantage for infants, children, and the elderly. Furthermore, a suspension helps improve the taste of some poor-tasting drugs, because the dispersion medium can be sweetened and flavored. A suspension is often a suitable dosage form in dermatology and for the parenteral administration of insoluble drugs. Suspension dosage forms are given orally, injected intramuscularly or subcutaneously, applied to the skin in topical preparations, and used topically in the eye and in the ear. Many antibiotic agents are unstable in solution for a period of time; therefore, packaging as a dry powder to be reconstituted by adding the dispersion medium when filling the prescription is a good approach.

Altogether, common pharmaceutical suspensions belong to three groups: oral suspensions (syrups), external suspension (lotions), and parenteral suspensions. The most common application of suspensions is for oral dosage form, which will be discussed in detail here. The following are examples of some common oral suspensions:

- Alumina, magnesia, and simethicone oral suspension (Mylanta®)
- Magnesia and alumina oral suspension (Maalox®)
- Sulfamethoxazole and trimethoprim oral suspension (Septra®, Bactrim®)
- Amoxicillin for oral suspension (Amoxil®)
- Ampicillin for oral suspension (Omnipen®)

- Cefaclor for oral suspension (Ceclor®)
- Cefixime for oral suspension (Suprax®)

The last four preparations consist of specific amounts of dry powder mixtures or granules, which are intended to be suspended in a specific quantity of water or some other vehicle prior to oral administration to produce a specific concentration.

In addition to the desired qualities of all pharmaceutical preparations (therapeutic efficacy, chemical stability of the components of the formulation, and esthetic appeal), specific requirements of a pharmaceutical suspension are:

- It should settle slowly and should be readily redispersed upon gentle shaking of the container.
- The particle size of the suspended drug should remain fairly constant during undisturbed standing for a long time.
- It should pour readily and evenly from its container.

C. Flocculation and Deflocculation

Zeta potential (ξ) is a measure of the potential existing at the surface of a particle. When ξ is relatively high ($\geq$25 mV), the repulsive forces between two particles exceed the attractive forces. Accordingly, the particles remain dispersed individually and are said to be deflocculated. Even when brought close together by random motion or agitation, deflocculated particles resist collision because of their high surface potential.

The addition of a preferentially adsorbed ion with a sign opposite to that on the particle leads to a progressive lowering of ξ. At some concentration of the added ion the electrical forces of repulsion are lowered sufficiently that the forces of attraction dominate. Under these conditions the particles may approach each other more closely and form loose aggregates termed floccules. Such a system is called a flocculated system. Table 6.6 summarizes the characteristics of flocculated and deflocculated particles.

The continued addition of the flocculating agent can reverse the above process, if the ξ increases sufficiently in the opposite direction. Thus, the adsorption of anions onto positively charged deflocculated particles in suspension will lead to flocculation. The addition of more anions can eventually generate a net negative charge on the particles. When this has achieved the required magnitude, deflocculation may occur again.

D. Formulation of Suspension

The formulation of a suspension possessing optimal physical stability depends on whether the particles in suspension are to be flocculated or to remain deflocculated. The approaches commonly used in the preparation of physically stable suspensions fall into two categories: the use of a

TABLE 6.6
Relative Properties of Flocculated and Deflocculated Particles in Suspension

Deflocculated	Flocculated
Particles exist in suspension as individual entities.	Particles form loose aggregates or floccules.
Rate of sedimentation is slow.	Rate of sedimentation is high.
The sediment eventually becomes very closely packed and the resulting hard cake is difficult, if not impossible, to redisperse.	The sediment is loosely packed and easy to redisperse.
Even after settling, the supernatant remains cloudy.	After rapid settling, a clear boundary exists between the sediment and the supernatant.

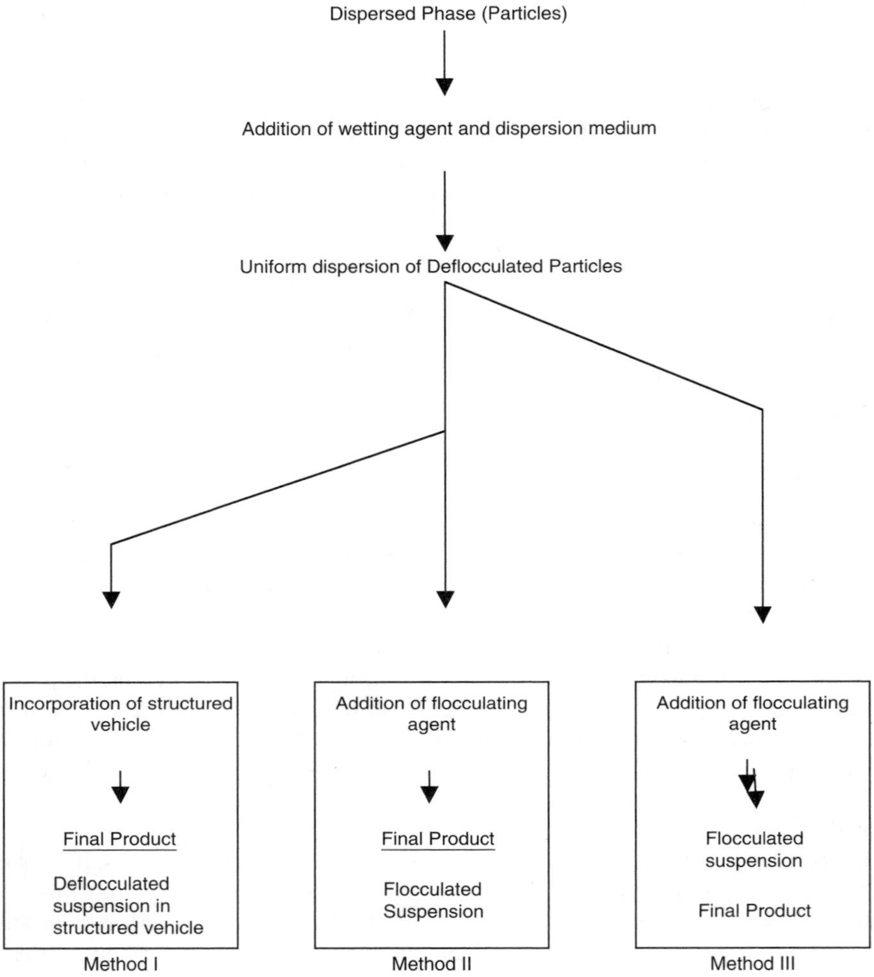

FIGURE 6.6 Different approaches to suspension formulation.

structured vehicle to maintain deflocculated particles in suspension and the application of the principles of flocculation to produce flocs that, although they settle rapidly, are easily resuspended with a minimum of agitation. Whatever the approach may be, to start, the particles (dispersed phase) are wetted by the addition of a minimum amount of wetting agents and dispersed in the dispersion medium as deflocculated particles.

Figure 6.6 represents three approaches commonly used to formulate a suspension. The first approach is to incorporate a structured vehicle to keep deflocculated particles in suspension for a long period. Structured vehicles are generally aqueous solutions of polymeric materials such as the hydrocolloids, which are usually negatively charged in aqueous solution. Typical examples are methylcellulose, carboxy-methylcellulose, bentonite, and carbopol. They function as viscosity-building suspending agents and reduce the rate of sedimentation of dispersed particles. The second approach utilizes principles of controlled flocculation as a measure to prevent cake formation. In this approach, the formulator takes the deflocculated, wetted dispersion of particles and attempts to bring about flocculation by the addition of a flocculating agent; most commonly, these are either electrolytes, polymers, or surfactants. The aim is to control flocculation by adding an appropriate amount of flocculating agent that results in the maximum sedimentation volume. The third approach is a combination of first two, in which flocculated particles are suspended in a structured vehicle that results in a product with optimum stability. The process involves dispersion of the particles and their subsequent flocculation. Finally, a lyophilic polymer is added to form the structured vehicle.

However, care must be taken while using this approach to avoid any incompatibility between the flocculating agent and the polymer used. A limitation is that virtually all the structured vehicles carry a negative charge. Therefore, an incompatibility arises if the charge on the particles is originally negative. Flocculation in this case requires the addition of a positively charged flocculating agent or ion. In the presence of such a material, the negatively charged suspending agent may coagulate and lose its suspendability. This situation does not arise with positively charged particles, as the required negatively charged flocculating agent is compatible with the negatively charged suspending agent. In the case of suspending negatively charged particles, it becomes necessary to use a protective colloid (fatty acid amine, gelatin, etc.) to change the sign of the particle from negative to positive. It is then possible to use the anionic flocculating agent and the negatively charged suspending agent as usual.

E. Evaluation of Suspension

During formulation, the formulator needs to evaluate a formulation in terms of the amount of flocculation in the suspension and compare this with the amounts found in other formulations. The two parameters commonly used for this are described below:

1. Sedimentation Volume (*F*)

This is the ratio of the equilibrium volume of the sediment, V_u, to the total volume of suspension, V_o.

$$F = V_u/V_o$$

The value of F ranges from 0 to 1. In the ideal suspension where the particles will be uniformly distributed without any separation of the dispersion medium, the value of F will be 1, as the volume of the sediment is equal to the total volume of the suspension. In the system where $F = 0.5$, 50% of the total volume of the suspension is occupied by the loose, porous flocs forming the sediment. The formulator's goal is to formulate a suspension with F equal to 1.

2. Degree of Flocculation (β)

A better parameter for comparing flocculated systems is to use, β, which relates the sedimentation volume of the flocculated suspension, F, to the sedimentation volume of suspension when deflocculated, F_∞, as expressed below:

$$\beta = F/F_\infty$$

With two formulations with β values 4 and 6, the formulation with β value 6 will be preferred, as the aim of a formulation is to produce as flocculated a product as possible. As the degree of flocculation in the system decreases, β approaches unity, the theoretical minimum value.

F. Preparation of Suspension

In the preparation of a suspension, the pharmacist must be familiar with the characteristics of both the dispersed phase and the dispersion medium. In some cases, the dispersed phase (powder) is readily wetted when added to the medium. However, sometimes the particles clump together and float on top of the vehicle and cause an uneven suspension. In these cases, the powder must be wetted with *wetting agent* first. Once the powder is wetted, the dispersion medium is added and mixed thoroughly before addition of the vehicle. The final product is then passed through a colloid mill or blender to insure uniformity. A preservative may be added to protect against bacterial contamination. During *extemporaneous compounding of suspensions*, the pharmacist puts the contents of the capsule into a mortar; if it is in tablet form, it will be crushed in a mortar with a pestle. Then the selected vehicle is slowly added to the powder and mixed well to create a paste before being diluted to the desired volume. Again, the difficulty that confronts the pharmacist is the stability of the drug when it is incorporated into a liquid vehicle. Drugs in liquid form have faster decomposition rates than those in solid form. To overcome this, the pharmacist can contact the pharmaceutical manufacturer to obtain stability information.

G. Stability of Suspension

1. Physical Stability and Stokes' Law

Settling of the suspended particles leading to separation of dispersed particles and dispersion medium is the most important factor associated with the physical stability of a suspension. Many factors involved in the rate of settling of the particles of a suspension are summarized in the equation below (Stokes' law):

$$\frac{dx}{dt} = \frac{d^2(\rho_i - \rho_e)g}{18\,\eta}$$

where dx/dt is the sedimentation rate, d is the diameter of the particles, ρ_i is the density of particles, ρ_e is the density of medium, g is the gravitational constant, and η is the viscosity of the medium.

The equation is valid only in an ideal situation in which uniform, perfectly spherical particles settle in a very dilute suspension (i.e., no collision of the particles), without effecting turbulence in the medium and without chemical or physical attraction between the particles and the medium. While in practical cases, conditions are not in strict accord with the assumptions of Stokes' law, the above equation does give the factors that influence the rate of settling. It is apparent from the equation that rate of sedimentation (dx/dt) will be reduced by

- Decreasing the particle size (d), provided the particles are kept in a deflocculated state
- Minimizing the difference in densities between the particles and the dispersion medium ($\rho_i - \rho_e$)
- Increasing the viscosity of the dispersion medium (η)

The first factor may be achieved by communition of the particles using appropriate techniques (pestle and mortar, colloid mills, etc.). Control of settling by minimizing the difference in densities between the particles and the dispersion medium is not too practical. However, viscosity of the dispersion medium can always be increased by the addition of viscosity building polymers (methyl cellulose, carboxy methylcellulose, etc.). However, too high a viscosity is undesirable, as it may affect the redispersability and pourability of the suspension.

2. Chemical Stability

Suspended particles are unlikely to undergo most chemical reactions that lead to degradation. However, most drugs in suspension have a finite solubility even though this may be very little. As a result, the material in solution may be susceptible to degradation by various ways described in Chapter 9. For the same reason, the preparation may be stabilized against microbial attack through the addition of antimicrobial preservatives.

VII. EMULSIONS

A. Definition of Emulsion

Emulsification is the most important phenomenon caused by the adsorption of surfactant at liquid–liquid interfaces. Interest in emulsions has grown steadily, and increasing numbers of people work with emulsions. Some are engaged in scientific research, others manufacture or use emulsions, and still others produce emulsifying agents or design machinery for emulsification.

An emulsion is a dispersion of two immiscible liquids, one of which is finely subdivided and uniformly distributed as droplets (the dispersed phase) throughout the other (the continuous phase) with the help of a third substance known as emulsifying agent. In emulsions, the diameter of the droplets is usually greater than 0.1 μm, which gives it a milky appearance. Very few emulsion droplets are smaller than 0.25 μm in diameter. An emulsion system has two phases: the internal phase and the external phase. One of the liquids appears in the emulsion in the form of separate globules, and its continuity is interrupted and accordingly called the discontinuous phase. The other liquid, which surrounds the droplets, remains uninterrupted and therefore it is referred to as the continuous phase. The liquids used in the emulsion are generally as water (or aqueous solutions of various salts and organic or colloidal substances) and oil (or resins, waxes, hydrocarbon, etc.) phases. There are two types of emulsion. One is a water-in-oil system (W/O), where oil is the continuous phase containing water droplets. The other system is oil-in-water (O/W), in which oil droplets are dispersed in a continuous phase of water. Whether the emulsion will be O/W or W/O can be determined from *Bancroft's rule,* which states that the phase in which the emulsifying agent is more soluble will be the external phase.

It is also possible to take an emulsion system and emulsify it further, such that a multiple emulsion is formed. For instance, a multiple emulsion would be water in oil in water (W/O/W), which would consist of very small drops of an aqueous phase dispersed in small drops of oil, which then are dispersed in another, larger aqueous phase. This can have advantages for separating incompatible components or adjusting drug release rates.

Emulsions offer advantages in drug delivery, such as providing a way of administering oil in a more palatable aqueous system. This helps the drug reach the site of action more easily. Furthermore, many drugs are administered in emulsions as a way of increasing the chemical stability of the drug (compared with that in solution) or of improving bioavailability over solid dosage forms. In addition, it can improve penetration and spreading. Based on the intended application, liquid emulsion may be employed orally, topically, or parenterally.This chapter emphasizes orally administered emulsions only.

Medicinal emulsions for oral administration are usually of the O/W type. Some examples of orally administered emulsions are liquid petrolatum emulsion used for lubricating cathartic, castor oil emulsion used as a laxative, and simethicone emulsion used as a defoaming agent for the relief of painful symptoms of excess gas in the gastrointestinal tract.

B. Theories of Emulsification

Many theories have been proposed to explain how emulsifying agents act in promoting emulsification and in maintaining the stability of the resulting emulsion. Some of the most prevalent theories are the *surface-tension theory*, the *oriented-wedge theory*, and the *plastic-film theory*. According to the surface tension theory, some surfactants or wetting agents are used to lower the interfacial tension of the two immiscible liquids, reducing the repellent force between the liquids. Thus, the surfactants facilitate the breaking up of large globules into smaller ones, which then have lesser tendency to coalesce. The oriented-wedge theory assumes monomolecular layers of emulsifying agent curved around the droplet of the internal phase. The theory is based on the presumption that certain emulsifying agents orient themselves around and within a liquid in a manner reflective of their solubility. Depending upon the shape and size of the molecules, their solubility, and thus their

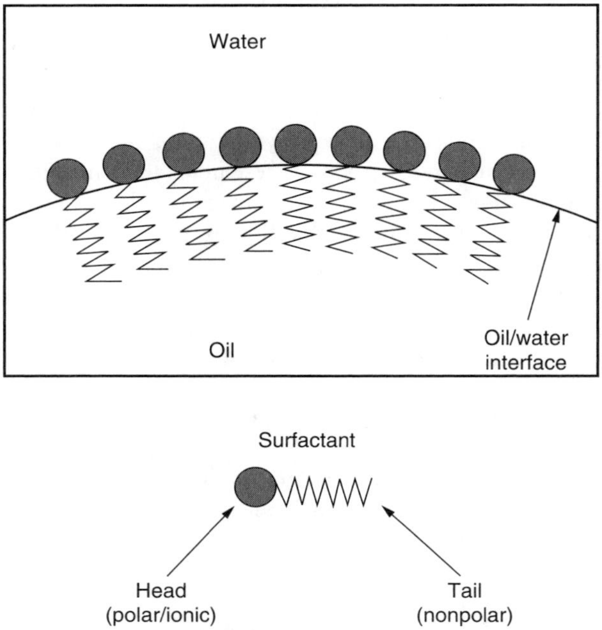

FIGURE 6.7 Emulsifier at the interface of water and oil.

orientation, the wedge-shaped arrangement causes the surrounding of either oil globules or water globules. Generally an emulsifying agent with a greater hydrophilic character than hydrophobic character promotes an oil-in-water emulsion, and a water-in-oil emulsion results from the use of an emulsifying agent that is more hydrophobic. The plastic-film theory places the emulsifying agent at the interface between the oil and water, surrounding the droplets of the internal phase as a thin layer of film adsorbed on the surface of the drops. The film prevents the contact and coalescing of the dispersed phase. Depending on the required task for each emulsion system, different theories are applied.

C. Emulsifiers: Mechanism and Types

Emulsifying agents, or emulsifiers, play an important role in stabilizing an emulsion system. Some desirable properties of an emulsifying agent are:

- Surface-active and reduces surface tension to below 10 dyn/cm
- Adsorbs quickly around the dispersed drops as a condensed, nonadherent film that will prevent coalescence
- Imparts to the droplets an adequate electric potential so that mutual repulsion occurs
- Increases the viscosity of the emulsion
- Effective in a reasonably low concentration

From a pharmaceutical point of view, an acceptable emulsifier must also be stable and compatible with other ingredients in the mixture. In addition, it has to be nontoxic and should not interfere with the stability or efficacy of the active agents.

Emulsifiers can be classified as surfactants and have hydrophilic (polar, water-loving) and lipophilic (nonpolar, oil-loving) structural portions within their molecular structures. Therefore, they serve as a bridge between the water and oil interface as shown in Figure 6.7. There are four major types of surfactants: anionic, cationic, amphoteric, and nonionic. Various types of materials that have been used in pharmacy as emulsifying agents are as follows:

- *Carbohydrate materials* such as acacia, tragacanth, agar, chondrus, and pectin form hydrophilic colloids when added to water and generally produce O/W emulsions. To prepare extemporaneous emulsions, acacia is the most frequently used emulsifier. Tragacanth and agar are commonly employed as thickening agents.
- *Protein substances* such as gelatin, egg yolk, and casein usually produce O/W emulsions. The disadvantage of using gelatin as emulsifier is that the system becomes more fluid upon standing.
- *High molecular weight alcohols* such as cetyl alcohol, stearyl alcohol, and glyceryl monostearate are used as thickening agents and stabilizers for externally used O/W emulsions such as lotions and ointments.
- *Wetting agents*, which contain both hydrophilic and lipophilic groups, can be anionic, cationic, or nonionic surface-active agents depending on their application for each system.
- *Colloidal clays* such as bentonite, magnesium hydroxide, and aluminum hydroxide generally form O/W emulsions.

The emulsifiers act as a barrier at the oil–water interface to minimize or prevent droplet–droplet contact and subsequent coalescence. The barrier may be a physical barrier or an electrostatic barrier or both. The barrier may or may not have much effect interfacial tension. The barrier is in the form of a film at the interface and may be of three general types:

- *Monomolecular films*: Some emulsifying agents (synthetic organic materials, e.g., the anionic, cationic, and nonionic surfactants described above) stabilize an emulsion by forming a monolayer of adsorbed molecules or ions at the oil–water interface and thereby reduce interfacial tension. The droplets are now surrounded by a coherent monolayer that also prevents coalescence between approaching droplets. These emulsifying agents are used in both O/W and W/O types emulsions.
- *Multimolecular films*: This type of emulsifying agent (hydrated hydrophilic macro molecules such as acacia, gelatin, lecithin, cholesterol, etc.) form multimolecular films around droplets of dispersed droplets. They do not cause an appreciable lowering in surface tension. Rather, their efficiency depends on their ability to form strong, coherent multimolecular films. These act as a coating around the droplets and render them highly resistant to coalescence, even in the absence of a well-developed surface potential. They form primarily O/W type emulsions.
- *Solid particle films*: This type of emulsifying agent (e.g., bentonite, magnesium hydroxide, etc.) consists of small solid particles that are wetted to some degree by both aqueous and nonaqueous liquid phases. They do not markedly affect interfacial tension but rather act as a physical barrier. They may form either O/W or W/O type emulsions.

D. Preparation of Emulsions

The first step in preparing an emulsion is the selection of the right emulsifier or blend of emulsifiers. To obtain stable emulsions against coalescence, emulsifiers must adsorb efficiently at the interfaces between the water and oil phase. An emulsifier that is too hydrophilic dissolves into the water phase, and one that is too hydrophobic dissolves into the oil phase. In either case, the emulsifier molecules may not adsorb efficiently at the interfaces. Consequently, the hydrophilic and hydrophobic nature must be balanced in the surfactant molecule for it to work well as a good emulsifier. The **hydrophilic-lipophilic balance (HLB)** number indicates the polarity of the molecules in an arbitrary range of 1 to 40, with the most commonly used emulsifiers having a value between 1 and 20. The HLB number increases with increasing hydrophilicity. HLB is given to each surfactant and used as a guiding criterion to select suitable emulsifiers. Table 6.7 shows the HLB ranges for some surfactant functions.

TABLE 6.7
Surfactant Functions According to HLB Range

Function	HLB Range
Antifoaming agent	0–3
Emulsifier, W/O	4–6
Wetting agent	7–9
Emulsifier, O/W	8–18
Detergent	13–15
Solubilizer	10–18

Oils are also given HLB values, but this HLB is relative as to whether an O/W or W/O emulsion is to be prepared and stabilized. Emulsifiers should have HLB values similar to those of the respective oils in order to achieve maximum stabilization. Mineral oil has an assigned HLB number of 4 when a W/O emulsion is desired and a value of 10.5 when an O/W emulsion is to be prepared. Accordingly, the HLB number of the emulsifiers should also be around 4 and 10.5, respectively. Examples of required HLB values of some other oils are given in Table 6.8.

TABLE 6.8
Required HLB Values of Oils and Waxes

	W/O	O/W
Beeswax	4	12
Cetyl alcohol	—	15
Mineral oil	4	10.5
Soft paraffin	5	12
Wool fat	8	10

The desired HLB numbers can also be achieved by mixing lipophilic and hydrophilic surfactants. The overall HLB value of the mixture is calculated as the sum of the fraction multiplied by the individual HLB.

EXAMPLE

In a mixture of 30% Span 80 (HLB = 4.3) and 70% Tween 80 (HLB = 15) what is the overall HLB value?

$$\text{HLB} = (0.3 \times 4.3) + (0.7 \times 15) = 11.8$$

The relative volume of the internal and external phases of an emulsion is also important, regardless of the type of emulsifier used. As the concentration of the internal phase increases, the viscosity of the emulsion also increases sharply up to a point beyond which a further increase in the internal phase concentration will cause a sharp decline of viscosity. At this point, the emulsion is said to have undergone a phase inversion; that is, it has changed from an O/W emulsion to a

W/O, or vice versa. In practice, emulsions may be prepared without inversion with as much as about 75% of the volume of the product being internal phase.

E. Methods of Emulsion Preparation

In the small scale extemporaneous preparation of an emulsion, the following three methods are used. However, the selection of the method depends on the nature of the emulsion components and the equipment available.

1. Continental or Dry Gum Method

This method is also referred to as the 4:2:1 method because for every four parts (volumes) of oil, two parts of water and one part of gum are added in preparing the initial or primary emulsion in a perfectly dry Wedgewood or porcelain mortor. The gum is added slowly to the oil first with continuous trituration. After the oil and the gum have been mixed, the two parts of water are added all at once and the mixture is triturated rapidly and continuously until a creamy white emulsion that produces a crackling sound to the movement of the pestle forms. Other liquid formulative ingredients that are soluble in or miscible with the external phase may then be added to the primary emulsion with mixing. Preservatives, stabilizers, colorants, and flavoring agents are usually dissolved in a suitable volume of water (assuming water is the external phase) and added as a solution to the primary emulsion. The emulsion is transferred to a graduated cylinder and made to volume with water previously swirled about in the mortar to remove the last portion of the emulsion. An electric mixer or blender may also be used for making the primary emulsion.

2. English or Wet Gum Method

In this method, the same proportions of oil, water, and gum are used as in the dry gum method, but the order of mixing is different. A mucilage of the gum is prepared by triturating the gum with twice its weight in water in a mortar. The oil is then added slowly in portions, and the mixture is triturated to emulsify the oil. After all of the oil has been added, the mixture is thoroughly mixed for several minutes to ensure uniformity. Then the other formulative ingredients are added, and the emulsion is transferred to a graduated cylinder and made to volume with water.

3. Bottle Method

In this method, powder gum is placed in a dry bottle, two parts of oil are then added, and the mixture is thoroughly shaken in the capped container. A volume of water approximately equal to the oil is then added in portions, the mixture being thoroughly shaken after each addition. When all the water has been added, the primary emulsion thus formed may be diluted to the proper volume with water or an aqueous solution containing other formulative agents. The method is useful for the extemporaneous preparation of emulsions from volatile oils or oleaginous substances of low viscosities.

F. Identification of Types of Emulsion

The general properties of an emulsion are determined mainly by the external phase. For instance, the O/W emulsion has water as the external phase; consequently, it behaves like a water system. It can be diluted with water but not with oil or oil-like liquids. It will conduct an electric current, since there is an uninterrupted layer of aqueous solution present, and it can be colored by water-soluble dyes. However, a W/O emulsion has oil as the external phase. It can be thinned down with oil or oil-like fluids but not with water. If water is added to W/O emulsion, it will enter the internal phase and increase its viscosity. Furthermore, a W/O emulsion will not conduct the electric current as readily as an O/W emulsion, and water-soluble dyes will not develop their tinting strength, since they can hardly reach the enclosed water droplets.

G. Stability of Emulsions

Generally speaking, an emulsion is considered to be physically unstable if the internal phase forms aggregates of globules and these globules fall to the bottom of the emulsion to form a concentrated layer of the internal phase. The distribution of droplet size through time shows the stability of that emulsion system. Also, an emulsion may be adversely affected by microbial contamination and growth.

It is easy to understand why water and oil are largely immiscible, as oils consist of long chain hydrocarbons, which have no affinity for small polar water. If water and oil are mixed vigorously, the oil will disperse as small drops in the water. However, these oil drops will tend to fuse with each other and eventually separate out as a separate phase, or **aggregates**, because the dispersion of a liquid into small drops greatly increases the interphase area and interfacial free energy. Aggregation of the drops reduces the interphase area, and thus reduces the total interfacial free energy. Thus, aggregates lead to increased sizes of the dispersed particles, which tend to accelerate into a total separation of the two phases with a tendency to rise to the top of the emulsion if the internal phase is oil or fall to the bottom if it is water. Such a separation of the two phases is known as **creaming** and is considered reversible. That means that the two phases can be redispersed with a mechanical force such as vigorous shaking. The creaming phenomena is not desirable, as it may lead to improper dosing if the emulsion is not mixed properly as well as as lack of esthetic appeal to the consumer.

Of greater concern than creaming when making an emulsion is the coalescence of the globules of the internal phase and the separation of that phase into a permanent layer thant can never be redispersed. That phenomena is known as **cracking** or **breaking** of the emulsion. Once that happens, attempts to reestablish the emulsion are generally unsuccessful.

To stabilize an emulsion and reduce coalescence, the particle size should be reduced to as fine as possible; the density difference between the internal and external phases should be minimal, and the viscosity of the external phase should be high. All these suggestions are according to Stokes' law described earlier in the section on suspension stability. In addition, choice of the proper emulsifying agent in the appropriate amount or concentration is of utmost importance.

Another problem associated with emulsion stability is **inversion**, which is the change of an emulsion from O/W to W/O or vice versa. Inversion is generally brought about by the addition of an electrolyte or by changing the phase volume ratio.

Another factor that can greatly effect the stability of an emulsion system is temperature. Excess heat or freezing result in the coarsening of an emulsion. Therefore, pharmaceutical manufacturers must consider temperature when products need to be shipped to locations with varying climates. Other environmental conditions such as the presence of light, air, and contaminating microorganisms can adversely affect the stability of an emulsion. Appropriate formulative and packaging steps can minimize these problems. For instance, light-resistant containers can be used for light-sensitive emulsions. For emulsions susceptible to oxidative decomposition, antioxidants may be included in the formulation. A combination of methylparaben, propylparaben, and alcohol can be added as preservatives to prevent contamination by bacteria, mold, yeast, etc.

VIII. MICELLES: CRITICAL MICELLAR CONCENTRATION AND MICELLAR SOLUBIZATION

As we add surfactant to a system, the surfactant molecules orient themselves at the interface, with the polar head pointing toward the aqueous phase and the nonpolar tail oriented toward the nonpolar

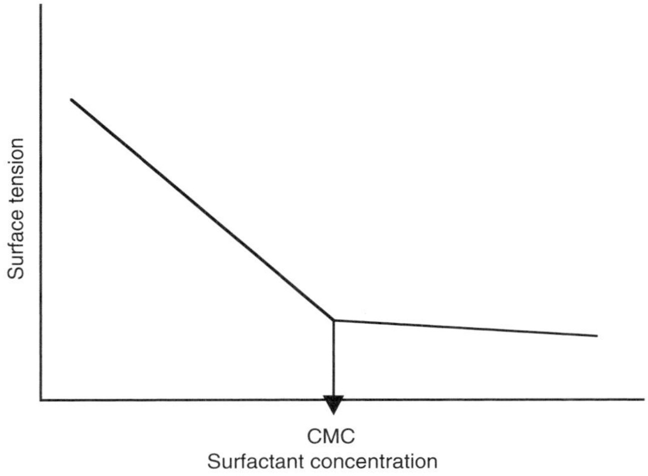

FIGURE 6.8 Demonstration of critical micelle concentration (CMC).

phase or air. With an increase in dissolved surfactant concentration, the surfactant molecules occupy more and more space at the interphase, and the surface tension of the liquid continues decreasing. At a certain concentration, the surface will be saturated with surfactant molecules, and the addition of more surfactant will consequently cause no further reduction in surface tension.

At this point the surfactant molecules form aggregates known as *micelles;* the concentration at which this happens is called the critical micelle concentration (CMC) (Figure 6.8). The micelles may contain 50 or more monomers. When micelles begin to form, the concentration of monomers will have reached a maximum in the solution. Any further increase in the total concentration of the surfactant will result in an increased number of micelles but no significant change in the monomer concentration. This concept is similar to the solubility limit of a dissolved material, which would normally result in crystallization of a solute once the limit is exceeded. The CMC represents the solubility limit for the monomer surfactant, and above this concentration the excess surfactant is displaced into micelles. Some factors affecting the CMC are the hydrophilic head group of the surfactant (depending on whether it is ionic and neutral), temperature (higher temperature raises the CMC). Micellar solubilization is one of the several utilities of forming micelles. Above CMC, some water-insoluble drugs can be trapped in the lipophilic core of the micelles and can be dispensed as aqueous solution. Examples of compounds used in this manner include volatile oils and fat soluble vitamins.

IX. CONCLUSION

Surface phenomena have a great impact on the formulation of pharmaceutical products, as they are directly related to solubility and dispersion of the active substance in a suitable medium or vehicle. As described in this chapter, two important pharmaceutical dosage forms are suspensions and emulsions because they are expected to demonstrate improved oral bioavailability compared to the same drug formulated as tablets or capsules. However, these expectations cannot be met if the suspension or emulsion is not formulated as a stabilized system, which requires a through knowledge and understanding of the surface phenomena.

HOMEWORK

1. Molecules on the surface of a liquid are subjected to intermolecular forces in the direction of the interface and toward the bulk of the phase. Therefore:
   (a) These forces tend to increase the surface.
   (b) These forces tend to decrease the surface.
   (c) These forces are compensated by adhesion (attraction to unlike molecules).
   (d) These forces are compensated by cohesion (attraction to like molecules).
2. The general term for molecules and ions that tend to accumulate at interfaces, and thereby reduce surface or intefacial tension, is:
   (a) Antifoaming agents
   (b) Detergents
   (c) Wetting agents
   (d) Surfactants
3. How does the increase in the viscosity of the liquid affect the rate of sedimentation in a suspension? (Assume that the density of the particles is greater than the density of the dispersing liquid.)
   (a) The sedimentation rate will not change.
   (b) The sedimentation rate will be higher.
   (c) The sedimentation rate will be slower.
   (d) Particle sedimentation will not take place.
4. Which one of the following phenomena is undesirable in pharmaceutical suspensions:
   (a) Slow settling of particles.
   (b) Particles agglomerate to dense crystals.
   (c) Particles readily redisperse upon agitation.
   (d) The suspension pours readily and evenly.
5. You have been asked to prepare a suspension with the following formula. You prepare this prescription and notice that the suspension settles extremely fast, such that within 30 seconds it has completely settled.

Tracmicin:	250 mg/ml
2% methylcellulose:	10 ml
Vanillin (flavoring agent):	0.2%
Water qs ad:	100 ml

   Which of the following techniques could be used to decrease the rate of settling:
   (a) You could increase the concentration or volume of methycellulose used in the formulation.
   (b) You could add a viscous liquid, such as glycerin, to this suspension.
   (c) You could decrease the particle size by levigating the Tracmicin in a mortar and pestle.
   (d) All of the above.
6. Emulsions are not stable systems; coalescence of droplets will occur. The process of coalescence can be reduced to insignificant levels by:
   (a) Decreasing the difference between the density of the dispersed phase and the density of the medium
   (b) Adding an agent that reduces the viscosity of the medium
   (c) Increasing the droplet size of the dispersed phase
   (d) All of the above

7. In oil-in-water (O/W) types of emulsions:
   (a) Water or an aqueous solution is the medium and oil is the dispersed phase.
   (b) Oil is the medium and water or an aqueous solution is the dispersed phase.
   (c) Oil is the medium and the emulsion of oil is the dispersed phase.
   (d) Oil is the medium and the emulsion of an aqueous solution is the dispersed phase.
8. The emulsion that permits the palatable administration of an otherwise distasteful oil, such as castor oil, is:
   (a) An oil-in-water emulsion
   (b) A water-in-oil emulsion
   (c) Both water-in-oil and oil-in-water emulsions
   (d) None of the above
9. According to the concept of the hydrophilic-lipophilic balance (HLB) system, the emulsifying agents in dosage forms should be selected to have:
   (a) The same or nearly the same HLB value as the aqueous phase of the intended emulsion
   (b) The same or nearly the same HLB value as the oily phase of the intended emulsion
   (c) An HLB value that is the average of the HLB values of the oily and aqueous phases of the intended emulsion
   (d) An HLB value that is the product of the HLB values of the oily and aqueous phases of the intended emulsion
10. Increasing the surfactant concentration above the critical micellar concentration will result in:
   (a) An increase in surface tension
   (b) A decrease in surface tension
   (c) A change in the number of micelles
   (d) All of the above

ANSWERS

1. (b)
2. (d)
3. (c)
4. (b)
5. (d)
6. (a)
7. (a)
8. (a)
9. (b)
10. (c)
11. A new formulation of a drug in the form of an externally applied medication (ointment) employs hydrophilic ointment as a base. The label lists the following inactive ingredients: Sodium laurylsulfate, butylparaben, stearyl alcohol, white petrolatum, water, and propylene glycol.

Which ingredient can be classified into the categories below:
(a) Aqueous (W) phase of the ointment: ________________
(b) Oily (O) phase of the ointment: ________________
(c) Emulsifying agent: ________________

Answers:
(a) Water and propylene glycol
(b) White petrolatum
(c) Sodium lauryl sulfate and stearyl alcohol

12. SILVADENE Cream 1% is a soft, white, water-miscible cream containing the antimicrobial agent silver sulfadiazine in micronized form, which has the following structural formula:

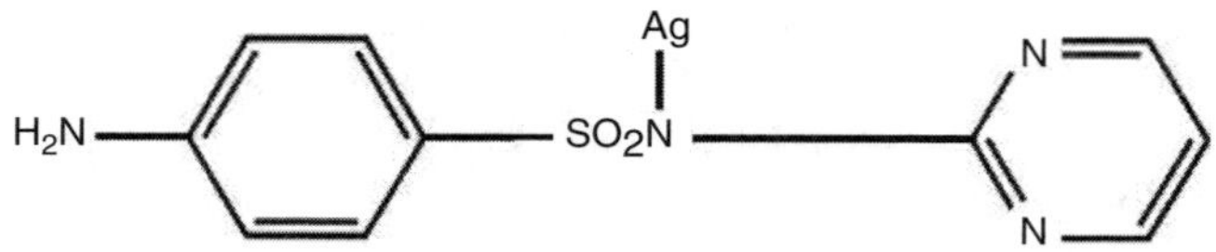

Each gram of SILVADENE Cream 1% contains 10 mg of micronized silver sulfadiazine. The cream vehicle consists of white petrolatum, stearyl alcohol, isopropyl myristate, sorbitan monooleate, polyoxyl 40 stearate, propylene glycol, and water, with methylparaben 0.3% as a preservative. SILVADENE Cream 1% (silver sulfadiazine) spreads easily and can be washed off readily with water.

Which ingredient of the vehicle can be classified into the categories below:
(a) Aqueous (W) phase of the ointment ________________
(b) Oily (O) phase of the ointment ________________
(c) Emulsifying agent ________________
(d) The cream base is an emulsion. Based on the above description, is it a water-in-oil (W/O) or an oil-in-water (O/W) emulsion?

Answers:
(a) Water and propylene glycol
(b) White petrolatum and isopropyl myristate
(c) Sorbitan monooleate and polyoxyl 40 stearate
(d) Oil in water (O/W) emulsion

13. How many grams of Span 80 (HLB 4.3) and Tween 80 (HLB 15) must be mixed to make 30 g of surfactant mixture with an HLB of 12.0?

Answers: 8.4 g of Span 80 and 21.6 g of Tween 80

14. Calculate the required HLB for the oil phase of the following O/W emulsion:

Ingredient	**Amount**	**Required HLB**
Cetyl alcohol	15.0 g	15
White wax	1.0 g	12
Anhydrous lanolin	2.0 g	10
Emulsifier, qs.	—	—
Glycerin	5.0 g	—
Distilled water ad	100.0 g	—

Answer: 14.3

CASES

CASE I

You are required to prepare an oral antacid suspension preparation with the following ingredients:

Aluminum hydroxide
Magnesium hydroxide
Syrup USP
Glycerin
Sorbitol solution
Methylparaben
Propylparaben
Flavor, color
Purified water

1. Explain how you will prepare the suspension.
2. What will be the nature of the suspension?
3. Explain the purpose of each ingredient in the formulation.

ANSWERS:

1. The parabens should be dissolved in a heated mixture (≈60°C) of the sorbitol solution, glycerin, syrup, and a portion of the water. The mixture is then cooled and the aluminum hydroxide and magnesium hydroxide added slowly with constant stirring. The flavor and color are added and finally volume is made up with sufficient purified water. The suspension is then homogenized, using a hand homogenizer or colloid mil.
2. As the formulation does not have any flocculating agent, it will be a deflocculated system stabilized in a viscous medium.
3. Aluminum hydroxide and magnesium hydroxide are active drug substances. Syrup USP, glycerin, and sorbitol solution impart viscosity to the vehicle as well as sweet taste to the preparation. Methylparaben and propylparaben are antimicrobial preservatives. Flavor and color add flavor and color to the preparation.

CASE II

You have been asked to prepare an emulsion with the following ingredients:

Mineral oil	500 ml
Acacia	125 g
Syrup	100 ml
Vanillin	30 m
Alcohol	50 ml
Purified water, qs	1000 ml

1. Explain how you will prepare the emulsion.
2. What will be the nature of the emulsion?

Answers:

1. By looking at the quantity (volume) of the aqueous phase, oil phase, and acacia, it appears that the dry gum method using the proportion of 4:2:1 will be the method of choice. The oil should be mixed with acacia first, followed by the addition of 250 ml of water all at once to form the primary emulsion. To this slowly add with trituration the remainder of the ingredients, with the vanillin dissolved in the alcohol and make up the final volume with water.
2. It will be an O/W emulsion, as acacia is soluble more in water.

CASE III

A powder with a density of 1.3 g/cc and an average particle diameter of 2.5 μm settles in water (viscosity 1 cps) at a rate of 1.02×10^{-4} cm/sec.

1. If the particle size of the powder is reduced to 0.25 μm, what will be the rate of settling through water?
2. If the same powder (2.5 μm) is now allowed to settle through a different dispersion medium such as glycerin (density 1.25 g/cc and viscosity 400 cps), what will be the rate of settling?
3. If the powder with smaller particle size (0.25 μm) is now allowed to settle through glycerin (density 1.25 g/cc and viscosity 400 cps), what will be the rate of settling?
4. What conclusion can you draw from the above results?

Answers:

1. 1.02×10^{-6} cm/sec
2. 4.25×10^{-8} cm/sec
3. 4.25×10^{-10} cm/sec
4. Particle size reduction and change in dispersion medium result in lowering the settling rate of the particles and thereby stabilizing the formulation.

REFERENCES

1. Adamson A. W., Gast A. P., *Physical Chemistry of Surface*, 6th ed., John Wiley & Sons, New York, 1997.
2. Ansel H. C., Allen L. V. Jr., Popovich N. G., *Pharmaceutical Dosage Form and Drug Delivery Systems*, 7th ed., Lippincott Williams & Wilkins, Philadelphia, 1999.
3. Becher P., *Emulsion: Theory and Practice*, Krieger, New York, 1977.
4. Buckton G., Interfacial phenomenon in *Drug Delivery and Targeting,* Harwood Academic Publishers, 1995.
5. Davies J. T., Rideal E. K., *Interfacial Phenomena*, 2nd ed. Academic Press, New York, 1963.
6. Gemnaro A. R., Ed., *Remington: The Science and Practice of Pharmacy*, 20th ed., Lippincott Williams & Wilkins, Philadelphia, 2000.
7. Martin A, Bustamante P., *Physical Pharmacy*, 4th ed., Lippincott Williams & Wilkins, Philadelphia, 1993.
8. Sutheins G. M., *Introduction to Emulsions*, Chemical Publishing Co., New York, 1947.
9. Thompson J. E., *A Practical Guide To Contemporary Pharmacy Practice*, Williams & Wilkins, Baltimore, 1998.
10. Tsujii K., *Surface Activity: Principles, Phenomena, and Applications,* Academic Press, New York, 1998.

7 Theory and Applications of Diffusion and Dissolution

Xiaoling Li and Bhaskara Jasti

CONTENTS

Learning Objectives

After finishing this chapter the student will have thorough knowledge of:

- Fick's law and diffusion of molecules across single layer barriers
- Diffusion of molecules through multiple layers
- Experimental methods to determine diffusion
- Partition coefficient and its role in the diffusion of molecules
- Dissolution phenomenon
- Noyes–Whitney and Hixon–Crowell equations
- Application of the theory of diffusion and dissolution to drug transport and dosage form design

I. INTRODUCTION

Diffusion governs the transport of the great majority of drugs across various biological barriers after administration. The theory of diffusion has been used in investigating the mechanism of drug transport as well as applied to the design and development of various controlled or sustained release

0-4152-8863-0/05/$0.00+$1.50
© 2005 by CRC Press LLC

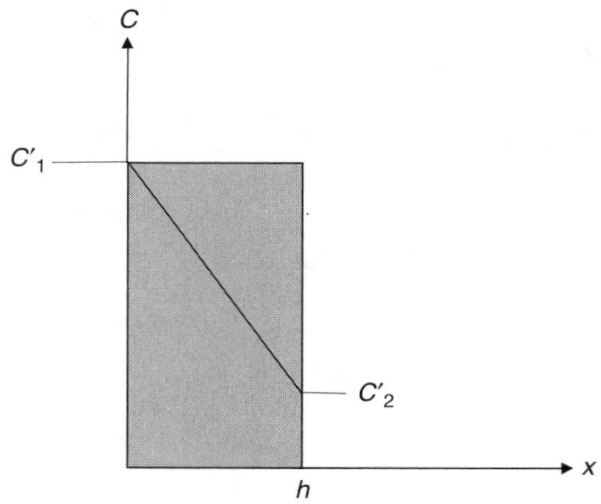

FIGURE 7.1 Schematic representation of a diffusion system: concentration gradient in a cross section of diffusional barrier.

drug delivery systems. Understanding of the theoretical principles provides a fundamental knowledge of drug transport across the biological membranes and developing controlled drug delivery systems.

II. DIFFUSION

Diffusion can be defined as a process by which molecules transfer spontaneously from a region of higher concentration to a region of lower concentration. Diffusion is a result of random molecular motion. Although diffusion of molecules with a wide spectrum of physicochemical properties occurs in various conditions and situations, the diffusion process can be abstracted to a simple system involving the molecules of interest, a diffusional barrier, and a concentration gradient within the barrier. The molecules that migrate from one location to another location are termed *diffusants*, also called *permeants* or *penetrants*. The medium in which the diffusant migrates is called the *diffusional barrier*. The *concentration gradient* is the concentration profile of the diffusant in the diffusional barrier. The concentration gradient is the driving force for diffusion. Molecules move from a high concentration region to a low concentration region in all directions. In Figure 7.1, the diffusion process is schematically described for a given direction, along the x-axis. In this system, the molecules with a concentration of C_1' diffuse through a diffusional barrier with a thickness of h. The concentration gradient of the diffusant is represented by the concentration profile, the line between C_1' and C_2'.

The number of molecules that diffuse through a unit area of the diffusional barrier in a given time is termed *flux*, J. Flux is a measure of the rate of molecular diffusion through the diffusional barrier.

A. Fick's Law

In 1800 Bertholot discovered a proportional relationship between the flux and the concentration gradient. This relationship was expressed as

$$J \propto \frac{dC}{dx} \tag{7.1}$$

In 1855, Fick described diffusion of molecules in quantitative terms, analogous to the conduction of heat. The quantitative description of diffusion through a given unit area, expressed as follows, is called *Fick's first law.*

$$J = -D\frac{dC}{dx} \tag{7.2}$$

where D is the diffusion coefficient, or diffusivity, and dC/dx is the concentration gradient. The negative sign in the equation indicates that the direction of molecular movement is opposite to the increase in the concentration of diffusants. When the area for diffusion (A) is not a unit area, Fick's first law is written as

$$J = -\frac{D}{A}\frac{dC}{dx} \tag{7.3}$$

The unit for flux is the amount of diffusant per unit area per unit time (grams/square centimeters/seconds). The units for concentration and distance are the amount of diffusant per unit volume (grams/cubic centimeters) and length (centimeters). Hence, the unit for diffusion coefficient is $(\text{length})^2/\text{time}$. A typical unit for the diffusion coefficient is square centimeters/seconds.

When the concentration changes of diffusant at a given location in the diffusional barrier is of interest, Fick's first law is unable to describe the diffusion process. To describe the time-dependent diffusion process, the changes in concentration of diffusant with time (dC/dt) in a unit volume element along the x-direction can be expressed by the following partial differential equation:

$$\frac{dC}{dt} = D\left(\frac{d^2C}{dx^2}\right) \tag{7.4}$$

For a concentration change in the three-dimensional system shown in Figure 7.2 (concentration change in a specific location labeled as x,y,z), the time-dependent diffusion is expressed by the general form of Fick's second law:

$$\frac{\partial C}{\partial t} = D\left(\frac{\partial^2 C}{\partial x^2} + \frac{\partial^2 C}{\partial y^2} + \frac{\partial^2 C}{\partial z^2}\right) \tag{7.5}$$

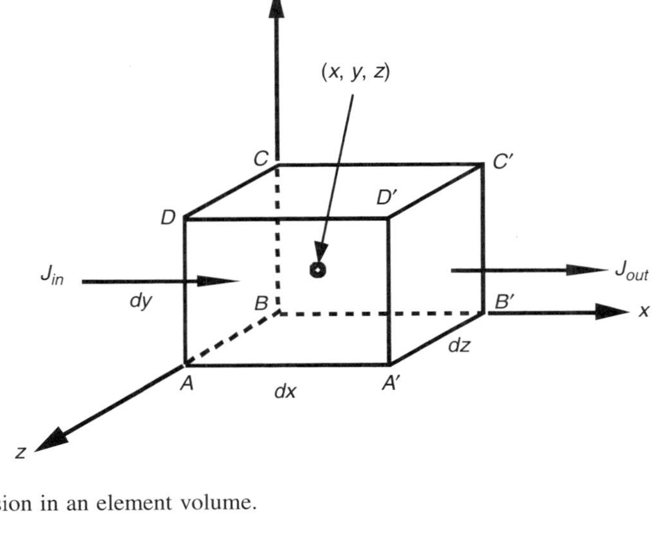

FIGURE 7.2 Diffusion in an element volume.

A general solution of Fick's second law for diffusion along the x-direction can be obtained by solving Equation 7.4:

$$C = \sum_{m=1}^{\infty} \left(A_m \sin \lambda_m x + B_m \cos \lambda_m x \right) e^{-\lambda_m^2 Dt} \tag{7.6}$$

where A_m, B_m, and λ_m are constants and can be determined by the initial and boundary conditions of a specific diffusion system. A solution of Equation 7.4 as expressed in Equation 7.6 gives the concentration as a function of the location (x) and time (t). The solution can be used to calculate the concentration of diffusant at a given time and location or time required to reach a specific concentration for a given location.

B. Nonsteady State and Steady State Diffusion

Nonsteady state diffusion refers to a diffusion process in which the concentration of diffusant in a given space (ABCD-A′B′C′D′ in Figure 7.2) is a function of time or the concentration of diffusant in the diffusional barrier varies with time. In this case, one should use Fick's second law to study the diffusion process. Mathematically, this condition can be described as

$$\frac{dC}{dt} \neq 0 \tag{7.7}$$

When the amounts of diffusant enter and leave the given space (ABCD-A′B′C′D′ in Figure 7.2) at the same rate, the concentration of the diffusant in the given volume is a constant. The concentration of diffusant in the given space is independent of time. The diffusion process that meets this condition is considered a *steady state* diffusion. The mathematical expression of this condition is

$$\frac{dC}{dt} = 0 \tag{7.8}$$

Applying this condition to Equation 7.4, we get

$$\left(\frac{d^2C}{dx^2} \right) = \frac{d}{dx} \left(\frac{dC}{dx} \right) = 0 \tag{7.9}$$

When a derivative of a variable is equal to zero, this variable must be a constant. In other words, the concentration gradient is a constant for diffusion at the steady state. Therefore, diffusion at the steady state gives a constant flux.

III. DISTRIBUTION OR PARTITION COEFFICIENT

Distribution or partition coefficient is a measure of the ability of a compound to distribute in two immiscible phases. The partition phenomenon is of paramount importance for the diffusion across skin and other epithelia. Since many other pharmaceutical processes, such as absorption from the gastrointestinal tract after oral administration, protein binding, drug distribution following entry into systemic circulation, extraction and isolation of pure drugs after synthetic manufacturing or from crude plant sources, formulation of a stable dosage form (emulsion, suspensions, etc.), and assay of plasma concentrations, are also based on the partition principles, it is important to study this phenomenon.

The ability of drugs to penetrate a biological membrane has traditionally been evaluated using its partition in an octanol (representing lipid materials) and water system. Occasionally, other organic solvents such as chloroform, ether, and hexane have been used as a lipid vehicle instead of octanol to evaluate drug partition behavior. When a drug is placed below its saturation in an immiscible system composed of octanol and water, the drug distributes in each solvent and eventually reaches equilibrium. The ratio of drug concentration in each phase is termed its *distribution coefficient* or *partition coefficient* (*K*) and can be expressed as:

$$K = \frac{\text{Concentration in Octanol}, C_o}{\text{Concentration in Water}, C_w} \quad (7.10)$$

Practically, concentration in an aqueous phase is determined by chemical assays, such as high performance chromatography, ultra violet spectroscopy, gas chromatography, gas chromatography–mass spectroscopy, etc., and the concentration in octanol or other organic phases is deduced by subtracting the aqueous amount from the total amount placed in the solvents.

Example: Succinic acid (0.15 g) dissolved in 100 ml of ether was shaken with 10 ml of water at 37°C. After equilibrium was achieved, the water layer contained 0.067 g of succinic acid. What is the partition coefficient (K) of succinic acid?

$$\text{Concentration in water} = \frac{0.067\text{ g}}{10\text{ mL}} = 0.0067\text{ g/mL and}$$

$$\text{Concentration in ether} = \frac{(0.15 - 0.067)\text{ g}}{100\text{ mL}} = 0.083\text{ g/mL}$$

$$\text{So, } K = \frac{0.083\text{ g/mL}}{0.0067\text{ g/mL}} = 0.124$$

Note: If both solvents are used in equal volume, *K* can be expressed as a ratio of the amount in each phase instead of a concentration.

IV. DIFFUSION COEFFICIENT AND PERMEABILITY COEFFICIENT

A diffusion coefficient represents the mobility of a molecule in a specific medium, the diffusional barrier in our discussion. The mobility of a substance in a diffusional barrier is determined by the physicochemical properties of the diffusant and the diffusional barrier and temperature. The relationship of the diffusion coefficient and these factors is expressed quantitatively in the Strokes–Einstein equation, which states the diffusion coefficient as a function of temperature (T), viscosity (η), and size of the diffusant (r):

$$D = \frac{k\text{T}}{6\pi\eta r} \quad (7.11)$$

where k is the Bolzmann constant.

TABLE 7.1
Diffusion Coefficients and Permeability of Various Chemicals in Different Media

Substances	Diffusional Barriers	Diffusion Coefficients D (cm^2/sec)	Permeability P (cm/sec)	References
Water	Human skin (37°C)	2.8×10^{-10}	2.8×10^{-7}	1
Oxygen	PVC[a] (25°C)	1.2×10^{-8}		2
	LDPE[b] (25°C)	4.6×10^{-7}		2
Glucose	Water (25°C)	6.7×10^{-6}		3
Ethanol	Water (25°C)	1.24×10^{-5}		3
	Human skin (32°C)		7.9×10^{-4}	4
Siquinavir	Caco-2 cell		1.5×10^{-6}	5
Scopolamine	Pig skin (37°C)		5.1×10^{-7}	6
Estradiol	Human skin (32°C)		2.8×10^{-7}	7
Testosterone	Human buccal (37°C)		1.8×10^{-6}	8
Nicotine	Human skin (37°C)		2.2×10^{-8}	9
Clonidine	Rat ileum (37°C)		4.3×10^{-5}	10
Fentanyl	Human skin (37°C)		2.9×10^{-6}	11
Nitroglycerine	Mouse skin (32°C)		5.6×10^{-6}	12
Piroxicam	Human jejunum (*in vivo*)		4.6×10^{-4}	13
Peptide (GMDP)	Rabbit ileum (37°C)		1.7×10^{-6}	14

[a] Unplasticized poly (vinyl chloride)
[b] Low density poly (ethylene)

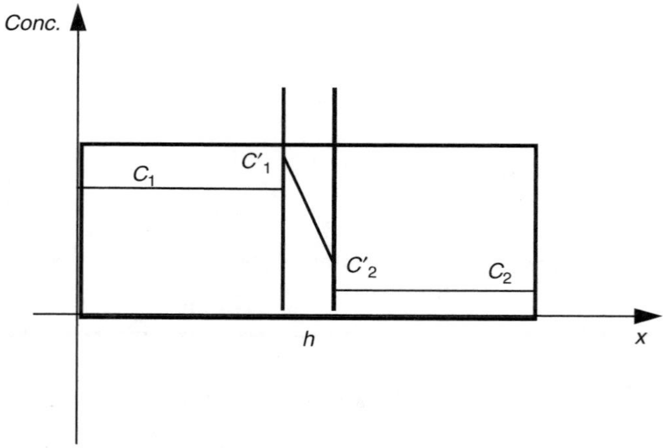

FIGURE 7.3 Diffusion across a membrane.

The diffusion coefficient in Equations 7.3 to 7.5 is treated as a constant. This is valid for homogeneous or isotropic diffusional barriers. The diffusion coefficients of the same diffusant vary in different diffusional barriers. Examples of diffusion coefficients of chemicals of pharmaceutical interest are given in Table 7.1.

Diffusion coefficients can be determined experimentally or calculated using theoretical or empirical models. A set-up for studying diffusion through a diffusional barrier is schematically illustrated in Figure 7.3. The set-up has two compartments divided by the diffusional barrier, for example, a membrane with a thickness of h. The concentrations of diffusion outside of a diffusional barrier or the concentrations of diffusant in each compartment are denoted C_1 and C_2. The concentrations of diffusant in the diffusional barrier are denoted C_1' and C_2' for each side of the barrier. C_1 and C_2 can be determined experimentally, but C_1' and C_2' are usually not known. To apply Fick's laws, one must know C_1' and C_2'. The concentrations of a diffusant in the diffusional barrier and the adjacent medium can be related by using partition coefficients (K) as follows:

$$K = \frac{\text{Conc. in membrane}}{\text{Conc. in solution phase}} = \frac{C_1'}{C_1} = \frac{C_2'}{C_2} \tag{7.12}$$

Therefore, the concentrations of diffusant in the diffusional barrier can be expressed as

$$C_2' = KC_1 \tag{7.13}$$

$$\text{and } C_2' = KC_2 \tag{7.14}$$

By replacing the concentrations of diffusant in the barrier by using these relationships, Fick's first law can be rewritten as

$$J = -\frac{DK}{h}\left(C_2 - C_1\right) \tag{7.15}$$

Since D, K, and h are constant for a given diffusional barrier, a constant *permeability* or *permeability coefficient* (P), is defined as

$$P = \frac{DK}{h} \tag{7.16}$$

The dimension of permeability coefficient is length per unit time, for example, centimeters per second. Permeability coefficient is commonly used in pharmaceutical sciences to determine or compare the permeation of a drug through various biological barriers. Examples of permeability coefficients of drugs permeating through different barriers are given in Table 7.1.

V. EXPERIMENTAL METHODS

The diffusion process can be studied by using various methods, such as the permeation method, sorption and desorption kinetics, determination of concentration profile, etc. The most commonly used method in pharmaceutical research is the permeation method. The experimental set-up for this method consists of two chambers separated by a diffusional barrier. Two types of diffusion cells used in the permeation studies are shown in Figure 7.4. A drug solution is charged to the donor chamber. The solution of the receiver chamber is removed partially or replenished with solvent or buffer solution at predetermined time intervals. When the receiver chamber is replenished with a solvent or a solution without a drug, the concentration of drug in the receiver chamber is maintained at a minimum level ($C_2 \approx 0$). This is called *sink condition*. To maintain sink condition, the concentration of a permeant in the receiver chamber is generally kept below 10% of its concentration in the donor chamber. In sink condition, Fick's first law can be simplified as

$$J = \frac{DK}{h} C_1 \tag{7.17}$$

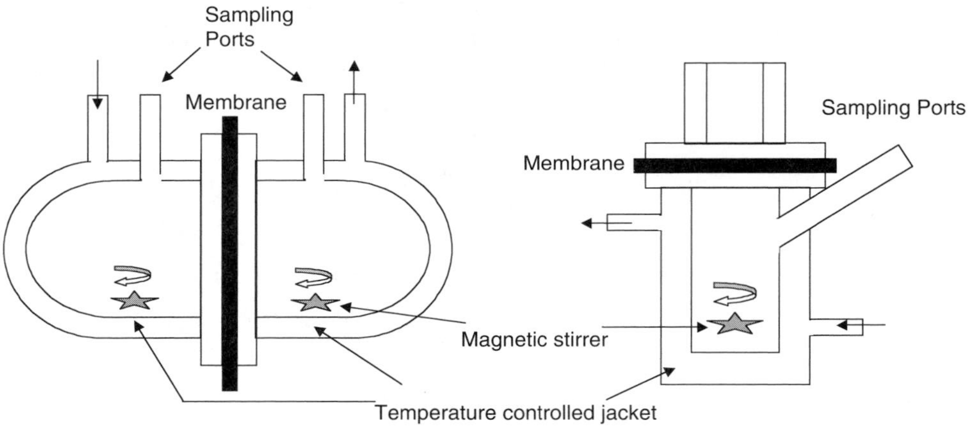

FIGURE 7.4 Side-by-side and vertical diffusion cells.

or

$$J = PC_1 \tag{7.18}$$

The amount of diffusant permeating the diffusional barrier (the membrane in this case) is determined quantitatively by chemical analysis, as described previously. Mathematically, the amount of cumulative permeation of diffusant (Q) can be derived from integration of Equation 7.19 over the time of diffusion.

$$J = \frac{dQ}{dt} \cdot \frac{1}{A} = \frac{DK}{h} C_1 \tag{7.19}$$

$$Q = PAC_1 t \tag{7.20}$$

Since the cumulative amount permeating the barrier (Q) at a given time (t) can be quantified and the concentration of donor chamber and the diffusional area are usually known, the permeation coefficient (P) can be obtained from the slope of a plot of cumulative permeation of diffusant vs. time. A typical plot of permeation study is shown in Figure 7.5.

As shown in Figure 7.5, the cumulative permeation curve has two portions. The initial portion of the curve represents nonsteady state diffusion, and the linear portion corresponds to steady state diffusion. The nonsteady state portion of the curve can be described mathematically by Fick's second law with defined initial and boundary conditions, while the linear portion can be expressed by Fick's first law. Figure 7.6, illustrates the diffusion process in a membrane by visualizing the concentration profile in the diffusional barrier at a different time. As time increases, the diffusant (dark color) diffuses into the membrane and eventually permeates through the membrane reaching the receiver chamber. When the diffusant appears in the receiver chamber, its amount increases gradually, which results in an increase in the cumulative amount of diffusant as shown in the Figure 7.5. Once the diffusion reaches steady state, the concentration gradient remains constant and the amount entering the receiver chamber is also constant, which results in a linear increase in the cumulative amount of diffusant permeated.

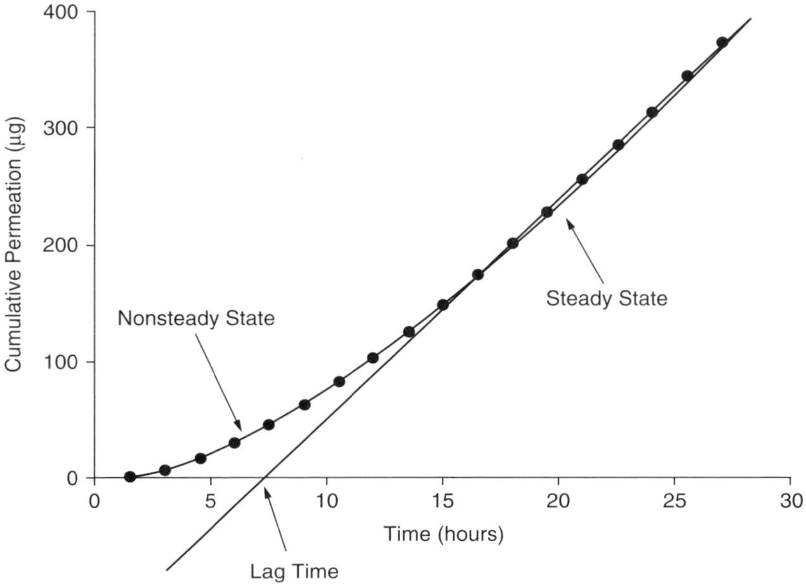

FIGURE 7.5 Diffusion of lidocaine through poly(vinyl alcohol acetate) membrane.

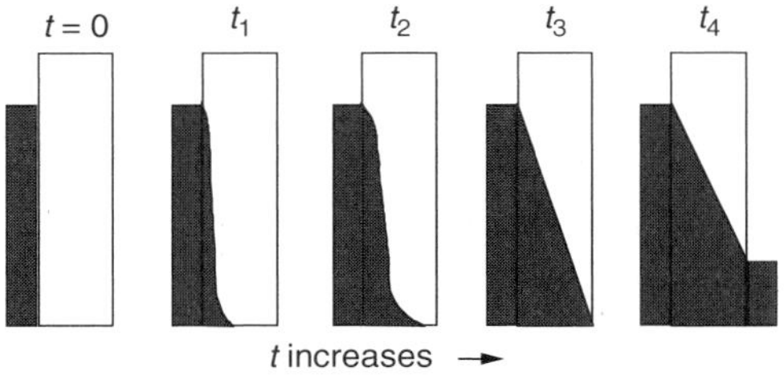

FIGURE 7.6 Diffusion through membrane at different times.

The time required to reach steady state is called the *lag time* (t_L). The lag time can be determined by extrapolating the linear portion of permeation vs. the time curve to the time axis. With lag time, Equation 7.20 is rewritten as

$$Q = \frac{DKAC_1}{h}\left(t - t_L\right) \tag{7.21}$$

or

$$Q = PAC_1\left(t - t_L\right) \tag{7.22}$$

The lag time can be calculated by

$$t_L = \frac{h^2}{6D} \tag{7.23}$$

Example: Diffusion of lidocaine through a 9% EVA membrane (0.051 mm) was studied by using a side-by-side diffusion apparatus with a diffusional area of 1 cm² at 32°C. The donor chamber was filled with a saturated lidocaine aqueous solution (3.9 mg/ml). Samples were withdrawn every 90 minutes from the receiver chamber and quantified by HPLC analysis. The cumulative permeation at different time points was determined as shown in the following table.

Time (hours)	Cumulative Permeation (μg)
1.5	1.29
3.0	5.82
4.5	15.97
6.0	29.49
7.5	45.19
9.0	62.86
10.5	82.38
12.0	103.32
13.5	125.32
15.0	148.61
16.5	174.38
18.0	200.89
19.5	228.08
21.0	256.08
22.5	284.49
24.0	313.37

1. Calculate the permeability of lidocaine across the EVA membrane. Cumulative permeation vs. time is plotted as shown in Figure 7.5. Data points at steady state are identified. A linear regression of these data points gives the following equation:

$$Q = 18.3t - 127.9$$

$$\text{Slope} = PAC_1 = 18.3$$

$$P = \frac{\text{Slope}}{AC_1} = \frac{18.3\,\mu\text{g/hr}}{1\,\text{cm}^2 \times 3.9\,\text{mg/cm}^3} = \frac{18.3\times10^{-3}\,\text{mg}/3600\,\text{sec}}{1\,\text{cm}^2 \times 3.9\,\text{mg/cm}^3} = 1.3\times10^{-6}\,\text{cm/sec}$$

2. Calculate the time lag for permeation of lidocaine through the EVA membrane.

$$t_L = \text{intercept of x axis} = \frac{127.9}{18.3} = 7.0\,\text{hr.}$$

3. Calculate the diffusion coefficient of lidocaine in the EVA membrane.

$$\text{Since } t_L = \frac{h^2}{6D},$$

$$D = \frac{h^2}{6t_L} = \frac{0.051\times10^{-2}\,\text{cm}}{6\times7\times3600\,\text{sec}} = 3.4\times10^{-9}\,\text{cm/sec}$$

4. Calculate the partition coefficient of lidocaine between water and the EVA membrane.

$$\text{Since } P = \frac{DK}{h},$$

$$K = \frac{Ph}{D} = \frac{1.3 \times 10^{-6}\,\text{cm/sec} \times 0.051 \times 10^{-2}\,\text{cm}}{3.4 \times 10^{-9}\,\text{cm/sec}} = 0.2$$

It is essential to control the experimental conditions to obtain a reproducible result in the permeation studies. Factors that should be maintained constant or kept consistent are temperature, medium or solvent, and agitation of liquid in contact with the diffusional barrier. Owing to the formation of stagnant liquid layers or diffusion layers adjacent to the diffusional barrier or membrane, the effective thickness of the diffusional barrier increases as shown in Figure 7.7. The thickness of a stagnant layer can be altered by the hydrodynamic movement of liquid surrounding the membrane. The stagnant layer can be minimized or diminished by the movement of liquid contents next to the diffusional barrier. In an *in vitro* permeation study, this is achieved by stirring. In a biological system, such as the gastrointestinal tract, gastrointestinal movement can effectively minimize the stagnant layer in the drug absorption process.

VI. DIFFUSION THROUGH BARRIERS WITH PARALLEL PATHWAYS OR MULTIPLE LAYERS

When two or more independent diffusional pathways are present in a diffusional barrier in a parallel format (Figure 7.8a), the total flux or apparent flux (J_T) is the sum of fluxes of each individual pathway.

$$J_T = J_1 + J_2 + \cdots + J_n \tag{7.24}$$

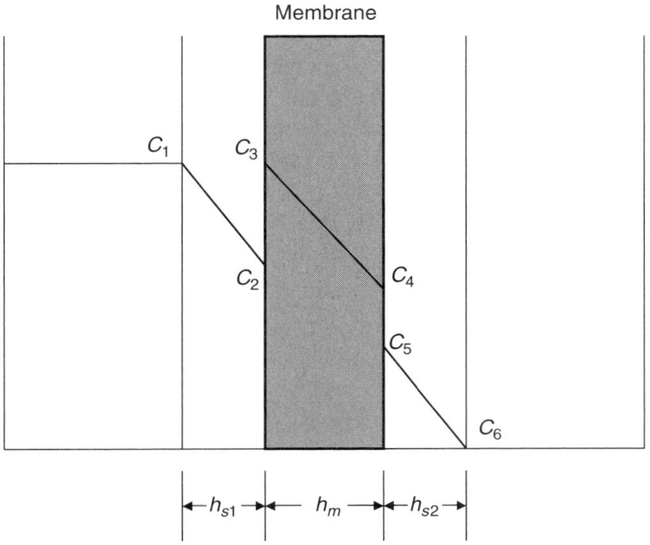

FIGURE 7.7 Stagnant layers in a membrane permeation system.

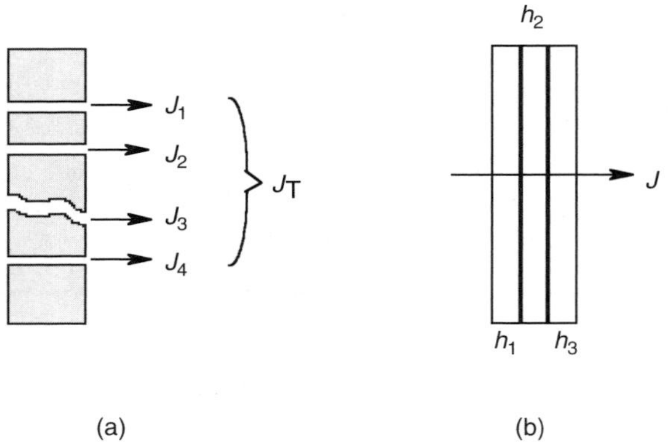

FIGURE 7.8 (a) Diffusional barrier with parallel independent pathway, (b) Multi-layer diffusional barrier.

The above equation can be written in an extended form by replacing J with Fick's first law as follows:

$$J_T = f_1 D_1 \frac{dC_1}{dx} + f_2 D_2 \frac{dC_2}{dx} + \cdots + f_n D_n \frac{dC_n}{dx} \tag{7.25}$$

where the f_1, f_2, ... and f_n are the fractional areas of each route or pathway.

Another type of diffusional barrier commonly seen in drug delivery systems and biological systems consists of more than one physically distinct layer. The diffusants travel through each layer in serial (Figure 7.8b). The permeability of a triply diffusional barrier can be written as

$$P_T = \frac{D_1 D_2 D_3 K_1 K_2 K_3}{h_1 D_2 K_2 D_3 K_3 + h_2 D_1 K_1 D_3 K_3 + h_3 D_1 K_1 D_2 K_2} \tag{7.26}$$

where h, D, and K are thickness, diffusivity, and partition coefficient, respectively. Each layer is denoted by subscripts. The flux of diffusant across the triply diffusional barrier can be calculated by using Fick's law with the permeability coefficient (P_T).

For parallel pathway diffusion or multiple layer diffusion, one of the pathways or one of the layers may play a dominant role, as a diffusional barrier and the diffusant will take the least resistant pathway, which becomes a dominant pathway in the transport process. When the dominant pathway in diffusion is responsible for the diffusion of diffusant, contributions from other pathways may be negligible and can be omitted. Therefore, Equation 7.25 can be simplified as

$$J_T = f_D D_D \frac{dC_1}{dx} \tag{7.27}$$

for a parallel diffusion system. Subscript D denotes the parameters for the dominant parallel pathway.

The roles of each layer in multilayer diffusion can be illustrated by using permeation of diffusant through a membrane with a stagnant layer. As shown in Figure 7.7, the diffusional barrier consists of a membrane layer and two aqueous stagnant layers. By using appropriate symbols for this specific case, Equation 7.26 is written as

$$P_T = \frac{D_s D_m D_s K_m}{h_{s1} D_m K_m D_s + h_m D_s D_s + h_{s2} D_s D_m K_m} = \frac{D_s D_m K_m}{h_m D_s + \left(h_{s1} + h_{s2}\right) D_m K_m} \tag{7.28}$$

where h, D, and K represent thickness, diffusion coefficient, and partition coefficient, respectively. Subscripts s and m denote stagnant layer and membrane.

The diffusion boundary layers can be minimized by stirring. So, $h_m >> (h_{s1} + h_{s2})$, and the permeation coefficient becomes

$$P_T = \frac{D_s D_m K_m}{h_m D_s} = \frac{D_m K_m}{h_m} \tag{7.29}$$

Therefore, the membrane is a dominant barrier for diffusion. If the stagnant layers play a significant role as a diffusional barrier, the contribution of the stagnant layers cannot be ignored. The permeation coefficient should be calculated based on Equation 7.28. In the extreme case $[(h_{s1} + h_{s2}) >> h_m]$, the permeation coefficient becomes

$$P_T = \frac{D_s}{\left(h_{s1} + h_{s2}\right)} \tag{7.30}$$

Under this condition, aqueous stagnant layers are the primary barrier for diffusion.

VII. DISSOLUTION

Dissolution is the process by which molecules leave the solid phase and enter into solution. In 1897, Noyes and Whitney established a quantitative description for dissolution as follows

$$\frac{dM}{dt} = \frac{DA}{h}\left(C_s - C_t\right) \tag{7.31}$$

where dM/dt is the rate of dissolution, D is the diffusion coefficient in the diffusion layer, h is the thickness of the diffusion layer, A is the surface area of the drug particles, C_s is the saturation concentration of the drug in the diffusion layer, and C_t is the concentration of drug in bulk fluids at time t. A schematic illustration of the dissolution of a solid is shown in Figure 7.9.

Equation 7.31 is based on an assumption that the diffusion layer, also called stagnant liquid film, present at the surface of the solid phase is a stationary layer with a thickness of h. The diffusion layer acts as a barrier for the diffusion of a solute or a drug before they enter solution. Solute is in saturated solution at the interface of a solid phase and a diffusion layer as indicated by Cs in Figure 7.9. However, the concentration of solute at the interface of solution and a diffusion layer is very low and indicated as Ct at time t. Thus, the concentration difference of solute between two sides of a diffusion layer constitutes a concentration gradient. Because both the Noyes–Whitney equation and Fick's first law are derived from diffusion of a diffusant through a diffusional barrier, one can easily identify the similarity of mathematical expressions of these two processes.

When the dissolution occurs under sink condition and controlled agitation, Equation 7.31 is written as

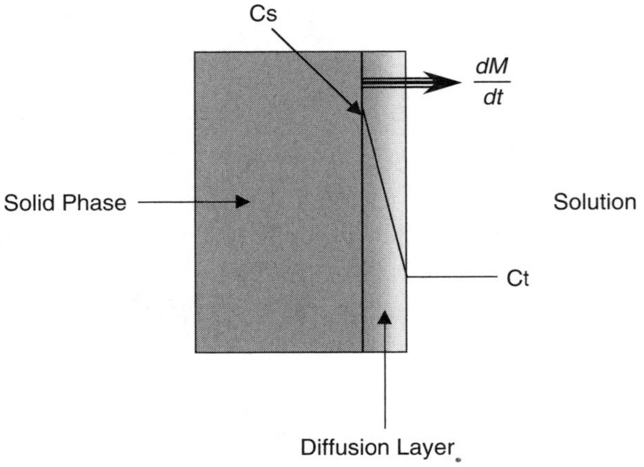

FIGURE 7.9 Dissolution of a drug from a solid.

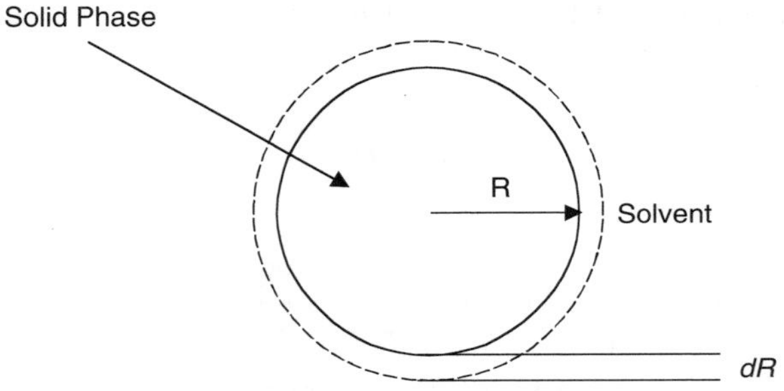

FIGURE 7.10 Dissolution of drug from a drug particle.

$$\frac{dM}{dt} = kAC_s \tag{7.32}$$

where k is the dissolution rate constant, which is a constant under a given temperature and agitation condition in a defined solvent.

Since the dissolution process is often a rate-limiting step in the absorption of drugs from the gastrointestinal tract, the dissolution rate of a drug from the dosage forms has been used as a biopharmaceutical parameter for the evaluation of effectiveness of drug release from a solid drug preparation, such as tablet, capsule, and powder. Following disintegration of a solid dosage form, small particles of drug are formed in liquid. These small particles undergo dissolution and are eventually dissolved in the solution. The official methods of dissolution tests can be found in the USP.

For particles with uniform size distribution, the rate of dissolution can be derived based on a particle, as shown in Figure 7.10.

Assuming the particle in Figure 7.10 is a sphere, it has a radius of R and a surface area of $4R^2$. As dissolution proceeds, the radius of the particle is reduced by dR. The volume of solid lost because of the reduction is

$$dV = 4\pi R^2 dR \tag{7.33}$$

and the weight loss corresponding to the radius reduction is ρdV, where ρ is the density of the solid. The weight loss in the dissolution process can also be described by the Noyes–Whitney equation; therefore,

$$\rho dV = -dM \tag{7.34}$$

or

$$\rho 4\pi R^2 dR = -kAC_s dt \tag{7.35}$$

When N particles are taken into account, the total surface area for N particles is $4\pi R^2 N$. Equation 7.35 then becomes

$$4\pi R^2 N\rho dR = -4\pi R^2 NkC_s dt \tag{7.36}$$

This equation can be simplified as

$$\rho dR = -kC_s dt \tag{7.37}$$

By integrating of Equation 7.37 with an initial condition of $R = R_0$ when $t = 0$, we obtain the relationship between the radius of the particle and time.

$$R = R_0 - \frac{kC_s t}{\rho} \tag{7.38}$$

Since the volume of a sphere is $\frac{4}{3}\pi R^3$, the mass of N spheres at time t is

$$M = \frac{4}{3} N\rho\pi R^3 \tag{7.39}$$

Replacing the radius in Equation 7.38 with the mass of undissolved particles (Equation 7.39) and rearranging the result to express mass as a function of diameter (d) gives

$$M^{1/3} = M_0^{1/3} - M_0^{1/3} \frac{2kC_s}{d\rho} t \tag{7.40}$$

A constant, the cubic root dissolution rate constant κ, was introduced to the above equation; thus

$$M^{1/3} = M_0^{1/3} - \kappa t \tag{7.41}$$

where

$$\kappa = M_0^{1/3} \frac{2kC_s}{d\rho}.$$

The cubic root dissolution rate constant has a dimension of $(\text{mass})^{1/3}/\text{time}$. Equation 7.41 is known as the Hixson–Crowell cubic root law. This expression gives a quantitative description of dissolution in terms of amount dissolved and time.

VIII. APPLICATIONS OF DIFFUSION AND DISSOLUTION

Scientists in the pharmaceutical industry and practice frequently encounter the principles of diffusion and dissolution. For example, absorption of drugs across many biological barriers and membranes, such as the gastrointestinal epithelium, skin, buccal mucosa, nasal mucosa, etc., takes place mainly by passive diffusion. If one observes the fate of tablets after administration, they first break down into small pieces (disintegration), followed by dissolution into the gastrointestinal fluid. Similarly, when capsules and powders are administered orally, dissolution occurs first before the drug is available for absorption. Several sustained and controlled release dosage forms are designed so that the drug release from the dosage form is limited by either diffusion or dissolution.

Two types of drug delivery systems, namely reservoir and matrix systems, have been designed in which diffusion is the rate limiting step in the drug release. In reservoir systems, the drug forms the core of the device surrounded by a polymeric membrane that controls its release. Such systems can be administered by different routes, for example, intrauterine devices (e.g., *Progestasert*®), implants (e.g., *Norplant*®), transdermal patches (e.g., *Transderm-Scop*®), and oculars (e.g., *Ocusert*®). In a matrix system, which is often described as a monolithic device, the drug is uniformly dispersed or dissolved within a polymer as a solid block. The release of the drug from the matrix system is controlled by its diffusion through the polymer matrix. Controlled or sustained release of the drug from delivery systems can also be designed by enclosing the drug in a polymer shell or coat where the dissolution of the polymer limits the drug release. After the dissolution or erosion of the coating, the drug molecules are liberated and become available for absorption. Release of the drug at a controlled rate is accomplished by controlling the thickness of the coating. *Spansule*® systems are a classic example of such dosage forms, where drug molecules are enclosed in beads with varying thickness of coating material to control the timed drug release. The drug enclosed within the thin coatings dissolves first, whereas drug release from the thicker coating will be slower. A complete description of these devices can be found in Chapter 11.

HOMEWORK

1. The diffusion coefficient of a permeant depends upon:
   (a) Diffusional medium
   (b) Diffusional length
   (c) Temperature
   (d) All of the above
   (e) None of the above
2. To simplify the experimental design and obtain reproducible results, a permeation study should be maintained with the following conditions:
   (a) Saturation concentration
   (b) Sink conditions
   (c) No stagnant layer
   (d) (b) and (c)
   (e) (a), (b), and (c)
3. The partition/distribution coefficient value of a drug will help in understanding:
   (a) Its solubility
   (b) Miscibility of different solvents
   (c) Its permeability across biological membranes
   (d) All of the above

4. Which of the following conditions will increase the rate of drug dissolution from a solid dosage form such as tablet?
   (a) Increase in the particle size of the drug
   (b) Increase in the surface area of drug particles
   (c) Increase in disintegration time
   (d) Increase in the amount of excipients to dilute the drug
5. The permeability coefficient of a weak electrolyte through the cell membrane will increase, if:
   (a) The particle size of the drug increases.
   (b) The surface area of drug particles increases.
   (c) The partition coefficient increases by adjusting the pH of media.
   (d) The rate of drug dissolution increases.
   (e) None of the above.
6. A steroid permeates through the EVA membrane with an area of 10.25 cm^2 and a thickness of 0.075 cm at 25°C. The concentration of steroid in the donor chamber is 0.004 mmol/ml. The amount of the steroid across the membrane at steady state is 3.5×10^{-3} mmole in 4.5 h. The lag time is 0.4 h.
   (a) Calculate the permeability coefficient.
   (b) Calculate the diffusion coefficient.
7. Zalcitabine, an anti-HIV agent, was investigated in transbuccal drug delivery studies. The permeabilities of zalcitabine across buccal mucosa with or without permeation enhancer were 5.2×10^{-6} and 1.75×10^{-7} cm/sec, respectively. The target dose of zalcitabine for anti-HIV therapy is 0.0825 mg/h. A reservoir type of transbuccal delivery system was designed to deliver zalcitabine with a reservoir drug concentration of 50 mg/ml. What will be the size (in square centimeters) of transbuccal delivery systems with and without enhancer?
8. A newly developed compound has a density of 1.05, aqueous solubility of 0.01 mg/ml, and a uniform particle size of 120 μm. In a dissolution study, 150 mg of the compound with a total surface area of 72 cm^2 was placed in 1 l of water with a stirring speed of 250 rpm at 37°C. After 20 minutes of dissolution, the undissolved solid was separated and dried to determine the weight. The remaining solid was 50% of initial amounts. The sink condition was maintained throughout the study.
   (a) Calculate the cubic root dissolution rate constant.
   (b) Calculate the dissolution rate constant.
   (c) Calculate the dissolution rate under the experimental conditions.
   (d) How long will it take to dissolve 75% of the compound?

ANSWERS

1. (d)
2. (d)
3. (c)
4. (b)

5. (c)
6. (a) 5.3×10^{-3} cm/sec, (b) 6.5×10^{-7} cm^2/sec
7. With enhancer 0.09 cm^2, without enhancer 2.6 cm^2
8. (a) 0.055 $mg^{1/3}$/min, (b) 0.11 cm/sec, (c) 0.08 mg/sec, (d) 35.7 min

CASES

CASE I: ETHANOL INTOXICATION FROM DIFFUSION THROUGH SKIN

When a mother gave a bath to her 15-month old baby girl, she forgot to check the temperature of the water and splashed water on the baby's forearm (50 cm^2 area). A year after the skin was burnt, the baby was scheduled for plastic surgery to mend the scar with the rest of the skin and prevent cicatricial contraction of the elbow. The night before the surgery, the baby girl, now 2 years, 3 months old was admitted into the hospital, and ethyl alcohol soaked wrappings were put in place on the injury site. Later it was discovered that the nurse applied boric acid into the wrapping accidentally.

On the morning of surgery (8 h later), the baby girl was found in a comatose condition. Physical examination found that her heart rate was 120 to 140 beats/min and body temperature was 38°C. Chemical analysis of blood indicated a sugar level of 15 mg/100 ml with high alcoholic content (0.8 mg/ml). When the wrappings were removed, the skin underneath blistered and partially detached. The child was promptly infused with 10% glucose and recovered in a few hours and was operated upon a year later without further complication.

The child's pediatrician called an on-call pharmacist to find out more about the intoxication. If you happened to be on-call that day, what information could you provide the physician on the following aspects?

1. How much alcohol is present in the child's blood?
2. Calculate the rate of alcohol permeation through the skin.
3. What is the flux of alcohol entering the child's blood circulation? If the wrappings were applied to twice the area in this case, would it increase the total alcohol content in blood?
4. If the concentration of alcohol in the wrappings was 1 mg/ml, what is its permeability coefficient?

Background information: While the reasons for the skin blisters and partial detachment under the wrappings are unknown, high alcohol levels indicated absorption of alcohol. The barrier property of skin is believed to reside entirely in the stratum corneum, the outermost layer of epidermis. The permeability coefficient of ethanol in intact human skin at 32°C is 7.9×10^{-4} cm/sec. It is assumed that ethanol absorption in damaged skin as described in this case is 50 to 100 times greater. The volume of fluids in this child was 1 l (based on body weight calculations).

Since no other form of exposure was mentioned, assume that all the alcohol entered the blood through the skin.

1. The body volume is 1 l, and the total volume of alcohol present in the child was (assuming no elimination or metabolism)

$$1 \text{ l} \times 0.8 \text{ mg/ml} = 800 \text{ mg}$$

2. 800 mg of alcohol permeated the skin in 8 h, so the rate of permeation into the body is 800 mg/8 h = 100 mg/hour.
3. As defined in Equation 7.2, flux is the rate of permeation per unit area. Since the total area of wrapping is 50 cm^2, alcohol flux across the skin in the child was 100 mg/h/50 cm^2 = 2.0 mg/h/cm^2.
4. $J = DK/h\ (C_2 - C_1)$, but $P = DK/h$ and under sink conditions C_1 is negligible, rearranging these equations, $P = J/C$. Thus, the permeability coefficient of ethanol is 2.0 mg/h/cm^2/1 mg/ml and, upon simplification, this would result in $P_{ethanol}$ to be 2 cm/h = 3.33×10^{-2} cm/sec. Thus, the permeability coefficient of ethanol through burnt skin is about 42 times that of intact skin.

REFERENCES

1. Bosman, I., Ensing, K., de Zeeuw, R., *Int. J. Pharm.,* 169, 65, 1998.
2. Brandrup, J., Immergut, E. H., and Grulke, E. A., *Polymer Handbook,* 4th ed., Wiley, New York, 1999, VI–545, 550.
3. Crank, J., *The Mathematics of Diffusion,* 2nd ed., Oxford University Press, Oxford, 1975.
4. Crank, J., and Park, G. S., *Diffusion in Polymers,* Academic Press, London, 1968.
5. Dowty, M., Dietsch, C., *Pharm. Res.,* 14, 1792, 1997.
6. Flynn, G. L., Yalkowsky, S. H., and Roseman, T. J., Mass transport phenomena and models: theoretical concepts. *J. Pharm. Sci.,* 63, 479–510, 1974.
7. Kikkoji, T., Gumbleton, M., Higo, N., Guy, R., and Benet, L., *Pharm. Res.,* 8, 1231, 1991.
8. Lide, D. R., *CRC Handbook of Chemistry and Physics,* 80th ed., CRC Press, Boca Raton, FL, 1999. 6–191.
9. Lyons, K., Charman, W., Miller, R., and Porter, C., *Int. J. Pharm.,* 199, 17, 2000.
10. Martin, A. N., Diffusion and dissolution, in *Physical Pharmacy,* 4th ed., Lea & Febiger, Marlver, 1993, 324–361.
11. Neilson, H., and Rassing, M., *Int. J. Pharm.,* (2000) 194:155.
12. Pershing, L., Parry, G., and Lambert, L., *Pharm. Res.,* 10, 1745, 1993.
13. Roy, S., Flynn, G., *Pharm. Res.,* 7, 842, 1990.
14. Scheuplein, R. J., *J. Invest. Dermatol.*, 45, 334, 1965.
15. Takamatsu, N., et al., *Pharm. Res.,* 14, 1127, 1997.
16. Wilschut, A., ten Berge, W. F., Robinson, P., McKone, T., Estimating skin permeation: the validation of five mathematical skin permeation models, *Chemosphere,* 30, 1275, 1995.
17. Williams, G., Sinko, P., Oral absorption of the HIV protease inhibitors: a current update, *Adv. Drug Del. Rev.,* 39, 211, 1999.
18. Zorin, S., Kuylenstierna, F., Thulin, H., *In vitro* test of nicotine's permeability through human skin. Risk evaluation and safety aspects, *Ann. Occup. Hyg.,* 43, 405, 1999.

8 Chemical Kinetics and Stability

*Tapash K. Ghosh**

CONTENTS

* No official support or endorsement of this article by the FDA is intended or should be inferred.

0-4152-8863-0/05/$0.00+$1.50
© 2005 by CRC Press LLC

Learning Objectives

After finishing this chapter the student will have thorough knowledge of:

- The rate and order of a reaction
- The methods to determine the order of a reaction
- The characteristics of reactions encountered commonly in drug degradation
- How to calculate half-life and shelf-life of a drug in a formulation
- Complex reactions
- The factors affecting the stability of a drug in a formulation and pathways of degradation
- How to predict the shelf-life of drug in a formulation

I. INTRODUCTION

The stability of a drug molecule in a marketed or to-be-marketed dosage form is always a concern to the people associated with the process. These people include a core of personnel including but not limited to developing scientists, regulatory agencies, prescribing physicians, dispensing pharmacists, and delivering nurses. Everybody has a prime goal in mind, and that is patient safety. The scientists involved in developing a dosage form keep several physicochemical factors of the drug in mind in terms of making a stable dosage form that in most cases should have a shelf life of at least 2 years. The regulatory agencies carefully scrutinize every submitted document by the sponsors before approving a product for the market. The dispensing pharmacist checks a product very carefully before dispensing it to the patient and explains to the patient all the measures ought to be taken to maintain the long-term stability of the product after it leaves the pharmacist's counter and sits in the patient's medicine cabinet. The pharmacist exercises special caution when dispensing extemporaneous preparations. Generally, extemporaneous preparations contain an expiration date not more than 2 weeks from the date of preparation. Still, pharmacists should make sure patients follow proper storage and handling procedures to maintain the potency of the drug within that stated period of time. At the end of the process, the delivering nurse should check for any visual incompatibility before administering the drug to a patient.

Stability is often expressed in terms of shelf life. The shelf life of a drug dosage form depends chiefly on formulation, environmental conditions, and packaging. The shelf life of a dosage form can be defined under three categories: chemical, physical, and aesthetic. Some measurement parameters of the three shelf lives of a dosage form are listed in Table 8.1.

TABLE 8.1
Three Drug Shelf Lives and Their Measurement Parameters

Category	Measurement Parameters
Chemical	Remaining percentage of labeled strength of active drug(s)
	Content of specified degradation products
	Solution pH
Physical	Drug dissolution rate of solid oral dosage forms
	Fully intact integrity of dosage forms
	Particle size and homogeneity in emulsions, suspensions, and suppositories
Aesthetic	Color, odor, and texture of drug product
	Color and clarity of label

Physical stability mostly refers to visual and organoleptic appearances, whereas chemical stability goes further to identify the pathways (mechanism) of degradation. Excessive temperature, moisture, pH, and oxidation are the four greatest threats to chemical and physical drug lives. The effects of temperature, pH, and light are the most scientifically predictable influences on drug stability. Most of the time, change in physical stability is an indirect effect of chemical stability and surfaces after chemical degradation takes place for some time. Often chemical degradation takes place without any change in the visual appearance but with loss of potency in the dosage form. Administration of a less potent dosage form to a patient means the patient may not only be deprived of the required pharmacological benefit of the drug, but may also receive an unwanted degradation product of the active drug with an entirely different pharmacological effect. Therefore, study of degradation kinetics of a drug in a given dosage form is a key step in the formulation development and optimization process.

Evaluation of degradation kinetics helps to stabilize a drug in a dosage form with the help of suitable excepients in an appropriate physicochemical environment. In addition, the shelf life of a product can be predicted from the kinetic study of a dosage form. Therefore thorough knowledge of kinetics is an invaluable tool for a formulation scientist. It is also important for a retail or hospital pharmacist to understand the degradation process in order to explain it to patients or physicians. Practicing pharmacists and pharmaceutical scientists play a very important role in the stability aspect of a drug molecule. Therefore a sound knowledge of the subject is considered very important. The following sections will explain step by step the various aspects of kinetics and stability.

II. KINETICS OF CHEMICAL DECOMPOSITION IN SOLUTION

To define stability and to predict the shelf life of a dosage form, it is essential to determine the kinetics of the breakdown of the drug in the dosage form under projected usage conditions. To understand the concept of kinetics the reader needs to be familiar with the concepts described in the following section.

A. Reaction Rate and Half-Life ($t_{1/2}$)

For a hypothetical reaction,

$$aA \rightarrow \text{Product } (P)$$

the rate of reaction of A into product (P) is defined by the derivative

$$-dA/dt$$

where the minus sign means the concentration (or amount) of A is decreasing with time. For P, the rate of production is defined by

$$+dP/dt$$

Here the concentration (or amount; amount = concentration × volume) of P always increases with time. According to the law of mass action, the rate of chemical reaction is proportional to the product of the molar concentration of the reactants each carried to the power equal to the number of moles of the substance undergoing reaction.

Therefore,

$$-dA/dt = k[A]^a \quad \text{or} \quad dA/dt = -k[A]^a$$

where k is the proportionality constant, also known as the rate constant, $[A]$ is the molar concentration of A as a function of time, and a is the number of moles of A undergoing reaction.

Half-life ($t_{1/2}$) is defined as the time required for half of the initial concentration (or amount) of reactants to form products.

B. Order of Reaction

The order of a chemical reaction refers to the way in which the concentration of the reactant influences the rate. Most commonly, zero-order and first-order reactions are encountered in pharmacy and will be discussed in detail below. However, the concept of second order will also be introduced.

1. Zero-Order Reactions

Again, for a hypothetical reaction,

$$A \rightarrow P$$

The rate equation for a zero-order reaction is defined as

$$dA/dt = -k[A]^0$$

Since $[A]^0 = 1$, the rate equation can be simply rewritten as

$$dA/dt = -k_0$$

The rate equation can be integrated as

$$\int_{Ao}^{At} dA = -k_0 \int_{o}^{t} dt$$

and solved to give the integrated rate equation for zero-order kinetics:

$$A_t - A_0 = -k_0 t$$

or

$$A_0 - A_t = k_0 t$$

where A_t is the amount of A at any time t, A_0 is the initial amount of A, and k_0 is the zero-order rate constant with the unit of concentration (or mass)/time.

The half-life ($t_{1/2}$) of a zero-order reaction can be deduced by the fact that at $t_{1/2}$

$$A_t = A_0/2$$

Substituting into the integrated equation for zero-order reactions will give

$$A_0/2 = A_0 - k_0 t_{1/2}$$

or

$$A_0 - A_0/2 = k_0 t_{1/2}$$

and

$$A_0/2 = k_0 t_{1/2}$$

Therefore, the half-life of zero-order reactions is defined by

$$t_{1/2} = 0.5A_0/k_0 \text{ or } A_0/2\, \boldsymbol{k}_0$$

Therefore, the half-life of zero-order reactions is directly proportional to the initial concentration of the reactants.

Drug X degrades by a zero-order process with a rate constant of 0.05 mg ml^1 $year^{-1}$ at room temperature. If a 1% weight/volume (w/v) solution is prepared and stored at room temperature:

1. What concentration will remain after 18 months?
2. What is the half-life of the drug?

Answers:

1. C_0 = 1% w/v = 10 mg/ml; t =18 months = 1.5 year; k_0 = 0.05 mg ml^{-1} $year^{-1}$
   $C = C_0 - k_0t = 10 - (0.05 \times 1.5) =$ **9.91 mg / ml**
2. $t_{1/2} = 0.5C_0/k_0 = (0.5 \times 10)/0.05 =$ **100 years**

---

2. First-Order Reactions

The rate equation for first-order kinetics is given by

$$dA/dt = -k_1[A]^1$$

or simply

$$-k_1[\mathrm{A}]$$

The rate equation can be integrated to give

$$\int_{Ao}^{At} dA/[A] = -k_1$$

which is solved to give the integrated rate equation for first order kinetics:

$$\ln A_t - \ln A_0 = -k_1t$$

which can be rewritten as

$$\ln A_t = \ln A_0 - k_1t \quad \text{or} \quad A_t = A_0e^{-k_1t}$$

Converting into base 10 log,

$$\log = \ln/2.303$$

$$\log A_t = \log A_0 - k_1t/2.303$$

The half-life of a first-order reaction is defined as the time ($t_{1/2}$) when $A_t = A_0/2$. Substituting this into the integrated equation:

$$\log A_0/2 = \log A_0 - k_1 t_{1/2}/2.303$$

$$\log\left[\frac{A_o}{\frac{A_o}{2}}\right] = k_1 t_{1/2}/2.303$$

which will give

$$\log 2 = k_1 t_{1/2}/2.303$$

or

$$t_{1/2} = 0.693/k_1$$

and

$$k_1 = 0.693/t_{1/2}$$

Note that the half-life of first-order reactions is independent of the initial concentration of the reactants. The units of k_1 will be (1/time) or time^{-1}.

The characteristic of a first-order reaction is such that over the same time period, the fraction of unchanged drug remaining will always be the same as from

$$A_t = A_0 e^{-k_1 t},$$

$$\frac{A_t}{A_0} = e^{-k_1 t}$$

which is a constant.

---

1. Ten (10) ml aqueous solutions of drug A (10% w/v) and drug B (25% w/v) are stored in two identical test tubes under identical storage conditions at 37°C for 3 months. If both drugs degrade by first-order, which drug will retain the highest percentage of initial concentration?
   (a) Drug A
   (b) Drug B
   (c) They will be the same.

   Answer: (c) Both drugs will have the same *percentage of initial concentration*, as the degradation process is first order.

2. The concentration of drug X in aqueous solution drops by 10% per month when stored at room temperature. If the degradation occurs by first order, what concentration will remain if a 5 mg/ml solution of the drug is stored under the same conditions for 3 months?

   Answer: Remaining concentration = $5 \times 0.9 \times 0.9 \times 0.9 = \boxed{\mathbf{3.65\ mg/ml}}$

3. A 5 gm/100 ml solution of drug X is stored in a closed test tube at 25°C. If the rate of degradation of the drug is 0.05 day^{-1}, calculate the time required for the initial concentration to drop to (a) 50% (half-life) and (b) 90% (shelf-life) of its initial value.

   Answer: From the unit of rate constant (day^{-1}), it is obvious that the degradation process is first-order. Therefore, from $A_t = A_0 e^{-k_1 t}$

(a) $t_{1/2\ (50\%)} = 0.693/K = 0.693/0.05 =$ **13.9 days**

(b) $t_{0.9\ (90\%)} = 0.105/K = 0.105/0.05 =$ **2.1 days**

Please practice the derivation of $t_{0.9\ (90\%)} = 0.105/K$ using the same format used for the calculation of $t_{1/2\ (50\%)} = 0.693/K$ above. Also, it should be noted that the initial concentration was not used anywhere to answer the above questions, as first-order processes are independent of initial or starting concentration.

4. A 5 gm/100 ml solution of drug X is stored in a closed test tube at 25°C. If the rate of degradation of the drug is 0.05 day^{-1}, calculate the time for the drug concentration to degrade to 2.5 mg/ml.

Answer: Using $\log A_t = \log A_0 - k_1 t/2.303$

$$\log 2.5 = \log 5 - (0.05 \times t)/2.303$$

$$t = \boxed{\textbf{13.9 days}}$$

The student should note that the time asked about here is $t_{1/2}$. Therefore, the problem could also be solved by using $t_{1/2\ (50\%)} = 0.693/K$.

---

a. *Apparent (or Pseudo) Zero-Order Reactions for Suspensions*

Suspensions (e.g., Amoxil®, Mylanta®, and Maalox®) are dosage forms in which the concentration of a drug exceeds the solubility. Suspension formulations, therefore, have solid particles suspended in a solution of drug. The decomposition of drugs in suspensions depends on the concentration of the drug in solution as shown below.

$$A_{\text{solid}} \rightarrow A_{\text{soln}} \rightarrow \text{Product}$$

The rate of decomposition of A in solution, therefore, is given by the first-order expression

$$dA/dt = -k[A]_{\text{soln}}$$

Since there is excess solid drug present, and it dissolves continuously to replace the portion of drug in solution, which is being converted to product, the concentration of drug in solution ($[A]_{\text{soln}}$) remains constant, and we can write

$$k[A]_{\text{soln}} = k_0'$$

which converts the decomposition process an apparent zero-order process, with k_0' being the apparent zero-order rate constant, although the actual decomposition of the drug from the solution may be first order. The above rate equation, therefore, can be rewritten as

$$dA/dt = -k_0'$$

and the integrated form of the apparent zero-order rate equation is expressed as

$$A_t = A_0 - k_0' t$$

An aqueous suspension of drug X contains 200 mg of drug X per teaspoon (5.0 ml). The solubility of drug X at 25°C is 1 g/350 ml in water, and the first-order rate constant for degradation of drug X in solution at 25°C is 3.9×10^{-6} sec^{-1} . Calculate (a) the zero-order rate constant and (b) the shelf life of the liquid preparation.

Answer:

(a) From $k_0' = k[A]_{soln} = (3.9 \times 10^{-6}\ \text{sec}^{-1})(1\ \text{g/350 ml}) = \boxed{\mathbf{1.11 \times 10^{-8}\ gm / ml\ sec^{-1}}}$

(b) $t_{90} = 0.1[A_0]/\ k_0' = (0.1)(0.04\ \text{g/ml})/\ 1.11 \times 10^{-8}\ \text{g/ml sec}^{-1} = 3.6 \times 10^5\ \text{sec} = \boxed{\mathbf{4.2\ days}}$.

The summary of the characteristics of zero-order and first-order reactions is presented in Table 8.2.

3. Second-Order Reactions

The rates of bimolecular reactions, which occur when two molecules come together, are frequently described by second-order equations. When the rate of reaction depends on the concentrations of A and B, with each term raised to the first power, the reaction is second order.

$$\underset{(a-x)}{A} + \underset{(b-x)}{B} \xrightarrow{k} \underset{(x)}{P}$$

$$-\frac{d[A]}{dt} = -\frac{d[B]}{dt} = k\,[A][B] = \frac{dP}{dt}$$

$$\frac{dx}{dt} = k(a-x)(b-x)$$

The symbols a and b are customarily used to replace A_0 and B_0, respectively, x is the decrease of concentration in time t, and $(a - x) = A_t$., $(b - x) = B_t$. If $a = b$, then

$$\frac{dx}{dt} = k(a-x)^2 \rightarrow \int_0^x \frac{dx}{(a-x)^2} = k\int_0^t dt$$

$$\downarrow$$

$$\frac{1}{a-x} - \frac{1}{a-0} = k(t-0)$$

$$\downarrow$$

$$\frac{A_0 - A_t}{A_0 A_t} = kt \leftarrow \frac{x}{a(a-x)} = kt$$

The half-life ($A_t = 0.5\ A_0$) can be obtained by the usual manner: $t_{1/2} = 1/(A_0\ k)$.

TABLE 8.2
Summary of the Characteristics of First- and Zero-Order Reactions

Parameter	Zero-Order	First-Order
Rate of disappearance of a drug from a site	Independent of remaining concentration at that site	Proportional to or dependent on remaining concentration
Plot of rate of disappearance with remaining concentration	$-dC/dt$ 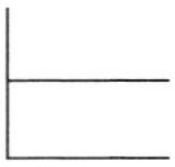 C	$-dC/dt$ 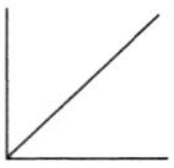 C
Rate equation (dC/dt)	$-dC/dt = K_o$	$-dC/dt = KC$
Integrated forms of above equation (C vs. time, t):	$C = C_o - K_o\,t$	$C = C_o e^{-Kt}$ $\ln C = \ln C_o - K\,t$ $\log C = \log C_o - (K/2.3)\,t$
Typical time concentration data	**Time** / **Concentration** 0 / 100 2 / 90 4 / 80 6 / 70 8 / 60 10 / 50 12 / 40	**Time** / **Concentration** 0 / 3200 2 / 1600 4 / 800 6 / 400 8 / 200 10 / 100 12 / 50
Shape of plot of integrated equation on Cartesian graph paper	C 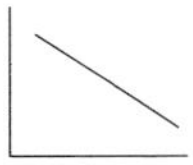 Time	C 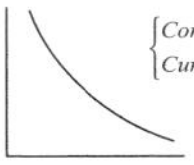 Time
Shape of plot of integrated equation on semilog graph paper (or shape of plot of logarthym of above equation):	log/ln C 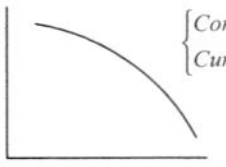 Time	log/ln C 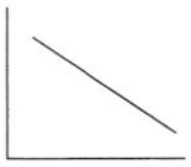
Slope of linear plot	Slope = $-K_o$	Slope = $-K$ (for ln plot) Slope = $-K/2.3$ (for log plot)
Rate constant and its units:	K_o (mg/mL • h, etc.)	K (h^{-1}; min^{-1}; day^{-1}, etc.)
$t_{\frac{1}{2}}$ (units)	$t_{\frac{1}{2}} = C_o/2K_o$ (hours, minutes, etc.)	$t_{\frac{1}{2}} = 0.693/K$ (hours, minutes, etc.)
Effect of concentration on $t_{\frac{1}{2}}$	$t_{\frac{1}{2}}$ increases with increasing starting concentration	$t_{\frac{1}{2}}$ is independent of starting concentration

The hydrolysis of sucrose is an example of a second-order reaction, since the rate also varies with the concentration of water. The amount of water required for the hydrolysis of sucrose is so small, however, relative to the large quantity present, that there is no significant change in the concentration of water. For practical purposes, therefore, the concentration of water is constant. Therefore, the reaction can be transformed into a pseudo first-order reaction by introducing $k_1 = k[H_2O]$.

III. METHODS OF DETERMINING REACTION ORDER

A. Substitution Method

The data collected from a kinetic study can be substituted into the integrated rate equations, for example:

Time (h)	Drug Conc. (mg/ml)
0	1000
2	950
4	900
6	850

For zero order, $A_t = A_0 - k_0 t$.

$$950 = 1000 - k_0(2)$$

$$900 = 1000 - k_0(4)$$

Solving both the equations shows that k_0 is the same (i.e., 25 mg/ml/h). Therefore, this is an example of zero-order reaction kinetics. The same conclusion could also be drawn by looking at the data as the amount lost every 2 hours being constant.

B. Graphical Method

The kinetic data can be plotted on linear or semi-log graph paper to determine the order of reaction.

C. Half-Life Method

This method is based on the relationship between the initial concentration of the reactant, the half-life, and the reaction order. For zero-order reactions, $t_{1/2}$ increases with increasing concentration, whereas for first-order reactions, $t_{1/2}$ does not change with change in concentration.

The concept of zero-order and first-order reactions will be made clearer by the following tutorial section.

TUTORIAL

Ten (10) g of a drug was dissolved in 100 ml of water maintained at 30°C. Samples of this solution were withdrawn at various intervals and assayed. The results are shown below.

Time	Concentration (mg/mL)
0.5	88.25
1.0	77.88
2.0	60.65
3.0	47.24
4.0	36.79
5.0	28.65
6.0	22.31
8.0	13.53

I. Plot the time points on the x-axis and their corresponding concentrations on the y-axis on regular (Cartesian) graph paper (spread the points to cover almost the entire axes). Connect the points and answer Questions 1 to 4 below.

1. What is the shape of the plot?
   (a) Inclining straight line
   (b) Declining straight line
   (c) Convex curve (declining and arching upward)
   (d) Concave curve (declining and arching inward)
2. What is the slope of the plot?
   (a) 0.25 h^{-1}
   (b) –0.25 h^{-1}
   (c) –10.00 mg/ml/h
   (d) –0.109 h^{-1}
   (e) The slope cannot be determined from the plot.
3. What type of reaction describes the data shown in the above table?
   (a) Zero-order
   (b) First-order
4. What is the rate constant of this process?
   (a) 0.25 h^{-1}
   (b) –0.25 h^{-1}
   (c) –10.00mg/ml/h
   (d) –0.109h^{-1}
   (e) The rate constant cannot be determined from the plot.

II. Plot the time points on the x-axis and their corresponding concentrations on the y-axis on two-cycle semilogarithmic graph paper (spread the points to cover almost the entire axes). Connect the points and answer Questions 5 to 8 below.

5. What is the shape of the plot?
   (a) Inclining straight line
   (b) Declining straight line
   (c) Convex curve (declining and arching upward)
   (d) Concave curve (declining and arching inward)
6. What is the slope of the plot?
   (a) 0.25 h^{-1}
   (b) –0.25 h^{-1}
   (c) –10.00 mg/ml/h
   (d) –0.109h^{-1}
   (e) The slope cannot be determined from the plot.
7. What type of reaction describes the data shown in the above table?
   (a) Zero-order
   (b) First-order
8. What is the rate constant of this process?
   (a) 0.25 h^{-1}
   (b) –0.25 h^{-1}
   (c) –10.00mg/ml/h
   (d) –0.109 h^{-1}
   (e) The rate constant cannot be determined from the plot.

III. Take the natural logarithm of the concentrations only and plot them on the y-axis and their corresponding times on the x-axis on regular (Cartesian) graph paper (spread points to cover almost the entire axes). Connect the points and answer Questions 9 to 12.

9. What is the shape of the plot?

(a) Inclining straight line
(b) Declining straight line
(c) Convex curve (declining and arching upward)
(d) Concave curve (declining and arching inward)

10. What is the slope of the plot?
(a) $0.25h^{-1}$
(b) $-0.25\ h^{-1}$
(c) –10.00mg/ml/h
(d) $-0.109\ h^{-1}$
(e) The slope cannot be determined from the plot.

11. What type of reaction describes the data shown in the above table?
(a) Zero-order
(b) First-order

12. What is the rate constant of this process?
(a) $0.25\ h^{-1}$
(b) $-0.25\ h^{-1}$
(c) –10.00 mg/ml/h
(d) $-0.109\ h^{-1}$
(e) The rate constant cannot be determined from the plot.

Ten (10) g of a drug was dissolved in 100 ml of water maintained at 30°C. Samples of this solution were withdrawn at various intervals and assayed. The results are shown below.

Time	Concentration (mg/ml)
0.5	95
1.0	90
2.0	80
3.0	70
4.0	60
5.0	50
6.0	40

IV. Plot the data, as I above, on regular (Cartesian) graph paper, connect the points, and answer Questions 13 to 16 below.

13. What is the shape of the plot?
(a) Inclining straight line
(b) Declining straight line
(c) Convex curve (declining and arching upward)
(d) Concave curve (declining and arching inward)

14. What is the slope of the plot?
(a) $0.25\ h^{-1}$
(b) 10.00 mg/ml/h
(c) –10.00 mg/ml/h
(d) $-0.109\ h^{-1}$
(e) The slope cannot be determined from the plot.

15. What type of reaction describes the data shown in the above table?
(a) Zero-order
(c) First-order

16. What is the rate constant of this process?

(a) 0.25 h^{-1}
(b) –10.00 mg/ml/h
(c) 10.00 mg/ml/h
(d) –0.109 h^{-1}
(e) The rate constant cannot be determined from the plot.

V. Plot the data, as you did in II above, on two-cycle semilogarithmic graph paper, connect the points, and answer Questions 17 to 20 below.

17. What is the shape of the plot?
(a) Inclining straight line
(b) Declining straight line
(c) Convex curve (declining and arching upward)
(d) Concave curve (declining and arching inward)

18. What is the slope of the plot?
(a) 0.25 h^{-1}
(b) 10.00 mg/ml/h
(c) –10.00 mg/ml/h
(d) –0.109 h^{-1}
(e) The slope cannot be determined from the plot.

19. What type of reaction describes the data shown in the above table?
(a) Zero-order
(b) First-order

20. What is the rate constant of this process?
(a) 0.25 h^{-1}
(b) 10.00 mg/ml/h
(c) –10.00 mg/ml/h
(d) –0.109 h^{-1}
(e) The rate constant cannot be determined from the plot.

VI. Take the natural logarithm of the concentrations only, plot them as you did in III above, as a function of their corresponding times on regular (Cartesian) graph paper, connect the plotted points, and answer Questions 21 to 24 below.

21. What is the shape of the plot?
(a) Inclining straight line
(b) Declining straight line
(c) Convex curve (declining and arching upward)
(d) Concave curve (declining and arching inward)

22. What is the slope of the plot?
(a) 0.25 h^{-1}
(b) 10.0 mg/ml/h
(c) –10.00 mg/ml/h
(d) –0.109 h^{-1}
(e) The slope cannot be determined from the plot.

23. What type of reaction describes the data shown in the above table?
(a) Zero-order
(b) First-order

24. What is the rate constant of this process?
(a) 0.25 h^{-1}
(b) 10.0 mg/ml/h
(c) –10.00 mg/ml/h
(d) –0.109 h^{-1}
(e) The rate constant cannot be determined from the plot.

Elimination half-lives of two drugs were determined at three doses on the same patients on separate occasions. The results are tabulated below. Use these data to answer Questions 25 to 30.

Elimination Half-Life ($t_{\frac{1}{2}}$) (h)

Dose (mg)	Drug 1	Drug 2
100	12.00	15.00
200	12.00	30.00
400	12.00	60.00

25. The elimination rate process of:
    I. Drug 1 is first order
    II. Drug 2 is first order
    III. Drug 1 is zero order
    IV. Drug 2 is zero order
    (a) I and II only
    (b) II and IV only
    (c) II and III only
    (d) I and IV only
26. The elimination rate constant of drug 1 is:
    (a) 0.058 mg/h
    (b) 0.058 h^{-1}
    (c) 3.33 h^{-1}
    (d) 3.33 mg/h
27. The elimination rate constant of drug 2 is:
    (a) 0.058 mg/h
    (b) 0.058 h^{-1}
    (c) 3.33 h^{-1}
    (d) 3.33 mg/h
28. If a 30-mg dose of drug 1 is given to a patient, how long will it take to eliminate 10 mg of the drug from the patient's body?
    (a) 3.0 h
    (b) 4.9 h
    (c) 5.7 h
    (d) 7.0 h
29. If a 30-mg dose of drug 2 is given to a patient, how long will it take to eliminate 10 mg of the drug from the patient's body?
    (a) 3.0 h
    (b) 4.9 h
    (c) 5.7 h
    (d) 7.0 h
30. What is the half-life of these drugs when 100 mg of each are administered to the same patient?
    (a) $t_{\frac{1}{2}}$ of both drug 1 and drug 2 is 12 h.
    (b) $t_{\frac{1}{2}}$ of drug 1 are 12 h and $t_{\frac{1}{2}}$ of drug 2 are 30 h.
    (c) $t_{\frac{1}{2}}$ of drug 1 are 15 h and $t_{\frac{1}{2}}$ of drug 2 are 60 h.
    (d) $t_{\frac{1}{2}}$ of drug 1 are 30 h and $t_{\frac{1}{2}}$ of drug 2 are 60 h.
    (e) $t_{\frac{1}{2}}$ of drug 1 are 12 h and $t_{\frac{1}{2}}$ of drug 2 are 15 h.

ANSWERS

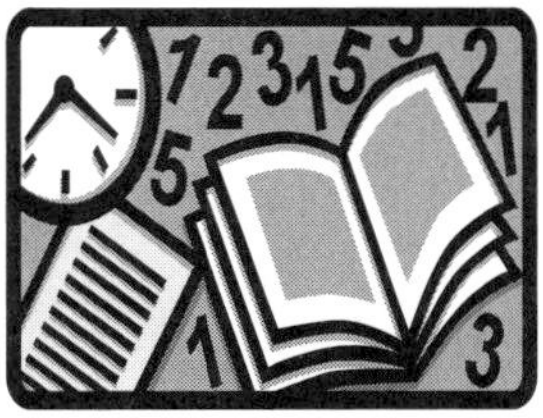

1. (d) (Refer to Figure 8.1.)
2. (e) (because the plot is not linear; refer to Figure 8.1)
3. (b) (Concentration–time data show concave curve in regular graph paper for first-order process.)
4. (e) (Rate constant cannot be determined in the absence of slope; refer to Figure 8.1.)
5. (b) (Refer to Figure 8.2.)
6. (d) [slope = (log 77.88 – log 36.79)/(1-4) = -0.109 h^{-1})
7. (b) (Concentration–time data show straight line in semilog graph paper for first-order process.)
8. (a) (slope = -0.109 h^{-1} = $-(K/2.303)$; $K = 0.25$ h^{-1})
9. (b) (Refer to Figure 8.3.)

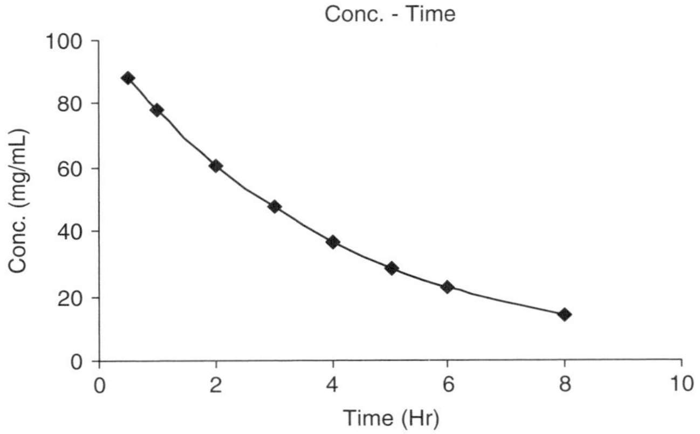

FIGURE 8.1 Conc.-time profile in regular graph paper.

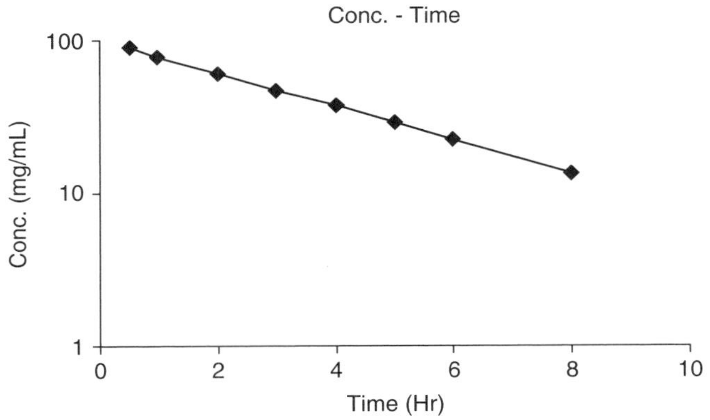

FIGURE 8.2 Conc.-time profile in semi-log graph paper.

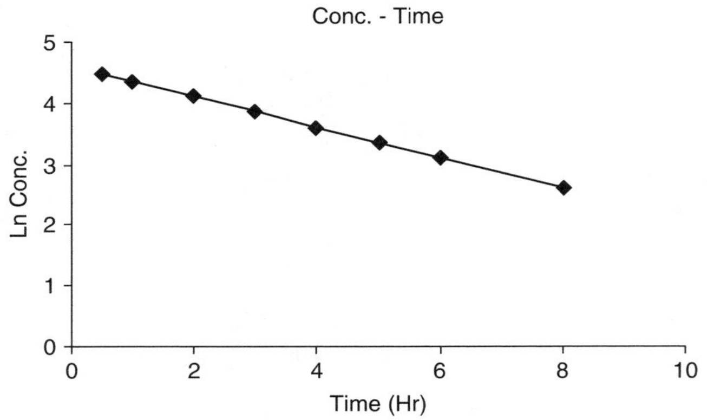

FIGURE 8.3 Conc.-time profile in regular paper.

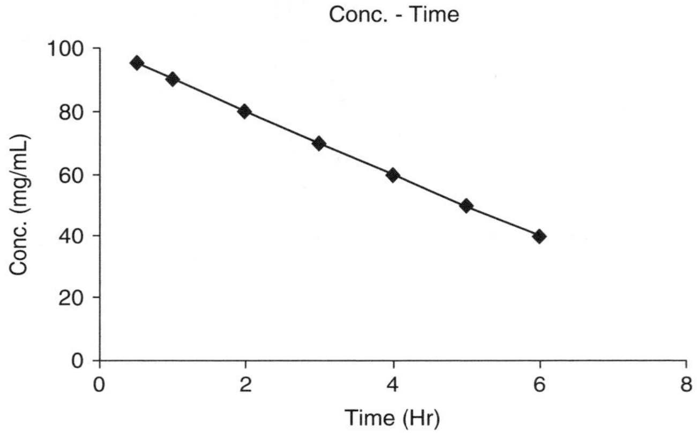

FIGURE 8.4 Ln Conc.-time profile in regular graph paper.

10. (b) (Slope = (Ln 77.88 – Ln 36.79)/(1-4) = –0.25 h^{-1})
11. (b) (Ln concentration–time data show straight line in regular graph paper for first-order process.)
12. (a) (K = –slope = 0.25 h)
13. (b) (Refer to Figure 8.4.)
14. (c) [slope = (90-80)/(1-2) = –10 mg/ml/h]
15. (a) (Concentration–time data show straight line on regular graph paper for zero-order process.)
16. (c) (K = –slope = 10 mg/ml/h)
17. (d) (Refer to Figure 8.5.)
18. (e) (because the plot is not linear; refer to Figure 8.5)
19. (a) (Concentration–time data show convex curve on semilog graph paper for zero-order process.)
20. (e) (Rate constant cannot be determined in absence of slope; refer to Figure 8.5)
21. (d) (Refer to Figure 8.6.)
22. (e) (because the plot is not linear; refer to Figure 8.6)
23. (a) (Ln Concentration–time data show concave curve on regular graph paper for zero-order process.)
24. (e) (Rate constant cannot be determined in absence of slope; refer to Figure 8.6)

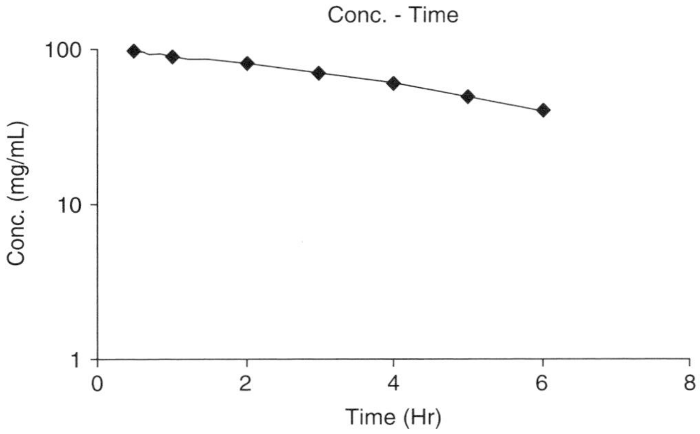

FIGURE 8.5 Conc.-time profile in semilog paper.

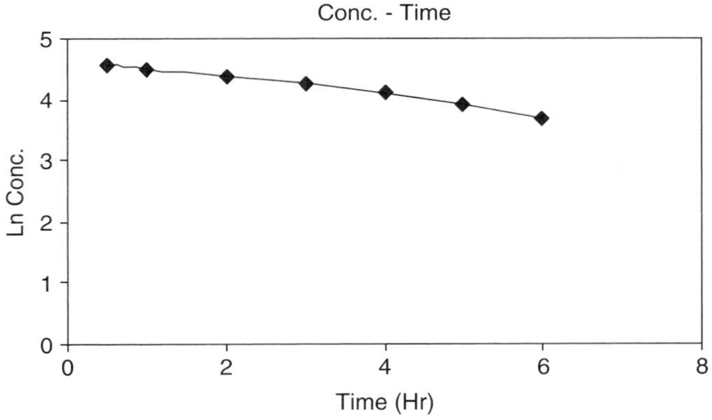

FIGURE 8.6 Ln Conc.-time profile in regular paper.

25. (d) (Elimination half-life does not change with dose for first-order process, whereas it changes proportionately for zero-order process. For zero-order reaction, $t_{1/2} = 0.5C_0/K$. Therefore, $t_{1/2} \propto C_0$ (or dose). So I and IV are correct options.
26. (b) (For first-order process, $K = 0.693/t_{1/2}$; $K = 0.693/12 = 0.058$ h^{-1}.)
27. (d) (For zero-order process, $K = 0.5\ C_0/t_{1/2}$; $K = (0.5*100)/15 = 3.33$ mg/h.)
28. (d) (For first-order process, $\ln C = \ln C_0 - Kt$; $\ln 20 = \ln 30 - 0.058t$; $t = 7.0$ h.)
29. (a) (For zero-order process, $C = C_0 - Kt$; $20 = 30 - 3.33t$; $t = 3$ h.)
30. (e)

IV. COMPLEX REACTIONS

In many cases, decomposition of drugs cannot be expressed by simple zero-, first-, second-, or higher-order equations. They involve a more than one-step reaction and accordingly are known as complex reactions. There are three types of complex reactions where the elementary reactions can be described with first-order kinetics:

1. Parallel (or side) reactions
2. Reversible reactions
3. Consecutive (or series reactions)

A. Parallel (or Side) Reactions)

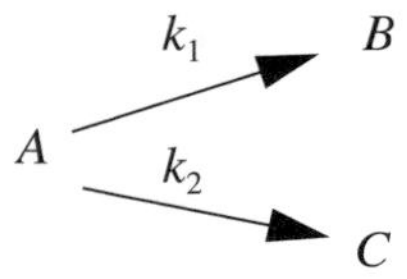

In a parallel reaction, the decomposition of drugs involves two or more pathways, the preferred route of reaction being dependent on reaction condition. As shown in the above scheme, drug A is simultaneously degraded into species B and C, with the first-order rate constants k_1 and k_2, respectively. The rate law for the decomposition of A is expressed as

$$-dA/dt = k_1[A] + k_2[A] \text{ or } -dA/dt = (k_1 + k_2)\,[A]$$

Degradation of nitrazepam tablets in the presence of moisture occurs via two pseudo first-order parallel reactions (please consult medicinal chemistry book). The ratio of two decomposition products depends on the amount of water in the environment.

B. Reversible Reactions

$$A \underset{k_f}{\overset{k_f}{\rightleftarrows}} B$$

In reversible reactions, drug A is converted into B with the forward rate constant k_f. At the same time B can also reverse back into A with a reverse rate constant k_r. This type of reaction is commonly seen with self-association of drugs (e.g., insulin hexamers), which leads to biological inefficacy (please consult medicinal chemistry book). The rate equation for A, therefore, is given by

$$-dA/dt = k_f[A] - k_r[B]$$

C. Consecutive (or Series) Reactions

$$A \xrightarrow{k_1} B \xrightarrow{\mathrm{k}_2} C$$

In a consecutive reaction, drug A is converted into an intermediate B, which further decomposes into C. This type of reaction mechanism is very important in biopharmaceutics and pharmacokinetics (e.g., absorption and elimination of drugs).

The rate equations are

$$-dA/dt = k_1[A]$$

$$dB/dt = k_1[A] - k_2[B] \text{ and}$$

$$dC/dt = k_2[B]$$

The hydrolysis of chlordiazepoxide follows a similar scheme. The neutral or cationic chlordiazepoxide is first transformed into lactam and, finally, in acidic solutions, to the yellow benzophenone (please consult medicinal chemistry book).

V. FACTORS AFFECTING THE RATE OF CHEMICAL REACTION (DRUG STABILITY)

In order to determine ways to prevent degradation of drugs in pharmaceutical formulations, it is important to identify the factors that accelerate the decomposition processes. In this section, pathways of drug degradation in pharmaceutical formulations will be examined by identifying the chemical groups present in classes of drugs that are particularly susceptible to chemical breakdown. Once the route of degradation has been identified, precautions can be taken to minimize the loss of activity. Some of the stability-affecting factors are common whether the drug is in solid, liquid, or semisolid dosage form. However, because the majority of stability problems are encountered in liquid dosage forms, the next section focuses on stability issues pertaining to liquid dosage forms. The previous discussion of the various orders of degradation processes and methods to calculate the rate constant for a reaction will help the reader to make shelf life predictions. Mechanistic analysis of drug degradation is discussed in detail in medicinal and pharmaceutical chemistry classes. Factors that affect drug degradation in a formulation are described below.

A. Temperature

The effect of temperature on drug degradation or stability has been studied extensively because it is directly related to determining the expiration date to go on the label. The effect of temperature on the reaction rate is described by the Arrhenius equation:

$$k = Ae^{-Ea/RT}$$

where k is the reaction rate constant; E_a is the activation energy, which is described as the energy barrier that has to be overcome if a reaction is going to occur when two reactant molecules collide; A is the frequency factor, which is assumed to be independent of temperature for a given reaction; R is the universal gas constant; and T is absolute temperature in the Kelvin scale (i.e., 273 + °C). The universal gas constant, R, is equal to 1.987 cal/mol/degree.

The Arrhenius equation can be rewritten in a linear form as

$$\ln k = \ln A - Ea/RT$$

or

$$\log k = \log A - Ea/2.303\ RT$$

Plotting log k as a function of 1/T on linear graph paper or k as a function of 1/T on semilog graph paper will give a straight line, with the slope equal to (–Ea/2.303R) and the y-intercept equal to log A or A, respectively, as shown below.

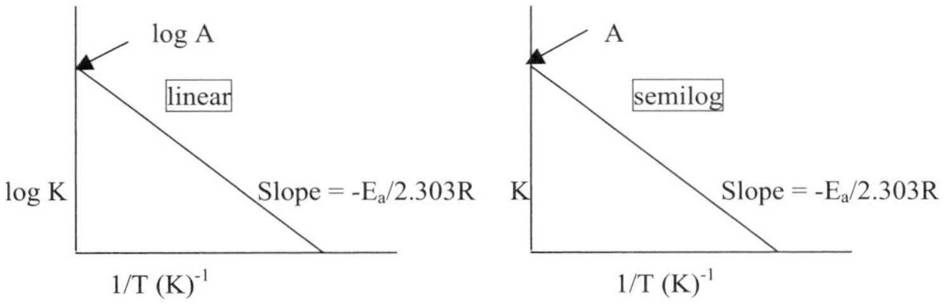

If the rate constants (k_1 and k_2) at two different temperatures (T_1 and T_2) are known, one can easily determine the activation energy and the frequency factor using the following equations.

$$\log k_1 = \log A - Ea/2.303\ RT_1$$

and

$$\log k_2 = \log A - Ea/2.303\ RT_2$$

Subtracting the second from the first,

$$\log\left[k_2/k_1\right] = -Ea/2.303\ R\left[\left(1/T_2\right)-\left(1/T_1\right)\right]$$

$$\text{or } \log\left(\frac{k_2}{k_1}\right) = \frac{E_a}{2.303R}\left(\frac{T_2 - T_1}{T_2T_1}\right)$$

If the activation energy and frequency factor at one temperature are known, one can easily determine the rate constant at any other temperature. This is important in the determination of expiration date of a new pharmaceutical product. However, some of the important underlying assumptions of this method are that (a) breakdown is purely thermal, (b) the substance does not coagulate on heating, and (c) there is no change in the mechanism of degradation with a change of temperature.

Example: The rate constant for first-order degradation of a drug (activation energy is 20.0 Kcal/mol) in a solution (5.0 mg/ml) at 70°C is $1.50 \times 10^{-3}\ h^{-1}$.

1. Calculate the amount remaining after storage for one week at room temperature (25°C).
2. Calculate the amount remaining after one week if the drug is kept at 70°C.
3. Calculate the shelf life of the drug at room temperature.
4. Calculate the half-life of the drug at room temperature.
5. Calculate shelfclife of the drug at 70°C.
6. Calculate the half-life of the drug at 70°C.
7. Discuss what you learned from this exercise.

Answer 1

$$T_1 = 25°C = 25 + 273 = 298°K;\ k_1 = ?$$

$$T_2 = 70°C = 70 + 273 = 343°K;\ k_2 = 1.50 \times 10^{-3}\ h^{-1}$$

$$E_a = 20.0\ Kcal/mol = 20{,}000\ cal/mol$$

$$R = 1.987\ cal/mol/degree$$

Plugging these values into the above equation,

$$\log\left(\frac{0.0015}{k_1}\right)=\frac{20000}{(2.303)(1.987)}\left(\frac{343-298}{(343)(298)}\right)=\frac{900000}{467737.50}=1.92$$

$$\left(\frac{0.0015}{k_1}\right)=\text{Antilog }(1.92)=10^{1.92}=83.18$$

$$k_1=\frac{0.0015}{83.18}=1.80\times10^{-5}\ \text{h}^{-1}$$

This is the first-order rate constant at 25°C.

The amount remaining after storage (C) for one week at room temperature (25°C) can be found from

$$C=C_0\,e^{-K_1t}=\left(5\ \text{mg/ml}\right)e^{-(0.000018)(7)(24)}=\boxed{\mathbf{4.98\ mg/ml}}$$

Answer 2

The amount remaining after storage (C) for one week at room temperature (70°C) can be found from

$$C=C_0\,e^{-K_2t}=\left(5\ \text{mg/ml}\right)e^{-(0.0015)(7)(24)}=\boxed{\mathbf{3.89\ mg/ml}}$$

Answer 3

Shelf life of the drug at room temperature can be found from

$$\text{Shelf life }(t_{90})=\frac{0.105}{k_1}=\frac{0.105}{0.000018}=5833.33\text{ hrs}=\boxed{\mathbf{243.1\ days}}$$

Answer 4

Half-life of the drug at room temperature can be found from

$$\text{Half-life }(t_{50})=\frac{0.693}{k_1}=\frac{0.693}{0.000018}=38500\text{ hrs}=1604.2\text{ days}=\boxed{\mathbf{4.40\ yrs}}$$

Answer 5

Shelf life of the drug at 70°C can be found from

$$\text{Shelf life }(t_{90})=\frac{0.105}{k_2}=\frac{0.105}{0.0015}=70\text{ hrs}=\boxed{\mathbf{2.92\ days}}$$

Answer 6

Half-life of the drug at 70°C can be found from

$$\text{Half-life }(t_{50})=\frac{0.693}{k_2}=\frac{0.693}{0.0015}=462\text{ hrs}=\boxed{\mathbf{19.25\ days}}$$

Answer 7

Degradation at 70°C occurs 83 times $\left(\frac{k_2}{k_1} = \frac{0.0015}{0.000018} = 83.33\right)$ faster than at room temperature. The implication is obvious in the longer half-life and shelf life at room temperature compared to 70°C.

B. pH AND HYDROLYSIS

The apparent rate constant for many reactions is affected markedly by pH. pH is perhaps the most important and widely examined parameter that affects hydrolysis of drugs in liquid formulations. If the drug is a derivative of a carboxylic acid or contains functional groups based on this moiety, for example, an ester, amide, lactone, lactam, imide, or carbamate, then that drug is liable to undergo hydrolytic degradation. Hydrolysis is frequently catalyzed by hydrogen ions (*specific acid catalysis*) or hydroxyl ions (*specific base catalysis*). Under specific acid catalysis and specific base catalysis, the reaction tends to go faster in acidic (low pH) and basic (high pH) medium, respectively, than they would otherwise in a neutral system. If the drug solution is buffered, the decomposition may not be accompanied by an appreciable change in the concentration of acid or base; however, it may be catalyzed by other acidic and basic species that are commonly encountered as components of buffers. This type of catalysis is referred to as *general acid-base catalysis*.

Acid-base catalysis is the most important type of catalysis when drug stability is considered. Solutions of many drugs undergo accelerated decomposition upon the addition of acids and bases. When the rate equation for such an accelerated reaction contains the term involving the concentration of H^+ or the concentration of OH^-, for example,

$$-\frac{d[A]}{dt} = \left\{k + k_{\mathrm{H}}\left[\mathrm{H}^+\right] + k_{\mathrm{OH}}\left[\mathrm{OH}^-\right]\right\}[A]$$

the reaction is said to be subjected to *specific acid-base catalysis* (k is the specific rate constant in water alone, k_H and k_{OH} are the catalytic coefficients for specific acid and base catalysis, respectively). In addition to the effect of pH, there may be catalysis by one or more species of the buffer components, which is called *general acid* or *general base catalysis* depending on whether the catalytic components are acidic or basic. For example, the decomposition of glucose in sodium acetate buffers can be written as follows:

$$-\frac{d[\mathrm{Glucose}]}{dt} = \left\{k + k_{\mathrm{H}}\left[\mathrm{H}^+\right] + k_{\mathrm{OH}}\left[\mathrm{OH}^-\right] + k_A[\mathrm{AcOH}] + k_B\left[\mathrm{Ac}^-\right]\right\}[\mathrm{Glucose}]$$

$$-\frac{d[\mathrm{Glucose}]}{dt} = k_{obs}[\mathrm{Glucose}]$$

where k_A and k_B are the catalytic coefficients for acetic acid (AcOH) and acetate (Ac^-, base), respectively, and the reaction is subjected to both *specific and general acid-base catalysis*, where

$$k_{obs} = \left\{k + k_{\mathrm{H}}\left[\mathrm{H}^+\right] + k_{\mathrm{OH}}\left[\mathrm{OH}^-\right] + k_A[\mathrm{AcOH}] + k_B\left[\mathrm{Ac}^-\right]\right\}$$

Several methods are available to stabilize the solution of a drug that is susceptible to acid-base catalyzed hydrolysis. In the development phase, prototype formulations are subjected to a wide

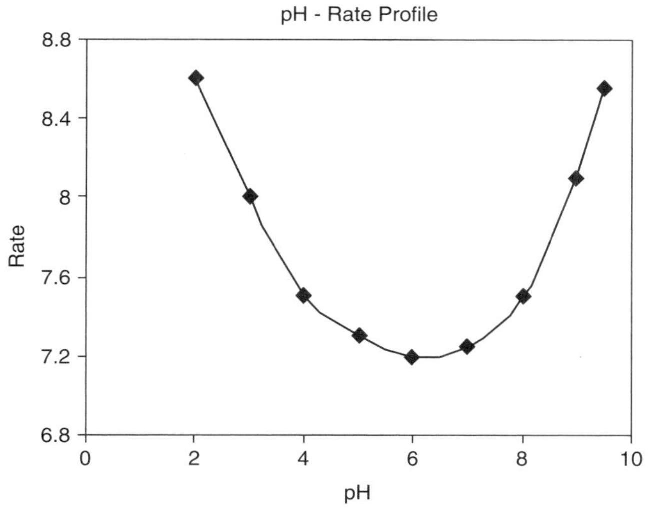

FIGURE 8.7 pH-Rate Profile.

range of pH, and the pH-rate profile is constructed to identify the pH of optimum stability. The pH of the final formulation is maintained around that pH. Various other chemical species, such as metal ions, may also exert catalytic effects on decomposition of drugs in solution.

Figures 8.7 and 8.8 demonstrate two hypothetical pH-rate profiles. They will help the reader interpret such a profile to identify the pH of maximum stability.

Figure 8.7 represents a pH-rate profile involving specific acid and specific base catalysis. The maximum stability occurs around pH 6.

The principal features of the pH-rate profile shown in Figure 8.8 are the specific acid catalysis below pH 1.8, the specific base catalysis above pH 7.8, and the sigmoid portion leading to a pH-independent plateau from pH 4 to 7.8. The maximum stability occurs around pH 2.

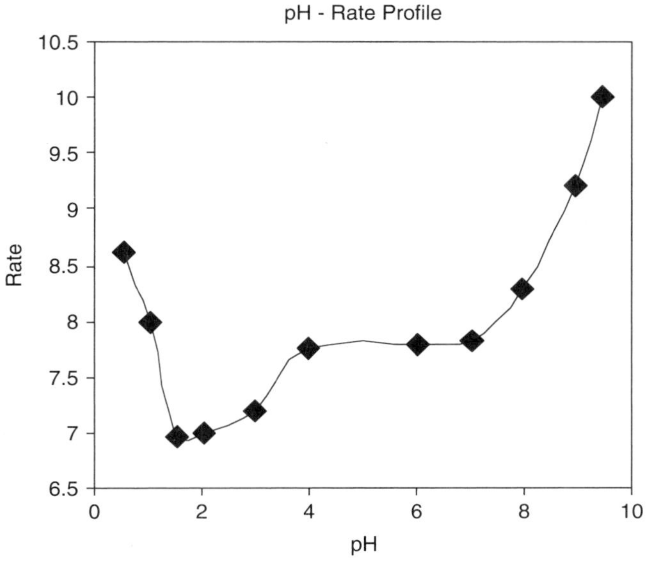

FIGURE 8.8 pH-Rate Profile (2).

Sometimes it becomes necessary to compromise between the optimum pH for stability and that for pharmacological activity. For example, many local anesthetics are most stable at acid pH, whereas for pharmacological activity they need to be in neutral or alkaline pH. Students should also keep pH-partition theory in mind, in which permeation through various biological membranes is a function of the state of ionization in that pH.

Alteration of dielectric constant by the addition of nonaqueous solvents such as alcohol, glycerin, or propylene glycol may in some cases reduce hydrolysis. The control of drug stability by modifying chemical structure using appropriate substituents has been suggested for drugs for which such a modification does not reduce therapeutic efficacy. In addition, other formulation ingredients, especially nature and amount of surfactants used in a liquid formulation can have a modifying effect on hydrolysis of drugs.

C. Oxidation

In a liquid formulation, a drug substance is often subjected to oxidative degradation. After hydrolysis, oxidation is the next common pathway for drug breakdown. Drugs that are affected by oxidative degradation include phenolic compounds such as morphine and phenylephine, catecholamines such as dopamine and adrenaline, steroids, antibiotics, vitamins, oils, and fats. Stabilization of drugs against oxidation involves observing various precautions during manufacture and storage. The best method to stabilize drugs against oxidation is to replace oxygen in the storage containers with nitrogen or any other inert gas. However, that may not be always feasible. In some cases the initiation of the oxidation process is initiated by catalysts that include heavy metals such as iron, cobalt, and nickel, so containers made from these metals should not be used during storage or manufacture of oxidation-susceptible drug–containing formulations. Alternatively, a very small amount of a compound or combination of compound called antioxidants, which can prevent, or at least delay, the propagation of the chain reaction, can be added in the preparation.

D. Ionic Strength

Electrolytes that are often added to the formulation to control tonicity increase the ionic strength of the formulation and may cause to a stability problem. Therefore, investigations of the influence of ionic strength on reaction rate are important. In fact, stability experiments should be conducted at constant electrolyte concentration to avoid any confusion arising from possible differences in electrolyte effects between different systems. The following Bronsted-Bjerrum equation describes the influence of electrolytes on the degradation rate constant:

$$\log k = \log k_0 + 2Az_Az_B\sqrt{\mu}$$

where μ is the ionic strength of the solution, k_0 is the rate constant in an infinitely dilute solution in which $\mu = 0$, z_A and z_B are the charge numbers of the two interacting ions, and A is a constant for a given solvent and temperature. For water at 25°C, A is approximately 0.51. Therefore, the above equation reduces to

$$\log k = \log k_0 + 1.02z_Az_B\sqrt{\mu}$$

It is evident from the above equation that a plot of log k vs. $\sqrt{\mu}$ should give a straight line with a slope of $1.02z_Az_B$, and k_0 will be evaluable from the intercept of log k_0. If one of the reactants is a neutral molecule, z_Az_B is 0 and the rate constant (k) should then be independent of the ionic strength in dilute solutions.

E. Solvent

The solvent (mainly the dielectric constant $\in$) in a liquid formulation plays a major role in the hydrolysis of the drug in the formulation. The equation that predicts the effect of solvent on the hydrolysis rate is

$$\log k = \log k_{\in=\infty} - Kz_A z_B/\in$$

where k is the rate of hydrolysis, K is a constant for a given system at a given temperature, and z_A and z_B are the charge numbers of the two interacting ions. It is evident from the above equation that a plot of log k as a function of the reciprocal of the dielectric constant ($\in$) of the solvent should be linear with a gradient of $Kz_A z_B$ and an intercept of log $k_{\in=\infty}$ (logarithm of the rate constant in a theoretical solvent of infinite dielectric constant). If the charges on the drug ion and the interacting species are the same, the gradient of the line will be negative. In this case, replacing water with a solvent of lower dielectric constant will achieve the desired effect of reducing the reaction rate. If the charges on the drug ion and the interacting species are opposite, the gradient of the line will be positive and water should not be replaced with a nonpolar solvent.

F. Isomerization

Isomerization is the conversion of a drug into its optical or geometric isomers, and such a conversion may be regarded as a form of degradation, often resulting in a serious loss of therapeutic activity. Several molecules including adrenaline, tetracyclines, pilocarpine, cephalosporins, and vitamin A undergo isomerization in one form or the other. The process of degradation following this type of reaction is reported to be pH dependent and catalyzed by various components of the buffers used in the formulation. Control of pH and various formulation components can reduce degradation from isomerization.

G. Photochemical Decomposition

Many pharmaceutical compounds, including hydrocortisone, prednisolone, riboflavine, ascorbic acid, and folic acid are light sensitive. Pharmaceutical products can be adequately protected from photo-induced decomposition by colored glass containers and storage in the dark. Amber glass excludes light of wavelength <470 nm and so affords considerable protection of compounds sensitive to ultraviolet light.

H. Polymerization

Polymerization is the process by which two or more identical molecules combine to form a complex molecule. Polymerization has been demonstrated during storage of concentrated aqueous solutions of amino-penicillins such as ampicillin sodium. Such polymeric substances have been shown to be highly antigenic in animals and they are considered to play a part in eliciting penicilloyl-specific allergic reactions to ampicillin in humans. Polymerization reactions may be prevented by the addition of methanol or similar solvents.

VI. KINETICS OF CHEMICAL DECOMPOSITION IN SOLID AND SEMISOLID DOSAGE FORMS

In spite of most dosage forms on the market being solid or semisolid, there have been relatively few attempts to evaluate the detailed kinetics of the decomposition of such dosage forms.

The major factors responsible for drug stability in **solid dosage forms** (tablets, capsules, caplets, etc.) include adsorbed moisture, storage temperature, the presence of light, and the presence of

TABLE 8.3
Major Factors Affecting Physical and Chemical Stability of Dosage Forms

Dosage Form	Factors Affecting Physical and Chemical Stability
Tablets and capsules	Excessive heat and moisture (entrapped from multiple use)
Oral and topical liquids	Excessive heat and cold
Injectable solutions	Excessive heat and cold
Powders for reconstitution	Excessive heat; liquid state after reconstitution
Semisolid dosage forms	Excessive heat and cold

oxygen. The chemical stability of active ingredients incorporated into **semisolid dosage forms** (creams, ointments, gels, etc.) frequently depend on the nature of the base used in the formulation. Nonetheless, like liquid dosage forms, the presence of excipients can cause stability problems in semisolid and solid dosage forms. Table 8.3 summarizes the major factors affecting physical and chemical stability of various dosage forms.

VII. STABILITY STUDIES AND EXPIRATION DATE

The purpose of stability testing is to provide evidence on how the quality of a **drug substance (active ingredient)** or **drug product (formulation)** varies with time under the influence of a variety of environmental factors such as temperature, humidity, and light. Stability testing permits the establishment of recommended storage conditions, retest periods, and shelf lives. Information on both drug substance (active ingredient) and drug product (formulation) are required by the Food and Drug Administration (FDA) in reviewing the application for new approval.

Although all regulated pharmaceutical products have to satisfy governmental regulatory authorities, surprisingly there are no nationally or internationally standardized storage conditions. However a draft guidance entitled "Stability Testing of Drug Substances and Drug Products" was published by the FDA in June 1998 to replace the 1987 guidance for submitting documentation on the stability of human drugs and biologics. The recommendations in this guidance are effective upon publication of the final guidance and should be followed in preparing new applications, resubmissions, and supplements. The guidance represents the FDA's current thinking on the design of stability studies for drug substances and dosage forms that should result in a statistically acceptable level of confidence for the established retest or expiration dating period for each type of application. The choice of test conditions defined in this guidance is based on an analysis of the effects of climatic conditions in the E.U., Japan, and the U.S. It references and incorporates substantial text from several International Conference on Harmonization (ICH) guidances (ICH Q1A, ICH Q1C, ICH Q1B, and ICH Q5C).

The guidance provides recommendations for the design of stability studies for drug substances and drug products that should result in a statistically acceptable level of confidence for the established retest or expiration dating period for each type of application. Information on the stability of a drug substance under defined storage conditions is an integral part of the systematic approach to stability evaluation. Stress testing helps determine the intrinsic stability characteristics of a molecule by establishing degradation pathways to identify the likely degradation products and to validate the stability-indicating power of the analytical procedures used. The design of the stability protocol for the drug product should be based on the knowledge of the behavior, properties, and stability of the drug substance and the experience gained from clinical formulation studies. The changes likely to occur during storage and the rationale for the selection of drug product parameters to be monitored should be stated.

The guidance provides recommendations regarding the design, conduct, and use of stability studies that should be performed to support:

- Investigational new drug applications (INDs)
- New drug applications (NDAs) for both new molecular entities (NMEs) and non–NMEs
- Abbreviated new drug applications (ANDAs)
- Supplements and annual reports
- Postapproval formulation, batch size, process, and manufacturing site changes
- Biologics license applications and product license applications

For a new drug, the expiration date is usually set at the time when 10% of the initial (labeled) dose is degraded (i.e., 90% of the dose is still active) at room temperature. If A_0 is the initial amount of drug in the dosage form, we can determine the time required for a drug to decompose to 90% of its original activity

$$A_0 \rightarrow 0.9\, A_0$$

To determine the decomposition of drugs at room temperature for a year or more is both time consuming and expensive. Instead, companies have adopted an accelerated stability method that involves examination of the drug decomposition at higher temperatures and extrapolating the data to room temperature. The following sequence is used:

- Measure the degradation rate constant (k) of the same formulation at several higher temperatures.
- Use an Arrhenius plot to determine the activation energy (Ea) and frequency factor (A). Using the data, calculate the rate constant at room temperature.
- Alternatively, plot log k vs. 1/T according to the Arrhenius equation, and extrapolate k at room temperature from that plot.
- Using k at room temperature, determine the time required for a drug to decompose to 90% of its original amount.

Pharmaceutical scientists must recognize that the order of a reaction may change during the period of the study, especially at an elevated temperature. Therefore, prediction of stability at room temperature based on results obtained from an elevated temperature study may not be always practical.

Another method that allows reasonable estimates of shelf life where an exact Ea value is not available is using the Q_{10} value. Q_{10} is defined as the factor by which the rate constant increases for a 10°C temperature increase and is represented by the expression

$$Q_{10} = \frac{K_{(T+10)}}{K_T}$$

Q_{10} values of 2, 3, or 4 can be used to represent, respectively, low, average, and high estimates of Q_{10} around room temperature (20 to 30°C) when Ea is unknown. For more simplification, the value of Q_{10} is chosen as either 2 or 5, according to Table 8.4, to calculate a conservatively shorter than actual value of t_{90}. Expiration (t_{90}) of a formulation at a higher or lower temperature can be predicted from the t_{90} of another temperature by using the expression

TABLE 8.4
Simplified Q_{10} Values

When Chemical Life Is	Use Q_{10} of
Known at a cooler temperature and to be estimated at a warmer temperature	5
Known at a warmer temperature and to be estimated at a cooler temperature	2

$$t_{90(T2)} = \frac{t_{90(T1)}}{Q_{10}^{\frac{\Delta T}{10}}}$$

where $t_{90\,(T2)}$ and $t_{90\,(T1)}$ are expiration times at temperatures $T2$ and $T1$, respectively. This estimate is independent of order of reaction. It is evident that an increase in temperature will shorten the shelf life, whereas lowering the temperature will prolong the shelf life of a formulation. Generally, the expiration date at room temperature is available from the standard monograph or from the product labels.

Example: The shelf life of a liquid drug is 21 days at 5°C. Approximately how long will the drug be stable at 37°C?

$$\text{Life at } 37°\text{C} = 21 \text{ days}/5^{[(37-5)/10]} = 21/5^{3.2} = 21/172.47$$
$$= 0.12 \text{ day or } 2.92 \text{ h (using the above table)}$$

Alternatively, using Q_{10} values as 2, 3 or 4

$$\text{Life at } 37°\text{C} = 21 \times 2^{-3.2} = 2.28 \text{ days } (Q = 2, \text{ possible})$$
$$= 21 \times 3^{-3.2} = 0.62 \text{ days } (Q = 3, \text{ likely})$$
$$= 21 \times 4^{-3.2} = 0.25 \text{ days } (Q = 4, \text{ conservative})$$

Example: A drug in solution is stable for 2 years at room temperature (25°C). How long may the drug be theoretically stable at refrigerator temperature (5°C)?

Using the table value, life at 5°C = 2 years/$2^{[(5-25)/10]}$ = $2/2^{-2}$ = 8 years

Alternatively, using Q_{10} values of 2, 3, or 4

$$\text{Life at } 5°\text{C} = 21 \times 2^{2} = 8 \text{ years } (Q = 2, \text{ conservative})$$
$$= 21 \times 3^{2} = 18 \text{ years } (Q = 3, \text{ likely})$$
$$= 21 \times 4^{2} = 32 \text{ years } (Q = 4, \text{ possible})$$

The student should note that to calculate the exact expiration date at a given temperature, we need to know the activation energy of the drug. All the above modes of calculation give an approximate figure. Students should also note the interchangeability of the terms *likely* and *conservative* based on the condition.

Example: The ampicillin monograph states that the reconstituted suspension is stable for 14 days in a refrigerator. If the product is left at room temperature for 6 h, what is the reduction in the expiration period?

The question to be addressed here is:

$$\text{Life at } 25°\text{C for 6 h (0.25 days)} = \text{how many hours at } 5°\text{C?}$$

This can be estimated by using the previously mentioned relationship and various Q_{10} values as follows:

$$0.25 \times 2^2 = 1 \text{ day } (Q = 2, \text{ possible})$$

$$0.25 \times 3^2 = 2.25 \text{ days } (Q = 3, \text{ likely})$$

$$0.25 \times 4^2 = 4 \text{ days } (Q = 4, \text{ conservative})$$

The above values describe the reduction in expiration period in various terms. Subtraction of these values from 14 will give possible, likely, and conservative estimates of new expiry dates.

Most manufacturers normally assign a shelf life of at least 2 years from the date of manufacture. Prescriptions requiring extemporaneous compounding by the pharmacist generally do not require that extended shelf life, as they are intended to be used immediately upon receipt by the patient within a limited period of time.

A. Accelerated Stability Testing Protocols

The objectives of accelerated stability tests may be defined as:

1. Quick evaluation of degradation in initial formulations of a product in selecting the best formulation from a series of possible choices.
2. The prediction of shelf life.
3. Identification of a rapid quality control tool, which ensures that no unexpected change has occurred in the stored product.

The first objective is accomplished by choosing from a series of possible choices the best formulation that exhibits the least amount of decomposition in a given time under the influence of a reasonably high stress. The second objective is achieved by using the results obtained from an accelerated test to predict the amount of decomposition in a product after a longer period of storage under normal conditions. The third objective can be accomplished by developing a sensitive, stability-indicating assay procedure to quantitate the drug substance and degradation products accurately when subjected to common stresses to predict the fate of the drug substance or drug product in long-term use conditions. Common high stresses or challenge factors considered in stability testing protocols are temperature, humidity, and light.

TABLE 8.5
Representative Study Design of Long Term and Accelerated Stability Study on a Drug Substance

Months	−20°C (Deep Freezer)	5°C (Refrigerator)	25°C/60%RH (Temperate and Subtropical)	30°C/70%RH (Tropical)
1.5	[X]	—	X	X
3	[X]	X	X	X
6	[X]	X	X	X
9	[X]	X	—	—
12	X	X	—	—
18	[X]	X	—	—
24	X	X	—	—
36	[X]	X	—	—
48	[X]	X	—	—
60	X	X	—	—

[X], samples put on stability but only analyzed on a need basis; X, analysis planned; —, analysis not planned

Accelerated stability testing requires the careful design of protocols that must define clearly the following or more:

1. The temperature and humidity for storage
2. Time points and frequency of sampling
3. The number of batches to be sampled
4. The number of replicates within each batch
5. A suitable light challenge
6. Details of assay

Excerpts from information specific to stability testing of drug products in a new drug application is presented in the following section. Procedures involved in stability testing of drug substances are similar and are summarized in the following section. However, the readers should find them in detail in the guidance (Ref. 7).

The purpose of the stability testing of drug substance is to determine for what period of time and under what conditions the drug substance can be stored and transported without relevant changes in quality. Long-term testing and accelerated testing are designed to cover storage in a refrigerator and storage in temperate and tropical climates, respectively. Additional testing includes but is not limited to stress testing under light in solid state and in solution, testing at solid state at different temperature and relative humidity and atmospheric conditions, and testing under forced decomposition under the influence of water (0.1 N HCl, 0.1 N NaOH, and 10% H_2O_2). The possibility of any racemization or isomerization of the drug substance is also investigated.

A representative study design of long-term and accelerated stability study on drug substances is presented in Table 8.5.

The design of the stability protocol for drug products should be based on knowledge of the behavior, properties, and stability of the drug substance and the experience gained from clinical formulation studies. Stability information from accelerated and long-term testing should be provided on three batches of the same formulation of the dosage form in the container and closure proposed for marketing. The long-term testing should cover at least 12 months. The test parameters should cover

TABLE 8.6
Long-Term and Accelerated Testing Conditions

	Conditions	Minimum Time Period at Submission
Long-term testing	25 ± 2°C, RH ±5%	12 months
Accelerated testing	40 ± 2°C, RH ±5%	6 months

those features susceptible to change during storage and likely to influence quality, safety, or efficacy. Analytical test procedures should be fully validated, and the assays should be stability indicating. The range of testing should cover not only chemical and biological stability, but also loss of preservative, physical properties and characteristics, organoleptic properties, and where required, microbiological attributes.

The length of the studies and the storage conditions should be sufficient to cover storage, shipment, and subsequent use (e.g., reconstitution or dilution as recommended in the labeling). It is now required that product labeling of official formulations provide recommended storage conditions and an expiration date. Official storage conditions are defined as follows: *Cold* is any temperature not exceeding 8°C, and *refrigerator* is a cold place where the temperature is maintained thermostatically between 2 and 8°C. A *freezer* is a cold place maintained between –20 and –10°C. *Cool* is described as any temperature between 8 and 15°C. *Room temperature* is the temperature prevailing in a working area, and *controlled room temperature* is a temperature maintained thermostatically between 15 and 30°C. *Warm* is any temperature between 30 and 40°C, and *excessive heat* is any temperature above 40°C. Table 8.6 describes recommended accelerated and long-term storage conditions and minimum times. Other storage conditions are allowable if justified.

Deviation of any stability parameter from the specification under the above testing conditions calls for additional testing in separate conditions. Results of the stability study ultimately govern the information presented on the label in terms of defining the expiration dating under correct storage conditions.

The design of a stability study is intended to establish, based on testing a limited number of batches of a dug product, an expiration dating period applicable to all future batches of the drug product manufactured under similar conditions. The stability protocol for both the drug substance and the drug product should be designed to allow storage under specifically defined conditions. For the drug product, the protocol should support a labeling storage statement at current room temperature (CRT), refrigerator temperature, or freezer temperature as defined earlier. A properly designed stability protocol should include the following information:

- Technical grade and manufacturer of drug substance and excipients
- Type, size, and number of batches
- Type, size, and source of containers and closures
- Test parameters
- Test methods
- Acceptance criteria
- Test time points
- Test storage conditions
- Container storage orientations
- Sampling plan
- Statistical analysis approaches and evaluations
- Data presentation
- Retest or expiration dating period

TABLE 8.7
Representative Study Design of a Long-Term and Accelerated Stability Study of a Drug Product

		Testing Intervals (Months)									
Code	**Storage Conditions**	**0**	**3**	**6**	**9**	**12**	**18**	**24**	**36**	**48**	**60**
B1	25°C/60% RH	X	X	—	X	X	—	X	X	[X]	[X]
	30°C/70% RH	X	—	X	X	—	X	X	X	[X]	[X]
	40°C/75% RH[a]	X	X	X	—	—	—	—	—	—	—
B2	25°C/60% RH	X	X	X	—	X	X	—	X	[X]	[X]
	30°C/70% RH	X	X	—	X	X	—	X	X	[X]	[X]
	40°C/75% RH[a]	X	X	X	—	—	—	—	—	—	—
B3	25°C/60% RH	X	—	X	X	X	—	X	X	[X]	[X]
	30°C/70% RH	X	X	X	—	X	X	—	X	[X]	[X]
	40°C/75% RH[a]	X	X	X	—	—	—	—	—	—	—

[X] = optional testing
X = definite testing

[a] Additional testing after 6 weeks of storage

Batches selected for stability studies should optimally constitute a random sample of the population of production batches. The sample time points should be chosen so that any degradation can be adequately profiled (i.e., at a sufficient frequency to determine with reasonable assurance the nature of the degradation curve). Stability testing for long-term studies generally should be performed at 3-month intervals during the first year, 6-month intervals during the second, and yearly thereafter.

A representative study design of a long-term and accelerated stability study of a semisolid drug product with three batches (B1, B2, and B3) is presented in Table 8.7.

Storage conditions during stability testing vary from manufacturer to manufacturer and even within a single company. Often different types of products are given different challenges. Products are usually stored in their final container. If at this stage the final pack has not been confirmed, a range of packs and pack materials must be tested. It is important that stability studies are performed at all stages of product development, namely during preformulation, formulation, scale-ups, clinical trials, and finally postmarketing.

B. Development Steps: Formulation to Expiration

While in the accelerated stability protocols, both the drug substance and drug product are challenged under common high stresses such as temperature, humidity, and light; however, for drug products this comes in the later phase of formulation development. The initial phases deal with finalizing the right environment of the formulation, that is, identifying the right pH, ionic strength, buffer species, buffer concentration, etc. Based on evaluation of several formulation factors, a few potential formulations are then subjected to accelerated stability testing to finalize the best formulation for further clinical studies.

One of the most important factors studied, especially for liquid dosage forms, is optimum pH or pH range for maximum stability. For that purpose it is necessary to evaluate the catalytic coefficients for specific acid and base catalysis and to determine the catalytic coefficients of possible buffers that might be useful in the formulation.

As a first step, it is important to achieve a buffer-independent pH-rate profile, since this will show at what pH the stability is greatest. As a hypothetical example with drug X, Figure 8.9 was generated from experiments carried out at constant temperature and constant ionic strength using

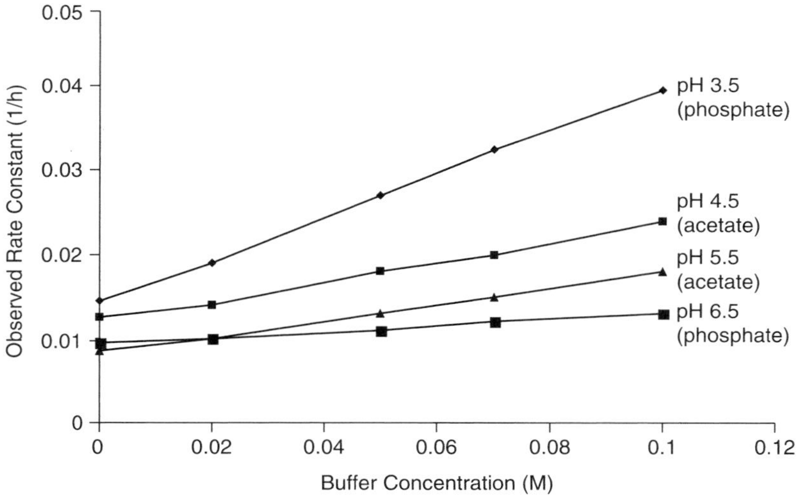

FIGURE 8.9 Effect of buffer concentration on the hydrolytic rate constant for drug X at 65°C as a function of pH.

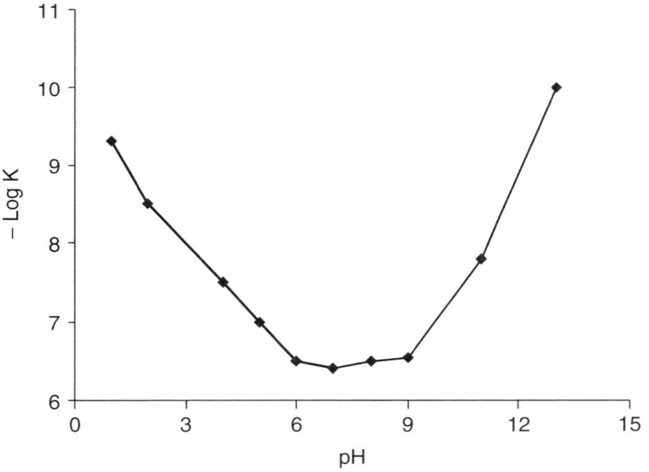

FIGURE 8.10 Log rate constant-pH profile for the degradation of Drug Y in buffer-free solution at 65°C.

a series of different buffers over the pH range 3.5 to 6.5. It is evident from the graph that an increase in buffer concentration had a marked effect on the hydrolysis rate. Comparison of the phosphate buffer profiles at pH 3.5 and 6.5 reveals that the effect of phosphate buffer concentration on this system was more pronounced at pH 3.5 (steeper curve) than at pH 6.5. Also, it appears that the effect of acetate buffer concentration did not change much between pH 4.5 and 5.5 (parallel curves). Therefore, for a stable liquid formulation of drug X, it is recommended to buffer the solution with phosphate buffer at pH 6.5. In selecting the optimum buffer concentration, careful attention has to be exercised to minimize the rate of degradation while maximizing the buffer capacity of the buffer system.

To remove the influence of the buffer, the reaction rate should be measured at a series of buffer concentrations at each pH and the data extrapolated back to zero concentration as shown in Figure 8.10. If these extrapolated rate constants are plotted as a function of pH, the required buffer-independent pH-rate profile will be obtained. Figure 8.10 illustrates the simple pH-rate profile obtained with a hypothetical drug Y. As we can see from the figure, this drug is very stable in unbuffered solution over a wide pH range (6 to 9) but degrades relatively rapidly in the presence of strong acids or bases. Since the influence of buffer components has been removed, this plot allows us to calculate the rate constants for specific acid and base catalysis. Removing the terms for the effect of buffer from equation we have

$$K_{obs} = k + k_{H}\,[H^{+}] + k_{OH}[OH^{-}]$$

Consequently a plot of measured rate constant k_{obs} against the $[H^{+}]$ at low pH will have a gradient equal to the rate constant for acid catalysis. Similarly, in the plot of k_{obs} against the $[OH^{-}]$ at high pH, the gradient will be the rate constant for base-catalyzed hydrolysis. The degradation of many drugs is particularly susceptible to the effects of buffers. Their hydrolysis rates at a specific pH in a buffer solution can sometimes be several times faster than in unbuffered solution at this pH. Therefore, the influence of buffer components and buffer concentration on the breakdown rates of these drugs has to be critically measured.

A complex pH-rate profile exists for many drugs as they undergo hydrolysis at different pH with differing susceptibility of the ionized and un-ionized forms of the drug molecules. An example is the pH-rate profile illustrated in Figure 8.8. It is evident that the drug exists at different ionic forms at different pH levels (cation, zwitterion, or anion). Even though it is not always possible to explain the pH-rate profile completely, it is possible to choose the pH for maximum stability. In the case of Figure 8.8, this would be between pH 4 and 7.5.

Once the pH of maximum stability is identified, an Arrhenius plot at that pH can be constructed to determine the shelf life at that particular pH. Sometimes in order to save time, experiments are designed to observe simultaneous effects of pH and temperature to determine the pH of maximum stability as well as shelf life. Figure 8.11 shows Arrhenius plots for the breakdown of a drug at several pH values. It is evident from the plots that the drug is most stable at pH 6.0. From the plots, rate of breakdown can be determined at any temperature. Activation energy of the drug also can be calculated from the slope of the plots.

Even after finalizing the pH or the buffer system, sometimes it becomes necessary to subject several potential formulations at that pH with different excipient at varying concentrations at a fixed elevated temperature for a fixed amount time. The purpose of this study is to investigate the effects of various formulation ingredients and their concentrations in order to choose the best stable

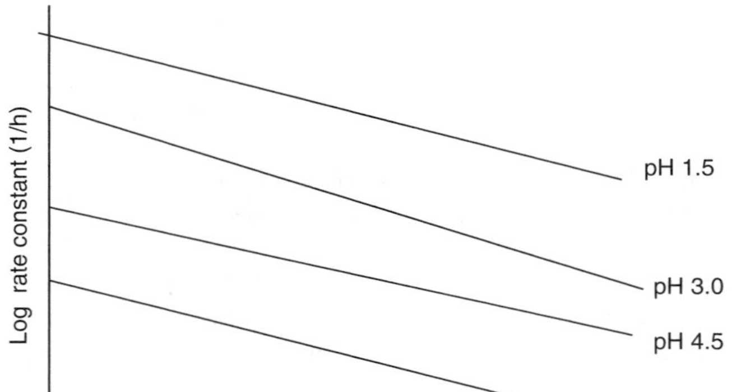

FIGURE 8.11 Sample log rate constant vs. (1/T) profile (Arrhenius plots) for the degradation of a drug at various pH conditions.

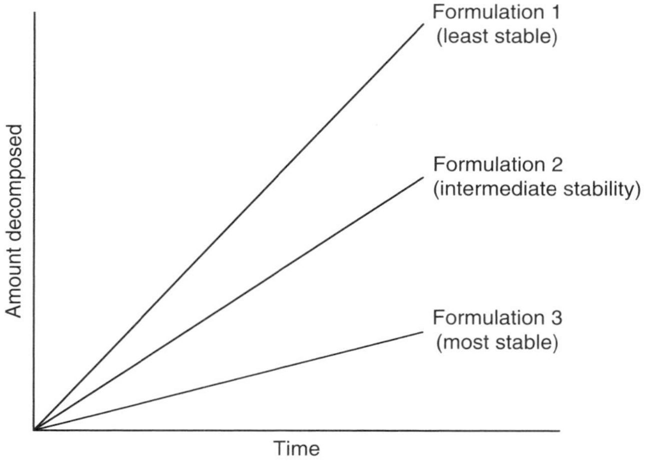

FIGURE 8.12 Amount decomposed–time profiles of various formulations of a drug at elevated temperature.

formulation as illustrated in Figure 8.12. Formulations 1, 2, and 3, with minor differences in composition in terms of formulation ingredients, are subjected to degradation at an elevated temperature. After a certain period of time, formulation 3 turns out to be the most stable formulation with the least steep degradation profile.

VII. CONCLUSION

Often the two terms *reaction kinetics* and *stability* are used interchangeably. Students should recognize that the goal of chemical kinetics is to elucidate degradation reaction mechanisms, whereas that of stability studies is to establish an expiration date. A competent pharmacist should have adequate knowledge to address patients' questions related to the stability of a finished product, and that demands a basic understanding of the possible modes or pathways of degradation and incompatibility of a drug molecule in a given formulation.

HOMEWORK

1. The expiration period for a reconstituted product is 24 h at room temperature. Estimate the expiration period when the product is stored in the refrigerator. *Answer:* 4 days
2. A newly reconstituted product is labeled to be stable for 48 h in a refrigerator. What is the estimated shelf life at room temperature? *Answer:* 1.9 hours
3. The expiration date for a product is 6 years from the current date when stored in a refrigerator. The product has been stored for one month at room temperature. If the product is now returned to the refrigerator, what is the new expiration date? *Answer:* 7.6 months
4. Phenethicillin degrades at pH 7.0 in aqueous solution at 25°C by first-order kinetics with a half-life of 517 h. How many hours will it take for a solution of phenethicillin to degrade to 90% of its starting concentration? *Answer:* 78.3 hours

5. Degradation of a new antiepileptic drug in an aqueous solution at 30°C shows the following concentration of the drug remaining as a function of time. The initial concentration of the drug was 100 mg/ml.

Time (days)	**Concentration Remaining (mg/ml)**
0.5	96.01
1.0	92.19
2.0	84.99
4.0	72.24
8.0	52.19
12.0	37.70
16.00	27.23
20.00	19.67
24.00	14.21
30.00	8.73
36.00	5.36
48.00	2.02

(a) Plot the data on suitable graph paper and determine the order of degradation. *Answer:* first-order

(b) From the slope of the straight line, calculate the rate constant and the half-life for the hydrolysis of this drug. *Answer:* $K = 0.081$ days^{-1}; $t_{1/2} = 8.5$ days

(c) If the drug is considered to be expired after it degrades by 10% or more of the original concentration at 30°C, what would be the expiration time of this compound. *Answer:* $t_{90} = 1.3$ days

6. Samples of a new anticancer drug in aqueous solution were stored at 45°C, 60°C, and 75°C. The first-order rate constants for the decomposition at different temperatures were determined as follows:

Temperature (°C)	45	60	75
Rate Constant (day^{-1})	0.052	0.069	0.089

Considering that the drug expires once the concentration decreases to 90% of the original, calculate the expiration time of the anticancer drug at 25°C (room temperature). *Answer:* 3.1 days

7. The following data show the degradation profile of ampicillin, an antimicrobial drug.

Time (Days)	**Concentration Remaining (mg/ml)**
0.50	993.4
1.00	986.8
2.00	973.6
4.00	947.1
6.00	920.7
8.00	894.2
12.00	841.3
18.00	762.0
24.00	682.7
32.00	576.9
40.00	471.1
48.00	365.3
72.00	48.0

(a) Plot the data on suitable graph paper and determine the order of degradation. Also calculate the rate constant and the half-life of this drug at 25°C. *Answer:* zero-order; $K = 13.2$ mg/ml.hr; $t_{1/2} = 37.9$ days

(b) Calculate the amount remaining after 36 days. *Answer:* 524.0 mg/ml

(c) If the drug is found to be ineffective after it has decomposed by 10% or more, calculate the shelf life of this product at 25°C. *Answer:* $t_{90} = 7.6$ days

8. An aqueous suspension of drug X was formulated containing 100 mg of the drug per teaspoon (5 ml). When the suspension was stored at room temperature (25°C), the half-life was found to be 3.0 days. The half-life increased to 20 days when the suspension was stored in the refrigerator (5°C). Assuming pseudozero-order degradation, calculate the rate constant of drug X at each of these temperatures. *Answer:* $k'_{0(25)} = 3.33$ mg/ml/day^{-1}; $k'_{0(5)} = 0.5$ mg/ml/day

CASES

CASE I

Freshly prepared Amoxicillin pediatric suspension was given an expiration time of 14 days and was directed to be kept refrigerated (4°C) after each use. After 3 days of use, the antibiotic was left at room temperature for one full day in July when the average room temperature stays around 30°C. Can the antibiotic be safely used for another 7 days?

1. What is the usual expiration time written on the label?
2. What may happen if the medication is not kept in the refrigerator for the entire recommended period?
3. What should be the proper advice for the pharmacist to give the patient's parents?
4. How did the pharmacist draw this conclusion?

Hint: Consult problems related to expiration date in the previous section.

CASE II

A bottle of aspirin tablets was left loosely closed in the medicine cabinet located in the bathroom. After a few months, the user complained about an unusual odor upon opening the container and brought the bottle back to the pharmacist. The pharmacist was asked to advise whether the patient should continue the medication or not.

1. What is the usual expiration time written on the label?
2. What might have happened to aspirin?
3. What might have caused the reaction?
4. What should be the proper advice for the pharmacist to give the patient?
5. How did the pharmacist draw this conclusion?

Hints: Aspirin (acetylsalicylic acid) is susceptible to hydrolysis. It was found that the bottle was not closed properly in bathroom's high humidity and also that the material of the bottle was not impervious to moisture. Salicylic acid and acetic acids are formed on hydrolysis. The acetic acid is detectable by its odor (vinegarlike).

CASE III

An extemporaneous preparation of procainamide hydrochloride solution (50 mg/ml) was used to treat a neonate for approximately 3.5 days as the recommended therapy. However, no efficacy was observed. Upon investigation, neither procainamide nor its metabolite was detected in the serum of the patients. When the product was assayed, it was found to contain less than 3 mg/ml of procainamide hydrochloride.

1. What might have happened to the procainamide hydrochloride?
2. What might have caused the reaction?
3. What should the pharmacist have done to prevent this?
4. What did we learn from this adverse experience?

Answers:

1. Procainamide rapidly degraded in the extemporaneous preparation.
2. A sample of the syrup that was used in compounding the preparation (imitation wild cherry syrup) was investigated. It was found to have a pH of 3.0. It is well known that amide bonds are subject to both acid- and base-catalyzed hydrolysis. The low pH accounted for the rapid disappearance of procainamide from the preparation.
3. To prevent this , the pH must be adjusted to 4 to 6, as recommended in Procainamide hydrochloride for Injection, USP.
4. It is prudent for the pharmacist to (1) make sure that the method to be used in compounding a formulation has been thoroughly studied and (2) be as rigorous as possible in duplicating the conditions specified.

CASE IV

During administration of a total nutrient admixture (TNA) formulation, the nurse practitioner observed separation of phases in the mixture.

1. What is a total nutrient admixture formulation?
2. What can cause separation of phases in such a formulation?
3. What is the safety issue of administering destabilized TNA?
4. What are the factors influencing TNA stability?

Answers:

1. Total nutrient admixture (TNA) is a formulation where lipid emulsion is combined with all other parenteral nutrition components in one container. Traditional parenteral nutrition formulations contain a solution of dextrose, amino acids, and micronutrients. Intravenous lipid emulsion is an oil-in-water emulsion, with triglyceride particles from soybean oil or soybean and safflower oils dispersed in water for injection, with egg yolk phospholipids added as emulsifier.
2. Creaming, aggregation, coalescence, cracking, etc. (discussed in full in emulsion chapter)

3. Administration of a destabilized total nutrient admixture is likely to cause major lipid emboli throughout the capillary system.
4. The conditions most likely to destabilize total nutrient admixture formulations are a final pH <5.3; final concentration of amino acids <2.5%; final concentration of dextrose <10%; final concentration of lipid emulsion <2%, storage at freezing or above room temperature, order of admixture, etc.

CASE V

A colorless 10% aqueous solution of sodium sulfacetamide was brought back to the pharmacist because it turned yellowish brown.

1. What can cause this color formation?
2. What can be done to prevent similar things from happening in future?
3. Should the solution be discarded?

Answers:

1. Sulfacetamide undergoes oxidative decomposition to develop color. Sulfacettamide solution, when stressed with heat in the presence of oxygen, develops a yellow color. The sulfanilamide produced by hydrolysis of sulfacetamide, in the presence of light and oxygen, produces a brown color. Therefore, the observed color formation can be a combination of hydrolysis, oxidation, heat, and light.
2. Hydrolysis is not a serious problem in sulfacetamide solutions at room temperature; however, it is a concern during sterilization by autoclaving at 121°C. Oxidative decomposition can be prevented by removing oxygen from solutions and sealing them in ampules. Antioxidants are often added to sulfacetamide solutions. However, sodium metabisulfite, added to reduce color development, also acts as a catalyst of hydrolysis. Moreover, the metabisulfite is not very effective in preventing color formation. Under normal storage conditions, 0.1% sodium thiosulfate almost completely stops color formation. The most effective inhibitor of color development appears to be sodium edetate, used in the 0.01 to 0.5% concentration range. This chelating agent probably works by complexing with trace metal ions that serve as catalysts for oxidation reactions. Apart from considering the above formulation aids, the solution should also be dispensed in amber ampules.
3. Yes, as a different compound has formed in course of reaction.

REFERENCES

1. Aulton, M.E., *Pharmaceutics: The Science of Dosage Form Design*, 6th ed., Churchill Livingstone, New York, 1988.
2. Cartensen, J.T., *Drug Stability: Principles and Practices*, 2nd ed. Marcel Dekker, New York, 1995.
3. Connors, K.A., Amidon, G.L., Stella, J.V., *Chemical Stability of Pharmaceuticals*, 2nd ed., Wiley, New York, 1986.
4. *Current Good Manufacturing Practice*, 21 CFR 211, 2004.
5. *DRAFT Guidance for Industry: Stability Testing of Drug Substances and Drug Products,* FDA, Center for Drug Evaluation and Research and Center for Biologics Evaluation and Research, June 1998.
6. Florence, A.T., and Attwood, D., *Physicochemical Principles of Pharmacy*, 3rd ed., Macmillan Press, Hampshire, UK, 1998.

7. *Guideline for Submitting Documentation for the Stability of Human Drugs and Biologics*, FDA, Center for Drugs and Biologics, Office of Drug Research and Review, February 1987.
8. Gennaro, A.R. (ed.), *Remington: The Science and Practice of Pharmacy*, 20th ed., Lippincott Williams & Wilkins, Philadelphia, 1995.
9. Martin, A., *Physical Pharmacy*, 4th ed., Lea & Febiger, Philadelphia, 1993.
10. Newton, D.W., Three drug stability lives, *Int. J. Pharm. Compounding,* 4(3), 2000.
11. Zatz, J.L., *Pharmaceutical Calculations*, 3rd ed., John Wiley, New York, 1995.

9 Drug and Dosage Form Development: Regulatory Perspectives

*Edward Dennis Bashaw**

CONTENTS

* The comments, suggestions, and interpretation contained in this document represent the personal view of the author and under no circumstances should the contents, in part or in whole, be interpreted as the view or recommendation of the U.S. Food and Drug Administration.

0-4152-8863-0/05/$0.00+$1.50
© 2005 by CRC Press LLC

Learning Objectives

After finishing this chapter the student will have a knowledge of:

- A general history of the FDA and the approval process
- An appreciation of how the current approval process has been shaped by regulation
- The different phases of drug development
- The investigational new drug application/new drug application process
- How the different scientific disciplines interact during the approval process

I. INTRODUCTION

The process for bringing a new drug to the market in the U.S. is a long and arduous one that attempts to balance the need for thorough testing of a potential drug, in both animal and human models, to ensure drug safety and efficacy. This chapter will review the history of the drug approval process and its evolution in the U.S. Although not the subject of this chapter, it should be noted that the FDA is divided into various centers including the Center for Biologic Evaluation and Research (CBER), the Center for Drug Evaluation and Research (CDER), the Center for Devices and Radiological Health (CDRH), and the Center for Food Safety and Nutrition (CFSAN). Each of these centers have their own regulatory authority under the Code of Federal Regulations (CFR) and have evolved from the drug approval process as the technology has grown in these areas. Because of this we will focus our discussion in this chapter on the drug approval processes used in CDER, as those used in the other centers either parallel or are related to the processes used in CDER. For those interested in the roles of the various centers in the FDA, the FDA website provides a link to each of the centers and their mission statements (http://www.fda.gov).

II. THE DRUG APPROVAL PROCESS: FROM DISCOVERY, TO MARKET, AND BEYOND

The discovery, approval, and successful marketing of a pharmaceutical product is a process that encompasses many years and millions of dollars. According to a February 1993 report by the Congressional Office of Technology Assessment, it costs, on average, $359 million to develop and launch a new drug. Because of the rising costs of pharmaceutical development and the fact that these costs are ultimately borne by the consumer through direct costs at the pharmacy and indirectly through increasing health plan costs, much effort has been expended recently in an attempt to streamline the drug approval process. Prior to reviewing these initiatives, one must first, however, review how we arrived at the current drug approval process. Only then, once a proper historical context has been laid out, can one fully appreciate the safeguards that are currently in place and the will of the public not to relive the tragedies of the past and peer into the future of the drug approval process.

A. A Brief History of Drug Regulation in the U.S.

1. The Early Years: Pre-1880

In the U.S. prior to 1880 the regulation of drugs and drug marketing was nonexistent or at best a state function. Although laws were in place regarding the importation of drugs and attempts were made to keep poisonous substances out of the marketplace, these were easily evaded due to the lack of a centralized authority charged with drug regulation. During this time the first rudimentary attempts were made in both medicine and pharmacy to establish standards of drug purity. In 1820 under the auspices of the U.S. Pharmacopeial Convention, the first recognized pharmacopoeia in the U.S. was published. This provided to the practitioner a list of drugs that the convention considered to be useful, and provided rudimentary information on their preparation and dosing. Later in 1888, the American Pharmaceutical Association established the National Formulary as another reliable source of drug information for the professional. While these first steps in establishing standards of drug purity were being taken, the vast expanse of the U.S. and the lack of trained practitioners throughout the country provided opportunities for patent medicines and entrepreneurs to thrive among a medically underserved populace.

2. The 1880s to 1906: Patent Medicines

Following the Civil War, patent medicines went into their heyday. The availability of cheap lithography for advertising, rapid transit via the railway, and a large number of veterans suffering from wounds were a huckster's delight. From about 1870 to 1910 opiate and cocaine laced products were widespread in the market as teething syrups, cough cures, brain tonics, and nerve foods. Probably no other product is more typical of this era than Vin Marini. Vin Marini was bordeaux wine laced with cocaine. It was sold throughout the world promising a return of vigor and strength. One advertisement claimed over 32,000 testimonials from the Pope down to the common man. While the exact amount of cocaine in Vin Marini is unknown today, one analysis from 1886 indicates that an 18-oz bottle contained just over 2.1 grains of cocaine (approximately 140 mg). As the labeled dose was one bottle per day, clearly sufficient amounts of cocaine were present to have quite an effect on an individual. Reports from that time indicate that addiction to Vin Marini was widespread. Curiously, opposition to Vin Marini was not raised because of its inclusion of cocaine, but from temperance groups because of its alcohol content.

Partially in response to the widespread availability of narcotics and partially due to a drug scandal arising from the purchase by the government of adulterated quinine during the Spanish-American War, the first Pure Food and Drug Act was passed in 1906. While the intent of the 1906 act was to establish standards for drug purity and labeling, the act itself was not well enforced and was actually used by patent medicine makers as a selling point. Under the provisions of the act, sponsors were required to list a product's ingredients. In complying with the law, many manufacturers added the phrase that their formulation was "guaranteed" under the Food and Drug Act of 1906, thus implying government recognition of the medicinal properties of their product. While these statements were eventually removed from the labels as a result of government pressure, the fact that the manufacturers felt bold enough to attempt such labeling demonstrates the general lack of enforcement power of the government under the original 1906 act. Effective enforcement power had to wait for a tragedy.

3. The 1930s: Elixir of Sulfanilamide and the 1938 Food Drug and Cosmetic Act

The first assertive attempt to regulate drugs in this country came about as an outgrowth of the Elixir of Sulfanilamide scandal. Briefly, in the early 1930s sulfa drugs were developed as the first effective agents to treat bacterial infections. A problem common to all sulfa drugs was their poor solubilty in aqueous media. In 1937 a chemist working for S.E. Massengill discovered that diethylene glycol was an excellent solvent for sulfanilamide. This new preparation was sold and distributed as Elixir of Sulfanilamide. About 110 people died from kidney failure after taking this product, many of them children. Under the laws in place at the time, the company was convicted only of misbranding their drug product, as an elixir, by definition, is a water-ethanol mixture. Public outcry for a stronger Food and Drug Administration resulted in the 1938 act that drastically increased the authority of the FDA and stated, for the first time that the safety of the drug product be proved prior to market. The major provisions of the Food, Drug, and Cosmetic Act of 1938:

- Required new drugs to be shown safe before marketing—starting a new system of drug regulation.
- Eliminated the Sherley Amendment requirement to prove intent to defraud in drug misbranding cases.
- Extended control to cosmetics and therapeutic devices.
- Provided that safe tolerances be set for unavoidable poisonous substances.
- Authorized standards of identity, quality, and fill-of-container for foods.
- Authorized factory inspections.
- Added the remedy of court injunctions to the previous penalties of seizures and prosecutions.

The 1938 act was a landmark piece of drug regulation and came at a time when the industrialization of pharmaceuticals was just beginning. No longer were drug manufacturers providing the raw materials to the community pharmacist for compounding. Instead they were providing more and more a finished product and transitioning the pharmacist from a small scale manufacturing laboratory into a dispensing and quality assurance role. The recognition of this change in the manufacture of pharmaceutical dosage forms partially led to the Durham–Humphrey Amendment of 1951. This amendment to the 1938 act clarified the vague line between prescription and non-prescription drugs already under the law. The amendment specifically stated that dangerous drugs, defined by several parameters, could not be dispensed without a prescription, witnessed by the prescription legend: "Caution: Federal law prohibits dispensing without a prescription." While the 1938 act and its amendments (see Table 9.1) did improve the process by which drugs reach the market, it still had significant shortcomings that became all too apparent with the marketing of thalidomide in the early 1960s.

4. The 1960s: Thalidomide and the Development of the Modern Drug Approval Process

Since the 1930s, drug development has become an industrialized process, with large corporations taking the lead. Under the 1938 Food Drug and Cosmetic Act all a sponsor had to demonstrate was that the drug itself was not harmful prior to marketing. Thalidomide was synthesized in West Germany in 1953 by Chemie Grünenthal as a mild sedative. Unfortunately, its reputation as a mild sedative with few side effects caused it to be used by pregnant women. It was marketed from October 1, 1957 (in West Germany) into the early 1960s. Thalidomide was available in at least 46 countries under many different brand names. While it was never marketed in the U.S., thalidomide was available under a sampling program in Canada in late 1959 and was licensed for prescription use in Canada on April 1, 1961. During this time a rise in a particular pattern of birth defects, called phocomelia, was noted among mothers who took thalidomide. The shortening or absence of hands and limbs was the most distinguishing feature. Although this is not the only birth defect seen with thalidomide, it is considered characteristic of the syndrome.

Thalidomide was not marketed in the U.S. due to the efforts of Dr. Frances Kelsey. Dr. Kelsey was the medical officer assigned to review the safety data for thalidomide prior to marketing. Her decision to obtain more data and to resist pressure to approve a drug that was already marketed in Europe spared the U.S. the brunt of this tragedy. In Canada, where it was available for approximately two years, an estimated 10 to 20 thousand babies were born with thalidomide related birth defects. Of these, an estimated 5000 are still alive today.

While the U.S. avoided the impact of thalidomide, thalidomide did have a major impact on the drug approval process in the U.S. A review of the FDA's role in keeping thalidomide off of the U.S. market shows how close it came to being licensed for use. As a result of ongoing congressional hearings into drug pricing, the Kefauver-Harris Amendment to the Food Drug and Cosmetic Act of 1938 was passed. This amendment heralded the birth of the modern drug approval process in

TABLE 9.1
An Abbreviated Timeline of Drug Regulation in the U.S.[a]

1900s	Pure Food and Drug Act of 1906
1930s	Federal Food, Drug, and Cosmetic Act of 1938
	• Requires demonstration of safety prior to marketing
1940s	Antibiotic Certification Program established
	• Requires FDA to certify the potency of each batch of antibiotics manufactured (repealed in 1983)
1950s	Durham–Humphrey Amendments of 1951
	• Delineate difference between prescription and over-the-counter drugs
1960s	Kefauver-Harris Amendments of 1962
	• Require demonstration of efficacy in addition to safety prior to marketing
	• Drug Efficacy Study Implementation (DESI) Program established with NAS
	• FDA forms the DESI to incorporate the recommendations of an NAS investigation of the effectiveness of drugs marketed between 1938 and 1962
1970s	FDA issues regulations on bioavailabilty and bioequivalence testing
	• Allows for the growth of the generic drug industry
1980s	Orphan Drug Act (1983)
	• Provides tax and other incentives to drug manufacturers to develop drugs for diseases with small target populations
	Drug Price Competition and Patent Term Restoration Act of 1984
	• Allows manufacturers to add up to 7 years of exclusive patent life to their products to compensate for time lost during FDA review
	Prescription Drug Marketing Act of 1988
	• Among other provisions, bans sale or trade of drug samples
1990s	Regulations published expanding access
	• Provide for parallel track access during drug development
	Prescription Drug User Fee Act of 1992
	• Drug manufacturers assessed a user fee to fund new review positions
	• FDA required to establish new tracking systems and put new performance goals in place.
	Generic Drug Enforcement Act of 1992
	Food and Drug Administration Modernization Act of 1997
	• Many structural changes to both the agency and the review process, including but not limited to:
	– Re-authorizing user fees for an additional 5 years
	– Establishing pediatric exclusivity
	– Attempting to regulate pharmacy compounding

[a] For a more detailed timeline see http://fda.gov/cder/about/history/

that it mandated demonstration of safety and efficacy testing of pharmaceuticals and placed an emphasis on preclinical animal testing for teratogens and carcinogenicity assessment.

In addition to looking forward to the safe development of new drugs, it was also realized that many drugs had been marketed between 1938 and 1962 without proper efficacy testing. In 1966 the FDA contracted with the National Academy of Sciences (NAS) and the National Research Council (NRC) to evaluate the effectiveness of 4000 drugs that had been approved on the basis of safety alone between 1938 and 1962. This led in 1968 to the Drug Efficacy Study Implementation (DESI) program in the FDA to incorporate the recommendations of the NAS/NRC investigation of the effectiveness of drugs marketed between 1938 and 1962. The DESI program evaluated over 3000 products and over 16,000 therapeutic claims. The last NAS/NRC report was submitted in 1969, but the contract was extended through 1973 to cover ongoing issues. The initial agency review of the NAS/NRC reports by the task force was completed in November 1970. By September

1981 final regulatory action had been taken on 90% of all DESI products. By 1984, final action had been completed on 3443 products; of these, 2225 were found to be effective, 1051 were found not effective, and 167 were pending final determination.

5. The 1980s to Today: Refinement of the Drug Evaluation Process

In the 1980s cracks began to be noticed in the drug approval process, while it was conceded that the FDA methods of review and approval generally resulted in safe drugs reaching the market. The FDA was also seen as hindering drug development by having the approval bar set too high. Critics of the FDA cited the availability of drugs first approved in other countries as an indication of a restrictive approval process. The vocal criticism of AIDS patient advocacy groups against the FDA for restricting their access to alternative therapies led to a revamping of the drug approval process to allow early access to drugs in cases of life-threatening diseases. Eventually Congress enacted two pieces of legislation that dramatically overhauled the drug approval process and how the FDA is funded: the Prescription Drug User Fee Act (PDUFA) of 1992 and the Food and Drug Administration Modernization Act of 1997 (FDAMA).

In PDUFA Congress recognized that the agency's funding was inadequate to the job of drug review. Instead of increasing funding through a larger appropriation (potentially resulting in increased taxes) Congress established a user fee system whereby drug companies paid the FDA a fee for their drug reviews. While the bulk of the agency's funding would still come from conventional sources, the funds generated via user fees would be earmarked for hiring new review staff and upgrade reviewer's information technology (IT) resources. Eventually over 600 additional reviewer and reviewer support positions were created by this system and the review timelines began to be shortened dramatically. As of 2004 the user fee for a new molecular entity was $573,000. For small companies, for whom the user fee would represent a financial burden, the FDA has the authority to waive the fee under certain conditions.

In exchange for the establishment of user fees the agency agreed to progressively shorter milestones in terms of the review itself. Ultimately the goal of the user fee program is to reduce, in stages, the time for review for a priority or standard drug approval to 6 months and 10 months, respectively. This represents a dramatic decrease in the time for review, which in the early 1990s was averaging almost 2 years. User fees have provided such a dramatic decrease in review timelines that the concept of user fees is being extended throughout the FDA.

Whereas PDUFA was designed to enhance the review process by funding additional reviewer positions, the Food and Drug Administration Modernization Act (FDAMA) of 1997 was much more broad reaching. Besides reauthorizing user fees (and establishing new guidelines for their implementation), FDAMA also codified a number of different legislative initiatives. These ranged from requiring the FDA to survey nonprescription drugs for levels of mercury to establishing a fast track access program for promising therapies for life-threatening diseases. Some of the features of FDAMA are summarized below:

- Reauthorizes PDUFA of 1992 for an additional 5 years. Reauthorization also establishes new performance goals for drug reviews and establishes response times for meetings with applicants among other structural changes in the review process.
- Establishes a mechanism for granting new drug application holders an additional 6 months of marketing exclusivity for completing trials in pediatric subjects.
- Requires the National Institutes of Health, to establish a registry of clinical trials for both federally and privately funded trials of experimental treatments for serious or life-threatening diseases
- Clarifies the FDA's authority and responsibility for conducting postmarketing surveillance of devices.

FDAMA has had a major impact on both the regulatory scope of the FDA and in its day to day operation. By serving as a vehicle for the reauthorization of user fees, Congress signaled to the agency a willingness to provide adequate resources to accelerate drug approval. Through the pediatric exclusivity provisions, Congress provided the FDA with a mechanism to obtain needed pediatric dosing data that is not normally generated during the approval process. Being such a disparate piece of legislation precludes a complete discussion of the provisions of FDAMA here. The law itself and the FDA's implementation of the various sections and subsections of the act can be found and tracked at the FDA's website.

III. THE DRUG APPROVAL PROCESS

Normally when one thinks of the drug approval process, one thinks of the three-box approach shown in Figure 9.1, that is, moving from discovery to an investigational new drug to a new drug application and to market. In reality, as shown in Figure 9.2, each of these phases is broken down into subphases, which have their own timelines and objectives.

A. Preclinical Research

Although we normally think of drug development as coming out of large industrial concerns, new drugs may be developed by a variety of different people or organizations, including independent researchers, university medical centers, government centers, or other organizations. At its inception, a target compound is identified either by synthesizing a new member of a known therapeutic class, such as an antibiotic, or as part of new research into how the body functions. From these studies, researchers develop new theories of how to treat illnesses and begin to search for compounds that will help achieve the desired effect on the body. As most pharmaceutical companies have links to the chemical industry, the generation of new compounds for development is a continuous process. While the ratio of synthesized compound to marketed pharmaceutical is less than 500:1, the potential payoffs are enormous.

Even with the vast payoffs of so-called blockbuster drugs, the cost of evaluating each of the synthesized compounds is a limiting factor in drug development. The development of low-cost, rapid screening tools is essential to the identification of a lead compound. Such screening tests can include adding target compounds to enzymes, cell cultures, or cellular substances grown in the laboratory to determine what effect, if any, the new compound has. This screening process can take a significant amount of time, and it is here that most of the attrition occurs in the drug development process, as most compounds tested either do not have the desired clinical profile or are not sufficiently safe to be usable.

Once a lead compound has been identified, more sophisticated *in vitro* animal models of the disease are tested to determine the compound's utility. This process involves using as few animals as possible, and researchers are careful to make sure that the animals are treated as humanely as possible during the studies. While animal models are being used less and less, there are often no

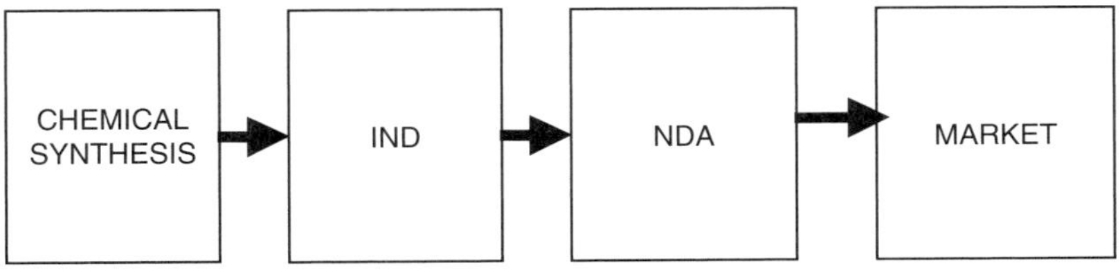

FIGURE 9.1 Traditional view of the drug approval process.

		PHASES of Drug Development				
	Preclinical Testing	**I**	**II**	**III**	**FDA**	**Post-Marketing Activities (PHASE IV)**
Years	3.5	1–2	2	3–5	Priority Drugs = 6 months Standard Drugs = 10 months	
Test Population	Animal studies Using both *in vivo* and *in vitro* models, also initial formulation and synthesis work	Healthy volunteers	Patient volunteers	Large Multicenter Clinical Trials in Patient volunteers	Review	Additional Post marketing testing required by FDA
Purpose	Initial assessment of safety and biological activity; also mechanism studies	Determine safety and dosage	Proof of concept, Initial safety evaluation in patients	Confirm effectiveness, monitor adverse reactions from long-term use	Review	Ongoing adverse event reporting
Success Rate	Thousands of compounds evaluated	<1% Enter trials			1 Approved	

IND FILED

NDA FILED

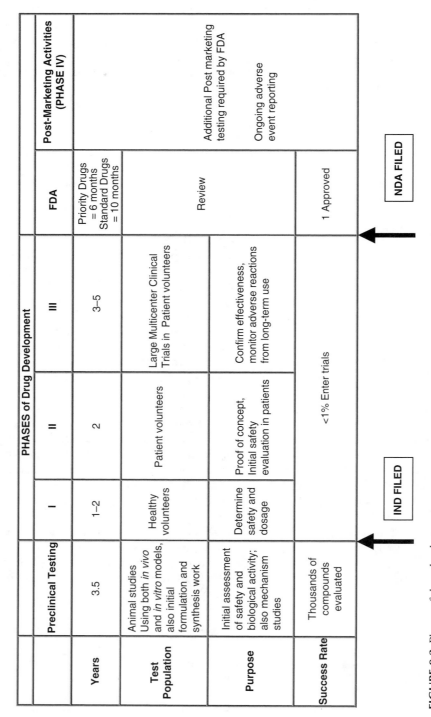

FIGURE 9.2 Phases of drug development.

alternatives. Often, more than one species is tested because compounds tend to act differently in different species. The goal of animal testing is to learn how the drug is absorbed by the body (i.e., its bioavailability), how the drug is metabolized in the body, the toxicity of the drug and its metabolites, and quickly the drug is excreted from the body. These studies are instrumental in predicting the dose for human testing via a methodology called allometric scaling. This method uses the exposure data from the best animal model, and attempts to correlate the known animal exposure to humans. From this correlation and initial starting dose can be derived that has an acceptable margin of safety.

In addition to this preclinical phase of testing to determine acute safety, both long-term and special safety studies are initiated to assess the risk of the drug being a carcinogen or teratogen. The experience with thalidomide resulted in the explosion of interest in the field of teratogenicity, and these tests include determining the NOEL or "no observed effect limit" or the dose at which no or minimal side effects are observed in the animals. Because of the time required to conduct these studies, they are often occurring or being "reported out" while the clinical studies are underway. Although no longer used, LD50 data (dose at which 50% of the test animals died) was routinely obtained until the mid to late 1970s. Since then the determination of the NOEL dose has been more commonly used in the animal safety assessment.

B. Formulation Development

It cannot be stressed too much that an early assessment of the desired dosage-form properties can contribute greatly to the speed of the drug development process. The lack of either an appropriate dosage form or a validated manufacturing process during the clinical trial can result in disaster. One must consider factors such as the target population (children, adults), the amount of drug to be given in each dose, stability of the drug product after manufacture, and the general design and cleanliness of the process. An in-house review conducted by one clinical division revealed that over 70% of all nonapproval letters issued in a 3-year period cited chemistry deficiencies as at least a partial reason for nonapproval.

Many reasons for nonapproval result from a lack of compliance with FDA regulations designed to ensure the quality of pharmaceutical manufacturing. These good manufacturing practice requirements govern quality management and control for all aspects of drug manufacturing. To enforce good manufacturing practice requirements, the FDA conducts field inspections where trained investigators periodically visit manufacturing sites to ensure that a facility is in compliance with the regulations. Good manufacturing practice requirements are updated regularly and represent state of the art procedures in pharmaceutical manufacturing to ensure a consistent and safe product.

Previously, the chemistry sections of an application were considered to be overly complex and drawn out. One of the reforms initiated by FDAMA is an effort to reduce the number of postmarketing manufacturing changes that require FDA approval. Under the new law, a company must obtain FDA approval before implementing any "major" manufacturing change. Such a change is one determined by the FDA to have substantial potential to adversely affect the identity, strength, quality, purity, and potency of a drug as they relate to safety or effectiveness. Other changes may be implemented either with or without submission of a supplemental new drug application. Changes not requiring a supplement may be reported in the annual report required to be filed with FDA or on another date as the agency may require.

Predating, but in the spirit of, FDAMA reforms, the FDA under its Scale-Up and Post-Approval Change (SUPAC) guidances allows various other types of changes to be completed by the sponsor with minimal FDA oversight. Currently the FDA has SUPAC guidances for topicals and solid oral dosage forms. In each guidance a classification system has been applied to specific types of changes, in either formulation or manufacturing processes, and various action levels from simple reporting to *in vivo* bio-studies are recommended for the changes. These changes combined with FDAMA sponsored initiatives have decreased the regulatory burden on the sponsor without affecting drug product quality.

C. Investigational New Drug Application

Once preclinical testing has been completed a decision has to be made about whether or not the drug has enough potential to proceed to *in vivo* studies in humans. There are three types of investigational new drug applications:

Commercial: most investigational new drug applications and what this discussion will primarily refer to.

Single Investigator or researcher: for one physician, at one study site, often university based. This investigational new drug application is most commonly used to test new dosing regimens of an approved drug.

Emergency/Compassionate Use: often used for patients with an otherwise incurable disease who do not meet enrollment criteria for an ongoing treatment protocol but may derive some benefit from the drug; can be used as part of the approval process in special circumstances as in development of antidotes where exposure to a toxic substance is required and is thus unethical.

Assuming that the company wishes to push the drug forward, the sponsor must file an investigational new drug application with the FDA for permission prior to using the drug on people in the U.S. One trend that has occurred over the past decade is the tendency for drug companies to do their initial work on humans outside of the U.S., in countries where the standard for initial testing is perceived to be lower. In response to this, Congress, through FDAMA and other legislation, has attempted to keep this initial development work in the U.S. by providing incentives to the sponsor. So far the success of these programs has been mixed, with some sponsors still preferring to do their research overseas. Whether or not this is truly a phenomena related to stringent FDA regulations, as its critics claim, or is just a reflection of the multinational nature of drug development is still hotly debated.

The investigational new drug application must contain summaries of all of the information the sponsor has generated in this country under their investigational new drug application or that they are aware of through the literature. By law, the investigational new drug application consists of a chemistry section, a preclinical (i.e., pharmacology) results section, and a medical review section in which planned first time in humans protocols are presented. Upon receipt, the FDA has 30 days to respond to this initial submission. If the sponsor has not heard from the FDA in this time, the sponsor may start its trials. Unless a sponser is placed on "clinical hold", this initial submission is the only time during the investigational new drug application process that the sponsor has to wait for FDA approval. After the initial submission a sponsor may start its trials the day after submitting its protocol to the FDA. Such a process of starting trials early without waiting for FDA comments does, from the sponsors point of view, speed up drug research. The sponsor also runs the risk that when the protocol is reviewed, the agency will find grounds to stop the trial via the issuance of a clinical hold letter. Such a letter identifies in detail the deficiences in the proposed clinical study plan and proposes an alternative trial design.

D. Clinical Trials

Ultimately all of the preclinical and formulation work is done to allow *in vivo* dosing in humans. These clinical studies (as outlined in Figure 9.3) have many subelements that allow for the rational and step-wise development of drugs.

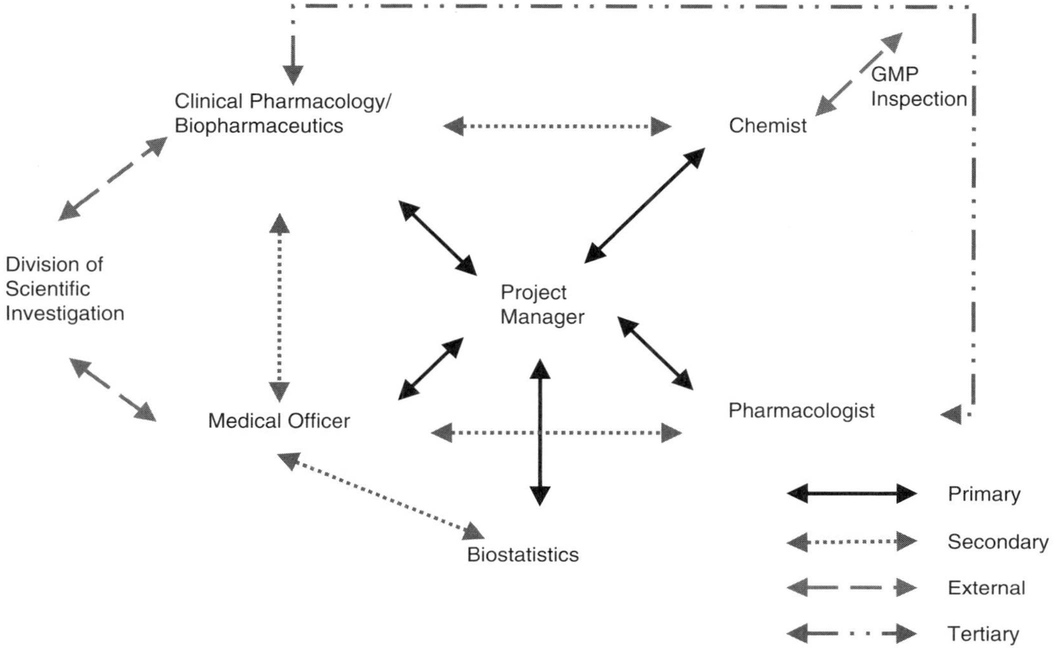

FIGURE 9.3 Representation of interaction between disciplines during the review process.

- *Phase I* trials usually involve administering the drug to a small number of healthy volunteers (typically about 20 to 100 people per trial) to learn how to administer the drug safely. Researchers closely monitor the participants' side effects profile in order to obtain initial dosing information and to adjust dosages if need be. At this time in addition to the initial clinical work, *in vivo* biopharmaceutic trials are also initiated.
- *Phase II* trials attempt to determine patients' responses to the drug; that is, these are proof of concept studies. Typically several hundred people with the disease participate in each Phase II trial. In Phase II trials a decision is made as to whether or not there is a useful therapeutic response evident that would warrant Phase III trials. If enough patients respond to the drug, the trial moves to Phase III.
- *Phase III* trials enroll a large number of participants (several hundreds to thousands). The FDA strongly recommends that drug sponsors meet with FDA officials before beginning Phase III trials to design the trials and their endpoints. Because of its timing, this meeting is often referred to as the end of Phase II, or EOP2, meeting. Traditionally these trials can involve thousands of subjects dispersed across many study sites (the so-called multicenter clinical trial). Normally the FDA required replication of clinical studies for approval, but as part of FDAMA, the agency is now willing to consider other sources of information such as *in vivo* biopharmaceutic studies as confirmatory information (provided a concentration-effect relationship has been previously demonstrated).

E. New Drug Application

Once all of the pivotal trials have been completed and the final reports are written, the sponsor can meet with the FDA and hold what is called a pre–new drug application meeting. At this meeting the sponsor provides the appropriate review division in the FDA with a first look at the proposed new drug application submission as a total package. Final discussions regarding organization of the reports and the preparation of reviewer-aids, either electronic or on paper, are proposed. If possible, the sponsor and the agency also agree that the package appears ready for submission. These pre–new drug application meetings are not required, but they provide the sponsor with additional interaction with the agency and improve their chances of approval by incorporating comments regarding document organization into their final printed or electronic package.

Once the sponsor submits the new drug application, the FDA will classify the filing to indicate the priority for review. Such classification can be determined based on the chemical nature of the drug molecule or therapeutic need. The following are frequently used classification types:

- Type S: standard review; other treatments are currently available
- Type P: priority review; therapeutic gain, new mechanism of action, etc.
- Type 1: new drugs; not available in U.S. market
- Type 2: derivative of active moiety approved in U.S.; can be ester, salt, etc.
- Type 3: new formulation of approved drug
- Type 4: new combination of drug with additional drugs
- Type 5: new manufacturer

An application can get a single designation such as P or S or multiple designations such as 1P, 3S, or 3 and 5.

One often unappreciated aspect of the investigational new drug application/new drug application development process is the amount of data and paper that are generated throughout the process. While the size of a new drug application can vary widely depending on drug class, indication, number of indications being sought, etc., the entire document for an average drug can run to 300 or more binders, each of which can contain up to 500 pages. From an electronic standpoint, a new drug application can easily fill 10 or more CD-ROMS with data. Clearly the investigational new drug application/new drug application process generates massive amounts of data in its quest to demonstrate safety and efficacy for a drug product, so much so that the days of a single reviewer commenting on the entire package are long gone. Instead, once a new drug application is submitted to a review division in the FDA, a multidisciplinary review team is formed to coordinate the review. While many of these people have been involved in the investigational new drug application process with this drug, it is not uncommon for different reviewers to be assigned to the new drug application because of competing priorities.

F. Drug Review

Once an applicant has submitted a new drug application to the FDA for review (and paid the appropriate user fee), the agency initiates a filing review. This is a screening process that allows the agency to reject applications that are incomplete. In filing a review, the agency is only concerned with the "type and number of studies submitted." The agency is not concerned, at this time, with the results of these trials — only that they are present. An example of this would be an applicant leaving out the results of teratogenicity testing. As the regulations call for adequate testing to determine the teratogenic potential of a drug, and the application cannot be acted upon without such information, the application would be returned to the sponsor with a refuse to file letter that lists the missing information. The agency has up to 60 days following submission to make a filing decision. If no action is taken by this time, the application is considered accepted by the agency for review.

It should be noted that there are at least three routes for drug approval: the 505 (b)(1), the 505 (b)(2), and the 505 subpart (j) application. Individually these represent the standard new drug application, the new drug application using a previous finding of safety and efficacy from another sponsor, and the abbreviated new drug application (i.e., for generic drugs), respectively. We will limit our discussion to the standard NDA review process.

G. The FDA Review Team

When a new drug application is submitted, the reviewing division forms a review team to coordinate the process of evaluating the application. The team normally consists of the following disciplines:

- *Medical:* The medical review is what most people envision when they think of the drug review process. The medical reviewer is primarily charged with determining the safety and efficacy of the drug product. This involves evaluation of the proper dosing regimen, analysis of the adverse event profile, and consideration of the reviews and issues raised by the other team members. Ultimately the medical reviewers are charged with answering the following questions:
  - Do the submitted clinical studies provide "substantial evidence" of the drug's effectiveness?
  - Do the submitted clinical studies demonstrate that the drug is safe when given as directed in the proposed labeling.
- *Chemistry:* The chemistry review is primarily concerned with the documentation claiming that the drug synthesis and manufacturing process are properly controlled. This is done by both review of the new drug application and by field inspection of the manufacturing site by one of the FDA district offices, often accompanied by the reviewing chemist. The chemist evaluates, among other things, the physical properties of the drug, the raw drug substance handling, specification setting, and final dosage form inspection and release procedures associated with manufacture. In doing so the chemist interacts with the clinical pharmcology/biopharmaceutics reviewer in the selection of appropriate *in vitro* release testing procedures.
- *Statistics:* The statistician primarily serves as a consultant to the medical review team to provide expert statistical consultation on trial design and evaluation. The statistician is intimately involved in the assessment of the efficacy and safety of the proposed treatment and in validating that the appropriate study design was used.
- *Clinical Pharmacology/Biophamaceutics*: The clinical pharmacology/biopharmaceutics reviewer is responsible for determining the disposition of the drug in humans along with its elimination (pharmacokinetics). Along with the medical officer, this reviewer evaluates the onset and timing of drug effect (pharmcodynamics). In addition, this reviewer is concerned with the *in vitro* release characteristics of the drug and dosage form (biopharmaceutics). There are overlapping responsibilities with the review chemist in terms of the setting of *in vitro* drug release testing procedures and specifications.
- *Pharmacology:* Unlike the clinical pharmacology/biopharmaceutics reviewer, the pharmacologist is concerned with the disposition and toxicity of the target drug substance in animals. The pharmacologist evaluates acute toxicity data, in addition to teratogenicity and carcinogenicity data. The pharmacologist works primarily with the medical officer in the evaluation of adverse effects and their prediction from animal models.
- *Project Management:* Project managers in the FDA represent the primary interface between the drug manufacturer and the review staff. They arrange, run, and document the decisions and agreements made at meetings. They route the documents to the appropriate review team member and are the primary liaison between the drug review team and agency management. It is the project management staff at the FDA that sees that the review team has what it needs when it needs it and keeps the administrative structure running.

Figure 9.3 shows an overview of the interactions between the individual review team members for a typical drug. Depending on the size of the application, more than one reviewer may be assigned to a new drug application. Sometimes, owing to specialization, one reviewer may be brought into a team to work on only one issue, such as the evaluation of *in vitro* drug interaction studies, to assist the primary reviewer.

H. Decision Making in the Review Process

A key point to realize about the drug approval process is that it is a bureaucracy, in that the final decision regarding approval is not the purview of a single individual but is the collective recommendation of the entire review team. While no one individual reviewer can approve a drug, an individual reviewer can, because of deficiencies in their section, recommend nonapproval. In cases where the decision is overly complex or is seen as precedent setting, the FDA has a mechanism in place to call together outside experts in the various specialty fields of medicine as an advisory committee. These committees meet at various times of the year to answer specific review related questions or to provide the FDA review staff with a public forum in which to discuss scientific issues related to drug approval and use. A typical committee might consist of three or four academic researchers, three or four practicing clinicians, a statistician, a clinical pharmacologist, and a lay member to represent the public. At these meetings both the FDA and the sponsor make presentations to the committee and respond to questions raised by the panel. The public is invited to attend advisory committee meetings, and some time at each open meeting is allowed for public comment. Once a committee meeting has been held, the minutesand copies of presentations are available on the World Wide Web.

The FDA usually completes its review of a standard drug in 10 to 12 months. One hundred and twenty days prior to a drug's anticipated approval, a sponsor must provide the agency with a summary of all the safety information in the new drug application, along with any additional safety information obtained during the review period.

Ultimately the drug approval process results in one of three outcomes:

- The drug is approved for marketing.
- The drug is approvable, but either additional information or clarification is needed or further labeling negotiation is required.
- The drug is not approved because of major deficiencies in the data (the sponsor can appeal this evaluation, withdraw the application, or resubmit an amended application at a later date).

As a condition of approval, the FDA may require a company to conduct postmarketing studies to either investigate a new dosing regimen or collect additional data on adverse events. Sponsors may also be required to set up a patient registry to collect data on who is really taking their compound and under what conditions. Or a company on its own may decide to undertake such a study to gather more safety information. A company may also undertake a study if it believes the reports of adverse drug reactions (ADR) it has received require such action. These studies may consist of new clinical trials or they may be evaluations of existing databases.

Immediately upon receipt of the final approval letter, the production, distribution, and marketing activities can begin. In an effort to make the review process transparent to industry and to the public at large, the FDA has begun posting the final approval letters and new drug application reviews for approved products on their website. Review of these documents provides the reader with significant insight into the drug approval process and is highly recommended for those seeking additional insight in this area.

I. Postmarketing Surveillance

Because of the numbers involved and the design of the clinical trials, adverse reactions that occur in fewer than 1 in 3000 to 5000 patients are unlikely to be detected in Phase I to III clinical trials and may be unknown when a drug is approved. An example of such unknown serious adverse event would be that of chloramphenicol induced aplastic anemia, which occurs at an estimated rate of 1 in 30,000. Unfortunately, with numbers like these, rare adverse reactions are more likely to be detected when the drug reaches large numbers of patients after it has been approved.

Because of well-publicized drug withdrawals in recent years because of adverse events (e.g., bromfenac, cisapride), Congress, as part of PDUFA-III, has proposed using some of the generated user fees to double the postmarketing reporting staff. This doubling of the staff is a tacit admission by Congress that postmarketing surveillance of drugs is both a labor intensive activity and one that is subject to the marketplace. Once on the market, physicians are free to use a drug in different doses, different dosing regimens, different patient populations, and in other ways they believe will benefit patients.

In addition to its required mechanisms of adverse event monitoring through the sponsors, the FDA also sponsors the MedWatch Partners program to collect reports of ADRs from health care professionals. Supported by more than 140 organizations including health professionals and industry, MedWatch Partners helps to ensure that safety information is promptly collected and that new information is rapidly communicated to the medical community. The reports submitted to the FDA come primarily from the new drug application holder, but individual physicians and other health care professionals also make significant contributions. The agency evaluates the reports for trends and implications and may require a company to provide more data, undertake a new clinical trial, revise a drug's labeling, notify health care professionals, or even remove a product from the market.

IV. INTERNATIONAL DRUG DEVELOPMENT AND APPROVAL

Up to this point we have been concerned with the development of and evolution of the drug approval process in the U.S. This is appropriate, as the U.S. Food and Drug Administration as a regulatory body predates most drug regulatory agencies and was the model for many of them. In Japan modern drug regulation in the U.S. mode was established in the 1950s as an outgrowth of the U.S. occupation. In Europe drug regulation was more of a registration process than regulation until thalidomide. In the wake of the thalidomide tragedy, and the observation that the U.S. did not suffer the consequences of it, most European countries established regulatory models based loosely on the FDA model. From a drug development standpoint, the need for individualized drug approval packages for the U.S., Europe, and Japan led to increased costs and was an active disincentive to the approval of some drugs in some countries.

Although not recognized as such at that time, the establishment of the European Community (now known as the European Union) as a single market for European goods opened up the door to the harmonization of regulatory requirements for pharmaceuticals. It was under this structure that in the 1980s a single market for pharmaceuticals was developed whereby commercial standards in a number of areas, including pharmaceuticals were brought into line. At the same time there were bilateral discussions between the U.S., Europe, and Japan about developing a single review package that could be used by any drug regulatory agency. Ultimately, these efforts coalesced in 1990 as the International Committee on Harmonization (ICH). This committee is made up of representatives from six member organizations and three observer parties:

Member Organizations
- European Union (E.U.)
- European Federation of Pharmaceutical Industries' Associations (EFPIA)
- Ministry of Health and Welfare, Japan (MHW)
- Japan Pharmaceutical Manufacturers Association (JPMA)
- U.S. Food and Drug Administration (FDA)
- Pharmaceutical Research and Manufacturers of America (PhRMA)

Observers

The World Health Organization (WHO)

The European Free Trade Area (EFTA), represented at ICH by Switzerland

Canada, represented at ICH by the Drugs Directorate, Health Canada

In addition to these organizations, the International Federation of Pharmaceutical Manufacturers Association (IFPMA) has been closely associated with ICH, to ensure contact with the research-based industry, outside the three primary ICH regions.

The stated purpose of the ICH is to develop one package of information that will be accepted for filing of a new drug application or its equivalent throughout Europe, Japan, and the U.S. Currently this document is known as the common technical document. The document was presented at the fifth international meeting of the ICH in 2000 and was published as a Step 4 document in November 2000. The specification for the electronic common technical document was signed for as a Step 2 document for testing in May 2001. The electronic common technical document specification provides information needed to prepare an electronic submission of the sections of the common technical document that are common to the ICH regions. A detailed discussion of the format and contents of the common technical document is beyond the scope of this chapter. For those interested, the FDA has published guidance on the organization and contents of the common technical document at their website and at the ICH organization website, which also has the latest news on the development and implementation of these initiatives.

While the common technical document will be an aide to the drug approval process, one additional aspect of drug approval must be considered from a multinational perspective. That is the regulation and inspection of drug manufacturing facilities. This part of the drug regulatory process is in some ways more important than the original approval of a drug, as it is responsible for the quality manufacture and control of a process that can theoretically be scaled up from producing a few hundred dosage units to millions of dosage units per batch.

To deal with these issues, and separate them from the ICH initiative, the U.S. and the E.U. negotiated a pharmaceutical mutual recognition agreement in 1997 to lower or eliminate regulatory barriers and promote trade between the two regions. The regions agreed to recognize each other's inspections of manufacturing facilities for human drugs and biologics. Prior to this, the FDA, for example, had to send its inspectors to Europe to conduct U.S. required inspections, something that caused no end of problems with the timing of reviews and foreign sensibilities regarding their sovereignty. For non-E.U. nations the U.S. still sends out its inspectors to make preapproval and routine inspections to ensure the delivery of quality pharmaceuticals to U.S. consumers. This both adds to the cost of drug development and presents timing issues concerning the availability of inspectors to visit foreign sites and the timing of the inspection within the approval schedule. As individual nations build up their internal drug review processes, other mutual recognition agreements may be negotiated on a case by case basis in regard to accepting foreign inspections.

V. CONCLUSION

The drug approval process in the U.S. has been shaped by both tragedy and legislation. The American public has come to expect safe drugs approved in a reasonable amount of time. The current drug approval process is in a state of transition both from a paper based to an electronic based system and from one tailored for the U.S. to a more global perspective. Further advances in the science of drug regulation are most likely to be evolutionary rather than revolutionary as the public's tolerance for approval of unsafe drugs makes massive experimentation with new standards of safety unlikely to be tolerated in the face of drug withdrawals. The process of drug approval, ultimately, is a direct reflection of the society it serves. As the needs and wants of society change, in terms of accelerated access to drugs and improved risk management, the drug approval process will adapt to meet these needs through both internal and external mechanisms.

HOMEWORK

Hints: Answers for the following questions may not be available in this chapter. Therefore, students are encouraged to visit the reference websites listed at the end of the chapter.

1. A number of important concepts have been discussed, but not explicitly defined in the previous sections. Based on your reading and knowledge, please define the following:
   Adulteration
   Legend drug
   Misbranding
   Patent medicine
   Teratogen
2. Compare and contrast misbranding vs. adulteration and give an example of each.
3. Give three examples of a grandfathered drug.
4. What was the Sherley Amendment?
5. Name the three official compendia recognized by the U.S.
6. Differentiate between a priority new drug application and a standard new drug application in terms of therapeutic potential and review clock times under PDUFA.
7. What heavy metal was the FDA required by Congress to evaluate in over-the-counter products as part of FDAMA?
8. Of the following time periods, 6 months, 9 months, or 1 year, how much patent life was granted under FDAMA for conducting pediatric studies?
9. What is the difference between pediatric exclusivity and the pediatric rule?
10. What are the major provisions of the Drug Price Competition and Patent Term Restoration Act of 1984?
11. What is an EOP2?
12. Healthy subjects are evaluated in which phase of drug development?
13. What is a proof of concept study, and in which phase of drug development does it occur?
14. Define what a lead compound is.

ANSWERS:

1. -**Adulteration** — The addition of a foreign substance to a drug product.
   -**Legend Drug** — A drug defined by the Federal Food, Drug and Cosmetic Act, as amended, and under which definition its label is required to bear the statement "Caution: Federal law prohibits dispensing without prescription." (see Durham-Humphrey Amendment of 1951)
   -**Misbranding** — The act of providing false information on a product label;
   -**Patent Medicine** — A drug sold without a prescription that is protected by a patent for that drug or combination of drugs.
   -**Teratogen** — A substance, drug or otherwise, that can cause birth defects in the developing fetus.
2. Simply put, the FDCA (Food, Drug and Cosmetic Act) requires that any information a manufacturer places on the label of a drug be truthful and not misleading. A product that contains mis-leading or inaccurate information in relation, but not limited to, dosing, side effects, or indications is misbranded. By comparison adulteration relates to the content of the drug product, whether or not it is contaminated by or processed in

unsanitary conditions or if it does not contain the proper drug substance in labeled amounts.

3. Morphine, digoxin, aspirin, acetaminophen
4. In 1911, in U.S. v. Johnson, the Supreme Court ruled that the 1906 Food and Drug Act did not prohibit false therapeutic claims, but only false and misleading claims about the ingredients or identity of a drug. The Sherley Amendment (named for Rep. Swagar Sherley) brought therapeutic claims within the jurisdiction of the Pure Food and Drugs Act, but required that the government to prove that false therapeutic claims for a drug were made with an intent to defraud before they would be judged as illegal.
5. The United States Pharmacopoeia, The National Formulary, The Homeopathic Pharmacopoeia
7. Mercury
8. 6 months
9. Pediatric Exclusivity-Under the Best Pharmaceuticals for Children Act of 2001 the FDA was authorized to extend patent protection on marketed drugs for six months, provided that the sponsor completed an FDA agreed upon series of trials in pediatrics (called the Written Request). The objective of both the act and the studies completed under the BPCA is to provide the prescriber with data gained through trials actually conducted in a pediatric population.. In contrast the Pediatric Rule-otherwise known as the Pediatric Research Equity Act of 2003 requires that when a new NDA is submitted to the FDA the NDA must contain data on the use of the drug in a pediatric population or request a deferral or a waiver of such studies if the use of the drug in a pediatric population is not indicated, would be impractical, or does not represent a meaningful therapeutic benefit (among other reasons).
10. -Authorizes approval of Abbreviated New Drug Applications or ANDA's for generic drugs.
    -Limits the FDA to only requiring bioavailability studies for ANDA's.
    -Establishes a 5 yr exclusivity period for a New Molecular Entity
    -Establishes a 3 yr exclusivity period for supplemental New Drug Applications
    -Allows a drug company to re-cover some of the patent time lost during the drug approval process, up to a period of 5 yrs.
11. EOP2 stands for End of Phase II meeting. The purpose of the EOP2 meeting is to provide an opportunity for the sponsor and reviewing division to (1) evaluate the results of the drug development program to date; (2) discuss the sponsor's plans and protocols relative to regulations, guidance's, and Agency policy; (3) identify safety issues, scientific issues, and/or potential problems and resolve these, if possible, prior to initiation of phase 3 studies; and (4) identify additional information important to support a marketing application.
12. Phase I
13. Proof of concept studies are small clinical studies done in patients with the disease of interest done in order to confirm the activity of the drug in the disease and to provide initial safety and dose ranging information.
14. A lead compound is an experimental drug that is the first compound in a new class that is introduced into testing. Often times a lead compound is discarded in favor of a related compound that has improved efficacy or safety characteristics. The "lead compound" can be said to have led the way for the drugs that followed in terms of defining clinical utilitiy.

ADDITIONAL RESOURCES AND REFERENCES

1. The website that provides a link to each of the centers and their mission statements of FDA is http://www.fda.gov.
2. The U.S. Food and Drug Administration's Center for Drug Evaluation and Research (CDER) provides information on its approval and regulatory processes at http://www.fda.gov/cder/.
3. The January 1995 FDA consumer report, "Benefit vs. Risk: How CDER Approves New Drugs," is available online at http://www.fda.gov/fdac/special/newdrug/benefits.html.
4. Thalidomide Victims Association of Canada (http://www.thalidomide.ca/english/index.htm)
5. Information on CDER's history can be obtained at http://www.fda.gov/CDER/about/history/.
6. Information on registry of clinical trials for both federally and privately funded trials of experimental treatments for serious or life-threatening diseases can be obtained at http://www.clinicaltrials.gov.
7. CDER's guidance documents are available online at http://www.fda.gov/cder/guidance/index.htm.
8. Information on advisory committee decisions can be obtained from http://www.fda.gov/cder/audiences/acspage/acslist1.htm.
9. Information on IFPMA/ICH can be obtained from http://www.ifpma.org/ich1.html.
10. For a more detailed explanation of FDAMA and its provisions, go to http://www.fda.gov/po/modact97.html.

Module III

10 Oral Conventional Solid Dosage Forms: Powders and Granules, Tablets, Lozenges, and Capsules

Melgardt M. De Villiers

CONTENTS

0-4152-8863-0/05/$0.00+$1.50
© 2005 by CRC Press LLC

Learning Objectives

After studying this chapter, the student will have a thorough knowledge of:

- The development process for conventional solid dosage forms
- The different types of conventional solid dosage forms
- The advantages and disadvantages of conventional solid oral dosage forms
- The methods of preparation and manufacture of conventional solid oral dosage forms
- Ingredients that are used to formulate conventional solid dosage forms
- The requirements for testing and assuring the quality of conventional solid oral dosage forms
- Ways to package conventional solid oral dosage forms
- Calculations involved in preparing and compounding solid oral dosage forms
- Formulation factors that affect the release of drugs in conventional solid oral dosage forms

I. INTRODUCTION

The purpose of this chapter is to give the student a sense of the importance of the basic knowledge and scientific principles required to understand the manufacturing, formulation, use, and biopharmaceutical aspects of conventional solid dosage forms.

Oral dosage forms are taken orally for a local effect in the mouth, throat, or gastrointestinal tract or for a systemic effect in the body after absorption from the mouth or gastrointestinal tract.

Oral dosage forms can be divided into two main groups based on the physical state of the dosage form during consumption, that is, solid oral dosage forms (tablets, capsules, or powders) and liquid oral dosage forms (solutions, syrups, emulsions, and powders for suspensions). This chapter focuses on conventional solid oral dosage forms and is divided into the following three sections:

1. Powder and granules
2. Tablets
3. Capsules

Throughout the chapter practical examples are given, and each section is followed by a tutorial. These problems are intended to test the student's progress in mastering the text up to that point. At the end of the chapter are case studies. These are practical problems that will demonstrate to the student the application of the knowledge acquired after completing this chapter.

II. TYPES OF CONVENTIONAL SOLID DOSAGE FORMS

In this chapter, *conventional oral solid dosage forms* will be defined as those solid dosage forms taken by or given orally to patients and intended to deliver the drug to the site of action without any time delay. The following definitions of solid dosage forms come from the United States Pharmacopoeia (USP). Examples of commercially available solid dosage forms are listed in Table 10.1.

A. Powders

Powders are dry mixtures of finely divided medicinal and nonmedicinal agents intended for internal or external use. Powders may be dispensed to a patient and used in bulk form such as powders measured by the spoonful to make a douche solution or they may be divided into single dosage units and packaged in folded papers or unit-of-use envelopes.

TABLE 10.1
Representative Commercially Available Conventional Oral Solid Dosage Forms

Drug	Trade Name	Dosage Form	Strengths
Allopurinol	Zyloprim	Tablets	100 mg, 200 mg
Alprazolam	Xanax		0.25 mg, 0.5 mg, 1 mg, 2 mg
Acebutalol	Sectral	Hard capsules	200 mg, 400 mg
Clomipramine	Anafranil		25 mg, 50 mg, 75 mg
Digoxin	Lanoxicaps	Soft capsules	0.1 mg, 0.2 mg
Ethosuximide	Zarontin		250 mg
Nabumetone	Relafin	Caplet	500 mg, 750 mg
Cefuroxime	Ceftin		125 mg, 250 mg, 500 mg
Amoxicillin	Amoxil	Chewable tablet	125 mg, 250 mg
Carbamazepine	Tegretol		100 mg
Cholestyramine	Questran	Powder	
Clotrimazole	Mycelex	Troche	10 mg
Isosorbide dinitrate	Isordil	Sublingual	5 mg
Sulfasalazine	Azulfidine EN	Enteric coated	500 mg
Aspirin	Alka-Seltzer	Effervescent	300 mg

B. Tablets

Tablets are solid dosage forms containing one or more medicinal substances with or without added pharmaceutical ingredients. Among the pharmaceutical agents used are diluents, disintegrants, colorants, binders, solubilizers, and coatings. Tablets may be coated for appearance, for stability, to mask the bitter taste of the medication, or to provide controlled drug release. Most tablets are manufactured on the industrial scale by compression, using highly sophisticated machinery. Punches and dies of various shapes and sizes enable the preparation of a wide variety of tablets of distinctive shapes, sizes, and surface markings.

Most tablets are intended to be swallowed whole. Some, however, are prepared to be chewable and have a pleasant taste and feel. Other tablets dissolve in the mouth (buccal tablets) or under the tongue (sublingual tablets), whereas effervescent tablets are intended to be dissolved in water before taking.

Tablets are formulated to contain a specific quantity of drug substance. To enable flexibility in dosing, manufacturers commonly make available various tablet or capsule strengths of a given medication. As required, a tablet may also be broken in half (many tablets are scored or grooved for this purpose) or more than a single tablet may be taken as a prescribed dose.

C. Capsules

Hard gelatin capsules are solid dosage forms in which one or more medicinal and inert substances are enclosed within small shells of gelatin. Capsule shells are produced in varying size, shape, thickness, softness, and color. Hard shell capsules, which have two telescoping parts — the body and the cap — are commonly used in extemporaneous hand filling operations as well as in small- and large-scale manufacture of commercial capsules. They usually are filled with powder mixtures and granules. After filling, the two capsule parts are joined for tight closure. They may also be sealed and bonded through a variety of special processes for added quality assurance and capsule integrity.

Soft-shell gelatin capsules, which are one bodied, are formed, filled, and sealed in the same process. Highly specialized and large-scale equipment is required, and thus soft gelatin capsules are only prepared commercially. They are rendered soft through the addition of a plasticizer to the capsule shell formulation. Soft gelatin capsules may be filled with powders, semisolids, or liquids.

Capsules are intended to contain a specific quantity of fill, with the capsule size selected to accommodate that quantity closely. In addition to their medication content, capsules usually contain inert pharmaceutical substances, such as fillers, disintegrants, solubilizers, etc. When swallowed, the gelatin shell is dissolved by the gastrointestinal fluids, releasing the contents.

Capsules containing only nontherapeutic materials are termed placebos and are used widely in controlled clinical studies to evaluate the activity of a drug compared to a nondrug in a group of human subjects.

D. Lozenges

Lozenges are solid preparations containing one or more medicinal agents in a flavored, sweetened base intended to dissolve or disintegrate slowly in the mouth, releasing medication generally for localized effects. Lozenges are prepared by molding or compression.

III. POWDERS AND GRANULES

A. Background

Powders represent one of the oldest dosage forms. However, with the increased use of highly potent compounds, powders as a dosage form have been largely replaced by capsules and tablets. In certain situations, powders still possess advantages and thus still represent a small portion of the solid dosage forms currently being employed. For example, because of their greater specific surface area, powders disperse and dissolve more readily than compacted dosage forms. Children and adults who have trouble swallowing tablets or capsules may find powders more acceptable. Immediately before use, oral powders are mixed with a beverage or applesauce.

1. Oral Powders

Oral powders generally can be supplied as finely divided powders or effervescent granules. The finely divided powders are suspended or dissolved in water or mixed with soft foods such as applesauce before administration. Antacids and laxative powders frequently are administered in this form. Powdered antibiotic syrups to be reconstituted before administration are also classified as oral powders.

2. Dentifrices

Dentifrices may be prepared in the form of a bulk powder, generally containing a soap or detergent, mild abrasive, and anticariogenic agent.

3. Douche Powders

Douche powders are completely soluble and are dissolved in water prior to use as antiseptics or cleansing agents for a body cavity. They most commonly are intended for vaginal use, although they may be formulated for nasal, otic, or ophthalmic use.

4. Insufflations

Insufflations are finely divided powders introduced into body cavities such as the throat. An insufflator (powder blower) usually is employed to administer these products. The Norisodrine Sulfate Aerohaler Cartridge (Abbott) is an example. In the use of this aerohaler, inhalation by the patient causes a small ball to strike a cartridge containing the drug. The force of the ball shakes the proper amount of the powder free, permitting its inhalation. Another device, the Spinhaler turbo-inhaler (Fisons), is a propeller-driven device designed to deposit a mixture of lactose and micronized cromolyn sodium into the lung as an aid in the management of bronchial asthma. However, the difficulty in obtaining a uniform dose has restricted their general use.

5. Oral Antibiotic Syrups

For patients who have difficulty taking capsules and tablets, for example, young children, a liquid preparation of a drug offers a suitable alternative, but many antibiotics are physically or chemically unstable when formulated as a suspension or solution. The method used to overcome this problem is to present the dry ingredients in a suitable container in the form of a powder or granules. When the pharmacist dispenses the product, a given quantity of water is added to constitute the solution or suspension.

Sometimes the amount of water added is varied to obtain nonstandard doses of the antibiotic as shown in the following example. If a prescription for an amoxycillin product calls for the addition of 80 ml of water to make 100 ml of constituted solution containing 125 mg amoxycillin per 5 ml, how should the instruction be changed to obtain 100 mg amoxycillin per 5 ml?

The volume of dry powder in the bottle is

$$100 \text{ ml} - 80 \text{ ml} = 20 \text{ ml}$$

Total drug present is

$$5\ \text{ml}/125\ \text{mg} = 100\ \text{ml}/x\ \text{mg}$$

$$x = 2500\ \text{mg or } 2.5\ \text{g}$$

The volume of water to add to get 100 mg/5 ml

$$2500\ \text{mg}/x\ \text{ml} = 100\ \text{mg}/5\ \text{ml}$$

$$x = 125\ \text{ml}$$

Since the powder occupies 20 ml, the volume of water needed is

$$125\ \text{ml} - 20\ \text{ml} = 105\ \text{ml}$$

6. Effervescent Granules

Effervescent granules contain sodium bicarbonate and either citric acid, tartaric acid, or sodium biphosphate in addition to the active ingredients. On solution in water, carbon dioxide is released because of the acid-base reaction.

Citric acid: $3\ NaHCO_3 + C_6H_8O_7.H_2O = C_6H_5Na_3O_7 + 3\ CO_2 + 3\ H_2O$
Tartaric acid: $2\ NaHCO_3 + C_4H_6O_6 = C_4H_4Na_2O_6 + 2\ CO_2 + 2\ H_2O$

The release of the water of crystallization makes the powder coherent and helps form the granules. The effervescence from the release of the carbon dioxide masks the taste of salty or bitter medications.

B. Advantages and Disadvantages of Powders and Granules

The advantages of powders and granules are the following:

1. Solid preparations are more stable than liquid preparations. The shelf life of powders for antibiotic syrups, for example, is 2 to 3 years, but once reconstituted with water it is 1 to 2 weeks.
2. Powders and granules are convenient forms in which to dispense drugs with a large dose. For example if the dose of a drug is 1 to 5 g it is sometimes not feasible to manufacture tablets to supply the drug to the patient.
3. Orally administered powders and granules of soluble medicaments have a faster dissolution rate than tablets or capsules, as these must first disintegrate before the drug dissolves.
4. Powders offer a lot of flexibility in compounding solids.

The disadvantages of powders and granules are as follows:

1. Bulk powders or granules are far less convenient for patients to carry than a small container of tablets or capsules.
2. The masking of unpleasant tastes may be a problem with this type of preparation.

3. Bulk powders or granules are not a good method of administering potent drugs with a low dose. This is because individual doses are usually extracted from the bulk using a 5 ml spoon, which is subject to variation in spoon fill (e.g., level or heaped spoonfuls).
4. Powders and granules are not a suitable method for the administration of drugs that are inactivated in the stomach; these should be presented as enteric-coated tablets.
5. Powders and granules are not well suited for dispensing hygroscopic or deliquescent drugs.

C. Preparation of Powders and Granules

During the manufacture and extemporaneous preparation of powders, the general techniques of weighing, measuring, sifting, and mixing are applied. The manually operated procedures usually employed by pharmacists for preparing powders are co-milling, trituration, pulverization by intervention, and levigation.

On a small scale, a pharmacist can reduce the particle size of powders by grinding with a mortar and pestle. This process is termed trituration. Another process, called pulverization by intervention, is also used for reducing the particle size of solids. A prime example is camphor, which cannot be pulverized easily by trituration because of its sticky properties; however, on the addition of a small amount of alcohol or other volatile solvent, this compound can be reduced readily to a fine powder because when the solvent is permitted to evaporate a fine powdered material is formed. Levigation is the process in which a nonsolvent is added to solid material to form a paste, and particle-size reduction then is accomplished by rubbing the paste in a mortar with a pestle or on an ointment slab using a spatula.

When blending two or more powders the method of geometric dilution is preferred, especially for unequal quantities of powders. This method ensures that small quantities of ingredients, usually potent drugs, are uniformly distributed throughout a powder mixture. The steps involved in geometric dilution are the following:

1. Weigh ingredients.
2. Place the ingredient with the smallest quantity in a mortar.
3. Combine this powder with an amount of the material present in the second largest quantity approximately equal to the amount already in the mortar.
4. Triturate the powders until a uniform mixture is formed.
5. Add another amount of the second ingredient equal in size to the powder volume already in the mortar and triturate well.
6. Continue adding powder to the mortar in this fashion until all the powder ingredients have been added.

D. Compounding Pharmaceutical Powders

When working with powders pharmacists should look out for efflorescent powders since they contain water of hydration, which may be released when the powders are triturated or are stored in an environment of low humidity. The water liberated can make the powder damp and pasty. Examples of efflorescent powders include caffeine, citric acid, codeine phosphate, ferrous sulfate, and atropine sulfate. Hygroscopic and deliquescent powders should also be handled with care since these substances become moist because of their affinity for moisture in the air. Double wrapping is desirable for further protection. Extremely deliquescent compounds cannot be prepared satisfactorily as powders.

The following tips can be useful for pharmacists who need to do some compounding that involves powders:

1. A coffee grinder will aid in the size reduction of small amounts of powders.
2. Mixing powders of similar particle size and density can be done in a plastic bag.
3. A dust mask can be used to protect the dispenser if the powder is very light.
4. Powders that are too fluffy can be compacted by the addition of a few drops of alcohol, water, or mineral oil.
5. Magnesium stearate less than 1% can be added to increase the flow and lubrication of powders.
6. Sodium lauryl sulfate up to 1% can be used to aid the wetting and dissolution of the powder or to reduce the electrostatic forces created by handling the powder.

E. Packaging of Powders and Granules

Oral powders may be dispensed in doses premeasured by the pharmacist, that is, divided powders or in bulk. Traditionally, divided powders have been wrapped in materials such as bond paper and parchment. However, the pharmacist may provide greater protection from the environment by scaling individual doses in small cellophane or polyethylene envelopes.

Divided powders (chartula or chartulae) are dispensed in the form of individual doses and generally are dispensed in papers, properly folded. They also may be dispensed in metal foil, small heat-sealed plastic bags, or other containers. Hygroscopic and volatile drugs can be protected best by using a waxed paper, double-wrapped with a bond paper to improve the appearance of the completed powder. Parchment and glassine papers offer limited protection for these drugs.

F. Size Classification of Powders

After preparation powders are classified according to their particle size. In order to qualify the particle size of a given powder, the USP uses the following descriptive terms:

Very coarse (No. 8) powder: All particles pass through a No. 8 sieve (2.38 mm) and not more than 20% pass through a No. 60 sieve.
Coarse (No. 20) powder: All particles pass through a No. 20 sieve (0.84 mm) and not more than 40% pass through a No. 60 sieve.
Moderately coarse (No. 40) powder: All particles pass through a No. 40 sieve (0.42 mm) and not more than 40 % pass through a No. 80 sieve.
Fine (No. 60) powder: All particles pass through a No. 60 sieve (0.25 mm) and not more than 40% pass through a No. 100 sieve.
Very fine (No. 80) powder: All particles pass through a No. 80 sieve (0.18 mm). There is no limit to greater fineness.

HOMEWORK

It will be helpful for students to find the following information from literature and other sources.

1. Name commonly available commercial drug products that are dispensed as:
   (a) Powders
   (b) Effervescent powders
   (c) Powders for reconstitution

2. What are the techniques for folding powder papers (chartulae)?
3. List examples of powders that are:
   (a) Efflorescent
   (b) Hygroscopic
   (c) Deliquescent
4. List sources for information on:
   (a) Tablets that can be crushed to make powders
   (b) Excipients that can be used in drug products
   (c) Suppliers of compounding materials for making powders

TUTORIAL

1. To aid patients breaking tablets when they have to take only a half or quarter of the tablet, tablets are
   (a) Scored
   (b) Hinged
   (c) Crushed
   (d) Squared
2. Powders disperse and dissolve more readily than compacted dosage forms because they
   (a) Contain disintegrants
   (b) Are better lubricated
   (c) Have a greater specific surface area
   (d) Flow better
3. Which of the following products are not classified as dispensed powder and granules
   (a) Dentifrices
   (b) Douches
   (c) Insufflations
   (d) Tinctures
4. If an ampicillin product for constitution calls for 75 ml of water to be added to give 100 ml of a 100 mg/5 ml syrup, how much water should be added to obtain a syrup containing 125 mg/5 ml.
   (a) 55 ml
   (b) 80 ml
   (c) 105 ml
   (d) 100 ml
5. Which is a major advantage of powders and granules?
   (a) Bulk powder is easily carried by patients.
   (b) They are more stable than liquid preparations.
   (c) They can mask unpleasant tastes.
   (d) They are good for dispensing hygroscopic and deliquescent drugs.
6. Which blending method is best suited for mixing powders of unequal quantities to best distribute a small amount of a very potent drug throughout a powder mix?

   (a) Trituration
   (b) Levigation
   (c) Geometric dilution
   (d) Granulation
7. The gas evolved when effervescent powders dissolve in water is
   (a) Oxygen
   (b) Carbon dioxide
   (c) Carbon monoxide
   (d) Water vapor
8. Packing divided powders in parchment chartulae is not suited for drugs that are
   (a) Anhydrous
   (b) Coarse
   (c) Very fine
   (d) Hygroscopic
9. Efflorescent powders when triturated or stored in low humidity release
   (a) Carbon dioxide
   (b) Water
   (c) Dust
   (d) Bad smells
10. Which of the following dosage forms are used for localized treatment in the mouth?
   (a) Chewable tablets
   (b) Sublingual tablets
   (c) Lozenges
   (d) Gelatin capsules

Answers: 1 (a); 2 (c); 3 (d); 4 (a); 5 (b); 6 (c); 7 (b); 8 (d); 9 (b); 10 (c).

IV. TABLETS

A. Background

Tablets have been a viable dosage form since well before William Brockendon's patent for a tablet machine in 1843. This very simple device consisted essentially of a hole (or die) bored through a piece of metal. The powder was compressed between two cylindrical punches; one was inserted into the base of the die at a fixed depth; the other was inserted at the top of the die and struck with a hammer. Today, this simple invention forms the basis for most modern tableting equipment. The term *tablet* clearly distinguished compressed tablets from pills. Pills were spherical or, less often, ovoid, solids prepared by rolling, cutting, and rounding wet powder masses into balls that were usually sugar coated. Pills are no longer used and have been replaced by tablets and capsules. The term *pill* however, is still used to describe tablets by many who are not familiar with the history of pharmacy.

B. Types of Tablets Dispensed by Pharmacists

Tablets may be classed, according to the method of manufacture, as compressed or tablets. The vast majority of all tablets manufactured are made by compression, and compressed tablets are the most widely used dosage form in the U.S. and most of the rest of the world. Tablets can be produced in a wide variety of sizes, shapes, and surface markings, depending upon the design of the punches and dies. Capsule-shaped tablets are commonly referred to as caplets. Boluses are large tablets intended for veterinary use, usually for large animals.

1. Multiple Compressed Tablets

These tablets are prepared by subjecting the tablet powder to more than one compression cycle. The result may be a multilayered tablet or a tablet-within-a-tablet. Multilayer tablets are mainly used for incompatible substances, for example, phenylephedrine hydrochloride and ascorbic acid in admixture with paracetamol. Paracetamol and phenylephedrine hydrochloride are contained in one layer, and paracetamol and ascorbic acid in the other.

2. Sugarcoated Tablets

Compressed tablets may be coated with a sugar layer that is colored or uncolored. Sugar coats are water soluble. The coat protects the drug from the environment and provides a barrier for bad tasting or smelling drugs.

3. Film Coated Tablets

Film coated tablets are compressed tablets coated with a thin layer of a polymer capable of forming a skin-like film over the tablet. These coats rupture in the gastrointestinal tract, exposing the drug.

4. Gelatin Coated Tablets

The innovator product gelcaps, is a capsule shaped compressed tablet coated with a gelatin layer. This allows the product to be smaller than an equivalent capsule filled with an equivalent amount of powder.

5. Enteric Coated Tablets

These tablets are intended to pass unchanged through the stomach to the intestines, where the tablets disintegrate and drug dissolution occurs. This helps protect drug molecules that are susceptible to degradation in gastric acid and drugs that can irritate the gastric mucosa. It can also be used to control the delivery of certain drugs to the intestines to enhance absorption.

6. Buccal or Sublingual Tablets

Buccal tablets are inserted in the buccal pouch, and sublingual tablets are inserted beneath the tongue. Where rapid drug availability is required such as in the case of nitroglycerin tablets these tablets are administered sublingually. They are sometimes also referred to as instant disintegrating or dissolving tablets. Examples are isoprenaline sulphate (bronchodilator) and glyceryl trinitrate tablets (vasodilator). These tablets are usually small and flat. Sometimes sweeteners are added.

7. Chewable Tablets

Chewable tablets, when chewed, produce a pleasant tasting residue in the mouth that when swallowed does not leave a bitter or unpleasant aftertaste. These tablets have been used in tablet formulations for children, especially multivitamin formulations, and for the administration of antacids and selected antibiotics. Chewable tablets prepared by compression usually contain mannitol, sorbitol, or sucrose as binders and fillers and colors and flavors to enhance their appearance and taste. Mannitol is sometimes preferred as a chewable base diluent, since it has a pleasant, cooling sensation in the mouth and can mask the taste of some objectionable medicaments.

8. Effervescent Tablets

These tablets are compressed effervescent powders. They are usually dissolved in a glass of water before administration. The resultant solution is usually a flavored, bubbling drink. This type of tablet offers quick dissolution of the active ingredient in water if the tablet is broken apart by the internal liberation of carbon dioxide. This also increases palatability.

9. Lozenges

Lozenges are compressed tablets that do not contain a disintegrant. Some lozenges contain antiseptics (e.g., benzalkonium) or antibiotics for local effects in the mouth. The second type of lozenge produces a systemic effect, for example, a lozenge containing vitamin supplements (multivitamin tablets). Lozenges must be palatable and slowly soluble. Flavors are normally included to make the lozenge more acceptable. Compressed lozenges are called troches in the USP.

10. Molded Tablets

Sometimes compounding prescriptions require that molded tablets be prepared from mixtures of medicinal substances and a diluent usually consisting of lactose and powdered sucrose in varying proportions. The powders are dampened with solutions containing high percentages of alcohol. The dampened powders are pressed into molds, removed, and allowed to dry. Solidification depends upon crystal bridges built up during the subsequent drying process and not upon the compaction force.

11. Immediate Release Tablets

These tablets are designed to disintegrate and release the drug absent of any rate-controlling features such as special coatings or other formulation techniques.

12. Extended Release Tablets

Extended release tablets, sometimes also called sustained release tablets, are designed to release a drug in a predetermined manner over an extended time. These will be discussed in Chapter 11.

C. Advantages and Disadvantages of Compressed Tablets

Compressed tablets have a number of advantages:

1. They enable an accurate dosage of medicament to be administered simply.
2. They are easy to transport in bulk and for the patient to carry.
3. The tablet is a uniform final product as regards weight and appearance and is usually more stable than liquid preparations.
4. The release rate of the drug from a tablet can be tailored to meet pharmacological requirements.
5. Tablets can be mass-produced simply and quickly, and the resultant manufacturing cost is therefore lower than other dosage forms.
6. Tablets are very versatile. Many variations of tablet formulations exist, and new ones are being developed continuously.

The major disadvantages of compressed tablets and their manufacturing are:

1. When the dose of the drug is large, tablets might be too big for children or even adults to swallow.
2. When drugs need to act very fast, the disintegration of the tablet and the dissolution of the drug from the tablet might be the rate-limiting step in determining the onset of drug action.
3. Compression can change the physical properties, particle size, and crystal form of the drug that can affect the proper action after administration.

4. Sometimes the chemical and physical properties of the drug make it difficult to overcome compression problems such as capping, lamination, and picking and sticking.
5. Tablet coating can be very a difficult process to master, and it can also be very expensive.

D. Oral Administration of Tablets

Solid dosage forms for oral administration are best taken by placing the dosage form upon the tongue and swallowing it with a glass of water or beverage. Taking solid dosage forms with adequate amounts of water is important. A scheme relating tablet breakup to drug dissolution and adsorption is shown in Figure 10.1.

Swallowing a tablet or capsule without water can be dangerous because the dry dosage form might lodge in the esophagus, causing esophageal ulceration, particularly when taken just before bedtime. Among the drugs of greatest concern in this regard are alendronate sodium, aspirin, ferrous sulfate, any nonsteroidal antiinflammatory drug (NSAID), potassium chloride, and tetracycline antibiotics.

In general, patients who suffer from gastro-esophageal reflux disease must take their medications with adequate amounts of water and avoid reclining for at least an hour to avoid reflux. The administration of oral medication in relation to meals is very important since the bioavailability and efficacy of certain drugs may be severely affected by food and certain drinks. The pharmacist should be knowledgeable about such instances and advise patients accordingly.

If a patient cannot swallow a solid dosage form, the pharmacist can suggest using an available chewable or liquid form of the drug. If these are not available, an ordinary tablet can be crushed or a capsule opened to facilitate ease of administration. Any unpleasant drug taste may be partially masked by mixing with custards, yogurt, rice pudding, other soft food, or fruit juice. The patient should be advised to consume the entire drug-food mixture to obtain the full drug dose, and the drug should not be premixed and allowed to set, because of stability considerations. However, oral

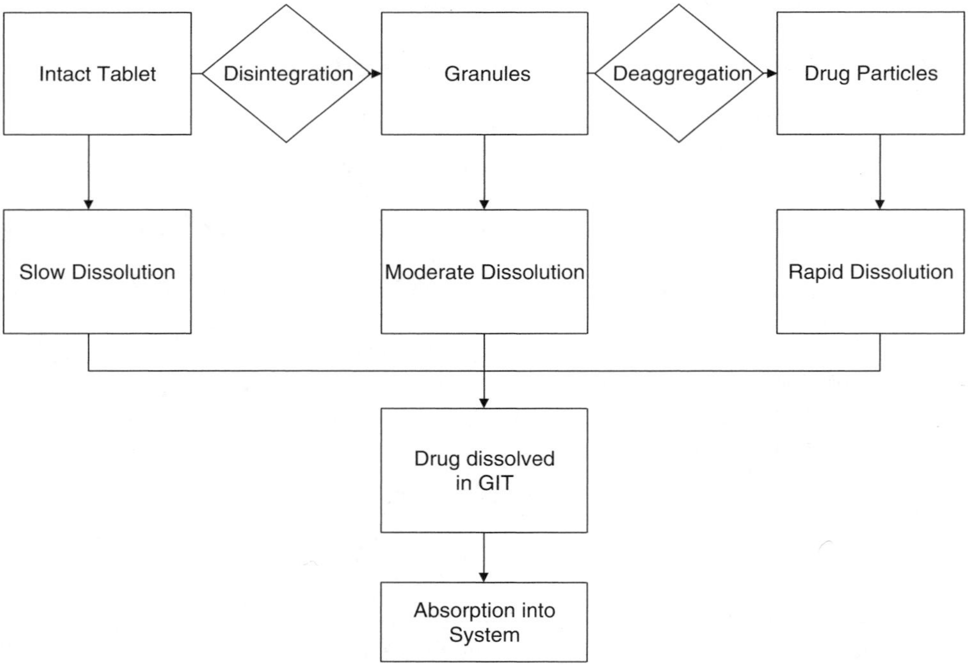

FIGURE 10.1 Schematic presentation of the disintegration and dissolution steps before absorption of a poorly soluble drug.

dosage forms that have special coatings (e.g., enteric) or are designed to provide controlled drug release must not be chewed, broken, or crushed to preserve their drug release features.

E. Design and Formulation of Compressed Tablets

Pharmaceutical manufacture is an important enterprise, and oral tablet manufacture is the most significant of all because more drugs are made as tablets than any other dosage form. For old and new drugs the tablet (or capsule) development cycle can be divided into five phases: concept, prototype, prepilot, pilot, and production. This paradigm is similar to the standard project management model consisting of project definition, project planning, project tracking and maintenance, and project closeout.

1. Concept Phase

The first step in standard project management models is the concept phase, in which a project team is appointed to formulate the new tablet in accordance with design control requirements.

2. Prototype Phase

During the prototype or feasibility phase various departments begin evaluating the feasibility of the tablet concept. The manufacturing group looks at production issues, and the marketing department may conduct focus panels to evaluate market acceptance.

3. Prepilot (Development) Phase

During this phase the manufacturing department works to develop the production process, while other departments develop the processes necessary to carry out functions such as packaging, distribution, and marketing. In design control language, this is the phase of design verification or the period of testing and proving the design features of the tablet.

4. Pilot (Scale-Up) Phase

Pilot plants are small manufacturing sites in which processes or techniques planned for full-scale production are tested in advance. Scale-up refers to the activity of moving from limited production to full production and typically takes place during the pilot phase of tablet development. In the pilot, or scale-up, phase design and process verification are completed and transferred into tablet specifications. The pilot phase culminates when the manufacturing process is confirmed and transfer to production facilities begins.

5. Production Phase

Production, or commercialization, is the final phase of product development. This is the phase in which the fully approved tablet is made available for marketing.

F. Essential Properties of Compressed Tablets

Tablets must provide an accurate dosage of medicament. Tablets must also be uniform in weight, appearance, and diameter. Another prerequisite of tablets for oral use is that when they are swallowed they should readily disintegrate in the stomach. This property represents a great paradox in formulation, since tablets should be produced with sufficient strength to withstand the rigors of processing, coating, and packing yet be capable of rapid breakdown when administered in order to release the drug rapidly. This disintegration involves the bursting apart of the compact by aqueous fluids penetrating the fine residual pore structure of the tablet.

Perhaps the most significant property of tablets is dissolution rate. The active ingredient must be available pharmacologically, and since drugs cannot be absorbed into the blood stream from the solid state, the active ingredient must first dissolve in the gastric or intestinal fluids before absorption can take place. Thus, dissolution of the drug from tablets into aqueous fluids is a very important property of solid dosage forms.

Tablets should also be stable in air and the environment over a reasonable period, and light and moisture should not affect their properties. Finally, tablets should be reasonably robust and capable of withstanding normal patient handling and handling during transport.

G. Tableting Excipients

Most tablets are not composed solely of the drug because drug powders are poorly compressible. Indeed, to ensure good compaction, powders intended for compression into tablets must possess two essential properties: powder flow and compressibility. Various excipients such as those listed in Table 10.2 are used to impart these and other essential properties to tablets. Most compressed tablets consist of the active ingredient and a diluent (filler), binder, disintegrating agent, and lubricant. Approved FD&C and D&C dyes or lakes (dyes adsorbed onto insoluble aluminum hydroxide), flavors, and sweetening agents may also be present.

1. Diluents

Diluents are added where the quantity of active ingredient is small or difficult to compress. Where the amount of active ingredient is small, the overall tableting properties are in large measure determined by the filler. Because of problems encountered with bioavailability of hydrophobic drugs of low water solubility, water-soluble diluents are used as fillers for these tablets.

2. Lubricants

Lubricants reduce friction during the compression and ejection cycle. In addition, they aid in preventing adherence of tablet material to the dies and punches. Because of the nature of this function, most lubricants are hydrophobic and as such tend to reduce the rates of tablet disintegration and dissolution (Figure 10.2). Consequently, excessive concentrations of lubricant should be avoided.

3. Disintegrants

A disintegrating agent assists in the fragmentation of the tablet after administration. Sometimes effervescent mixtures are used as disintegrating agents in soluble tablet systems.

4. Binders

Binders give adhesiveness to the powder during the preliminary granulation and to the compressed tablet. They add to the cohesive strength already available in the diluent. While binders may be added dry, they are more effective when added out of solution. The most effective dry binder is microcrystalline cellulose, which is commonly used for preparing tablets by direct compression. Binding agents can be added in two ways depending on the method of granulation:

1. As a powder as in slugging or in dry granulation methods.
2. As a solution to the mixed powders as in wet granulation.

5. Moisturizing Agents

In wet granulation, a moistening agent is required that is usually water. In cases where water cannot be used because the drug is hydrolyzed, alcohol is often substituted. Absolute alcohol is expensive,

TABLE 10.2
Pharmaceutical Excipients Used in the Manufacture of Tablets and Capsules

Excipient	Function	Concentration	Incompatibility	Therapeutic Effect
Lactose monohydrate NF EP, JP: powder, various mesh	Filler/binder	Maximum 65–85%	Maillard reaction primary amines	Laxative Lactose intolerance
Lactose monohydrate NF, EP, JP: DT, direct tableting	Filler/binder	Range depends on compressibility of drug	Same as for mactose monohydrate NF	Same as for lactose monohydrate NF
Lactose monohydrate NF modified sSpraydried	Filler/binder	Range depends on compressibility of drug	Same as for lactose monohydrate NF	Same as for lactose monohydrate NF
Lactose anhydrous NF: DT direct tableting	Filler/binder	Range depends on compressibility of drug	Same as for lactose monohydrate NF	Same as for lactose monohydrate NF
Lactose anhydrous NF powder	Filler/binder	Range depends on compressibility of drug	Same as for lactose monohydrate NF	Same as for lactose monohydrate NF
Microcrystalline cellulose (MCC) NF, EP, JP : 101, 102, 12 (DC), etc.	Filler/binder Disintegrant Absorbent Glidant	20–90% 5–15% 20–90% 5–20%	Strong oxidizing agents	Large quantities act as laxative
Cellulose powder NF	Filler/binder Binder Glidant Disintegrant	0–100% 5–25% 1–2% 5–15%	Same as for MCC	Same as for MCC
Starch NF	Filler Binder Disintegrant	Ranges 2–25% 3–15%		Rare allergic reactions
Starch pregelatinized NF	Filler/binder	Same as starch, NF but more compressible		Same as for starch
Sucrose NF	Filler Binder Coating	Not very compressible 2–20% (dry) 50–67% (wet) 50–67% syrup	Incompatibilities owing to impurities in sugar	Cariogenic, diabetes, or sugar intolerance
Sucrose compressible NF	Filler/binder	More compressible than sugar	Same as for sugar	Same as for sugar
Calcium phosphate, dibasic USP: direct compression	Filler/binder	Maximum 65–75%; needs lubricant	Tetracycline, indomethacin, aspirin, ampicillin, aspartame, etc.	Source calcium in nutritional supplements
Calcium phosphate anhydrous USP powder & DC grade	Filler/binder	Maximum 65–75%; always needs lubricant	Same as for calcium phosphate, dibasic USP	Same as for calcium phosphate, dibasic USP
Calcium carbonate USP, powder, DC, or with gum or starch	Filler/binder	Maximum 65–75%; always needs lubricant	Acids and ammonium salts	Used as antacid and calcium supplement
Mannitol USP, powder, DC grades, pyrogen free	Filler/binder	10–90%; chew tablets	None reported in dry state	Laxative in high doses; allergic reactions

TABLE 10.2 (continued)
Pharmaceutical Excipients Used in the Manufacture of Tablets and Capsules

Excipient	Function	Concentration	Incompatibility	Therapeutic Effect
Sorbitol NF, powder and DC grades	Filler	Both direct compressible and regular available	Forms chelates, water-soluble metal ions, PEG	Not unconditionally safe for diabetics; osmotic laxative
Dextrose NF	Filler/binder	Depends on compressibility of drug	Mallaird reaction with primary amines; strong oxidizing agents; warfarin sodium, etc.	Source carbohydrates
Dextrin NF	Filler/binder	Depends on compressibility of drug	Strong oxidizing agents	Source carbohydrates
Acacia NF, spraydried and low micro	Binder	1–5%	Amidopyrine, apomorphine, ethanol, etc.; oxidizing effect	Hypersensitivity
Methyl cellulose NF	Binder Disintegrant	1–5% 2–10%	Aminacrine HCl, chlorocresol, parabens	Flatulence and GIT distension, esophageal obstruction
Ethyl cellulose NF	Binder coating	1–3% Microencapsulation 10–20% Sustained release 3–20% Coating 1–3%	Paraffin waxes	
Alginic acid NF	Disintegrant Binder Control release	1–5%	Alkaline earth metals Strong oxidizing agents	Antacid Treats gastro-esophageal reflux
Sodium starch Glycolate NF, EP: JRS Vivastar	Super disintegrant	2–8%	Ascorbic acid	
Croscarmellose sodium NF, EP: JRS vivasol	Super disintegrant	Tablets 0.5–5% Capsules 10–25%	Strong acids Soluble iron salts	Large dose is laxative
Magnesium stearate NF, powder, regular and vegetable	Lubricant/glidant	0.25–5.0%	Strong acids, alkalis, and iron salts; strong oxidizing agents	Large dos is laxative and causes mucosal irritation
Calcium stearate NF powder	Lubricant/glidant	1% and less	Same as for magnesium stearate	
Zinc stearate NF powder	Lubricant/glidant	0.5–1.5%	Decomposed by strong acids	Fatal pneumonitis in infants after inhalation
Stearic acid triple pressed NF FCC powder, regular and vegetable	Lubricant/glidant	1–3%	Metal hydroxides and oxidizing agents; naproxen	
Hydrogenated vegetable oil NF	Lubricant/glidant	0.1–2%		
PEG, polyethylene glycol NF EP, all molecular weights	Lubricant/glidant Coating	Soluble tablets 5–15%	Oxidizing activity; softening of solids	Laxative

TABLE 10.2 (continued)
Pharmaceutical Excipients Used in the Manufacture of Tablets and Capsules

Excipient	Function	Concentration	Incompatibility	Therapeutic Effect
Silicon dioxide collodial NF	Lubricant/glidant	0.1–0.5%	Diethylstilbesterol preparations	Inhalation danger; granulomas
Talc USP, EP	Lubricant/glidant			
Mineral oil NF light	Lubricant/glidant	1–2%	Strong oxidizing agents	Large doses impair appetite and interfere with absorption of fat soluble vitamins
Polysorbate NF FCC, various grades	Wetting agent	0.1–3%	Reduces activity of paraben preservatives	Hypersensitivity
Sodium lauryl sulfate NF	Wetting agent Lubricant	1–2% 1–2%	Reacts with cations	Irratation of skin, eyes, lungs; pulmonary sensitization

Note: NF = National Formulary; EP = European Pharmacopeia; JP = Japanese Pharmacopeia; DT = Direct Tableting.

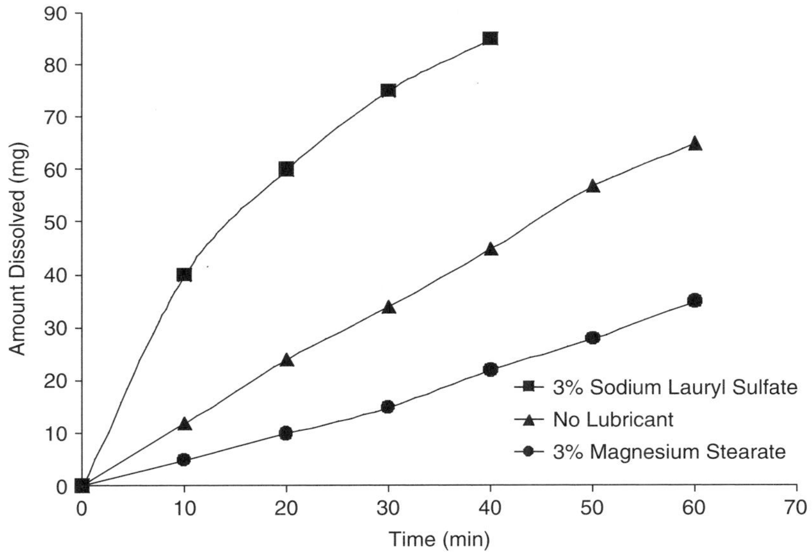

FIGURE 10.2 Effect of lubricant (magnesium stearate) and surfactant (sodium lauryl sulfate) on the dissolution rate of salicylic acid.

and thus industrial methylated spirits are used. Care must be taken to remove all traces of the solvent during drying or the tablets will possess an alcoholic odor. Isopropyl alcohol is an alternative moistening agent. It is difficult, however, to remove the last traces from granules, and it possesses an objectionable odor.

6. Adsorbents

Adsorbents are substances included in a formulation that are capable of holding quantities of fluids in an apparently dry state. Oil-soluble drugs, fluid extracts, or oils can be mixed with adsorbents and then granulated and compressed into tablets. Fumed silica, microcrystalline cellulose, magnesium carbonate, kaolin, and bentonite are examples of commonly employed adsorbents.

7. Glidants

Glidants are materials that are added to tablet formulations in order to improve the flow properties of the granulations. They act by reducing interparticulate friction.

8. Colorants

Colorants are often added to tablet formulation to add value or for product identification. Both FD&C and lakes are used. Most dyes are photosensitive when exposed to light. The Federal Food and Drug Administration (FDA) regulates the colorants employed in drugs.

H. Powder Flow and Compressibility

Powder flow, or fluidity, is required so the material can be transported through the hopper of a tableting machine. Inadequate fluidity this gives rise to arching, bridging, or rat-holing. This leads to variable die filling, which produces tablets that vary in weight and strength, and therefore steps must be taken to ensure that fluidity is maintained. Powder flow can be improved mechanically by the use of vibrators. However, the use of these devices can cause powder segregation and stratification.

A better method to enhance powder fluidity is to incorporate a glidant into the formulation. Materials such as fumed silicon dioxide are excellent flow promoters even in concentrations of less than 0.01%. Another way to improve powder flow is to make the particles as spherical as possible, for example, by spray drying or by the use of spheronization machines. The most popular method of increasing the flow properties of powders is by granulation. Most powdered materials can be granulated, and the improvement in flow can be quite startling. Icing sugar, for example, will not flow, but if it is granulated with water, it flows more easily.

Compressibility is the property of forming a stable, intact compact mass when pressure is applied. Acetaminophen is poorly compressible, whereas lactose compresses well. The physics of powder compression and why some materials compact better than others is a subject on its own and is not described in detail here. However, much research is in progress to characterize compaction behavior, but little is known about why some materials compress better than others. It is known, however, that in nearly all cases, granulation improves compressibility. Generally, materials that do not compress produce soft tablets; brittle crystalline materials yields brittle tablets.

There are at least three measurable hardness parameters (Hiestand indices) that can give a clue to the compatibility and intrinsic strength of powdered materials. The strain index is a measure of the internal entropy, or strain, associated with a given material when compacted. The bonding index is a measure of the material's ability to form bonds and undergo plastic transformation to produce a suitable tablet. The third index, the brittle fracture index, is a measure of the brittleness of the material and its compact. In general, the higher the bonding index, the stronger a tablet is likely to be, and the higher the strain index, the weaker the tablet. Since the two parameters are opposite in their effect on the tablet, it is possible for a material (such as microcrystalline cellulose) to have a relatively high strain index but superior compaction properties because of an extraordinary bonding potential. The higher the brittleness index, the more friable the tablet is likely to be.

The relationship between flow and compressibility is demonstrated by Carr's compressibility index (Table 10.3). This index evaluates the flow of a powder by comparing the poured (fluff) density and the tapped density.

TABLE 10.3
Carr's Compressibility Index

Carr's Index (%)	Type of Flow
5–15	Excellent
12–16	Good
18–21	Fair to possible
23–25	Poor
33–38	Very poor
>40	Extremely poor

$$\text{Carr's index} = [(\text{tapped density} - \text{poured density})/\text{tapped density}] \times 100$$

This simple index can be used on small quantities of powders according to the values listed below. Poured density is measured when the powder is poured under gravity into a calibrated cylinder and is related to the weight of powder occupying a specific volume. The tapped density of this same sample is measured when the sample is tapped until no further change in the powder volume is observed.

Another measure of the flowability of a powder is the angle of repose. This is the maximum angle at which a pile of unconsolidated powder can remain stable. It has generally been accepted that such slopes tend to have an angle varying from 25 to 40°. The exact angle depends upon conditions such as size, shape, and density of the powder particles, roughness of the powder surfaces, sorting or mixture of sizes, and height of fall of the grains. Studies of the effects of these various characteristics determining the angle of repose of loose material have produced diverse results. In general, increasing the particle size results in lowering the angle of repose and improving the flow.

I. Desirable Properties of Drugs and Excipients Necessary for Tableting

Irrespective of the type of tablet, general criteria for these raw materials are necessary. To produce accurate, reproducible dosage forms it is essential that each component be uniformly dispersed within the mixture and that any tendency for component segregation be minimized.

1. Particle Size and Segregation of Mixtures

In general, the tendencies of a powder mix to segregate can be reduced by maintaining similar particle size distribution, shape, and theoretically, density of all the ingredients. Flow properties are enhanced by using regular shaped, smooth particles with a narrow size distribution, together with an optimum proportion of fines (particles <50 μm). In contrast, decreasing particle size produces tablets of increased strength, as well as reduced tendency for lamination. An alternative approach aimed at reducing the segregation tendencies of medicaments and excipients involves milling the former to a small particle size and then physically absorbing it uniformly onto the surface of the larger particles of an excipient substrate. By these means ordered (interactive), as opposed to random, mixing is realized and dissolution is enhanced because of the fine dispersion.

2. Moisture Content

One of the most significant parameters contributing to the behavior of many tablet formulations is the level of moisture present during manufacture, as well as that residual in the product. In addition to its role as a granulation fluid and its potentially adverse effects on stability, moisture level may be critical in minimizing certain faults, such as lamination, that can occur during compression. Moisture levels can also affect the physicochemical properties of materials. For example, as the

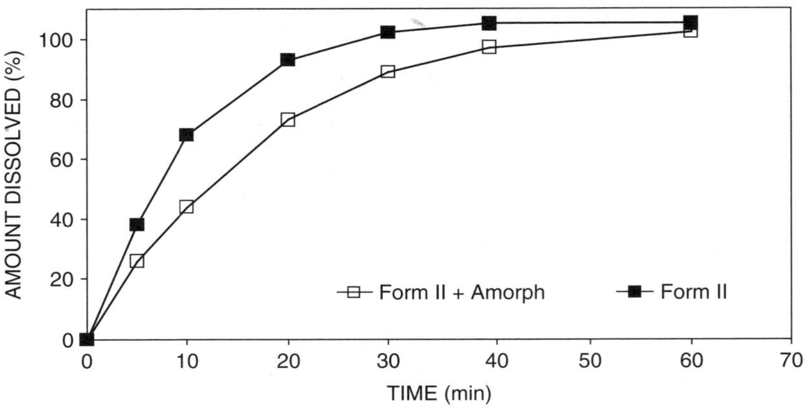

FIGURE 10.3 Dissolution profiles of rifampin from capsules showing the influence of a small amount (2 to 8%) of amorphous material on the dissolution rate of the drug.

moisture content of anhydrous lactose increases, it is absorbed by the lactose, thereby converting it from the anhydrous to the hydrous form. This transformation produces changes in compaction. Accelerated drug degradation and crystal transformation rates have also been traced to high residual moisture content.

3. Crystalline Form

Crystal transformation can occur during manufacturing processes such as tableting, mixing, and granulation. These polymorphic changes can have a profound effect on tablet performance in terms of processing, *in vitro* dissolution, and *in vivo* absorption as shown in Figure 10.3. Several researchers have concluded that polymorph, pseudopolymorph, amorphous content, and crystal habit influence the compactibility and mechanical strength of tablets prepared from polymorphic materials. Polymorphism is not restricted to active ingredients; different forms of lactose, sorbitol, etc. have been reported.

J. Drug Excipient Compatibility

Another area that should be considered when choosing the excipients to be used in the tablet formulation is that of drug–excipient interactions. There is still much debate about whether drug–excipient compatibility testing should be conducted before formulation. These tests most often involve the trituration of small amounts of the active ingredient with a variety of excipients. Critics of these small-scale studies argue that their predictive value has yet to be established and, indeed, they do not reflect actual processing conditions. Figure 10.4 shows a scheme for determining chemically compatible excipients using DSC; confirmatory analysis using HPLC, IR, or TLC is shown.*

K. Manufacturing Compressed Tablets

Tablets are prepared by three general methods: wet granulation, dry granulation (roll compaction or slugging), and direct compression (Figure 10.5). The purpose of both wet and dry granulation is to improve the flow of the mixture and to enhance its compression properties. Based on these three processes, the preparation of tablets can also be divided into dry methods and wet methods. Dry methods include direct compression, slugging, and roller compaction, and wet methods include wet granulation.

* DSC = differential scanning colorimetry; IR = infrared spectroscopy; HPLC = high performance liquid chromatography; TCL = TLC = thin layer chromatography.

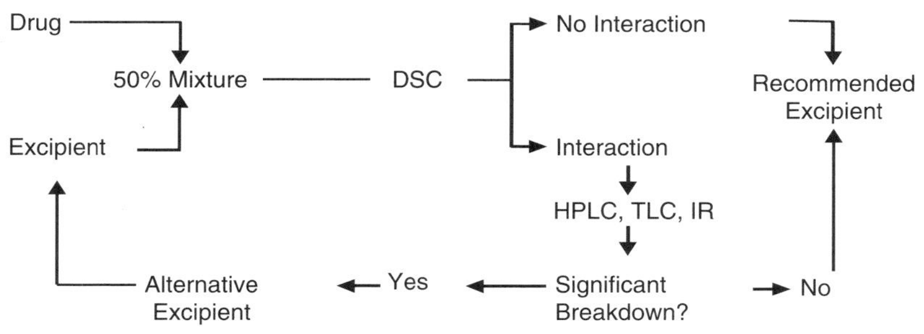

FIGURE 10.4 Scheme for determining chemically compatible excipients using DSC and confirmatory analysis using HPLC, IR, or TLC.

1. Powder Compaction

For simplicity, the physics of tablet compaction discussed here will deal with the single-punch press, in which the lower punch remains stationary. Initially, the powder is filled into the die, with the excess being swept off. When the upper punch first presses down on the powder bed, the particles rearrange themselves to achieve closer packing. As the upper punch continues to advance on the powder bed, the rearrangement becomes more difficult, and deformation of particles at points of contact begins. At first, the particle undergos elastic deformation, which is a reversible process, but as the continual pressure is applied, the particle begins to deform irreversibly. Irreversible deformation can be due to either plastic deformation, which is a major factor attributing to the tablet's mechanical strength, or brittle fracture, which produces poor quality compacts that crumble as the tablets are ejected. In general, as increasing pressure is applied to a compact, its porosity will be reduced.

2. Direct Compression

The possibility of compressing mixed powders into tablets without an intermediate granulating step is an attractive one. For many years, several widely used drugs, notably aspirin, have been available in forms that can be tableted without further treatment. Recently, there has been a growing impetus to develop so-called direct compression formulations, and the range of excipients, especially diluents, designed for this specific role has expanded dramatically. It is possible to distinguish two types of direct compression formulations: (a) those in which a major proportion is an active ingredient, and (b) those in which the active ingredient is a minor component (i.e., <10% of the compression weight). In the former, the inherent characteristics of the drug molecule, in particular the ability to prepare a physical form that will tablet directly, has profound effects on the tablet's characteristics.

Dry methods and, in particular, direct compression are superior to methods employing liquids, since dry processes do not require the equipment and handling expenses required in wetting and drying procedures and can avoid hydrolysis of water-sensitive drugs. Materials currently available as direct compression diluents may be divided into three groups according to their disintegration properties and their flow characteristics:

1. Disintegration agents with poor flow, for example, micro-crystalline cellulose, microfine cellulose, and directly compressible starch
2. Free-flowing materials that do not disintegrate, for example, dibasic calcium phosphate
3. Free flowing powders that disintegrate by dissolution, for example, spray-dried lactose, anhydrous lactose, spray-crystallized maltose, dextrose, sucrose, mannitol, and amylose

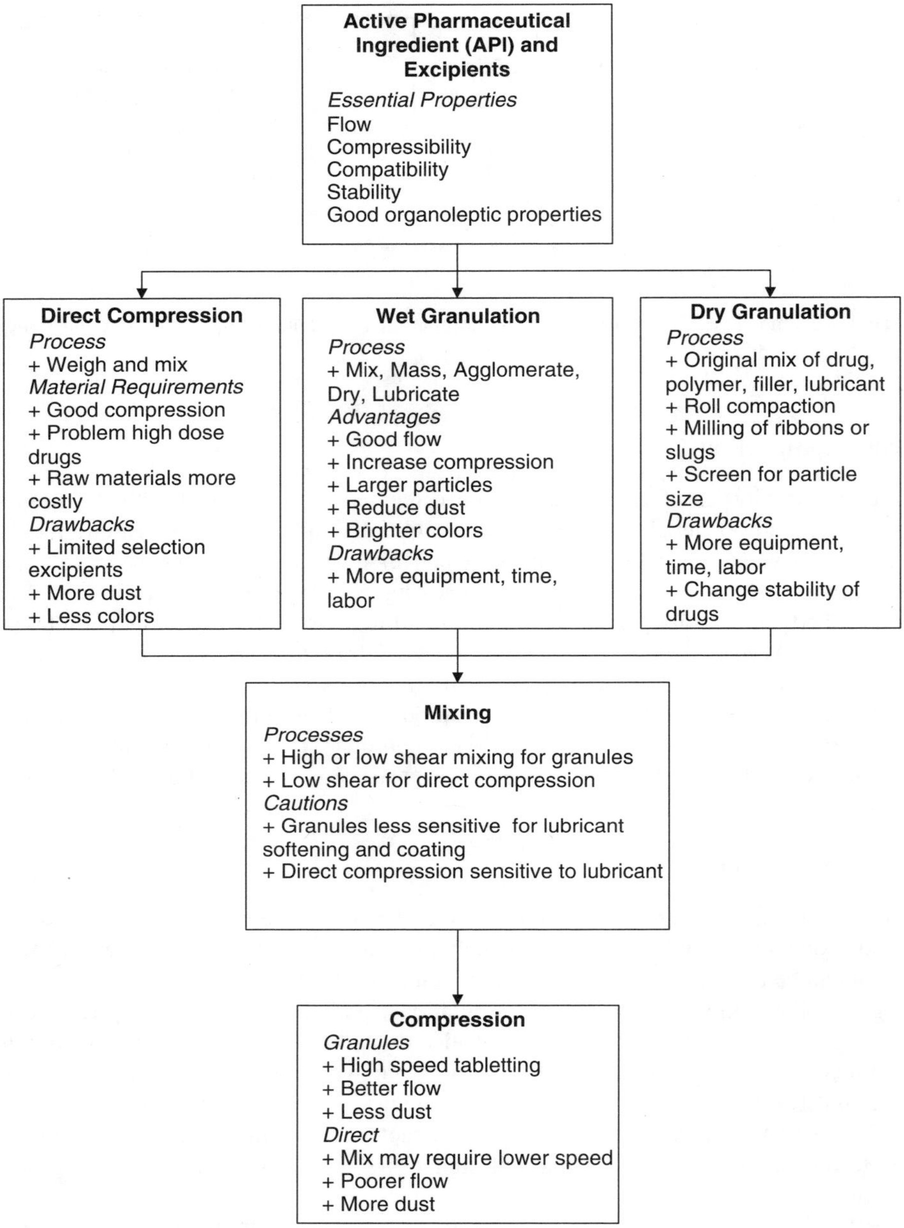

FIGURE 10.5 Schematic illustration of the most common tablet manufacturing processes.

Direct compression has some advantages:

1. For certain diluents, it was found that extremely hard tablets could be made with ease with no sign of lubrication difficulties.
2. Tablets prepared from soluble direct compression vehicles containing a disintegrant generally show faster dissolution times than those prepared by wet granulation.
3. When the right excipients are used, powder flow is very good and tablets exhibit excellent friability and rapid disintegration time.

Microcrystalline cellulose is perhaps the most widely used direct compression excipient. It exhibits the highest capacity and compressibility of all known direct compression vehicles, but its flow properties are relatively poor. Another popular direct compression excipient is dibasic calcium phosphate, a comparatively cheap insoluble diluent with good flow properties.

3. Granulation

Granulation is the process of particle size enlargement of powdered ingredients and is carried out to confer better flow and compressibility to powder systems. In addition to this, the ideal properties of a granule are that:

1. When compacted, a tablet granulation should confer physical strength and form to the tablets.
2. The granulation should be capable of being subjected to high compression pressures without defects forming.
3. A good granulation should have a uniform distribution of all the ingredients in the formulation.
4. The particle size range of the granulation should be log normally distributed with only a small percentage of both fine and coarse particles.
5. Granules should be as near spherical as possible and robust enough to withstand handling without breaking down.
6. The granulation should be relatively dust free, thus minimizing powder spread during tableting.

a. Dry Granulation

Granulation by compression or slugging is one of the dry methods that have been used for many years for moisture- or heat-sensitive ingredients. The blend of powders is forced into dies of a large heavy-duty tableting press and compacted. The compacted masses are called slugs. An alternative technique is to squeeze the powder blend into a solid cake between rollers. This is known as roller compaction. The slugs or roller compacts are then milled and screened in order to produce a granular form of tableting material that flows more uniformly than the original powder mix.

b. Wet Granulation

The wet granulation process has many advantages over the other granulation methods but is not readily suitable for hydrolyzable and thermo labile drugs such as antibiotics. In wet granulation, the binder is normally incorporated is a solution or mucilage. The choice of the liquid phase depends upon the properties of the materials to be granulated. Water is the most widely used binder vehicle, but nonaqueous granulation using isopropanol, ethanol, or methanol may be preferred if the drug is readily hydrolyzed. Changes in drug solubility resulting from a change in solvent affect granule strength owing to solute migration.

Granule growth is initiated by the formation of liquid bridges between primary particles. Soluble excipients dissolve in the binder solution to increase the liquid volume available for wet granulation; consequently, granules of large mean size are formed. The soluble components crystallize on drying to form more solid bridges. If one of the components absorbs water, this reduces the volume of binder liquid available to form wet granules, so smaller granules will result. Thus, the choice of binder vehicle can affect the characteristics of the dry granules.

The trend has moved toward using machines that can carry out the entire granulation sequence in a single piece of equipment: the mixer-granulator-dryer. The use of such machines may reduce granulation time by factors between 5 and 10.

4. Tableting Machines

With the exception of tableting presses designed to produce coated or layered tablets, the development of tableting equipment has continually evolved. In many areas, the incentives have come

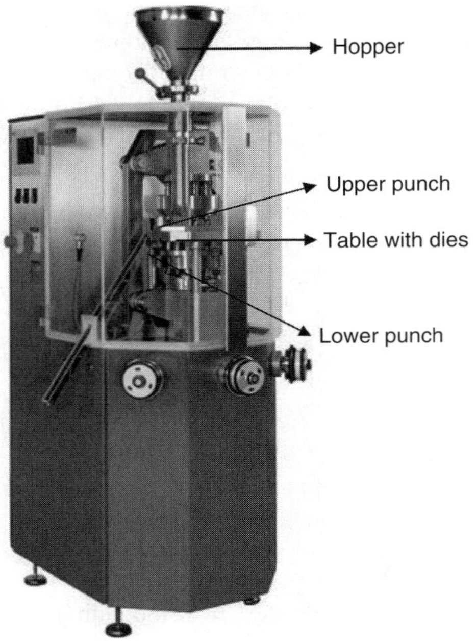

FIGURE 10.6 Multiple station rotary compacting press, XL 100, for producing pharmaceutical/vitamin tablets and compacts (Copyright by Korsch, America Inc., 18 Bristol Drive, South Easton, MA 02375).

from the pharmaceutical industry, rather than the tablet press manufacturers, because of certain trends in tableting operations. These include the desire for higher rates of production, direct compression of powders, stricter standards for cleanliness as part of an increasing awareness of GMP,* and a wish to automate, or at least continuously monitor, the process.

The ways in which individual manufacturers of tableting equipment (Figure 10.6) have sought to achieve higher output fall into four groups:

1. Increasing the effective number of punches (i.e., multitipped types)
2. Increasing the number of stations
3. Increasing the number of points of compression
4. Increasing the rate of compression (i.e., turret speed)

To produce an adequate tablet formulation, certain requirements, such as sufficient mechanical strength and desired drug release profile, must be met. At times, this may be difficult for the formulator to achieve because of poor flow and compatibility characteristics of the powdered drug. This is of particular importance when one has only a small amount of active material to work with and cannot afford to make use of trial-and-error methods. The study of the physics of tablet compaction using instrumented tableting machines enables the formulator to systematically evaluate the formula and make any necessary changes.

5. Coating

Tablets may be coated for a variety of reasons, including protection of the ingredients from air, moisture, or light. Tablets are also coated to protect the drug against decomposition or to disguise or minimize the unpleasant taste of certain medicaments. Coating also enhances the appearance of tablets and makes them more readily identifiable. In addition, coatings can be resistant to gastric

* GMP = good manufacturing practice.

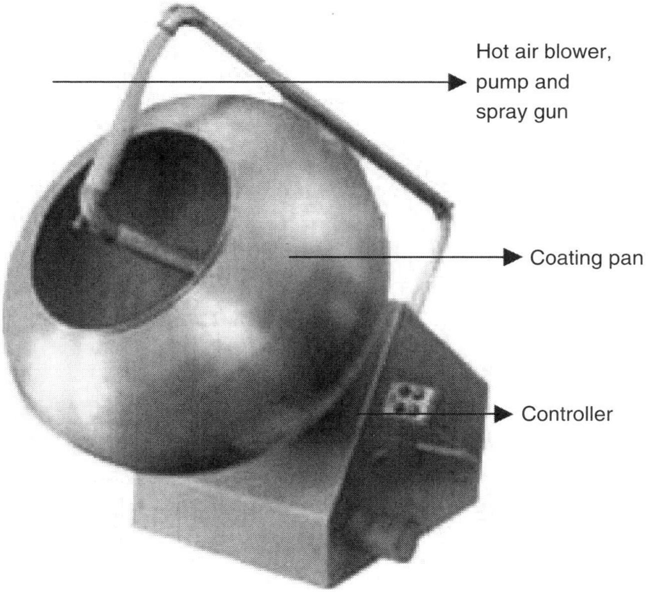

FIGURE 10.7 Tablet coating pan (Copyright by Grovers International, Sundervan Complex, Mumbai, India).

juices but readily dissolve in the small intestine. These enteric coatings can protect drugs against decomposition in the acid environment of the stomach.

a. *Sugarcoating*

The sugarcoating process involves building up layers of coating material on the tablet cores as they are tumbled in a revolving pan (Figure 10.7) by repetitively applying a coating solution or suspension and drying off the solvent. Traditionally, the cores were made using tooling with deep concave geometry to reduce problems associated with producing a sufficient coat around the tablet's edge. However, this shape may not be ideal for all products because of the inherently softer crown region exhibited in tablets manufactured from such tooling. In addition, deep concave tooling often produces tablets of poor mechanical strength.

During coating, care must be exercised to minimize penetration of coating solutions into the core itself, although the coat should adhere well to the tablet surface. Before sugarcoating, the core is coated with a sealing coat of shellac, PVP*-stabilized types of shellac, or other polymeric materials, such as cellulose acetate phthalate and polyvinyl acetate phthalate. The sealing coat must be kept to the minimum thickness consistent with providing an adequate moisture barrier. The next stage is to build up a subcoat that will provide a good bridge between the main coating and the sealed core, as well as round off any sharp corners. This step is followed by smoothing or grossing. The finishing stage is accomplished by again applying one or two layers of clear syrup. The tablets are then left for several hours before being transferred to the polishing pan. The polish is a dilute wax solution (e.g., carnauba or beeswax in petroleum spirit) applied sparingly until a high luster is produced.

b. *Film Coating*

Film coating has increased in popularity for various reasons. The film process is simpler and, therefore, easier to automate. It is also faster than sugarcoating, since weight gains of only 2 to 6% are involved, as opposed to more than 50% with sugarcoating. In addition, moisture involvement

* PVP = polyvinyl pyrolidone.

can be avoided (if absolutely necessary) by using nonaqueous solvents. Moreover, distinctive identification tablet markings are not obscured by film coats.

Two major groups of film coating materials may be distinguished: (a) those that are nonenteric and, for the most part, cellulose derivatives, and (b) those that can provide an enteric effect and are commonly esters of phthalic acid. Within both groups, it is general practice to use a mixture of materials to give a film with the optimum range of properties. Films may contain a plasticizer that prevents the film from becoming brittle with consequent risk of chipping. The choice of plasticizer depends on the particular film polymer.

The nature of the solvent system may markedly influence the quality of the film and, to optimize the various factors, mixed solvents are usually necessary. More specifically, the rate of evaporation and, hence, the time for the film to dry, has to be controlled within fine limits if a uniform smooth coat is to be produced. The solvent mixture must be capable of dissolving the required amount of coating material yet give rise to a solution within a workable range of viscosity.

Until recently, alcohols, esters, chlorinated hydrocarbons, and ketones have been among the most frequently used types of solvents. However, because of increasing regulatory pressures against undesirable solvents, there has been a pronounced trend toward aqueous film coating. Because of the need to develop a uniform color with minimum application, the colorants used in film coating are more likely to be lakes than dyes. In lakes, the colorant has been absorbed onto the surface of an insoluble substrate.

c. *Modified-Release Coatings*

A coating may be applied to a tablet to modify the release pattern of the active ingredient. Two general categories, enteric coating and controlled-release coating, are distinguished. The former are insoluble in the low pH environment of the stomach but dissolve readily in the small intestine with its elevated pH. They are used to minimize irritation of the gastric mucosa by certain drugs and to protect others that are degraded by gastric juices. Modified release coats are discussed in more detail in Chapter 11.

L. Packaging and Storing Tablets

Tablets are stored in tight containers in places of low humidity and protected from extremes in temperature. Products that are prone to decomposition by moisture generally are co-packaged with a desiccant packet. Drugs that are adversely affected by light are packaged in light-resistant containers. With a few exceptions, tablets that are properly stored will remain stable for several years or more.

In dispensing tablets, the pharmacist should use a similar type of container as provided by the manufacturer. The patient should maintain the drug in the container dispensed. Storage conditions, as recommended for the particular product, should be maintained by the pharmacist and patient and expiration dates observed.

The pharmacist should also be aware that the hardness of certain tablets might change upon aging, usually resulting in a decrease in the disintegration and dissolution rate of the product. The increase in tablet hardness can frequently be attributed to the increased adhesion of the binding agent and other formulation components in the tablet. Examples of increased tablet hardening with age have been reported for a number of drugs including aluminum hydroxide and sodium salicylate.

Certain tablets containing volatile drugs, such as nitroglycerin, may experience the migration of the drug between tablets in the container, resulting in a lack of uniformity among the tablets. Further, when packing materials such as cotton come in contact with nitroglycerin tablets it may absorb varying amounts of nitroglycerin, rendering the tablets subpotent. The USP directs that nitroglycerin tablets be preserved in tight containers, preferably of glass, at controlled room temperature. The USP further directs that nitroglycerin tablets must be dispensed in the original, unopened container, labeled appropriately.

M. Compendial Standards and Quality Assurance

Compressed tablets may be characterized or described by a number of specifications. These include the diameter, shape, thickness, weight, hardness, disintegration time, and dissolution characteristics. The diameter and shape depend on the die and the punches selected for the compression of the tablet. The tablets may be scored in halves or quarters to facilitate breaking if smaller doses are desired. The top or lower surface may be embossed or engraved with a symbol or letters that serve as an additional means of identifying the source of the tablets. These characteristics along with the color of the tablets tend to make them distinctive and identifiable with the active ingredient they contain.

The remaining specifications assure the manufacturer that the tablets do not vary from one production lot to another. It is convenient to divide the types of test procedures into two major categories: (a) those that are requirements in an official compendium (Table 10.4), and (b) those

TABLE 10.4
USP Recommended Tests and Specifications for Quality Control and Quality Assurance of Tablets

Property	Test	Specification
Uniformity of dosage units	Weight	When the tablet contains 50 mg or more of the drug substance or when the latter comprises 50% or more, by weight, of the dosage form. Twenty tablets are weighed individually, and the average weight is calculated. The variation from the average weight in the weights of not more than two of the tablets must not differ by more than the percentage allowed; no tablet differs by more than double that percentage. Tablets that are coated are exempt from these requirements but must conform to the test for content uniformity if it is applicable.
	Thickness	Tablet thickness is determined with a caliper or thickness gauge that measures the thickness in millimeters. Plus or minus 5% may be allowed, depending on the size of the tablet.
	Content	The content uniformity test has been extended to monographs on all coated and uncoated tablets and all capsules intended for oral administration where the range of sizes of the dosage form available includes a 50 mg or smaller size, in which case the test is applicable to all sizes (50 mg and larger and smaller) of that tablet or capsule.
Drug release	Disintegration	For compressed, uncoated tablets the testing fluid is usually water at 37°C, but in some cases the monographs direct that simulated gastric fluid TS be used. The conditions of the test vary somewhat for coated tablets, buccal tablets, and sublingual tablets. For most uncoated tablets the period is 30 min, although the time for some uncoated tablets varies greatly from this. For coated tablets up to 2 h may be required. For sublingual tablets, such as CT isoproterenol hydrochloride, the disintegration time is 3 min.
	Dissolution	For some tablets the USP monographs direct compliance with limits on dissolution rather than disintegration. Since drug absorption and physiological availability depend on having the drug substance in the dissolved state, suitable dissolution characteristics are an important property of a satisfactory tablet. Dissolution specifications state that a certain percentage of drug must dissolve in a specified time.
Other tests	Moisture	Must comply with minimum standard.
	Impurities	No impurities above required limits
	Packaging and storage	Must meet USP and FDA requirements.
	Tablet hardness	Must meet USP and FDA or in-house requirements.
	Friability	Must meet USP and FDA or in-house requirements.

that, although unofficial, are widely used in commerce. Here information will be largely restricted to tests that are mandatory in the USP and the National Formulary.

In general, the content uniformity USP test is designed to establish the homogeneity of a batch by assaying 10 tablets individually, after which the arithmetic mean and relative standard deviation (RSD) are calculated. The USP criteria are met if the content uniformity lies within 85 to 115% of the label claim, and the RSD is not greater than 6%. Provision is included in the compendium for additional testing if one or more units fail to meet the standards.

Example: Evaluation of Content Uniformity

The contents of 10 tablets containing 5 mg of a drug were measured to be 4.86 mg, 4.87 mg, 5.01 mg, 5.01 mg, 4.95 mg, 4.91 mg, 4.88 mg, 4.92 mg, 4.89 mg, and 4.91 mg. If the content uniformity must be between 85 and 115% of the label claim, with a RSD not greater than 6%, this data can be used to determine if the batch of tablets complies with USP specifications. Using a standard calculator, the mean of the 10 assay results was calculated to be 4.935 mg with a standard deviation of 0.106 mg. This correlates to a content of 98.7% of the label claim and a RSD of 2.15%. This means the batch of tablets complies with USP specifications because the content uniformity lies within 85 to 115% of the label claim and the RSD is not greater than 6%.

---

1. Dissolution Testing

The bioavailability of drugs from tablets can be markedly influenced by the rate and efficiency of the initial disintegration and dissolution process. Some of the reasons why dissolution testing is important include:

Product optimization: By conducting dissolution studies in the early stages of a product's development, differentiations can be made between formulations and correlations identified with *in vivo* bioavailability data.

Monitoring manufacturing processes: The conduct of such testing from the early product development through product approval and commercial batch production assures the control of any potential variables of materials and processes that could affect drug dissolution and the product's quality standards.

In vitro–in vivo correlation: In assessing such batch-to-batch bioequivalence, the FDA allows manufacturers to examine scale-up batches of 10% of the proposed size of the actual production batch, or 100,000 dosage units, whichever is greater.

Regulatory requirements: New drug applications contain *in vitro* dissolution data generally obtained from batches that have been used in pivotal clinical or bioavailability studies and from human studies conducted during product development. Once these specifications are approved they become official specifications for all subsequent batches and bioequivalent products.

The goal of *in vitro* dissolution testing is to provide as far as is possible, a reasonable prediction of, or correlation with, the product's *in vivo* bioavailability. The biopharmaceutical classification system has been developed, which relates combinations of a drug's solubility (high or low) and its intestinal permeability (high or low) as a possible basis for predicting the likelihood of achieving a successful *in vivo–in vitro* correlation. Drugs determined to have the following characteristics are considered:

1. High solubility and high permeability
2. Low solubility and high permeability

3. High solubility and low permeability
4. Low solubility and low permeability

For a high solubility, high permeability drug, an *in vivo–in vitro* correlation may be expected if the dissolution rate is slower than the rate of gastric emptying (the rate-limiting factor). In the case of a low solubility, high permeability drug, dissolution may be the rate-limiting step for drug absorption, and an *in vivo–in vitro* correlation may be expected. In the case of a high solubility, low permeability drug, permeability is the rate-limiting step, and only a limited *in vivo–in vitro* correlation may be possible. In the case of a drug with low solubility and low permeability, significant problems would be likely for oral drug delivery.

A number of formulation and manufacturing factors can affect the disintegration and dissolution of a tablet, including:

1. The particle size of the drug substance in the formulation
2. The solubility and hygroscopicity of the formulation
3. The type and concentration of the disintegrant, binder, and lubricant used (see example for salicylic acid in Figure 10.3)
4. The manufacturing method, particularly the compactness of the granulation and the compression force used in tableting
5. The in-process variables that may occur

The USP includes seven apparatus designs (Table 10.5) for drug release and dissolution testing of immediate release oral dosage forms, extended release products, enteric coated products, and transdermal drug delivery devices. Of primary interest here are USP Apparatus 1 and USP Apparatus 2, used principally for immediate release solid oral-dosage forms.

Dissolution testing should be carried out under physiological conditions, if possible. This allows interpretation of dissolution data on *in vivo* performance of the product. However, strict adherence to the gastrointestinal environment need not be used in routine dissolution testing. The testing conditions should be based on physicochemical characteristics of the drug substance and the environmental conditions the dosage form might be exposed to after oral administration. The volume of the dissolution medium is generally 500, 900, or 1000 ml. Sink conditions are desirable but not mandatory. Sink has been defined different ways in the past as either 10 to 20% of solubility necessary for dissolution. The USP prefers 33% of solubility. An aqueous medium with a pH range of 1.2 to 6.8 (ionic strength of buffers the same as in USP) should be used. To simulate intestinal fluid, a dissolution medium of pH 6.8 should be employed. A higher pH could be justified on a case-by-case basis and, in general, should not exceed pH 8.0.

To simulate gastric fluid, a dissolution medium of pH 1.2 should be employed without enzymes. The need for enzymes in simulated gastric fluid and simulated intestinal fluid should be evaluated on a case-by-case basis and should be justified. Recent experience with gelatin capsules indicates the possible need for enzymes (pepsin with simulated gastric fluid and pancreatin with simulated intestinal

TABLE 10.5
Types of Dissolution Apparatuses Most Commonly Used to Test Solid Dosage Forms

Type of Apparatus	When Best Used
Apparatus 2 (paddle)	Default apparatus
Apparatus 1 (basket)	Preferred over paddle for enteric coated tablets or beads
Apparatus 3 (reciprocating cylinder)	Low solubility compounds where surfactant concentration is excessive
Apparatus 4 (flow-through)	Enteric coated dosage forms, low solubility compounds, other strategies requiring a switchover in media

fluid) to dissolve pellicles, if formed, to permit the dissolution of the drug. Use of water as a dissolution medium is also discouraged because test conditions such as pH and surface tension can vary depending on the source of the water and may change during the dissolution test itself, owing to the influence of the active and inactive ingredients. For water insoluble or sparingly water soluble drug products, use of a surfactant such as sodium lauryl sulfate is recommended. The need for and the amount of the surfactant should be justified. Use of a hydro alcoholic medium is discouraged.

All dissolution tests for immediate release dosage forms should be conducted at 37 ± 0.5°C. The baskets or paddles are rotated between 50 and 150 rpm. The basket and paddle method can be used for performing dissolution tests under multimedia conditions (e.g., the initial dissolution test can be carried out at pH 1.2, and, after a suitable time interval, a small amount of buffer can be added to raise pH to 6.8). Alternatively, if addition of an enzyme is desired, it can be added after initial studies (without enzymes). Use of Apparatus 3 allows easy change of the medium. Apparatus 4 can also be adopted for a change in dissolution medium during the dissolution run. Certain drug products and formulations are sensitive to dissolved air in the dissolution medium and need deaeration. Apparatus suitability tests should be carried out with a performance standard (i.e., calibrators) at least twice a year and after any significant equipment change or movement. However, a change from basket to paddle or vice versa may need recalibration. The equipment and dissolution methodology should include the product related operating instructions such as deaeration of the dissolution medium and use of a wire helix for capsules.

There is growing recognition that where inconsistencies in dissolution occur, they occur not between dosage units from the same production batch, but rather between batches, or between products, from different manufacturers. However, since dosage forms within a batch are generally the same, the concept of *pooled dissolution testing* has emerged. The pooled samples come from the individual dissolution vessels in the apparatus or from multiple dosage units dissolved in a single vessel.

2. Mechanical Strength of Tablets

The resistance of the tablet to chipping, abrasion, or breakage during storage, transportation, and handling before usage depends on its hardness. In the past, a rule of thumb described a tablet to be of proper hardness if it was firm enough to break with a sharp snap when it was held between the second and third fingers, with the thumb as the fulcrum, yet did not break when it fell on the floor. In industry the force is measured in kilograms, and when used in production, a hardness of 4 to 6 kg is considered the minimum for a satisfactory tablet. The most widely used apparatus to measure tablet hardness or crushing strength is the Schleuniger apparatus. A number of attempts have been made to quantitate the degree of hardness. These are unofficial tests not prescribed in pharmacopoeias.

A tablet property related to hardness is friability, and the measurement is made with the Roche friabilator. Rather than measuring the force required to crush a tablet, the instrument is designed to evaluate the ability of the tablet to withstand abrasion in packaging, handling, and shipping. A number of tablets are weighed and placed in the tumbling apparatus, where they are exposed to rolling and repeated shocks resulting from freefalls within the apparatus. After a given number of rotations, the tablets are weighed, and the loss in weight indicates the ability of the tablets to withstand this type of wear.

Another approach taken by many manufacturers when they evaluate a new product is to send the package to distant points and back using various methods of transportation. This is called a shipping test. The condition of the product on its return indicates its ability to withstand transportation handling.

N. Common Tablet Manufacturing Problems

Problems often arise in tableting operations owing to a variety of moisture, pressure, temperature, and static charge factors. These problems can lead to:

1. Content uniformity variations owing to segregation of powder mixtures
2. Inability to meet blend uniformity requirements
3. Unreliable material flow to the tableting press
4. Tablet weight variations with increased compression rates
5. Difficulty in establishing proper sample collection locations, number of samples, and acceptance criteria

Generally failure of tablet formation during tableting causes one of three things to happen, that is, capping, lamination, and chipping. Capping describes the partial or complete separation of the top or bottom of the tablets from the main body. Lamination is the separation of the tablet into two or more distinct layers. Chipping is the splitting of small parts from the surface of the tablet. The cause of these problems is usually ascribed to entrapment of air among the powder particles. The air does escapes not during compression but afterward when the pressure is released. Bad punches and dies, not enough lubricant, too dry granules, or incorrect setup of the tableting press can also contribute to these tablet failures.

O. Examples of Tablet Formulations

1. Example 1

Direct Compression Tablet Containing a Low Dose, Highly Soluble Active Ingredient

Formula	%	Use
Diphenhydramine HCl powder	10.0	Active
Microcrystalline cellulose (Avicel; pH 102)	25.0	Filler (direct compression)
Lactose	62.2	Filler
Croscarmellose sodium (Ac-di-Sol)	2.00	Disintegrant
Colloidal silicon dioxide	0.20	Glidant
Magnesium stearate	0.30	Lubricant
Stearic acid	0.30	Lubricant

a. *Method*

1. Screen all the ingredients through a 20-mesh sieve.
2. Blend everything except the lactose, stearic acid, and magnesium stearate in a V-shaped blender for 3 min.
3. Add lactose to the batch and blend for 10 min.
4. Add stearic acid to the batch and blend for 3 min.
5. Add magnesium stearate to the batch and blend for 5 min.
6. Tablet using flat-faced bevel-edge tooling with appropriate size.

2. Example 2

Wet Granulated Tablet Containing a High Dose, Partially Soluble Active

Formula	%	Use
Theophylline anhydrous	30.0	Active
Anhydrous lactose	65.5	Filler (direct compression)
Croscarmellose sodium (Ac-di-Sol)	3.00	Disintegrant
Stearic acid	1.00	Lubricant
Granulating fluid (10% PVP solution)	Qs	Binder

a. Method

1. Screen all the ingredients through a 20-mesh sieve.
2. Blend everything except stearic acid in a V-shaped blender for 20 min.
3. Granulate powder with PVP solution.
4. Screen granules to appropriate size and dry.
5. Add stearic acid to dried granules and blend for 3 min.
6. Tablet using flat-faced bevel-edge tooling with appropriate size.

HOMEWORK

It will be helpful for students to find the following information from literature and other sources:

1. The top 20 commercially available tablet products.
2. Ten commercially available enteric coated tablets.
3. Available USP requirements and test conditions for tablet dissolution testing of products found in Questions 1 and 2.
4. Differences between immediate release and sustained or controlled release tablets.

TUTORIAL

1. Which method of tablet manufacture can be used to combine two incompatible substances in the same tablet:
   (a) Sugarcoating
   (b) Film coating
   (c) Enteric coating
   (d) Multilayer compression
2. Solidification of molded tablets depends on:
   (a) Compaction force
   (b) Crystal bridges formed during drying
   (c) Solubility of the drug
   (d) Powder flow
3. Lozenges usually do not contain the following tableting excipient:
   (a) Sucrose
   (b) Cross-linked povidone
   (c) Lactose
   (d) Gelatin
4. A major disadvantage of compressed tablets as a dosage form is:
   (a) Physical changes in the drug during compression
   (b) Tablets are difficult to transport.

   (c) The release of drug cannot be controlled.
   (d) They cannot be produced uniformly.
5. Patients who cannot swallow enteric coated tablets should:
   (a) Crush the tablet before taking it
   (b) Dissolve the tablet before drinking it
   (c) Try to swallow tablet without water
   (d) Consult pharmacist for alternative
6. For a drug to be absorbed from the gastrointestinal tract the drug needs to be:
   (a) Coated before tableting
   (b) In solution
   (c) Disintegrated
   (d) Kept in the tablet for as long as possible
7. Adequate powder flow ensures that after tableting:
   (a) Tablets of constant weight are produced.
   (b) The drug is released more quickly.
   (c) Drug molecules are crushed.
   (d) Smooth tablets are produced.
8. Which of the following excipients are added to tablet formulations to provide enough bulk for compression:
   (a) Glidants
   (b) Diluents
   (c) Lubricants
   (d) Disintegrants
9. Complete mixing of magnesium stearate with tablet granules will:
   (a) Decrease the crushing strength of tablets
   (b) Increase tablet hardness
   (c) Increase tablet dissolution
   (d) Increase tablet disintegration
10. Fumed silica is not used in tablets as a:
   (a) Diluent
   (b) Glidant
   (b) Lubricant
   (d) Adsorbent
11. The mixing process whereby micronized drug particles are uniformly distributed over diluent particles is called:
   (a) Reactive mixing
   (b) Ordered mixing
   (c) Ball mixing
   (d) Random mixing
12. According to Hiestand compaction indices, a high strain index for a compound means that:
   (a) Stronger tablets are formed.
   (b) Harder tablets are formed.
   (c) Smooth tablets are formed.
   (d) Weaker tablets are formed.
13. During the first part of tablet compaction the particles undergo:
   (a) Irreversible deformation
   (b) Elastic deformation
   (c) Brittle deformation
   (d) Plastic deformation

14. Which of the following is not an ideal property of a tablet granulation?
    (a) Uniform distribution of ingredients
    (b) Near spherical in shape
    (c) Contain large amounts of fines
    (d) Withstand high compaction pressures
15. Which of the following excipients can be used as a binder in granulation?
    (a) Magnesium stearate
    (b) Starch mucilage
    (c) Fumed silica
    (d) Isopropanol
16. To protect drugs from decomposition by moisture, tablets are co-packaged with:
    (a) Cotton wool
    (b) Fumed silica
    (c) A desiccant package
    (d) Aluminum foil
17. Which of the following factors does not influence the speed of drug dissolution from tablets?
    (a) Particle size of the drug
    (b) Solubility of the drug
    (c) Tablet hardness
    (d) Weight uniformity
18. What is considered an adequate crushing strength for uncoated compressed tablets?
    (a) 1 to 2 kg
    (b) 4 to 6 N
    (c) 4 to 6 kg
    (d) 70 to 90 N
19. Which formulation factors adversely affect the release rate of drugs from compressed tablets?
    (a) Poor chemical stability
    (b) Small effective surface area of the drug
    (c) Good aqueous solubility of the binder
    (d) The absence of polymorphism
20. Which formulation of the following tablet tests are not an official pharmacopoeial test?
    (a) Hardness
    (b) Disintegration
    (c) Dissolution
    (d) Content uniformity

Answers: 1 (d); 2 (b); 3 (b); 4 (a); 5 (d); 6 (b); 7 (a); 8 (b); 9 (a); 10 (a); 11 (b); 12 (d); 13 (b); 14 (c); 15 (b); 16 (c); 17 (d); 18 (c); 19 (b); 20 (a)

V. CAPSULES

A. Background

Capsules are solid dosage forms in which the drug substance is enclosed within either a hard or soft soluble shell, usually formed from gelatin. The capsule may be considered a container drug delivery system that provides a tasteless and odorless dosage form without need for a secondary coating step, as may be required for tablets. Swallowing is easy for most patients, since the shell is smooth and hydrates in the mouth, and the capsule often tends to float on swallowing in the liquid taken with it. Their availability in a wide variety of colors makes capsules aesthetically pleasing. There are numerous additional advantages to capsules as a dosage form, depending on the type of capsule employed.

The first capsule prepared from gelatin was a one-piece capsule patented in France by Mothes and Du Blanc in 1834. Although the shells of these early capsules were not plasticized, such capsules likely would be classified today as soft gelatin capsules based on shape, contents, and other features. Intended to mask the taste of certain unpleasant tasting medication, they quickly gained popularity.

Today capsules may be classified as either hard or soft, depending on the nature of the shell. Most capsules of both types are intended to be swallowed whole; however, some soft gelatin capsules are intended for rectal or vaginal insertion as suppositories.

B. Hard Gelatin Capsules

Most capsule products manufactured today are of the hard gelatin type. It is estimated that the utilization of hard gelatin capsules to prepare solid dosage forms exceeds that of soft gelatin capsules by about 10-fold.

1. Advantages and Disadvantages of Hard Gelatin Capsules

Some of the advantages of hard shell capsules as a dosage form include:

1. Hard gelatin capsules often have been assumed to have better bioavailability than tablets. Most likely, this assumption is derived from the fact that the gelatin shell rapidly dissolves and ruptures, which affords at least the potential for rapid release of the drug.
2. Hard shell capsules allow a degree of flexibility of formulation not obtainable with tablets. Often they are easier to formulate because there is no requirement that the powders be formed into a coherent compact that will stand up to handling. However, the problems of powder blending and homogeneity, powder fluidity, and lubrication in hard capsule filling are similar to those encountered in tablet manufacture.
3. Modern capsule filling equipment makes possible the multiple filling of diverse systems such as beads, granules, small tablets, powders, and even semisolids.
4. Capsules make possible the filling of bead-type modified release products since they are filled without a compression process that could rupture the particles.
5. Hard gelatin capsules are ideally suited for clinical trials and are widely used in preliminary drug studies. For comparative bioequivalence studies tablets can even be hidden in capsules to ensure the test being blinded.

Disadvantages of hard shell capsules include:

1. From a manufacturer's point of view, there is perhaps some disadvantage in the fact that the number of suppliers of shells is limited.
2. Filling equipment is slower than tableting, although that gap has narrowed in recent years with the advent of high-speed automatic-filling machines.
3. Generally, hard gelatin capsule products tend to be more costly to produce than tablets; however, the relative cost effectiveness of capsules and tablets must be judged on a case-by-case basis.
4. This cost disadvantage diminishes as the cost of the active ingredient increases or when tablets must be coated. Furthermore, it may be possible to avoid the cost of a granulation step by choosing encapsulation in lieu of tableting.

5. Highly soluble salts (e.g., iodides, bromides, and chlorides) generally should not be dispensed in hard gelatin capsules. Their rapid release may cause gastric irritation owing to the formation of a high drug concentration in localized areas.
6. Both hard gelatin capsules and tablets may become lodged in the esophagus, where the resulting localized high concentration of certain drugs (e.g., doxycycline, potassium chloride, indomethacin) may cause damage.

2. Manufacturing Hard Gelatin Capsule Shells

Hard gelatin shells are manufactured by a process in which stainless steel mold pins are dipped into warm gelatin solutions and the shells are formed by gelatin on the pin surfaces. Gelatin is the most important constituent of the dipping solutions, but other components may be present. There are three producers of hard shell capsules in North America (Shionogi Qualicaps, Indianapolis, IN; Capsugel Div. Warner-Lambert Co., Greenwood, SC; and Pharmaphil Corp, Windsor, Ontario).

Gelatin is prepared by the hydrolysis of collagen obtained from animal connective tissue, bone, skin, and sinew. This long polypeptide chain yields, on hydrolysis, 18 amino acids, the most prevalent of which are glycine and alanine. Gelatin can vary in its chemical and physical properties, depending on the source of the collagen and the manner of extraction. There are two basic types of gelatin. Type A, which is produced by an acid hydrolysis, is manufactured mainly from pork skin. Type B gelatin, produced by alkaline hydrolysis, is manufactured mainly from animal bones. The two types can be differentiated by their iso-electric points (4.8 to 5.0 for type B and 7.0 to 9.0 for type A) and by their viscosity-building and film-forming characteristics. Either type of gelatin may be used, but combinations of pork skin and bone gelatin are often used to optimize shell characteristics because bone gelatin contributes firmness, whereas pork skin gelatin contributes plasticity and clarity.

Commonly, various soluble synthetic dyes (coal tar dyes) and insoluble pigments are used to change the color of the capsule shell. Commonly used pigments are the iron oxides. Colorants not only play a role in identifying the product, but also may play a role in improving patient compliance. Titanium dioxide may be included to render the shell opaque. Opaque capsules may be employed to provide protection against light or to conceal the contents. When preservatives are employed, parabens are often used.

In general, information on capsules is printed before filling. Empty capsules can be handled faster than filled capsules and, should there be any loss or damage to the capsules during printing, no active ingredients would be involved.

3. Hard Shell Sizes and Shapes

For human use, empty gelatin capsules are manufactured in eight sizes, ranging from 000 (the largest) to 5 (the smallest) as shown in Figure 10.8. The volumes and approximate capacities for the traditional eight sizes are listed in Table 10.6. The largest size normally acceptable to patients is a No. 0. Size 0 hard gelatin capsules with an elongated body (size OE) also are available. They provide greater fill capacity without an increase in diameter. Three larger sizes are available for veterinary use: 10, 11, and 12, with capacities of about 30, 15, and 7.5 g, respectively. Although the standard shape of capsules is the traditional, symmetrical bullet shape, some manufacturers have employed distinctive proprietary shapes. Lilly's Pulvule is designed with a characteristic body section that tapers to a bluntly pointed end. Smith Kline Beacham's Spansule capsules exhibit a characteristic taper at both the cap and body ends.

In the wake of several incidents of tampering with over-the-counter capsules, sometimes with fatal consequences, much thought was given on how to make capsules safer. Attention was focused on sealing techniques as possible means of enhancing the safety of capsules by making them tamper-evident (i.e., so that they could not be tampered with without destroying the capsule or at least causing obvious disfigurement).

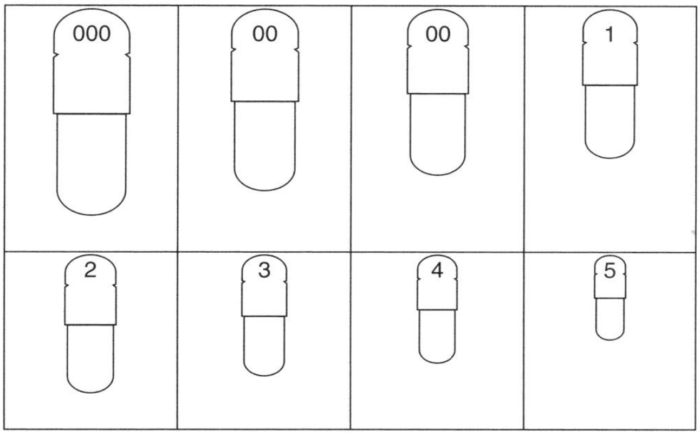

FIGURE 10.8 Hard gelatin capsule sizes.

TABLE 10.6
Approximate Capacities of Capsules

Human Sizes	Capacity (ml)	Fill Weight at 0.8 g/ml Density
5	0.12	0.10
4	0.20	0.19
3	0.27	0.24
2	0.37	0.30
1	0.48	0.40
0	0.67	0.54
00	0.95	0.76
000	1.36	1.10

These sealing and self-locking closures also help prevent the inadvertent separation of filled capsules during shipping and handling. This problem is particularly acute in the filling of noncompacted, bead, or granular formulations. Hard gelatin capsules are made self-locking by forming indentations or grooves on the inside of the cap and body portions. When fully engaged, a positive interlock is created between the cap and body portions. Examples of self-locking capsules include Posilok (Shionogi Qualicaps), Loxit (Pharmaphil), and Coni-Snap (Capsugel; Figure 10.9).

Hard gelatin capsules may also be hermetically sealed by banding (i.e., layering down a film of gelatin, often distinctively colored, around the seam of the cap and body). Parke Davis' Kapseal and Quali-Seal (Shionogi Qualicaps) are examples of banding capsules. Banding currently is the single most commonly used sealing technique.

Example: Determining Capsule Fill Weight

To determine the size of capsule to be used or the fill weight for a formulation the following practical relationship is used:

$$\text{Capsule fill weight} = \text{tapped bulk density of formulation} \times \text{capsule volume}$$

For example, if the formulation has a fill weight of 450 mg and a tapped density of 0.80 g/ml, then the volume occupied by fill weight = 0.45 g/0.80 g/ml = 0.56 ml. A size 0 capsule has a fill volume 0.54 ml, so this capsule should be used.

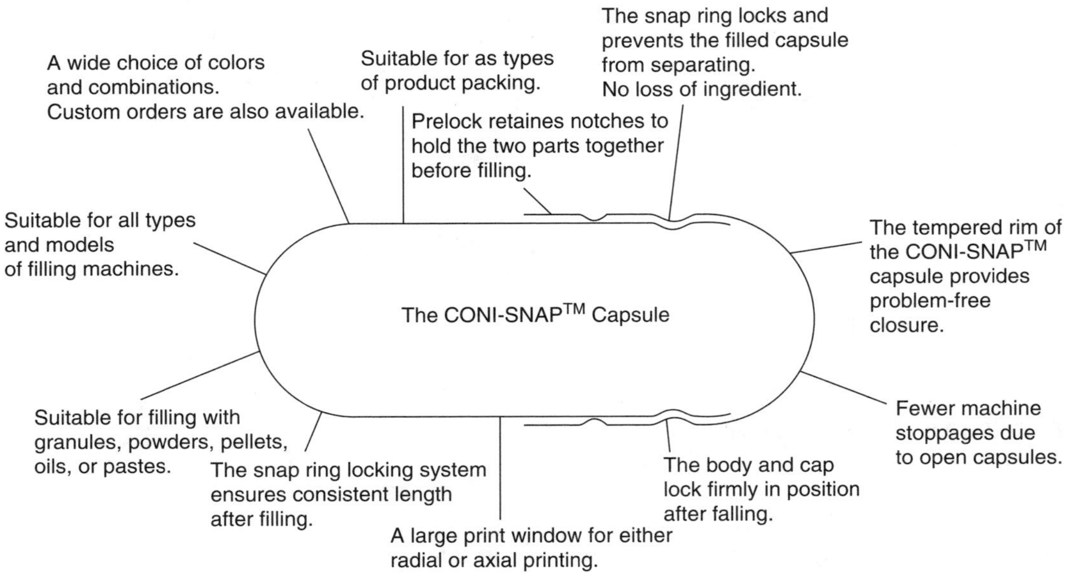

FIGURE 10.9 Coni-Snap® capsules from CAPSUGEL, the world's most popular brand of two-piece gelatin capsule (Copyright by CAPSUGEL, 201 Tabor Road, Morris Plains, NJ 07950).

4. Storage, Packaging, and Stability of Hard Shell Capsules

Finished capsules normally contain an equilibrium moisture content of 13 to 16%. This moisture is critical to the physical properties of the shells, since at lower moisture contents (<12%), shells become too brittle; at higher moisture contents (>18%) they become too soft. It is best to avoid extremes of temperature and to maintain a relative humidity of 40 to 60% when handling and storing capsules.

The bulk of the moisture in capsule shells is physically bound, and it can readily transfer between the shell and its contents, depending on their relative hygroscopicity. The removal of moisture from the shell could be sufficient to cause splitting or cracking, as has been reported for the deliquescent materials potassium acetate and sodium cromoglycate. Conditions that favor the transfer of moisture to powder contents may lead to caking and retarded disintegration or other stability problems. It may be useful to preequilibrate the shell and its contents to the same relative humidity within the acceptable range before filling.

Another problem is the loss of water solubility of shells, apparently because of sufficient exposure to high humidity and temperature or to exposure to trace aldehydes. Such capsules develop a skin, or pellicle, during dissolution testing, exhibit retarded dissolution, and may fail to meet the USP drug dissolution specifications. This decrease in solubility of gelatin capsules has been attributed to gelatin crosslinking caused by impurities such as formaldehyde.

Capsules can be individually protected by enclosure in strip or blister packs. In the former the units are hermetically sealed in strips of aluminum foil or plastic film. In the latter one of the films enclosing the units is formed into blisters. An ideal foil or film for these packs should be:

1. Heat stable
2. Impermeable to moisture, water vapor, air, and odors
3. Strong enough for machine handling
4. Reasonably easy for patients to tear and open

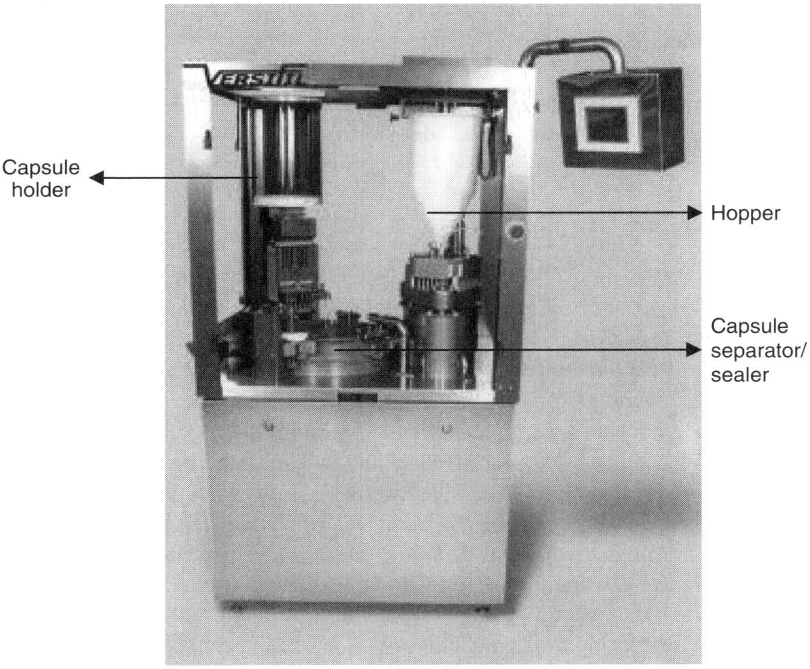

FIGURE 10.10 The Bohanan Versifil, a hard gelatin encapsulation machine designed to fill capsules at a rate of 160 to 700 capsules per minute (fill rate depends on tooling used) (Copyright by A.W. Bohanan Co., Inc., Dallas, NC).

5. Filling Hard Gelatin Capsules

The several types of filling machines (Figure 10.10) in use in the pharmaceutical industry have in common the following operations:

- *Rectification:* The empty capsules are oriented so that all point the same direction (i.e., body-end downward).
- *Separation of caps from bodies:*The rectified capsules are delivered body-end first and vacuum applied from below to pull the bodies down.
- *Dosing of fill material:* Various methods are employed including the Auger fill principle, vibratory fill principle, piston-tamp principle, and nonpowdered filling.
- *Replacement of caps and ejection of filled capsules:* Pins are used to push the filled bodies up into the caps for closure, and to push the closed capsules out of the bushings.

The machines used to fill capsules by this process may be either semiautomatic or fully automatic. Semiautomatic machines require an operator to be in attendance at all times and are capable of filling as many as 120,000 to 160,000 capsules in an 8-h shift. Fully automatic machines are rated to fill that many capsules in 1 h. There are four main powder-filling methods:

- *Auger Fill Principle:* The fill of the capsules in these machines is primarily volumetric.
- *Vibratory Fill Principle:* The capsule body passes under a feed frame that holds the powder in the filling section. Formulation requirements for this machine typically require the addition of stearate lubricants to prevent the binding of push rods and guides.

- *Piston-Tamp Principle:* Piston-tamp machines are fully automatic filters in which pistons tamp the individual doses of powders into plugs (sometimes referred to as slugs), which often resemble soft tablets in consistency, and eject the plugs into the empty capsule bodies. There are two types of piston-tamp fillers: dosator machines and dosing-disk machines.
- *Nonpowder Filling:* In addition to powder dosing, filling devices also are available that can feed beads or pellets, micro-tablets, tablets, and liquid or pasty materials into capsules.

6. Tips for Compounding Capsules

1. Sodium lauryl sulfate, up to 1%, can be added to powders to neutralize electrostatic forces.
2. Capsules may be colored by adding a dye to the powder before it is placed in a clear capsule. This helps to distinguish various strengths of powders and capsules. In addition, mixing bases and caps of different colored capsules customizes the colors. It is best to use two capsule machines for this process.
3. Liquids can be incorporated into capsules by mixing with melted polyethylene glycol 6000 or 8000 or a related concentration of this substance. The mixture can be poured into capsules, where it will solidify. The capsule can then be closed and dispensed.
4. Liquids can be dispensed in capsules by using a syringe to drop the liquid into the capsule base. Oils can be mixed with a fat or fatty acid, including cocoa butter, and slightly heated. The mixture can then be poured into capsules.
5. Locking-type capsules minimize the loss of their contents, whether powder, liquid, or semisolid; they also work well with hand-operated capsule-filling machines.

7. Capsule Powder Formulations and Choice of Excipients

Powder formulations for encapsulation should be developed in consideration of the particular filling principle involved. The requirements of the formulation imposed by the filling process, such as lubricity, compactibility, and fluidity, are not only essential to a successful filling operation, but also may be expected to influence drug release from the capsules. Indeed, the various filling principles themselves may influence drug release. This seems particularly evident for machines that form compressed plugs.

When immersed in a dissolution fluid at 37°C, hard gelatin capsules can rupture first at the shoulders of the cap and body where the gelatin shell is thinnest. As the dissolution fluid penetrates the capsule contents, the powder mass begins to disintegrate and deaggregate from the ends to expose drug particles for dissolution. The efficiency by which the drug is released depends on the wettability of the powder mass, how rapidly the dissolution fluid penetrates the powder, the rate of disintegration and deaggregation of the contents, and the nature of the primary drug particles. These processes, in turn, can be significantly affected by the design of the formulation and the mode of filling.

The excipients used to manufacture capsules are similar to those used to make tablets listed in Table 10.2. In addition formulations to be filled on dosator machines may sometimes benefit from the better compactibility of microcrystalline cellulose, particularly when drug dosage is large. Surfactants may be included in capsule formulations to increase the wetting of the powder mass and enhance drug dissolution. The waterproofing effect of hydrophobic lubricants may be offset by the use of surfactants. Numerous studies have reported the beneficial effects of surfactants on disintegration and deaggregation or drug dissolution. The most common surfactants employed in capsule formulations are sodium lauryl sulfate and sodium docusate in levels of 0.1 to –0.5% to overcome wetting problems (Figure 10.2).

C. Soft Gelatin Capsules

Soft gelatin capsules (sometimes referred to as softgels) are made from a more flexible, plasticized gelatin film than hard gelatin capsules.

1. Advantages and Disadvantages of Soft Gelatin Capsules

Several advantages of soft gelatin capsules derive from the fact that the encapsulation process requires that the drug be a liquid or at least dissolved, solubilized, or suspended in a liquid vehicle.

1. Since the liquid fill is metered into individual capsules by a positive-displacement pump, a much higher degree of reproducibility is achieved than is possible with powder or granule feed in the manufacture of tablets and hard gelatin capsules.
2. A higher degree of homogeneity is possible in liquid systems than can be achieved in powder blends. A content uniformity of <3% has been reported for soft gelatin capsules manufactured in a rotary die process.
3. Another advantage that derives from the liquid nature of the fill is rapid release of the contents with potentially enhanced bioavailability. The proper choice of vehicle may promote rapid dispersion of capsule contents and drug dissolution.
4. Soft gelatin capsules are hermetically sealed as a natural consequence of the manufacturing process. Thus, this dosage form is uniquely suited for liquids and volatile drugs. Many drugs subject to atmospheric oxidation may also be formulated satisfactorily in this dosage form.
5. Soft gelatin capsules are available in a wide variety of sizes and shapes. Specialty packages in tube form (ophthalmics, ointments) or bead form (various cosmetics) are also possible.

The disadvantages of soft gelatin capsules include:

1. Many times such products must be contracted out to a limited number of companies that have the necessary filling equipment and expertise. Materials must be shipped to the soft gelatin capsule facility and products must be shipped back to the pharmaceutical manufacturer for final packing and distribution. Additional quality control measures may be required.
2. Soft gelatin capsules are not an inexpensive dosage form, particularly when compared with direct compression tablets.
3. There is a more intimate contact between the shell and its liquid contents than exists with dry-filled hard gelatin capsules, which increases the possibility of interactions. For instance, chloral hydrate formulated with an oily vehicle exerts a proteolytic effect on the gelatin shell; however, the effect is greatly reduced when the oily vehicle is replaced with polyethylene glycol.
4. Drugs can migrate from an oily vehicle into the shell, and this has been related to their water solubility and the partition coefficient between water and the nonpolar solvent. The possible migration of a drug into the shell must be considered in the packaging of topical products in soft gelatin tube-like capsules, as this could affect drug concentration in the ointment, as applied. For other products, such as oral capsules or suppository capsules, both the shell and the contents must be considered in judging drug content when migration occurs.

2. Composition of the Shell

Similar to hard gelatin shells, the basic component of soft gelatin shells is gelatin. However, the shell has been plasticized by the addition of glycerin, sorbitol, or propylene glycol. Other components may include dyes, opacifiers, preservatives, and flavors. The ratio of dry plasticizer to dry gelatin determines the hardness of the shell and can vary from 0.3 to 1.0 for a very hard shell to 1.0 to 1.8 for a very soft shell. Up to 5% sugar may be included to give the shell a chewable. The basic gelatin formulation from which the plasticized films are cast usually consists of 1 part gelatin,

1 part water, and 0.4 to 0.6 parts plasticizer. The residual shell moisture content of finished capsules is in the range of 6 to 10%.

3. Formulation of Soft Gelatin Capsules

The formulation for soft gelatin capsules involves liquid rather than powder technology. Materials are generally formulated to produce the smallest possible capsule consistent with maximum stability, therapeutic effectiveness, and manufacture efficiency. Soft gelatin capsules contain a single liquid, a combination of miscible liquids, a solution of a drug in a liquid, or a suspension of a drug in a liquid. The liquids are limited to those that do not have an adverse effect on the gelatin walls. The pH of the liquid can be between 2.5 and 7.5.

Liquids with more acid pHs tend to cause leakage by hydrolysis of the gelatin. Liquids with pH's higher than 7.5 and aldehydes decrease shell solubility by tanning the gelatin. Other liquids that cannot be encapsulated include water (greater than 5% of contents) and low molecular weight alcohols such as ethyl alcohol. Emulsions cannot be filled because inevitably water will be released that will affect the shell. The types of vehicles used in soft gelatin capsules fall into two main groups:

1. Water-immiscible, volatile, or more likely nonvolatile liquids such as vegetable oils, aromatic and aliphatic hydrocarbons (mineral oil), medium-chain triglycerides, and acetylated glycerides
2. Water-miscible, nonvolatile liquids such as low molecular weight polyethylene glycol (PEG 400 and 600) have come into use more recently because of their ability to mix readily with water and accelerate dissolution of dissolved or suspended drugs.

All liquids used for filling must flow by gravity at a temperature of 35°C or less. The sealing temperature of gelatin films is 37 to 40°C. Typical suspending agents for oily bases are beeswax (5%), paraffin wax (5%), and animal stearates (1 to 6%). Suspending agents for nonoily bases include PEG 4000 and 6000 (1 to 5%) and solid nonionic or glycol esters (10%).

4. Manufacture of Soft Gelatin Capsules

The oldest commercial process, the semiautomatic plate process has been surpassed by more modern, continuous processes. The first continuous process was the rotary die process. Aside from its being a continuous process, the rotary die process reduced manufacturing losses to a negligible level and content variation to the ±1 to 3% range, both major problems with earlier processes.

In the rotary die process two plasticized gelatin ribbons (prepared in the machine) are continuously and simultaneously fed with the liquid or paste fill between the rollers of the rotary die mechanism. The forced injection of the feed material between the two ribbons causes the gelatin to swell into the left- and right-hand die pockets as they converge. As the die rolls rotate, the convergence of the matching dies pockets seals and cuts out the filled capsules.

Another continuous process for the manufacture of soft gelatin capsules filled with powders or granules involves a measuring roll, a die roll, and a sealing roll. As the measuring roll and die roll rotate, the measured doses are transferred to the gelatin-linked pockets of the die roll. The Globex Mark 11 Capsulator (Kinematics and Controls Corp., Deer Park, NY) produces truly seamless, one-piece soft gelatin capsules by a bubble method.

5. Soft or Liquid Filled Hard Gelatin Capsules

Perhaps the most important reason soft gelatin capsules became the standard for liquid-filled capsules was the inability to prevent leakage from hard gelatin capsules. The advent of such sealing techniques as banding and self-locking hard gelatin capsules, together with the development of high-resting-state viscosity fills, has made liquid/semisolid-filled hard gelatin capsules a feasible dosage form. Commercial examples include Vancocin HCl (Eli Lilly and Co., Indianapolis, IN). As with soft gelatin capsules, any materials filled into hard capsules must not dissolve, alter, or otherwise adversely affect the integrity of the shell, and generally, the fill material must be pumpable.

Three formulation strategies based on a high resting-state viscosity after filling have been described:

- *Thixotropic formulations:* Such systems exhibit shear thinning when agitated and thus are pumpable. Yet, when agitation stops, the system rapidly establishes a gel structure, thereby avoiding leakage.
- *Thermal-setting formulations:* For these, excipients are used that are liquid at filling temperatures, but that gel or solidify in the capsule to prevent leakage.
- *Mixed thermal-thixotropic systems:* Improved resistance to leakage may be realized for low or moderate melting-point systems. Above its melting point, the liquid can still be immobile because of thixotropy.

Materials that may be used as carriers for the drug include vegetable oils, various fats such as cocoa butter, and polyethylene glycols. For thixotropic systems, the liquid excipient is often thickened with colloidal silicas. Powdered drugs may be dissolved or suspended in thixotropic or thermal-setting systems. Since the more lipophylic the contents the slower the release rate, varying release rates may be achieved by selecting excipients with varying hydrophylic–lipophylic balances.

D. Examples of Capsule Formulations

1. Example 1

Sustained Release Matrix Capsule Dosage Form

Formula	%	Use
Pseudoephedrine HCl	24	Active
Hydropropylcellulose (Klucel HXF)	15	Binder
MCC	20	Filler
Pregelatinized starch	20	Filler
Dicalcium phosphate	20	Filler
Magnesium stearate	1	Lubricant

a. *Method*

1. Blend everything except magnesium stearate in a V-shaped blender for 10 min.
2. Fill the blend into size 0 gelatin capsules using an automatic capsule filling machine with dosator piston force of 200 N.

2. Example 2

Liquid Filled Formula for Sealed Hard Gelatin Capsules

Formula	%	Use
Acetaminophen	500 mg	Active
Glycerol esters of saturated C8-C18 fatty acids (Gelucire 33)	300 mg	Carrier

Stages of the hard gelatin capsule sealing process are:

1. Drug is mixed into the melted carrier and kept at a constant temperature of 40°C.
2. Capsules are moistened with a 50:50 water and ethanol mixture sprayed onto the joint, and capillary action draws the liquid into the space between the body and the cap.

3. Excess fluid is removed by suction because the melting point of gelatin is lowered by the presence of water.
4. The product is maintained at a constant temperature of 40°C while filling.
5. A homogeneous suspension is maintained in the product hopper.
6. The required volumes of liquid are accurately dosed into capsules.
7. The machine ejects a filled capsule body when the cap is missing.
8. Application of gentle heat of approxmately 45°C completes the melting over a period of about 1 min, and the two gelatin layers are fused together to form a complete 360° seal.
9. The gelatin setting or hardening process is completed while the product returns to room temperature. This process is best carried out on trays.

E. Compendial Standards and Regulatory Requirements for Capsules

As is the case for tablets, several requirements and specifications are set for capsules in the USP and other pharmacopoeia (Table 10.7). These specifications ensure the stability, quality, and release of drugs from capsules.

F. Process Optimization for Solid Dosage Form Manufacturing

Process optimization for tablet and capsule formulations are summarized in Table 10.8.

HOMEWORK

It will be helpful for students to find the following information from literature and other sources:

1. The top 20 commercially available capsule products.
2. Five commercially available soft gel capsules.
3. Available USP requirements and test conditions for dissolution testing of products found in Questions 1 and 2.
4. Suppliers of empty gelatin capsules.

TUTORIAL

1. When preparing hard gelatin capsules, the formulator is generally not concerned with:
   (a) Powder blending and homogeneity
   (b) Powder flow
   (c) Powder lubrication
   (d) Powder compaction
2. Why should highly soluble chloride salts not be dispensed in hard gelatin capsules?
   (a) Capsules will dissolve slower.
   (b) The salts will decompose.

TABLE 10.7
USP-Recommended Tests and Specifications for Quality Control and Quality Assurance of Capsules

Test	Specification
Containers' permeability and sealing	There are specifications listed in the USP prescribing the type of container suitable for the repackaging or dispensing of each official capsule and tablet. Depending on the item, the container might be required to be tight, well-closed, and light resistant.
Disintegration	The compendial disintegration test for hard and soft gelatin capsules follows the same procedure and uses the same apparatus described in this chapter for uncoated tablets. The capsules are placed in the basket-rack assembly, which is repeatedly immersed 30 times per minute into a thermostatically controlled fluid at 37°C and observed over the time described in the individual monograph. To fully satisfy the test, the capsules disintegrate completely into a soft mass with no firm core and only some fragments of the gelatin shell.
Dissolution	The compendial dissolution test for capsules uses the same apparatus, dissolution medium, and test as that for uncoated and plain coated tablets described in this chapter. However, in instances in which the capsule shells interfere with the analysis, the contents of a specified number of capsules can be removed and the empty capsule shells dissolved in the dissolution medium before proceeding with the sampling and chemical analysis.
Stability	Stability testing of capsules is performed to determine the intrinsic stability of the active drug molecule and the influence of environmental factors such as temperature, humidity, light, formulation components, and the container and closure system. The battery of stress-testing long-term stability and accelerated stability tests help determine the appropriate conditions for storage and the product's anticipated shelf life.
Moisture	The USP requires determination of the moisture-permeation characteristics of single-unit and unit-dose containers to assure their suitability for packaging capsules. The degree and rate of moisture penetration is determined by packaging the dosage unit together with a color-revealing desiccant pellet, exposing the packaged unit to known relative humidity over a specified time, observing the desiccant pellet for color change (indicating absorption of moisture) and comparing the pre- and postweight of the packaged unit.
Weight variation	The uniformity of dosage units may be demonstrated by determining weight variation or content uniformity. The weight variation method is as follows. *Hard capsules:* Ten capsules are individually weighed and the contents removed. The emptied shells are individually weighed and the net weight of the contents calculated by subtraction. From the results of an assay performed as directed in the individual monograph, the content of active ingredient in each of the capsules is determined. *Soft capsules:* The gross weight of 10 intact capsules is determined individually. Then each capsule is cut open with scissors or a sharp blade, and the contents are removed by washing with a suitable solvent. The solvent is allowed to evaporate at room temperature over a period of about 30 min, with precautions taken to avoid uptake or loss of moisture. The individual shells are weighed and the net contents calculated. From the results of the assay directed in the individual monograph, the content of active ingredient in each of the capsules is determined.
Content uniformity	Unless otherwise stated in the monograph for an individual capsule, the amount of active ingredient, determined by assay, is within the range of 85 to 115% of the label claim for 9 of 10 dosage units assayed, with no unit outside the range of 70 to 125% of label claim. Additional tests are prescribed when two or three dosage units are outside of the desired range but within the stated extremes.

(c) Rapid release may cause gastric irritation.
(d) The capsule shell will disintegrate.

3. What is the major difference between Type A and Type B gelatin?
   (a) A is produced by acid hydrolysis and B by alkaline hydrolysis.
   (b) A is produced by alkaline hydrolysis and B by acid hydrolysis.

TABLE 10.8
Process Optimization for the Manufacturing of Solid Dosage Forms

Process	Factors to Control
Granulation	Effect of granulation parameters
	Granulation time
	Speed of choppers (I and II) or mixer blades
	Solvent addition rate and overall amount
	Ratio of intragranulate disintegrant and binder agents
	Screen size for milling (e.g., 0.6 or 0.8 mm)
	Adjusting mill screen size up or down to fine tune hardness
	Evaluation of optimized granulate and tablet attributes
Drying	Fluidized bed drying temperature vs. target loss on drying and range limits and the effect on granulate and tablet properties (flow, capping, sticking)
Blending (mixing)	Blending times
	Lubricant split into two parts (preblending and final blending)
	The effect on content uniformity, granule lubrication, and dissolution profile.
	Evaluation of unit dose sampling vs. content uniformity
Compression	Effect of hardness on tablet properties (aging, dissolution, friability)
	Evaluation of hardness range limits
	Evaluation of stability results of optimized manufacturing process
Evaluation	Testing according to regulatory and pharmacopoeial specifications

   (c) A is made from animal bone and B is made from pork skin.
   (d) A contains mainly alanine and B contains mainly glycine.

4. If a formulation has a fill weight of 350 mg and a tapped density of 0.75 g/ml, what will the volume occupied by the fill weight be?
   (a) 1.00 ml
   (b) 0.50 ml
   (c) 4.60 ml
   (d) 0.47 ml
5. The ideal foil or film for individual packing of capsules in strips or blister packs should:
   (a) Allow air to permeate through it
   (b) Be heat insensitive
   (c) Be strong enough not to be torn by the patient
   (d) Melt around 30°C
6. The ideal powder characteristics for successful filling of capsules include:
   (a) Poor compactibility
   (b) Poor lubrication
   (c) Have excellent flow properties
   (d) Have low bulk density
7. The active ingredient in capsules often tend to make up a:
   (a) Low percentage of the content
   (b) High percentage of the content
8. Which of the following is not considered a mechanism whereby glidants improve powder flow?
   (a) Reducing particle roughness
   (b) Reducing particle cohesiveness
   (c) Increasing electrostatic charges
   (d) Removing moisture

9. The more general or increased use of soft gelatin capsules is restricted by:
   (a) A high degree of reproducibility
   (b) A high degree of homogeneity
   (c) Rapid release of contents
   (d) Substantially high cost
10. The main difference between soft and hard gelatin capsules is:
   (a) Dyes are added to the capsule shell.
   (b) Hard gelatin shells are not plastized.
   (c) Hard gelatin shells are plastized.
   (d) The basic composition of soft shells is not gelatin.
11. Leakage from soft gelatin capsules can be caused by:
   (a) Hydrolysis of gelatin at low pH
   (b) Addition of surfactants
   (c) Vegetable oils
   (d) Polyethylene glycol 6000
12. A thixotropic formulation can be filled into soft gelatin capsules because it:
   (a) Gels when agitation stops
   (b) Reacts with capsule shell
   (c) Remains liquid at room temperature
   (d) Gels at high temperature
13. Which of the following tests are not required for capsules in the USP?
   (a) Disintegration
   (b) Dissolution
   (c) Moisture permeation
   (d) Mean diameter
44. When can dissolution tests be performed on capsule contents removed from their shells?
   (a) If the shell does not dissolve.
   (b) If the capsules are too big.
   (c) If the dissolved shell interferes with analysis.
   (d) If the capsule shell dissolves.
15. The decrease in solubility of gelatin capsules has been attributed to:
   (a) Acid hydrolysis
   (b) Gelatin crosslinking
   (c) Trace amounts of glycine
   (d) Trace amounts of alanine

Answers: 1 (d); 2 (c); 3 (a); 4 (d); 5 (b); 6 (c); 7 (b); 8 (c); 9 (d); 10 (b); 11 (a); 12 (a); 13 (d); 14 (b); 15 (b)

CASES

After carefully studying this chapter the student should be able to handle and solve some problems associated with the formulation, evaluation, and application of tablet, capsule, and powder dosage forms. The following are theoretical problem scenarios. Try to address each problem and suggest answers based on the information given in this chapter.

CASE 1

The following is an example of a compounded formula for effervescent granules. If you were the pharmacist, how would you prepare and dispense this product?

Rx
Aspirin: 300 mg/5 g
Effervescent granules qs: 100 g
Sig. Dissolve one teaspoonful in one glass of water and drink. Repeat every 8 hours.

Answer:

If each dose weighs 5 g, then there will be 20 doses in 100 g. Each dose contains 300 mg of aspirin (0.3 g) so the total amount of drug needed is 6 g. The amount of effervescent powder needed is 100 – 6 = 94 g.

A good effervescent blend consists of both citric acid and tartaric acid in a 1:2 ratio. Using the reaction schemes given on page 279 and the molecular weights of the compounds, we know that 1 g citric acid reacts with 1.2 g sodium bicarbonate, and 1 g tartaric acid reacts with 1.12 g sodium bicarbonate. Since the ratio is 1:2 citric acid to tartaric acid, we can double the amount of tartaric acid to 2.24 g. In total we then need 3.44 g sodium bicarbonate for the reaction, and the ratio becomes 1:2:3.44 (citric acid:tartaric acid:sodium bicarbonate).

Since the acids also act as a flavor in the formulation, we always leave some unreacted acid, and for this purpose the ratio can be changed to 1:2:3.4. Since the formulation needs 94 g of the effervescent mix, the quantity of each ingredient can be calculated using the ratio as 14.67 g citric acid, 29.38 g tartaric acid, and 49.94 g sodium bicarbonate.

Next weigh out the ingredients and prepare the powder using the methods described in the text. If other ingredients such as flavors, lubricants, etc. are needed, calculate and determine the best options. The prepared powder then has to be stored under the right conditions of low humidity in a tightly sealed container with a desiccant package. In addition, advise the patient on how to correctly prepare and administer the effervescent solution.

CASE 2

A formulation scientist gets a 10-g sample of a highly potent drug from the drug discovery department in his company. He is asked to formulate a conventional solid dosage form for clinical trials that exhibits good dissolution and content uniformity. The drug is a weak acidic compound with a pKa of 4.1. The solubility of the compound is 0.01 mg/ml. It is photosensitive. The mean volume particle size of the powder is 6.2 μm, and the dose required per unit is 0.5 mg. Drug–excipient compatibility studies have indicated that this compound is incompatible with magnesium stearate, polyethylene glycol, lactose, and fumed silica. The compound is also hygroscopic.

Given the above information, how would the scientist formulate the drug in a solid dosage form?

Answer:

In setting up a design for this formulation, keep the following in mind:

1. The drug is poorly soluble, and the formulation should optimize dissolution. The addition of surfactants to the formulation might be required.
2. The dose is very low, and this would require special attention to the blending process since a large volume of diluent might be required. Content uniformity might be a problem. A special mixing process such as ordered mixing might be required.

3. Lubrication could be a problem that would preclude making tablets since the drug is incompatible with magnesium stearate and fumed silica. Addition of microcrystalline cellulose as a diluent and sodium lauryl stearate, as a surfactant and lubricant could overcome this problem.
4. Since the drug is light sensitive, tablets might be less stable than opaque hard gelatin capsules. Since the drug dose is very low, hygroscopicity in the formula should not be a problem. Bulk drug should, however, be protected during manufacturing.
5. The best solution seems to be preparation of fill formula containing the drug and a high concentration of microcrystalline cellulose with 0.5 to 1% of sodium lauryl sulfate added. These ingredients should be thoroughly mixed by an ordered mixing process, ensuring content uniformity. The fill formula should be filled in the correct size hard gelatin capsules. The capsule shells should be opaque to prevent exposure of the drug to light.

CASE 3

A drug firm was making sugarcoated tablets for many years. As a cost reduction step they decided to switch to film coating. Experimental batches made in R&D were found satisfactory on storage at all accelerated conditions. However, when the first production batch was made, the film coated tablets were rejected by quality control because of a pitted appearance (small holes on the surface).

What might have caused the pitted appearance of the tablet and how might that be overcome?

Answer:

The punch and dies used for making the core tablets were probably very rough. Film coating is always a much thinner coat than a sugar coat and cannot mask the rough surface of core tablets. New punches and dies should be used. It is essential for core tablets to have smooth surfaces for film coating.

CASE 4

A hospital pharmacist working closely with an oncology department receives many prescriptions for compounded formulas where experimental therapies are dispensed to patients in hard gelatin capsules. To be better prepared for handling these prescriptions he decides to buy some small scale manufacturing equipment, supplies, and excipients.

What would be the essential items he will need to improve his productivity and the quality of the products he makes?

Answer:

Capsule shells of different sizes
Diluents: lactose and microcrystalline cellulose
Lubricants: magnesium stearate and sodium lauryl sulfate
Glidant: fumed silica
A coffee grinder to reduce particle size
A small hand mixer
A hand-operated capsule-filling machine
Approved dyes for coloring
Desiccant packages
Suitable containers: good sealing and tamper proof

CASE 5

A pharmacist making hard gelatin capsules using a hand-filling machine decides to buy a semi-automatic machine because of increased demand for his product. When he tried the same size capsules on the new machine, he found that the capsule weight had to be increased. The formula was a powder granulation containing lactose and starch as the excipients. He decides to increase the amount of starch in the capsule because it increases flow, and it is cheaper. However, since more capsules were made with the machine, the supply lasted longer, and after three months' storage, the pharmacist observed that the capsules became brittle.

What might have caused that brittleness, and how it could be avoided?

Answer:

The reason for this is that starch was very good at removing the moisture from the capsule shell. Since the amount of starch was increased in the formula, more moisture was removed, which caused the capsule to become brittle. To solve the problem perhaps starch could be replaced by lactose.

Another option, depending on the stability of the drug in contact with moisture and the flow properties of the granules, is to leave some moisture in the granules before filling the capsules. However, that may lead to fungus growth. The best solution would be to replace the starch with lactose.

CASE 6

How would you approach the formulation of a new chemical entity in a solid dosage form if the drug candidate has low (sometimes extremely low) solubility or low permeability and relatively high molecular weight?

Answer:

The challenge is to select the most appropriate formulation for our project in terms of speed of development, cost-effectiveness, compliance, and suitability for the intended clinical use, whether it comes from the outside or from our own research.

First we select the best physical form of the active pharmaceutical ingredient, such as particle size, polymorphic form, crystal habit, and salt form, in terms of stability, dissolution profile, and compatibility with the intended formulation and mode of administration.

Then the formulation and process development group is responsible for developing the most appropriate dosage form, be it a tablet, capsule, oral solution, suspension, intravenous solution, lyophilized product, or cream. The choice of form is based on a number of factors, including *in vivo* results and targets, the physical characteristics and stability of the material, the patient population, and company strategy and standards.

If it is decided that an oral solid dosage form will be developed, we decide on the type:

1. Compressed tablets (wet granulation, roller compaction, direct compression, film coating)
2. Chewable tablets
3. Bilayer tablets
4. Capsules: powder filled, bead filled, liquid filled
5. Sustained release tablets

As a first step, the formulation and process development team carries out model formulation studies to help select appropriate excipients and processes. The careful selection of the excipients (anything that is not the active pharmaceutical ingredient) and manufacturing process determines the performance of the formulation in terms of bio-pharmaceutical behavior, stability, appearance, manufacturability, dissolution rate, and palatability. During development, we work with marketing to design the final market appearance of the product, which includes shape, color, and packaging. In summary, we design products to maximize the opportunities for drugs to get to the appropriate population of patients.

Based on the results of the model formulation study, the scientists prepare more refined formulations suitable for use in clinical trials. As the clinical trials progress, the researchers refine, optimize, and, finally, scale-up laboratory-scale formulations and processes to commercial production scale. At each step, formulation and process development works to ensure the quality of the experimental drug product and the reproducibility of the process.

In addition to developing new formulations, the group is working on new techniques to support formulation development, such as using NIR (Near Infrared Spectroscopy) to determine the end-point in powder blending and electric capacitance tomography to evaluate drying profiles in fluid-bed drying. The scientists are constantly trying to improve on current delivery systems and are at the forefront of new drug delivery technologies such as rapidly dissolving tablets, chewable tablets, devices for aerosol drug delivery, and needleless injections.

If low bioavailability is a problem, it may be due to a poor dissolution rate, poor permeability across the biological membranes, or because the drug is not present at a sufficient concentration for a sufficient period of time in the portion of the gastro-intestinal tract where it is absorbed. There are formulation technologies that can be used to address each of these problems. For instance, we can improve the dissolution rate of the drug by changing the physical properties of the active pharmaceutical ingredient (most frequently by reducing its particle size) or formulating it with excipients that will enhance solubility.

Other types of formulations are designed to improve permeability by affecting transport mechanisms through the gastrointestinal membrane or by improving stability during the absorption process. There are many technologies available to affect drug transit through the gastrointestinal tract including gastro-retention, controlled release, and immediate release. However, these approaches usually result in the development of a nonstandard formulation, which has an impact on development complexity, time, and cost.

REFERENCES

1. Allen, L.V., *The Art, Science, and Technology of Pharmaceutical Compounding,* American Pharmaceutical Association, Washington, DC, 1997.
2. Ansel, H.C., Allen, L.V., & Popovich, N.G., *Pharmaceutical Dosage Forms and Drug Delivery Systems,* 7th ed., Lippincott, Williams & Wilkens, Baltimore, MD, 1999.
3. Aulton, M.E., ed., *Pharmaceutics the Science of Dosage Form Design,* Churchill Livingstone, New York, 1988.
4. Banker, G.S., & Rhodes, C.T., eds., *Modern Pharmaceutics,* Marcel Dekker, New York, 1996.
5. Gennaro, A.R., ed., *Remington: The Science and Practice of Pharmacy,* 20th ed., Lippincott, Williams & Wilkens, Baltimore, MD, 2000.
6. Kibbe, A.H., ed., *Handbook of Pharmaceutical Excipients,* 3rd ed., American Pharmaceutical Association, Washington, DC, 2000.
7. Stoklosa, M.J., & Ansel, H.C., *Pharmaceutical Calculations,* Williams & Wilkens, Media, PA, 1996.
8. *United States Pharmacopoeia,* USP 25: NF 20, United States pharmacopoeial convention, Rockville, MD, 2002.

11 Oral Controlled Release Solid Dosage Forms

Emmanuel O. Akala

CONTENTS

0-4152-8863-0/05/$0.00+$1.50
© 2005 by CRC Press LLC

Learning Objectives

After studying this chapter the student will be able to:

- Describe the advances in drug delivery systems
- Describe various terms associated with modified release dosage forms
- Discuss advantages and disadvantages of controlled release dosage forms
- Describe the various factors affecting the design of controlled release dosage forms
- Classify oral controlled-release dosage forms with commercial examples for each class
- Describe the mechanisms of release and the equations for modeling the release of bioactive agents from each class of controlled oral dosage form
- Solve some numerical problems associated with the design of oral controlled-release dosage forms
- Describe the type of formulation science that forms the basis of the design of some commercially available oral controlled-release dosage forms

I. INTRODUCTION

A. Developments in Pharmaceutical Dosage Form Design

A drug is rarely administered to human being as a pure chemical compound. What is given is a drug product containing the drug. When a drug is prepared in a form suitable for administration, it is called a dosage form or a drug product or, in a modern term, a delivery system. Almost anything done to the dosage form may alter the availability (the rate and the amount) of the drug delivered to the desired place in the body. The following discussion of developments in pharmaceutical dosage form design is intended to provide an overview of the progress that has been made in the efforts to improve upon the delivery of bioactive agents (drugs). The divisions into various generations serve only to point out the transition from one major form of delivery to another based on advances in physical, chemical, and biological sciences. Some of the newer drug delivery systems will be handled in clinics by current pharmacy students.

Historically, humans developed primitive ways of introducing drugs into the body:

- Chewing leaves and roots of medicinal plants
- Inhaling soot from the burning of medicinal substances
- Primitive extracts from plants and animals
- Treatment of asthma by smoking medicinal cigarettes as recently as 1950s

These primitive approaches to the delivery of drugs lacked consistency, uniformity, and specificity.

1. The First Generation Drug Delivery Systems

The first generation drug delivery systems (pharmaceutical dosage forms) appeared toward the end of the nineteenth century and in the twentieth century, and they have consistency and uniformity. These drug delivery systems (conventional dosage forms) include tablets, capsules, elixirs, syrups, suspensions, emulsions, and solutions and topical administration of ointments, lotions and creams, suppositories or injection of suspensions and solutions. Though these conventional drug delivery systems are still with us, the need for more efficient drug delivery systems was realized with time. The aims of the anticipated drug delivery systems were to deliver the minimum amount of drug necessary to the site of action to produce the desired therapeutic response (spatial placement of the drug) and to deliver the drug at an optimal rate to maximize the beneficial response and to minimize unwanted side effects (temporal delivery of the drug).

2. The Second Generation Drug Delivery Systems

With advances in biopharmaceutics, pharmacokinetics, and human physiology and their applications to drug formulation and development, various modifications in drug molecules and dosage forms were introduced. These second generation drug delivery systems (dosage forms) represent attempts to improve upon conventional dosage forms. These dosage forms were supposed to protect drugs against hostile conditions along the gastrointestinal tract, prolong their action if necessary, and improve bioavailability. The materials of formulation include polymers, waxes, plastics, and oils. They are described by various terms such as *repeat action, prolonged action,* and *timed release*. These drug products were introduced in the 1950s. In the second generation drug delivery systems, repetitive, intermittent dosing of a drug occurs from one or more immediate release units incorporated into a single dosage form (e.g., enteric coated tablets). Though these drug delivery systems provide some degree of control, it is not complete. More often than not, the release of the drug is subject to the environmental conditions at the site of administration or drug release, and the drug products do not generally permit a long-term drug release. Further, they do not produce a uniform blood–drug concentration as a function of time.

3. The Third Generation Drug Delivery Systems

In the middle to late 1960s, the term *controlled drug delivery* was introduced to describe a new concept of dosage form design. Controlled drug delivery refers to the precise control of the rate at which a drug dosage is released from a delivery system, ideally in a constant or near constant manner over a long period of time. In order to permit an accurate, reproducible, and predictable drug therapy, the third generation drug delivery systems (controlled drug delivery systems) were designed to provide drug release that is dependent on the properties of the device and the physicochemical characteristic of the drug and the delivery system and independent of the environmental factors existing at the site of administration (i.e., pH of fluids and the presence of enzymes in the body).

The third generation drug delivery system moved beyond extension of the blood level within the therapeutic range, which is characteristic of the second generation drug delivery systems, to controlled drug release such that a constant blood drug level can be maintained for a long period of time. The design of the third generation drug delivery systems that revolves around the zero-order approach is based on the assumption that optimal clinical outcomes may be achieved by maintaining a constant drug plasma concentration and that the relationship between plasma drug concentration and therapeutic effect of a drug is invariant with time. This is the era of time when drug formulation scientists believe that the most desirable situation is that "the flatter the plasma drug concentration vs. time curve the better".

Based on the type of mechanism that triggers and eventually controls the release of drug incorporated into the system, the controlled-release drug-delivery systems are classified as osmotically controlled systems, swelling-controlled systems, magnetically controlled systems, chemically controlled systems, electrically controlled systems, and diffusion-controlled systems. The design of the vast majority of the third generation drug delivery systems involves the use of polymers. As there is an element of diffusion of drug molecules in most of the systems, polymers serve as permeable barriers that the drug must cross before reaching the body fluids. Polymers are particularly suitable for this purpose because their properties can be manipulated easily and diffusion rates of drug molecules through polymers are orders of magnitude less than the diffusion rates of the same molecules through water.

4. The Fourth Generation Drug Delivery Systems

Third generation drug delivery systems have no control over the fate of the drug once it enters the body. With the advent of highly potent drugs such as peptides, proteins, and low molecular weight anticancer drugs with narrow therapeutic indices, efforts were geared toward drug delivery systems that could exercise control on the time of availability and the localization of the drug in the body. Various drug delivery systems belong to this generation: targetable, modulated, pulsatile and self-regulated, or feedback controlled drug delivery systems.

a. *Drug Targeting or Site-Specific Drug Delivery*

This mode of delivery involves specific delivery of the active compound (drug) to its site of action and keeping it there until it is inactivated or detoxified. Drug targeting increases the therapeutic potential and reduces side effects of the drug. Targeted (site-specific) delivery systems by the parenteral route are, at present, in different stages of development, and most of them consist of the following components: an active moiety for the therapeutic effect, a carrier for protection and changing the disposition of the drug, and a homing device for selection of the assigned target (site-specificity). The systems are suitable for anticancer drugs and highly potent recombinant peptides and proteins in which site-specific delivery is needed to reduce side effects (particularly the paracrine and autocrine acting proteins). The occurrence of severe side effects with cytokines such as tumor necrosis factor and interleukin-2 limits their therapeutic potential. This problem can be circumvented by the delivery of these proteins at the proper site, rate, and dose. This problem also exists for anticancer drugs. Gastrointestinal targeting of peptide and protein drugs to a suitable site for absorption is also being investigated.

Commercial targetable drug products have reached the clinic in the field of immunotherapy. It is believed that new immunotherapies, such as monoclonal antibodies, antisense compounds, vaccines, and angiogenesis inhibitors, will revolutionize cancer treatment in the near future. Monoclonal antibodies bind to specific targets on cancer cells, then destroy the cells or mark them for destruction by the immune system. The specificity of the drugs is such that they do not destroy healthy cells; they have fewer side effects than traditional chemotherapies, and they can reduce the relapse rate among chemotherapy patients. Examples of monoclonal antibodies already approved by the Food and Drug Administration (FDA) are Rituxan (rituximab, made by Genentech) for B-cell non-Hodgkin's lymphoma, Herceptin (transtuzumab, made by Genentech) for breast cancer,

Mylotarg (gemtuzumab ozoganmicin, made by Wyeth) for acute myeloid leukemia, and Campath (alemtuzumab, made by Millenium) for chronic lymphocytic leukemia. Zevalin (90Y-ibritumomab tiuxetan, made by IDEC Pharmaceuticals) is a radioimmunotherapeutic agent that aims to combine the targeting power of monoclonal antibodies with the cancer-killing ability of radiation. It has been approved by the FDA.

Cancer vaccines have the potential to prevent or delay cancer recurrence by destroying residual cells following first-line therapy. Some of them have been approved in Australia and Canada, and some are in advanced clinical development in the U.S. Antisense compounds are complimentary small fragments of RNA. They bind to specific sequences of mRNA target sites and prevent translation and thereby inhibit the production of disease-causing proteins. ISIS' Vitraven has been approved by FDA for the treatment of cytomegalovirus in AIDS patients. The angiogenesis inhibitors block the development of new blood vessels that supply nutrients and blood to tumors. They can shrink tumors and prevent them from growing. None has made it to the clinic.

b. Pulsatile or Modulated Drug Delivery Systems

Twenty-four-hour ambulatory blood pressure monitoring has revealed a marked circadian rhythm in hypertensive patients. Receptor down-regulation has been observed with a long-term delivery of nitrates and hormones. Consequently, pulsatile drug delivery has been suggested as the best mode of delivery for certain drugs so that their delivery should mirror the physiological release profiles of endogenous peptides and proteins or human physiological needs. Efforts in this direction resulted in a new body of knowledge called chronotherapeutics, which is the integration of chronobiology, the science of biologic rhythm (i.e., the predictable cyclic variability of human biologic functions), physiology, and therapeutics.

c. Self-Regulated or Feedback-Controlled Drug Delivery Systems

The long-term complications of diabetes, such as cardiovascular and neuropathic problems, have been associated with the poor control over glucose levels that result from conventional therapy. This consideration has been an impetus for the development of self-regulated or feedback-controlled drug delivery systems. The types that are being investigated can be classified as follows:

- Closed loop systems that involve biosensor-pump combinations: The major requirement is that there should be a known relationship between plasma level and pharmacological effect. The components of the system are (a) a biosensor that determines the plasma level of the drug, (b) an algorithm to calculate the required input rate for the delivery system, and (c) a pump system capable of administering the drug at the required rate over a prolonged period of time.
- Closed loop systems that involve delivery by self-regulating systems: Drug release is controlled as a consequence of response to stimuli in the body. Efforts are directed toward insulin release in response to glucose concentration. The progress on the development of these self-regulating systems can be found in most text books on pharmaceutical biotechnology.
- Closed loop systems based on microencapsulated secretory cells: Reports have shown that clinical data obtained on human secretory islet of Langerhans cells encapsulated in alginate-based microspheres for restoring insulin production in a biofeedback fashion are very encouraging. Polymers are used to protect the secretory cells from the body's microenvironment.

5. The Fifth Generation Drug Delivery Systems

Gene therapy, intended to treat the cause of a disease rather than the symptoms, is essentially the redesign of cells by introducing therapeutic genes into genetically disabled cells to deliver therapeutic agents made by the gene. There are two methods of inserting genetic material into cells: viral and nonviral gene transfer methods. Such systems are in various phases of drug development.

II. CONTROLLED RELEASE ORAL DRUG DELIVERY SYSTEMS: SOLID DOSAGE FORMS

A. Various Terms Associated with Modified Release Dosage Forms

Modified release dosage forms are drug products designed to alter the time and rate of release of the bioactive agent (drug substance). They can achieve certain therapeutic or convenience objectives that conventional dosage forms cannot offer. Modified release dosage forms can be classified as follows:

Delayed (sustained) release dosage forms: In delayed-release dosage forms, repetitive, intermittent dosings of a drug occur from one or more immediate-release units incorporated into a single dosage form. Examples of types of delayed-release dosage forms are:
- Repeat-action (release of two doses successively)
- Prolonged action (initial dose plus a replenishment)
- Enteric coated tablets (timed release is achieved by a polymer barrier coating)
- Modified enteric coated tablets (additional drug is applied over the enteric coat and is released in the stomach, while the remainder is released further down the GI tract)

Controlled release dosage forms: This system releases the bioactive agent (drug) over an extended period of time. It avoids the undesirable sawtooth characteristics of the plasma concentration vs. time profiles of the conventional drug products. The dosage form can control the rate of release of the incorporated bioactive agents and maintaining constant drug concentrations at the biophase: target tissue or cells. The nature of the controlled release dosage form is such that the release is determined by the design of the system and the physicochemical properties of the drug and is independent of the external factors or the microenvironment in which the dosage form is placed. The design of controlled release dosage forms is based on the philosophy that optimum biological response occurs when the level and the time of availability of the drug to the target tissue are optimized. Since release-rate control complements drug molecular structure in determining both the selectivity and duration of action, it is believed that controlled-release drug-delivery systems permit more accurate, effective, reproducible, and predictable drug administration. Figure 11.1 shows a hypothetical serum drug concentration vs. time curves of various oral dosage forms.

Targeted dosage forms: Two types of targeted dosage forms are commonly described: site-specific targeting and receptor targeting. In site-specific targeting, drug release is at the diseased organ or site of absorption, while in receptor targeting, drug release is at the receptor for the drug action.

All the dosage forms described above can be grouped under the generalized term *controlled drug delivery. Controlled drug delivery* is now a generalized term that encompasses any system that is capable of spatial placement and temporal delivery of a drug. In spatial placement, the drug is targeted to a specific organ or tissue (this includes targeted dosage forms); in temporal delivery, the rate of drug delivery to the target tissue is controlled (this includes the delayed and controlled release dosage forms as well as pulsatile or modulated drug delivery systems because the rate of delivery is controlled based on human physiological need).

B. Rationale for the Development of a Drug as a Controlled Release Dosage Form

- Controlled release drug delivery systems increase efficacy by maintaining constant plasma drug level in the therapeutic window range for an extended period of time.
- Controlled release drug delivery systems reduce dosing frequency and eliminate drug accumulation in the body, thereby minimizing untoward effects.
- Controlled drug delivery increases patient compliance.
- The cost of treatment for chronic diseases may be reduced.

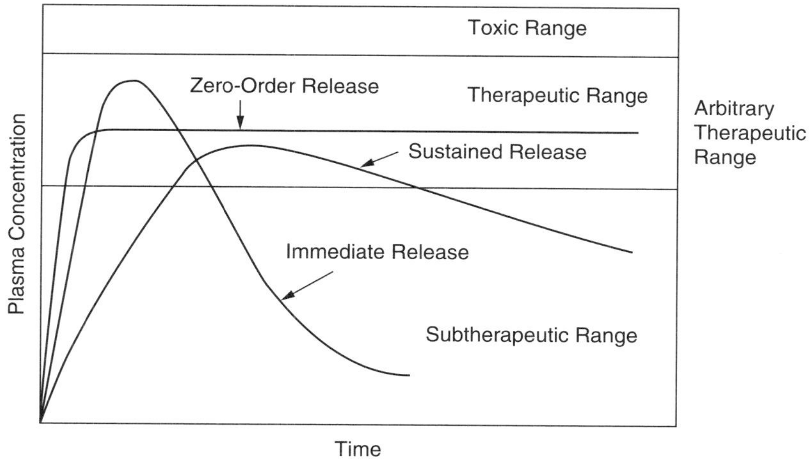

FIGURE 11.1 Hypothetical serum drug concentration vs. time curves of various oral dosage forms. (From Jantzen, G. M., and Robinson, J. R., Sustained and controlled release drug delivery systems, in *Modern Pharmaceutics,* 3rd ed., Banker, G. S. and Rhodes, C. T., Eds., Marcel Dekker, New York, 1996, p. 575, with permission.)

C. Disadvantages of Controlled Release Dosage Forms

It is important to consider some of the drawbacks of using controlled release dosage forms, especially as an aid in drug product evaluation and selection.

- Dose dumping can result from the failure of controlled release dosage forms. Dose dumping has been defined as either the release of more than the usual fraction of drug or as the release of drug at a greater rate than the customary amount of drug per dosing interval, such that potentially adverse plasma levels may be reached. Controlled release dosage forms are particularly prone to dose dumping if not properly designed and fabricated because they usually contain the equivalent of two or more drug doses present in a conventional dosage form. This problem is exemplified by delayed and enteric coated dosage forms: If poorly formulated, the enteric coating may be dissolved in the stomach, resulting in premature release of the drug with the concomitant irritation of gastric mucosa. Moreover, if the coating is poorly done, the enteric coating may not dissolve at the proper site along the gastrointestinal tract. Thus, the drug may not be available for absorption.
- Should the patient suffer from an adverse drug reaction or become accidentally intoxicated, the removal of the drug from the system may be difficult, if not impossible, for controlled release dosage forms.
- Orally administered controlled release dosage forms may yield erratic or variable drug absorption owing to interactions with the contents of gastrointestinal tract and changes in gastrointestinal motility. The result is drug ineffectiveness.

D. Biological and Physicochemical Considerations in the Design of Controlled Release Dosage Forms

1. Physicochemical Factors

a. *Dose Size*

Generally, 0.5 to 1.0 g is considered to be the maximum amount for a single dose of a conventional dosage form. For drugs requiring large doses (>500 mg) in conventional dosage forms, the size of the controlled release drug products for the same drugs will be too large for the patient to swallow since they may contain two or more times the amount of drug as in conventional dosage forms.

b. Stability

Since most drugs are weak acids or weak bases, they are prone to acid–base hydrolysis and enzymatic biotransformation in the gastrointestinal tract. The general principle of dosage form design for drugs that are unstable in the stomach is to place the drug in a system that releases its contents only in the intestine. Drugs that are not stable in the small intestine can also be put in a system that releases the drug in the stomach or can be targeted to the colon (a less hostile environment in terms of density and diversity of metabolizing enzymes). However, these principles are not possible for the design of controlled release dosage forms that release their contents uniformly over the length of the gastrointestinal tract. Because of the large surface area available for absorption in the small intestine, it is believed that drugs that are not stable in the small intestine should be administered through an alternative route (e.g., nitroglycerin).

c. Ionization

Most drugs are weak acids or bases. The principle that is applicable to conventional dosage forms is also true of controlled release dosage forms: The uncharged form of the drug has the best permeation across biological membranes. This principle is based on the pH-partition hypothesis, which states that the nonionized form of an acid or basic drug, if sufficiently lipid soluble, is absorbed, but the ionized form is not. The fraction of drug in solution that exists in the nonionized form depends on the both the dissociation constant of the drug and the pH of the dissolution medium.

d. Aqueous Solubility

Knowledge of the aqueous solubility of a drug is important in the design of controlled release dosage forms. This solubility is influenced strongly by the pH of the microenvironment in the gastrointestinal tract. Drugs with low aqueous solubility (<0.01 mg/ml) are not suitable for dosage forms based on diffusional systems: the driving force for diffusion through the system is the concentration of drug in the polymer or solution. However, drugs with very high solubility and rapid dissolution are often too difficult to formulate into controlled release dosage forms, as it is difficult to decrease the dissolution rate in order to modify the absorption of the drug. It is believed that buffering agents may be added to slow or modify the release rate of a fast dissolving drug in the formulation of a controlled release dosage form. The requirements for the effectiveness of the system are that the controlled release dosage form must be a nondisintegrating system and the buffering agent must be released slowly.

e. Partition Coefficient

Lipid solubility vis-à-vis drug absorption and disposition is very important in dosage form design and development. Polar drug molecules are often poorly absorbed after oral administration, while lipid soluble drugs with favorable partition coefficients are usually well absorbed and often result in improved pharmacological activity, as the drug must diffuse through a variety of biological membranes that act as lipid-like barriers. The oil–water partition coefficient (often experimentally determined as the 1-octanol–water partition coefficient) is often used as a guide to the lipophilic nature of a drug. Though drugs with high partition coefficients can penetrate biological membranes easily, they often have low aqueous solubility; moreover, drugs with low partition coefficients often have high aqueous solubility, but they cannot penetrate biological membranes easily for absorption. Thus a balance is required between aqueous solubility and partition coefficient for efficient penetration of biological membrane and activity of the drug. A log K (partition coefficient) between 1 and 3 is considered ideal for controlled release drug candidates.

Considering the materials of formulation of controlled release dosage forms, partition coefficient is a measure of the preference of a solute for the polymer relative to the surrounding release medium. Generally, if a solute has a chemical structure similar to that of the polymer, the partition coefficient will be high, whereas if the structures are different, the partition coefficient will be low. This principle is the basis for the release of lipophilic drugs (e.g., progesterone) from silicone rubbers as opposed to ionic drugs. Studies have shown that steroids with high partition coefficients permeate polyHEMA hydrogels crosslinked with high concentrations of ethylene glycol dimethacrylate by a solution diffusion mechanism (diffusion through the polymer). The permeability was found to be higher than that of hydrocortisone, which has a lower partition coefficient. However, polyHEMA gel fabricated without a crosslinker is more hydrophilic and contains a higher percentage of bulk water; consequently, the more water soluble hydrocortisone showed higher permeability in the gel.

f. Polymer Structure

Apart from biocompatibility, the diffusivity of a drug through a polymer is an important criterion in its selection for use in a dosage form. The physical structure and chemical nature (for biodegradable polymers) of the polymeric material are important parameters controlling drug diffusion coefficient and hence release rate. Knowledge of the diffusion coefficient and release rate dependence on structural characteristics of the polymeric material is of utmost importance. The main polymer structural parameters affecting drug release through diffusion coefficient include the average molecular weight between cross-links (inversely proportional to the cross-linking density), degree of hydration, degree of crystallinity and size of crystallites, glassy or rubbery state, porosity, and tortuousity of porous materials and relaxational phenomena observed in swelling-controlled systems. Fillers, plasticizers, and other additives are equally important.

g. Size of Solute

The release of solute or its diffusivity in a polymer is often a complex kinetic parameter that is determined by the properties of the solute such as size and shape and the properties of the polymer. Generally, if permeation occurs via the pore mechanism (i.e., through water-filled pores), the solute size will have an important effect on diffusivity. However, permeation via the partition mechanism is less solute-size dependent. Solute diffusion coefficients in cross-linked polyacrylamide and polyvinylpyrolidone gel show a logarithmic dependence on implant polymer concentration and solute molecular weight.

h. Complex Formation

Drug protein binding influences the distribution equilibrium of drugs. Plasma proteins exert a buffer function in the disposition of drugs, especially distribution; the elimination half-life of the drugs will be long, and they may not be qualified to be formulated into controlled release dosage forms. Only the free, nonprotein-bound fraction of the drug can diffuse into the tissue from the blood vessels. The equilibrium between free and bound drug acts as a buffer system and maintains a relatively constant concentration of the drug over a long period of time via the dissociation of the drug protein complex. Some of these physicochemical parameters are summarized in Table 11.1.

2. Biological Factors

a. Absorption

For a controlled release dosage form, the rate of release of the drug rather than the rate of absorption should be the rate-limiting step in the absorption process. Thus, it is strongly desirable to have a rate of release much less than the rate of absorption. A desirable absorption profile for a drug to be formulated into controlled release dosage form is as follows: If the absorption rate constant lies between 0.17 and 0.23 h^{-1}, then 80 to 90% will be absorbed; if the absorption rate constant lies between 0.25 and 0.35 h^{-1}, then more than 95% will be absorbed. It is assumed that there is a constant absorption rate over the entire length of the small intestine and that the absorption is a first-order process. Thus, drugs whose absorption is site specific (riboflavin and chlorothiazide) and those with erratic absorption (gentamicin and kanamycin) are not good candidates for oral controlled-release dosage forms.

TABLE 11.1
Physicochemical Parameters in the Design of Controlled Release Dosage Forms

Parameters	Comments
Dose size	Drugs administered in large single doses are not suitable for development as oral controlled-release products. Dose should be ≤500 mg.
Size of solute	Drug molecular weights less than 500 do not produce problems with drug absorption and hence may be suitable for oral controlled drug delivery. Diffusion of drugs through polymer by a pore mechanism is solute-size dependent.
Aqueous solubility	Drugs with very high solubility and rapid dissolution are often too difficult to formulate into controlled release dosage forms, as it is difficult to decrease the dissolution rate in order to modify the absorption of the drug. Absorption of poorly soluble drugs is often dissolution-rate limited, and they are not suitable for controlled delivery systems
Partition coefficient	Drugs with very high partition coefficients easily permeate the lipid portion of the biological membranes. However, they have difficulty. Afterwards, partitioning into systemic circulation. However, drugs with low partition coefficients do not even permeate the upper lipid portion of the biomembranes. A balance is therefore needed between oil and aqueous solubilities. Drug molecules with log K (partition coefficient) between 1 and 3 are good candidates for controlled release.
Stability	For complete and reproducible drug release in the body, a drug candidate for oral controlled release must be stable in the entire range of gastrointestinal pH and under the influence of the microenvironment (enzymes and microbes) at the site of absorption.
Ionization	The uncharged form of the drug has the best permeation across biological membranes.
Polymer structure	Polymer structural parameters of importance are cross-linking density, degree of hydration, degree of crystallinity, porosity and tortuousity of porous materials, fillers, plasticizers, and other additives.

b. Distribution

Dose calculation in the formulation of controlled release dosage forms is often based on the one-compartment pharmacokinetic model. The apparent volume of distribution for a one-compartment model is the ratio of the dose to the initial concentration immediately after an intravenous bolus injection, and this parameter affects the amount of drug in the plasma or in target tissues. Of all the apparent volumes of distribution for a two-compartment pharmacokinetic model, the apparent volume of distribution at steady state is considered to be the most appropriate.

c. Elimination

The relationship between clearance (Cl_T) and biological half-life ($t_{1/2}$) is obtained as in Equation 11.1:

$$t_{1/2} = 0.693V_D/Cl_T \tag{11.1}$$

The volume of distribution, V_D, and Cl_T affect $t_{1/2}$. A decrease in clearance will increase the biological half-life; changes in the volume of distribution will also cause similar changes in biological half-life. In the design of oral controlled-release dosage forms, a balance is required between the two extremes: short and long half-lives. The general belief is that a very short half-life (<2 h) requires a large amount of drug, as exemplified by furosemide and levodopa, while a very long half-life (>8 h) is not desirable since the drug resides in the body too long to need frequent dosing.

Since one of the reasons for the design of controlled release dosage forms is to maintain plasma drug concentration as constant as possible within the therapeutic range, biotransformation of the drug, especially hepatic clearance, is an important factor. Consequently drugs that are highly extracted are not good candidates for controlled release dosage forms. Moreover, enzyme induction and enzyme inhibition are of utmost importance. Drugs that induce or inhibit enzyme synthesis are believed to be poor candidates for the design of controlled release dosage forms.

TABLE 11.2
Biological Parameters in the Design of Controlled Release Dosage Forms

Parameters	Comments
Absorption	The drug absorption rate constant should be much higher than the drug release rate constant since the principle of oral controlled-release dosage form design is that the drug release rate should be the rate-limiting step.
Elimination	A very short half-life (<2 h) requires a large amount of drug, while a very long half-life (>8 h) is not desirable since the drug resides in the body too long to need frequent dosing.
Disease condition	Controlled release dosage forms are designed for the treatment of chronic rather than acute diseases.
Distribution	Dose calculation in the formulation of controlled release dosage forms is often based on the one-compartment pharmacokinetic model. Whenever it is desirable to apply a two-compartment pharmacokinetic model in dose calculation, the apparent volume of distribution at steady state is considered the most appropriate.
Margin of safety	Highly potent drugs are often characterized by a narrow therapeutic index and are very difficult to formulate into controlled release dosage forms from the point of view of adequate release rate and dose dumping arising from misuse by the patient.

d. Disease Condition

Generally, controlled release dosage forms are designed for the treatment of chronic rather than acute diseases.

e. Margin of Safety

Highly potent drugs are often characterized by a narrow therapeutic index. Thus, drugs with very narrow therapeutic indices are difficult to formulate into controlled release dosage forms because of the problem of ensuring appropriate release rates. While the risk of product failure and hence dose dumping can be minimized to some extent, it is difficult if not impossible to avoid patient misuse, which can lead to dose dumping and result in untoward effects with drugs of narrow therapeutic indices.

E. Classification of Oral Controlled-Release Drug Delivery Systems

The basic tenet in the design of oral controlled-release drug delivery systems is that the kinetics of drug release, rather than the kinetics of drug absorption, controls the availability of the drug. Many innovative systems for oral controlled delivery of drugs have been developed. Most of them belong to the second and third generation drug delivery systems. Some have reached the stage of use in clinical situations, but most are at the early stage of development. These oral controlled-release dosage forms are classified based on the mechanism that triggers and eventually controls the release of the drug incorporated in the system. Some of the biological factors are summarized in Table 11.2.

1. Osmotically Controlled Sytems

The osmotic device consists of a core containing a drug alone or together with an osmotic agent that is surrounded by a semipermeable polymer membrane equipped with an orifice for delivery of the drug. Examples of semipermeable membranes utilized to regulate the osmotic permeation of water are cellulose acetate, ethylcellulose, polyurethane, polyvinyl chloride, and polyvinyl alcohol. When in contact with body fluids (i.e., stomach fluid), the osmotic agent draws in water through the semipermeable membrane because of the osmotic pressure gradient and forms a saturated solution inside the device. The increase in volume caused by the imbibition of water leads to the development of pressure inside the device. The flow of saturated solution of drug out of the

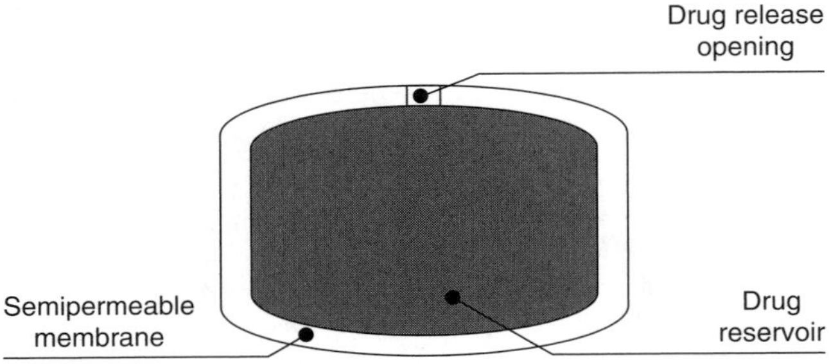

FIGURE 11.2 Elementary osmotic pump. (From Heilman, K. *Therapeutic Systems: Rate-Controlled Drug Delivery: Concepts and Development,* 2nd rev. ed., Georg Thieme Verlag Thieme-Straton, New York, 1984, p. 51, with permission.)

device through the delivery orifice relieves the pressure inside. This process continues at a constant rate (releasing a constant amount of drug per unit time) until the entire solid agent has been dissolved. The drug release rate is virtually unaffected by the pH of the environment and is essentially constant as long as the osmotic gradient remains constant (i.e., until the excess undissolved drug is depleted, at which time the release rate decreases to zero). This system is an elementary osmotic pump (Figure 11.2). Among the factors that affect drug release from an elementary osmotic pump are saturated solubility of the drug, osmotic pressure, size of the delivery orifice, and type of membrane.

Several variations of this basic osmotic pump design have been described. One system contains the drug in solution in an impermeable membrane (flexible bag) placed within the core containing the osmotic agent. The high osmotic pressure generated in the core causes the compression of the inner membrane, and the drug is pumped out through the hole. There is an osmotic device for implantation (osmotic minipump). The oral osmotic pump drug delivery system (Oros, or gastrointestinal therapeutic system) was introduced by Alza Corporation. Table 11.3 shows many oral controlled-release drug products based on osmotic delivery systems.

The rate at which drug is pumped out of the osmotic system through the orifice produced by a laser beam is the same as the volume flow rate of water into the core multiplied by drug concentration, as shown in Equation 11.2:

$$dM/dt = (dV/dt)\ C_s \tag{11.2}$$

where dM/dt is the amount of drug (milligrams) delivered per unit of time (hours), dV/dt is the volume flow rate [milliliters per unit of time (hours)], and C_s is the concentration of the saturated solution.

The volume flow rate of water into the osmotic device can be expressed in terms of membrane permeability (K), semipermeable membrane surface area (A), thickness (h) of the membrane, osmotic pressure difference ($\Delta\pi$), and hydrostatic pressure difference (Δp), as shown in Equation 11.3:

$$dV/dt = (AK/h)(\Delta\pi - \Delta p) \tag{11.3}$$

The orifice is usually big enough that $\Delta\pi >>> \Delta p$; then the equation reduces to

$$dV/dt = (AK/h)(\Delta\pi) \tag{11.4}$$

Therefore,

$$dM/dt = (AK/h)(\pi)\ C_s \tag{11.5}$$

TABLE 11.3
Osmotically Controlled Oral Drug Products

Products	Active Drug	Developed by/Marketed by
Procardia XL	Nifedipine	Alza Corp/Pfizer, Inc.
Calan SR	Verapamil	Alza Corp/G.D. Searle & Co.
Volmax	Albuterol	Alza Corp/Muro Pharmaceuticals
Ditropan XL	Oxybutynin	Alza Corp/Alza Corp.
Glucotrol XL	Glipizide	Alza Corp/Pfizer, Inc.
Minipress XL	Prazosin	Alza Corp/Pfizer, Inc.
Teczem	Enalapril and diltiazem	Merck & Co. Inc. and Hoechst Marion Roussel, Inc./Hoechst Marion Roussel, Inc.
Tiamate	Diltiazem	Merck & Co. Inc./Hoechst Marion Roussel

It is obvious from Equation 11.5 that the release rate will be zero-order as long as the concentration of drug solution (C_s) is above saturation. Further, since the osmotic pressure of the gastrointestinal fluids is negligible compared with that of the core, π can be substituted for $\Delta\pi$ in Equation 11.4 to get Equation 11.5. Both π and C_s will be constant in the presence of excess solid in the core. Other terms on the right hand side of Equation 11.5 can be maintained constant by a proper selection of the semipermeable membrane.

2. Ion-Exchange Resins

Crosslinked polymer networks containing pendantly attached ionizable groups are capable of exchanging ions attracted to the ionized groups with ions of the same charge present in solution. They are called ion-exchange resins and are often prepared in the form of beads or particles. They are insoluble in water. The resins may be either cation exchangers (the resin ionizable group is acidic: sulphonic, carboxylic, or phenolic groups) or anion exchangers (the ionizable group of the resin is basic: amine or quartenary ammonium groups).

In the design of controlled release dosage forms, ion-exchange resins are employed to form chemical complexes with drug substances. Binding of the drug to resin is effected by passing a solution of the drug through the resin in a chromatographic column. The drug-resin complex is washed to get rid of contaminating ions and dried to form beads. The drug-resin complex may be tableted, encapsulated, or suspended in an aqueous solution for oral administration. The drug-resin complex may be coated with a suitable polymer to further control the release of drug. The schemes describing the displacement and release of drug in the gastrointestinal fluids are as follows:

Cation-exchange resin:

$$\text{Poly}-\left(\text{SO}_{3^-}\right)\text{Drug}^+ + \text{HCl}/\text{NaCl} \leftrightarrow \text{Poly}-\left(\text{SO}_{3^-}\right)\text{H}^+/\text{Poly}-\left(\text{SO}_{3^-}\right)\text{Na}^+ + \text{Drug}^+\text{Cl}^-$$

Anion-exchange resin:

$$\text{Poly}-\text{N}\ \left(\text{CH}_3\right)_3 + \text{Drug}^- + \text{HCl}/\text{NaCl} \leftrightarrow \text{Poly}-\text{N}\ \left(\text{CH}_3\right)_3 + \text{Cl}^- + \text{Drug}^-\text{Na}^+/\text{Drug}^-\text{H}^+$$

The reaction involving HCl is expected to be predominant in the stomach, while the reaction involving NaCl is expected to be predominant in the small intestine. The majority of the ion-exchange resins for oral controlled-release delivery systems have sulfonic acid as the ionizable moiety. Consequently, cationic or basic drugs have been developed with ion-exchange as the mechanism that triggers drug release. Ions must diffuse into and out of the resin for exchange to occur. Thus, the following factors exert significant influence on the rate of release of basic drugs from cation-exchange resins: the diameter of the resin beads, the degree of crosslinking within the resin, the pK_a of the ionizable resin group, and electrolyte concentration in the microenvironment of drug release. Examples of drug products based on ion-exchange are Ionamin Capsules by Celltech (containing phentermine resin) and Tussionex and Pennkinetic Extended Release Suspensions by Celltech (containing hydrocodone polistirex and chlopheniramine polistirex). Moreover, resin complexes have been used to mask the taste of bitter drugs and to reduce the nausea produced by some irritant drugs.

3. Diffusion Controlled Systems

In diffusion controlled systems, polymers function as physical barriers to drug transport and hence control the rate of release of a drug when administered orally. Generally, diffusion controlled systems are of two types: reservoir devices and matrix devices.

a. Reservoir Devices

The drug core is encased by a water-insoluble polymeric material. The mesh (the space between macromolecular chains) of these polymers, through which drug penetrates or diffuses after partitioning, is of molecular level. Drug release is believed to involve the following processes: Water, gastric juices, or intestinal juices penetrate the coating and the drug dissolves to produce a saturated solution, assuming the amount of drug present is enough to achieve this condition; the drug diffuses though the polymeric coating (i.e., molecular diffusion through the polymer network is the mode of mass transport). A fairly constant rate of drug release occurs if the concentration of drug in the core (inside the polymeric coating) remains at saturation. The nature of the polymeric material governs the rate of release of drug from reservoir devices.

In amorphous polymer, the segmental mobility of polymer chains below the glass transition temperature (Tg) is low because of the entanglements between neighboring polymer chains. The result is low permeability of solutes. For example, ethylcellulose (Tg = 128°C) and methacrylate-ethacrylate copolymers (Eudragit RS and RL, Tg = 55° C) have low permeability at 37°C. It is the plasticizing effect of water and other fluids taken up by the polymers in the gastrointestinal tract that brings down the Tg to allow drug permeability through the polymer to a level to achieve adequate release. The degree of crystallinity is another polymer structural parameter that affects drug permeability. Crystallites act as physical crosslinks between polymer chains (crystallinity occurs when polymer chains have sufficient geometrical regularity that the molecules are able to align themselves together and crystallize to ordered structures). Normally polymers are not 100% crystalline, and the term *semicrystalline* is frequently used .

If the polymer coating results in a very low permeability, the problem can be circumvented as follows:

- Incorporation of plasticizers into the polymer, which may act to lower the glass-transition temperature or increase the uptake of fluids into the polymer coating. Further, depending on the solubility of the plasticizers, they may remain in the coating or be eluted to form micropores or macropores.
- The use of copolymers may lower the Tg and increase moisture uptake. Copolymers of methyl methacrylate and ethacrylate (Eudragit NE) have been reported to have a Tg of only about 5°C and absorb substantially more water than do polymethylmethacrylate homopolymers; the permeability of the copolymers is adequate without the incorporation of any additives.

- Water-soluble additives are often incorporated to form a disperse phase in the polymer coating. When placed in an aqueous environment, the water-soluble additives (often called pore-formers) in the polymer coating dissolve to leave pores. The coating becomes porous (depending on the type of pore-former used, it may contain micropores, down to molecular size, or macropores) during drug release and allows both lipophillic and hydrophilic drugs, including ions, to pass through. In this case the release rate is a function of the total concentration (both dissociated and undissociated) of drug within the core.

Mathematical expressions describing the release of solutes from polymeric drug delivery systems are based on Fick's laws of diffusion. Fick's first law states that the flux (J) or rate of solute transfer across a plane of unit area in the direction of decreasing concentration is proportional to the concentration gradient across the area:

$$J = -D\ (dc/dx) \tag{11.6}$$

The flux, J, has the units of amount/area-time; D is the diffusion coefficient of the drug in the polymer and has units of area/time; dc/dx is the concentration gradient or change in concentration (c) with respect to distance (x) in the polymer.

Fick's first law (Equation 11.6) can be simplified as shown in Equation 11.7. K is the partition coefficient that relates the concentration of the drug just inside the polymer membrane surface to the concentration in the adjacent region; d is the thickness of the diffusion layer (the diffusion path).

$$J = DK\Delta C/d \tag{11.7}$$

ΔC is the concentration difference across the polymer membrane. The type of equation used to estimate the amount of drug released is a function of the geometry of the reservoir device: slab, sphere, or cylinder. In the case of a slab in which drug release is restricted to only one surface, the steady-state release rate (dM_t/dt) is expressed as shown below:

$$dM_t/dt = ADK\Delta C/d \tag{11.8}$$

M_t is the amount of drug released after time t, and A is the surface area of the device. Theoretically, a constant rate of drug release is possible if all the terms on the right-hand side of Equation 11.8 are held constant. However, this is very difficult to achieve when the drug delivery device is exposed to water or gastrointestinal tract fluids, as some of the parameters will no longer be constant. Furthermore, for semicrystalline polymers, it is believed that the diffusion through the polymers can still be treated as Fickian, with little modification to account for reduction in solute mobility owing to tortuousity of diffusion paths between crystallites.

The design of oral controlled-release dosage forms of the reservoir type involves adjusting the release rate based on the pharmacokinetics and pharmacodynamic requirements. The following factors are considered to be important in the design of such systems. Generally, for drugs with molecular weights in the usual range of about 150 to 300, the effective permeability coefficient [which is a function of the diffusion coefficient (D) and partition coefficient (K)] is dependent primarily on the polymeric coating material. Further, polymer ratio in the coating formulation, polymer film thickness (drug release rate from an insoluble polymeric membrane increases as the membrane thickness decreases), and hardness of microcapsules (the harder the microcapsule, the longer the time for drug release) affect the release rate. The area of the coating film can also be manipulated. However, the area and thickness of the coating film can be manipulated over a limited range; the permeability coefficient is believed to be of prime importance.

TABLE 11.4
Examples of Oral Controlled-Release Dosage Forms Based on the Pellet-Type Reservoir-Diffusion System

Product	Manufacturer	Comment
Ornade Spansule capsule (phenylpropanol amine HCl and chlorpheniramine maleate)	SmithKline Beecham	Coated pellets in capsule
Theo-Dur (theophyline)	Key Pharmaceuticals Inc.	Pellets in tablets
Toprol-XL (metoprolol succinate)	AstraZeneca LP	Pellets in tablets
Indocin SR capsules (indomethacin)	Merck	Coated pellets in capsule
Compazine Spansule capsules (prochlorperazine)	SmithKline Beecham	Coated pellets in capsule

b. Methods for Developing Reservoir Devices

i. Coated Beads or Pellets

The first step is the coating of a drug solution onto preformed cores. The cores, also known as nonpareil seeds or beads, are prepared using a slurry of starch, sucrose, and lactose. Microcrystalline cellulose has also been used to form the cores. The second step involves the covering of the core (beads) by an insoluble but permeable coat, which provides sustained release of the drug. Drug crystals may comprise the core when it is necessary to minimize the size of the core. Pan-coating or air-suspension techniques are used for the coating process.

Some materials that have been used to coat the core containing the drug are cross-linked poly(vinyl alcohol), hydroxypropyl cellulose, polyvinyl acetate, ethyl cellulose, methacrylic ester copolymers, polyethylene glycol, beeswax, carnauba wax, and cetyl alcohol. Plasticizers and other additives such as pore-formers may be added to the coating solution. Further developments in coating technology have yielded low organic-solvent systems (coating emulsions) as well as polymer lattices and pseudolattices, which are aqueous dispersions. Aqueous-based coating systems such as Aquacoat and Surelease, which obviate the use of organic solvent-based systems, are commercially available. However, water evaporation and particle coagulation to make the formation of a homogeneous or continuous film are very important considerations, unlike the case with organic-solvent systems. To form an optimal film coating, it is believed that the minimum film-formation temperature (MFT) must be exceeded. Minimum film-formation temperature is affected by the type and amount of plasticizers added to polymer coatings made from Eudragit derivatives. The coating thickness may be varied for different pellets to achieve sustained release. Also, the usual tradition in the design of sustained drug delivery systems using the pellet technique is to leave about one quarter to one third of the beads (the core containing the drug) uncoated with the insoluble but permeable coat. These pellets will provide an immediate release of the drug. The pellets can be filled into a capsule or compressed lightly into tablets. Examples of products available commercially as coated beads or pellets are shown in Table 11.4.

ii. Microencapsulation

Microencapsulation is the technique used to encapsulate small particles of drug, solution of drug, or even gases in a coat, usually a polymer coat. Generally, any method that can induce a polymer barrier to deposit on the surface of a liquid droplet or a solid particle can be used for fabricating microcapsules. Examples are discussed below.

- *Coacervation*: In simple coacervation, the liquid or solid to be encapsulated is dispersed in a solution of polymer with which it is miscible. Then a nonsolvent for the polymer that is miscible with the dispersion medium is added to induce the polymer to form a coacervate (polymer-rich) layer around the disperse phase. The coacervate is then cooled and crosslinked to form capsule walls. Apart from the addition of the nonsolvent, coacervate deposition may be achieved by the addition of salts. Complex coacervation involves two polymer solutions in which the two polymers are oppositely charged under certain conditions. The interaction of the two polymers results in a decrease in the solubility of the polymers and consequently of the deposition of the polymer layer (coacervate) at the particle–solution interface. The coacervate is then crosslinked. Water-insoluble polymers can also produce a coacervate: Desolvation of the water-insoluble polymers in nonaqueous solvents results in the deposition of a coacervate around aqueous or solid disperse droplets. This desolvation can be induced thermally, and phase diagrams are often used to select conditions for phase separation to form a coacervate. Examples of polymers used for coacervation are gelatin, gum Arabic (acacia), hydroxyethylcellulose, poly(acrylic acid), starch, poly(vinyl alcohol), poly(vinylpyrollidone), and starch. Poly(methylmethacrylate), ethylcellulose, cellulose actetate phthalate, and cellulose nitrate have also been used for coacervation.

- *Interfacial polymerization*: This method of microcapsule formation can be carried out based on the reactions between oil-soluble and water-soluble monomers at the oil–water interface of water-in-oil or oil-in-water dispersions. The monomers diffuse to the oil–water interface, where they react to form a polymeric membrane; hence, the term *interfacial polymerization*. The drug is dissolved in the disperse phase. The size of the microcapsules is determined by the size of the emulsion droplets. Various polymers have been used as the wall materials for the preparation of microcapsules by interfacial polymerization: polyurethanes, polyamides, polysulfonamides, polyphthalamides, and poly(phenylesters). The most widely investigated is polyamide. Generally, the choice of polymer to be used is restricted to those that can be formed from monomers with preferential solubilities in one phase. The polyamide is formed by interfacial condensation of water-soluble alkyldiamines with oil-soluble acid dichlorides. The acid dichlorides are those with long carbon chains between the acid chloride groups, since such acid dichlorides hydrolyze slowly at the oil-in-water interface. Amines that have been used include ethylene diamine, diethylene triamine, hexamethylene diamine, and triethylene tetramine, while the dichlorides include terephthaloyl dichloride and trimesoyl chloride.
- *Solvent evaporation*: The polymeric wall material and the drug are usually dissolved or suspended in a water-immiscible volatile organic solvent. The drug–polymer–solvent solution or suspension is then dispersed in an aqueous solution, which may contain a surfactant, to form an emulsion. As the organic solvent evaporates, small polymer particles (micropcapsules) precipitate from the emulsion. Many process and formulation variables affect the size and quality of the microcapsules: speed of agitation, oil:water ratio in the emulsion, and the presence of additives. Other methods of fabricating microcapsules are thermal denaturation, hot melt, spray-drying, salting out, and electrostatic methods.

Microencapsulation has been used to prepare aspirin and potassium chloride (Mico-K Extecaps) and combines the property of extended release with improved gastrointestinal tolerability.

c. *Matrix Devices*

A matrix (monolithic) device consists of an inert polymeric matrix in which a drug is uniformly distributed. Drugs can be dissolved in the matrix or the drugs can be present as a dispersion (Figure 11.3). Further, the matrix may be homogeneous or granular (porous) with water-filled pores. The state of presentation leads to different release characteristics.

i. *Dissolved Drug*

When the system is a homogeneous matrix containing a dissolved drug, the release can be described by a solution to Fick's second law of diffusion, which states that the rate of change of concentration with time (t) at a particular level is proportional to the rate of change of the concentration gradient at that level. Mathematically, Fick's second law of diffusion is

$$dc/dt = D\ (d^2c/dx^2) \tag{11.9}$$

The symbols have the same meaning as described earlier for Fick's first law of diffusion. Equation 11.9, after appropriate manipulation, describes the release pattern as being dependent on square root of time for a homogeneous matrix of a certain geometry containing a dissolved drug. It has been used to model drug release from hydrogels containing a dissolved drug.

ii. Dispersed Drug

If a drug is dispersed as a solid in the homogeneous matrix (monolith) instead of being completely dissolved, the release kinetics are altered. The derivation of the release rate expression is based on Fick's first law of diffusion. Higuchi (18) developed a mathematical model for the analysis of such system: the release of dispersed drugs by diffusion from a stationary matrix with a planar geometry such as an ointment. The model has been applied to a homogeneous polymer matrix containing a suspended drug. The expression of the amount of drug released per unit area from a slab takes this form:

$$M_t = (C_sD(2A - C_s)t)^{1/2} \tag{11.10}$$

where M_t is the amount of drug released per unit area at time t, D and C_s are the diffusion coefficient and solubility of the drug in the polymer, respectively, and A is the concentration of drug initially present in the matrix. When $A >> C_s$, the equation reduces to

$$M_t = (2C_sDAt)^{1/2} \tag{11.11}$$

This is known as Higuchi's square root of t equation, which describes a linear relationship between the amount of drug released and $t^{1/2}$ provided D, A, and C_s are constant. It was indicated that the model would only be suitable for systems where A exceeded C_s by a factor of 3 or 4. Further, the simplifying assumption that $A >> C_s$ has been reported to be reasonable for almost all polymer–drug dispersions containing more than 5% w/w (weight-in weight) drug and is often valid for polymer–drug dispersions containing as little as 1% w/w drug. In this model, it is assumed that the solid drug dissolves from the surface layer and that this layer becomes exhausted of dispersed particles.

The derivation and development of the equation for the release of drug from a homogeneous matrix containing a dispersed drug requires certain assumptions, which have been reported as follows:

- Release is controlled by the matrix rather than via aqueous boundary layers at the surface of the matrix or channels formed by dissolved solute.
- Drug is present as finely divided particles uniformly disperse in the polymer matrix and release is not dissolution rate limited.
- The diffusion coefficient is constant.
- A pseudo-steady state exists.
- The concentration of the drug in the matrix is much greater than its solubility in the matrix.
- There is no volume change.
- The amount of drug released at the edges is negligible.
- Drug release occurs under sink conditions.

iii. Porous Matrix

The porous space in this system is often conceived as a collection of differently sized capillaries, randomly oriented and connected. Usually, the matrix is made of a hydrophobic polymer with negligible drug permeation. Release of drug is triggered by the ingress of water or gastrointestinal fluids into the pores and channels, followed by dissolution and diffusion of drug molecules. The interconnecting capillaries or channels filled with drug particles in the porous matrix are caused by high loading of drug in the polymer.

Higuchi (18) developed this model for drug release from a porous or granular matrix:

$$M_t = (C_{ss}D_s - \varepsilon/\tau(2A - \varepsilon C_{ss})t)^{1/2} \quad (11.12)$$

where M_t, A, and t have the same definitions as described previously for a homogeneous matrix containing a dissolved drug. C_{ss} and D_s are the solubility and diffusion coefficient, respectively, of the drug in the permeating fluid or release medium. The terms ε and τ are introduced to account for the porosity and tortuousity, respectively, in the matrix. Examples of polymers that have been used for the fabrication of porous matrix devices include copolymers of methylmethacrylate and methyl acrylate, poly(vinyl chloride), and polyethylene.

iv. Hydrophilic Matrix

In this system, the release of drug occurs by gelation and diffusion. The size and shape of the hydrophilic matrix change as water penetrates into the system. It is believed that if the polymer does not dissolve but only swells and if the drug has not completely dissolved in the solvent, then diffusion occurs from the saturated solution through the gel layer. This behavior is characteristic of polymers such as carboxymethylcellulose, sodium carboxymethylcellulose, and hydroxypropyl methylcellulose. The equation for modeling the drug release kinetics takes this form:

$$M_t = (C_{ss}D_s - \varepsilon/\tau(2(M_o/V) - \varepsilon C_{ss})t)^{1/2} \quad (11.13)$$

where M_o is the amount of the drug in the matrix, and V is the effective volume of the hydrated matrix, which varies with time of hydration.

A commercially available controlled delivery system for verapamil utilizes the hydrophilic matrix system: Calan SR oral caplets by Searle or Isoptin SR 240 mg oral tablets by Knoll Laboratories. Verapamil is incorporated into the matrix made of sodium aliginate (natural polysaccharide). Following contact with gastric fluid, the matrix becomes hydrated and drug diffuses through the gel layer. Reports have shown that Calan SR or Isoptin SR 240 mg once daily or in two equally divided doses caused a dose-dependent reduction in blood pressure that was equivalent to that of conventional verapamil, 80 to 160 mg every 8 hours; the controlled delivery system also showed a more favorable side effect profile. Further, niacin (nicotinic acid) is available commercially as Slo-Niacin tablet (Polygel controlled release niacin) made by Upsher-Smith. The release of niacin occurs through a water-soluble binder in the tablet. The hydration of the binder on the outer surface of the tablet results in a gel layer from which the drug diffuses out. The rate of dissolution and drug availability are controlled by the diffusion of the drug from the gel layer and the rate of tablet erosion. Sustained-release niacin has been compared with the immediate-release niacin. The sustained-release form was found to lower low density lipoprotein significantly more than the immediate-release preparation; however, the immediate release preparation increased the high density lipoprotein significantly more than the sustained-release preparation. The sustained release preparation of niacin reduced the vasodilatory side effect characteristic of the immediate release preparation of niacin. The extended release felodipine (Plendil ER made by AstraZenneca) also utilizes the hydrophilic gel mechanism: a reservoir of felodipine is surrounded by a second layer that also contains felodipine embedded in a gel-forming substance. The type and amount of the gel-forming

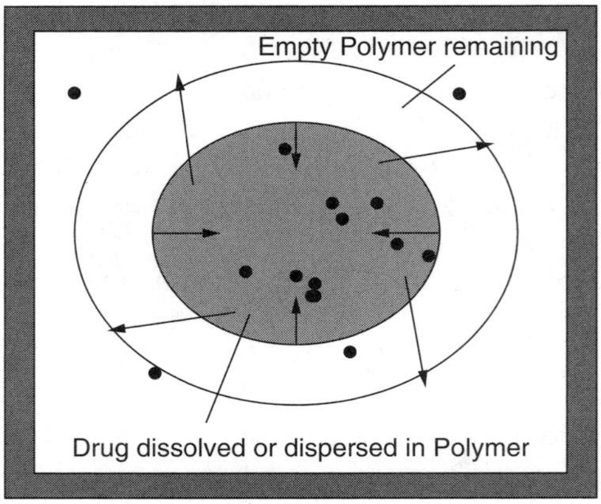

FIGURE 11.3 Schematic representation of a matrix diffusional system. (From Amidon, G., Choe, S., and Know, T., *Biopharmaceutics, Version 1.0,* College of Pharmacy, University of Michigan, Ann Arbor. PCCAL International, 1997, with permission.)

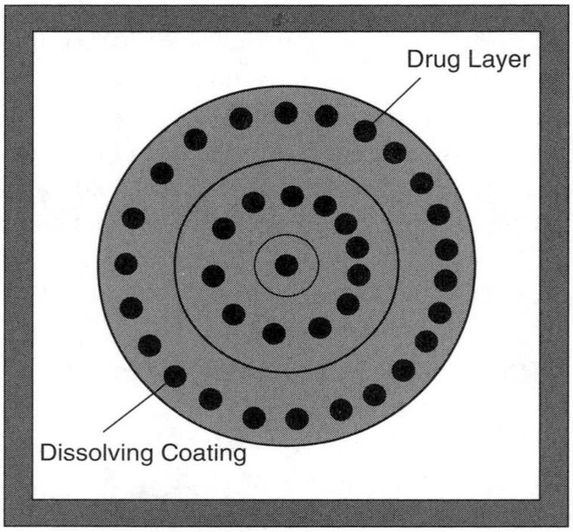

FIGURE 11.4 Schematic representation of a dissolution controlled system. (From Amidon, G., Choe, S., and Know, T., *Biopharmaceutics, Version 1.0,* College of Pharmacy, University of Michigan, Ann Arbor. PCCAL International, 1997, with permission.)

substance determines the drug-release rate. The mechanism of controlled release is by diffusion through the gel layer and attrition of the gel layer.

4. Dissolution Controlled Systems

a. *Matrix Dissolution*

The drug is dispersed or dissolved within the polymeric matrix. Alternatively, the matrix system may also be made by direct compression of a mixture of drug, a slowly soluble polymer, and other excipients into tablets (Figure 11.4). When a wax material is used in the preparation of the matrix dissolution systems, the drug is mixed with the wax material and either spray congealed or sprayed into water. The drug is released as the matrix is eroded. The rate of drug availability is controlled by the rate of penetration of the gastric fluid into the matrix. The rate of penetration depends on the porosity of the matrix among other factors. A typical commercial product that utilizes this principle is Nifedipine ER tablet for oral use (Adalat CC tablet made by Bayer), which is a once-daily controlled-release formulation. In this system, an outer core contains a slow-release formulation of nifedipine dispersed in a hydrophilic gel matrix, and the inner core contains a fast-release formulation of nifedipine. Following contact with gastric fluid, erosion of the matrix begins at the tablet surface, dissolution of nifedipine follows, and then drug absorption occurs as the tablet passes down the gastrointestinal tract. It is believed that as the erosion of the outer coat advances, the fast release of the inner core of nifedipine compensates for the decrease in the release of the drug in the outer core, thereby providing a fairly constant plasma level of nifedipine for 24 hours. Nifedipine ER has been reported to show comparable blood pressure control to Nifedipine GITS (Procardia XL manufactured by Pfizer Inc.). Other commercial products whose availability is controlled by matrix dissolution are Trilafon tablets (containing perphenazine and made by Schering) and Donnatal Extentabs (containing phenobarbital, hyoscyamine, atropine, and scopolamine and made by A. H. Robins).

b. Encapsulated Dissolution Controlled System

The design of this system involves coating the drug with slowly dissolving polymeric materials. Once this membrane has dissolved, the drug core is available for immediate release and absorption. Control over drug release is achieved by adjusting the thickness and hence the dissolution rate of the polymeric membrane. Consequently, by compressing multiple particles of drug with varying thickness, which results in varying erosion times, into a tablet, it is possible to obtain a uniform sustained release. Drug particles without a polymeric barrier are often included for immediate release of the drug. A commercial product based on this principle is Cardizem CD capsules (containing diltiazem hydrochloride and made by Hoechst Marion Roussel), available for the treatment of hypertension and it requires a twice daily dosing. The multiparticulate formulation consists of beads with a coating of variable thickness. The coating dissolves at varying times (3 to 12 h) in the gastrointestinal tract with the concomitant sustained release characteristics. The microencapsulation technique discussed earlier (under reservoir systems) can also be used for the preparation of encapsulated dissolution control systems. Examples of materials used for microencapsulation are cellulose acetate butyrate, cellulose acetate phtalate, ethylcellulose, shellac, gelatin, and carnuba wax. These microcapsules are often filled into capsules; they are not tableted so as to retain the integrity of the microcapsules. Other commercial products that depend on encapsulated dissolution control systems for the release of the drugs are: Diamox sequels (containing acetazolamide and made by Lederle laboratories), Dexedrine spansule (containing dextroamphetamine sulfate and made by SmithKline and Beecham), and Thorazine spansule (containing chlorpromazine and made by SmithKline and Beecham).

5. Dissolution and Diffusion Controlled Systems

In this system, the drug forms the core, which is surrounded by a partially water-soluble polymer. Drug release is triggered by the dissolution of the water-soluble part of the polymer membrane followed by the diffusion of the drug through the holes or pores in the polymer (Figure 11.5). This is achieved by incorporating water-soluble additives in the polymer coating. When placed in an aqueous environment, the water-soluble additives (often called pore-formers) in the polymer coating dissolve to leave pores. The process is also used to increase the permeability of drugs through reservoir delivery systems.

The kinetics of the release of drugs from dissolution-controlled systems is often described by the Noyes-Whitney equation:

$$dC/dt = k\ A/V\ (C_s - C) \tag{11.14}$$

where dC/dt is the dissolution rate; V is the volume of the solution; k is the dissolution rate constant, which is equal to the diffusion coefficient of the drug, D, divided by the thickness of the diffusion layer or the coating, h; and A is the surface area of the exposed solid. For the dissolution and diffusion dissolution controlled systems, D is the diffusion coefficient of the drug through the pores.

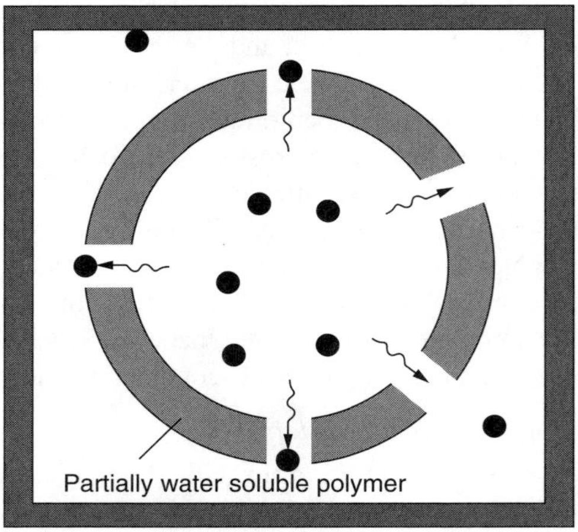

FIGURE 11.5 Schematic representation of dissolution and diffusion controlled system. (From Amidon, G., Choe, S., and Know, T., *Biopharmaceutics, Version 1.0,* College of Pharmacy, University of Michigan, Ann Arbor. PCCAL International, 1997, with permission.)

C_s is the saturated solubility of the drug, and C is the concentration of the drug in the bulk solution. Constant dissolution rate is assured if all the parameters defined above remain constant, which is difficult to achieve. Cube root law has been used to describe the *in vitro* release kinetics of spherical particles: The change in surface area can be related to the weight of the particle (sink condition is assumed). The equation is as shown below.

$$W_o^{1/3} - W^{1/3} = k^1 t \tag{11.15}$$

where W_o and W are initial weight and the weight of the amount remaining at time t, respectively. k^1 is the cube-root dissolution-rate constant.

F. Introduction to the Pharmacokinetics of Oral Controlled-Release Dosage Forms

For clinical justification, the pharmacokinetic characteristics of the active constituent should play a large role in the decision to develop a controlled release product, as discussed below.

The therapeutic range ratio (also called therapeutic index, TI) is defined as the ratio of the minimum toxic concentration to the minimum effective concentration. The principle applied to most drugs in the design of controlled release dosage forms is that the frequency of administration should be such that the ratio of maximum to minimum plasma concentrations at steady-state (dosage form index) is less than TI; however, the dose should be high enough to produce effective concentrations. The dosage form index depends on the first-order elimination rate constant, the administered dose, the volume of distribution, the dosing interval, and the absorption rate constant. In a conventional dosage form, a relatively large dosage form index is often observed because of rapid absorption of the drug into the body.

For a linear, one-compartment system with repetitive intravenous dosing (constant dose and constant dosing interval), the ratio of maximum to minimum drug concentration in the plasma at steady state is given by

$$\frac{C_{ss}\max}{C_{ss}\min} = \left[\left(\frac{D}{V_D}\right)\frac{1}{\left(1-e^{-k\tau}\right)}\right] \div \left[\left(\frac{D}{V_D}\right)\frac{e^{kt}}{\left(1-e^{-k\tau}\right)}\right] = e^{k\tau} \tag{11.16}$$

where k is the first-order elimination rate constant, D is the administered dose, V_D is the volume of distribution, and τ is the dosing interval. When the therapeutic index is known, the proper dosage interval is often dictated by

$$e^{k\tau} \leq TI \text{ and } \tau \leq (t_{1/2}) \times \frac{\ln TI}{\ln 2} \tag{11.17}$$

From the above mathematical relationships, if the therapeutic index is 2, then the dosing interval should be shorter than one elimination half-life. Furthermore, for drugs with short half-lives, (e.g., $t_{1/2} \leq 2 - 5$ h) and low therapeutic index (<3), it is believed that the proper dosing schedule to maintain desired blood drug levels in the body should be set at unreasonable frequency. One way to solve this problem is to select a drug that has a long enough half-life to necessitate once or twice daily administration, but most drugs have short half-lives, and they have to be formulated as controlled release products in which drug input into the body is slowed.

For oral dosage forms, if the drug confers a one-compartment open model on the body with first-order absorption, it is believed that for a given elimination-rate constant, $C_{ss}max/C_{ss}min$ decreases as the absorption rate decreases (Figure 11.6). Thus, if a pharmaceutical dosage form with an immediate release profile gives rise to a very high C_{ss},max/C_{ss},min, a controlled release preparation will solve the problem: Drug input into the body is slowed. $C_{ss}max/C_{ss}min = e^{k\tau}$ is the

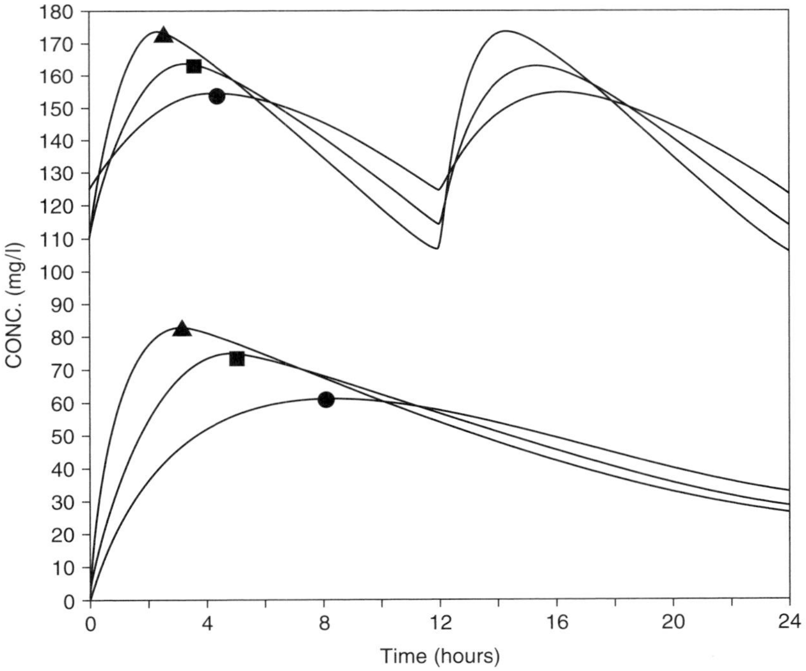

FIGURE 11.6 Effect of decreased apparent absorption rate on concentration profiles and C_{ss},max/C_{ss},min ratio, for a one-compartment model with first-order absorption. Single dose curves are found at the bottom of the figure and steady-state profiles with 12-h dosing intervals are represented at the top. Absorption half-lives are 45 min (▲), 1.5 h (■), and 3 h (●). Corresponding C_{ss},max/C_{ss},min ratios are 1.63, 1.44, and 1.25, respectively. The elimination half-life is kept at 12 h. (From Lippoid, B. C., Controlled release products: approaches of pharmaceutical technology, in *Oral Controlled Release Products: Therapeutic and Biopharmaceutic Assessment,* Gundert-Remy, U. and Moller, H., Eds., Wissenschaftliche Verlagsgesellschaft, Stuttgart, 1990, p. 39, with permission.)

dosage form index for the controlled release preparation, and the dosing interval must make it possible for the resulting ratio of the maximum to minimum plasma drug concentration to be smaller or equal to the therapeutic index. It is believed that the controlled release dosage form with the smallest dosage form index is the best design product.

The previous discussion makes it clear that the ideal goal of a controlled release drug delivery system is the maintenance of a constant plasma concentration; also, the C_{ss},max/C_{ss},min ratio must be as low as possible and it must be less than the therapeutic index. The rate of drug delivery should be zero-order to achieve this constant plasma concentration (a constant rate that is independent of the amount of drug remaining in the dosage form and that is constant over time). This is reminiscent of continuous intravenous infusion, where the rate of drug input is equal to the rate of elimination. It is known that at steady state, the rate of drug administration and the rate of drug elimination are equal. Thus, rate in is equal to rate out. Clearance is the proportionality constant that makes the average steady-state plasma concentration equal to the rate of administration.

$$R_{in} = (Cl)(Cp_{ss}\,ave) \tag{11.18}$$

$$R_{out} = (Cl)(Cp_{ss}\,ave) \tag{11.19}$$

These equations work for any single- or multiple-compartment model; consequently, they are often referred to as model-independent equations. They are the basis for the calculation of the zero-order release rate constant necessary to maintain a constant drug plasma level when drug elimination is by first-order kinetics as shown below:

$$k_{r0} = R_{in} = R_{out} = k_{el}\,Cp\,V_D \tag{11.20}$$

where k_{r0} is the zero-order rate constant for drug release (amount/time), k_{el} is the first-order rate constant for overall elimination of the drug (time^{-1}), Cp is the desired plasma drug concentration (amount/volume), and V_D is the volume of distribution. The usual practice is to design single-dose pharmacokinetic studies to calculate the pharmacokinetic parameters for the calculation of the zero-order rate constant for drug release. More often than not, non-zero-order release of the drug may be equivalent clinically to zero-order release.

The total dose (D_t) required for the design of a controlled release dosage form consists of an initial dose (D_i) to provide a blood level for therapy and the maintenance dose (D_m) that sustains the therapeutic plasma level

$$D_t = D_i + D_m \tag{11.21}$$

The maintenance dose can be expressed in terms of the zero-order release rate constant and the time for the sustained release from one dose as shown below:

$$D_t = D_i + k_{r0}t_r \tag{11.22}$$

The maintenance dose may be released after the initial dose has achieved a blood level equal to the therapeutic drug level (Cp). Alternatively, the design of the dosage form may be such that the maintenance dose begins to release the drug at time zero. To avoid what is often described as *topping,* the initial dose is often reduced as shown below:

$$D_t = D_i - k_{r0}t_p + k_{r0}t_r \tag{11.23}$$

where t_p is the time needed to achieve peak plasma concentration.

The equation below provides an estimation of the release rate (k_{r0}) required for the design of controlled release dosage forms.

$$k_{r0} = R_{in} = R_{out} = k_{el}\ Cp\ V_D \tag{11.24}$$

This equation can be expressed in terms of total clearance (Cl_T):

$$k_{r0} = R_{in} = R_{out} = CpCl_T \tag{11.25}$$

By definition, rate of administration is given as

$$R_{in} = \text{Dose}\ (D_0)/\tau \tag{11.26}$$

where τ is the dosing interval (the time needed to maintain a therapeutic concentration). If there is no built-in loading dose, then the dose needed for maintaining a therapeutic concentration for the dosing interval is

$$D_0 = CpCl_T\tau \tag{11.27}$$

G. Elements of Regulatory Assessment of Controlled Release Products

Most practicing pharmacists are involved directly or indirectly in drug product selection. Consequently, an introduction to the elements of regulatory assessment of controlled release products is necessary. This information is discussed extensively in most textbooks on pharmaceutical technology and on FDA web site are recommended for reading. Determination of the safety and efficacy of controlled release drug products and demonstration of the controlled release nature of the products are very important.

- For drugs that have been approved by the FDA as safe and effective in conventional dosage forms, the FDA has taken the position that controlled clinical studies may be required to demonstrate the safety and efficacy of the drugs in the controlled release formulation. However, if the pharmacokinetic–pharmacodynamic relationship is well defined, then pharmacokinetic data may be accepted in lieu of clinical studies. Bioavailability data for the drug in the controlled release formulation are always required.
- For drugs that have been previously approved as safe and effective in controlled release dosage forms, data are required to establish bioavailability comparability to an approved controlled release drug product.
- Single dose bioavailability studies are acceptable for determining the fraction of the amount absorbed, lack of dose dumping, and lack of food effects. Pharmacokinetic studies performed under steady-state conditions are acceptable to demonstrate comparability to an approved immediate release drug product, occupancy time within a therapeutic window, and percent fluctuation and are acceptable for supporting dosage administration labeling.
- The optimum single dose study would be a three-way crossover comparing a rapidly available dosage form and the controlled release dosage form under fasting conditions with the controlled release dosage form administered immediately after the ingestion of a high fat meal.
- The product should show consistent pharmacokinetic performance between individual dosage units.
- The product should allow for the maximum amount of drug to be absorbed while maintaining minimum patient-to-patient variation.

III. CONCLUSION

The developments in pharmaceutical dosage form design have been outlined to enable students to see the progressive nature of the efforts on delivery of bioactive agents (drugs). Various considerations in the design of oral controlled-release dosage forms have been presented to enable students to make informed decisions about the adequacy of the rationale for the design of some of the controlled release products they may come across in practice. The major classes of oral controlled-release products have been presented, with emphasis on the mechanisms that trigger the release of the drugs and the materials used in their preparations.

The tutorial questions together with the case studies will help develop lines of thoughts as students read about the oral controlled-release products and handle them in practice.

TUTORIAL

1. Write the equation of estimating drug release from dissolution-controlled systems.
2. The intrinsic dissolution rate constant of a drug is 4×10^{-5} cm/s. Calculate the rate of dissolution in milligrams per hour from a tablet of surface area 3 cm^2 under sink conditions. The solubility of the drug is 30 mg/ml.
3. $dC/dt = kA/V (C_s - C)$ or $dM_t/dt = kA (C_s - C)$ are valid equations for estimating the amount of drug released from a dissolution-controlled device. Comment on the situations or conditions under which either of the equations can be used.
4. It is believed that the type of equation used to estimate the amount of drug released is a function of the geometry of the reservoir device. Write down the equation that describes drug release from a reservoir device of slab geometry. Define fully all the parameters in the equation.
5. Calculate the rate of flux in milligrams per minute from a diffusion-controlled device when A = 1.8 cm^2, D = 10^{-6} cm^2/s, C = 18 mg/ml, d = 55 μm, and K = 6.
6. A controlled release device has a core drug concentration of 100 mg/ml, a surface area of 2.43 cm^2, and a thickness of 0.125 mm. Previous studies show that the film has the following properties:

   $$D = 1 \times 10^{-8} \text{ cm}^2/\text{s and partition coefficient, K, is 33.}$$

   Calculate the drug release rate through the film of the device

   $$dM/dt = \left(2.43 \text{ cm}^2\right)\left(10^{-8} \text{ cm}^2/s\right)\left(3600 \text{ s/mm}\right)\left[\left(100 \text{ mg/cm}^3\right)(33)\right]\big/0.0125 \text{ cm}$$

   $$= 23.10 \text{ mg/h}$$

7. The equation below is often used to estimate the rate of delivery of a solute from an elementary osmotic pump.

   $$dM/dt = (AK/h)(\Delta\pi)\, C_s$$

   Explain the meaning of all the terms in the equation.

The delivery rate of drug from an elementary osmotic pump may be by zero-order and non-zero-order. Give the condition under which these phenomenon can occur.

8. Given that the area of the semipermeable membrane of the elementary osmotic pump is 2.2 cm^2, its thickness is 0.025 cm, the permeability coefficient is 2.8×10^{-6} cm^2/atm h) and the osmotic pressure is 245 atm, calculate the rate of delivery of the solute under zero-order conditions if the concentration of the saturated solution at 37°C is 330 mg/cm^3.
9. An elementary osmotic pump has a total surface area of 3 cm^2 and a membrane thickness of 250 μm. C_s, which is the concentration of the saturated solution, was found to be 150 mg/ml. If the value of $K\pi$ was 5.0×10^{-2} cm^2/h, estimate the zero-order delivery rate in milligrams per hour.
10. The pharmacokinetic parameters of a drug, as determined from i.v. data are as follows: desired SS blood level = 10 mg/l; V_D = 15 l; k_{el} = 0.2 h^{-1}; F = 1. Estimate the rate of drug release for a zero-order release device that would maintain the desired blood levels in the body.
11. A drug is normally given at 10 mg four times a day. Suggest an approach for designing a 12-hour zero-order release product.
    (a) Calculate the desired zero-order release rate.
    (b) Calculate the concentration of the drug in an osmotic pump type of oral dosage form that delivers 0.5 ml/h of fluid.
12. An industrial pharmacist would like to design a sustained-release drug product to be given every 12 h. The active drug ingredient has an apparent volume of distribution of 10 l, an elimination half-life of 3.5 h, and a desired therapeutic plasma drug concentration of 20 mg/ml. Calculate the zero-order release rate of the sustained-release drug product and the total amount of drug.

ANSWERS

1. The Noyes-Whitney equation, which is shown below:

$$dC/dt = k\ A/V\ (C_s - C)$$

where dC/dt is the dissolution rate; V is the volume of solution; k is the dissolution rate constant, which is equal to the diffusion coefficient of the drug, D, divided by the thickness of the diffusion layer or the coating, h; and A is the surface area of the exposed solid.
2. $dM_t/dt = (4 \times 10^{-5}$ cm/s)(3600 s/h)(3 cm^2) (30 mg/cm^3) = 12.96 mg/h
3. $dC/dt = kA/V\ (C_s - C) \equiv dM_t/dt = kA\ (C_s - C)$. The equation on the left hand side is equivalent to the one on the right hand side because M_t (amount of drug) is equal to the product of C (concentration of drug) and V (the volume of solution of the drug). The parameters supplied in a particular problem will determine which of the equations to use in solving the problem.
4. $dM_t/dt = ADK\Delta C/d$. M_t is the amount of drug released after time t, A is the surface area of the device, D is the diffusion coefficient of the drug in the polymer, C is the concentration,

K is the partition coefficient that relates the concentration of the drug just inside the polymer membrane surface to the concentration in the adjacent region, and d is the thickness of the diffusion layer (the diffusion path).

5. $dM/dt = (1.8\ cm^2)(10^{-6}\ cm^2/s)(60\ s/min)\ [(18\ mg/cm^3)(6)]/\ 0.0055\ cm = 2.12\ mg/mm$
6. $dM/dt = (AK/h)(\pi)\ C_s$, where dM/dt is the amount of drug (milligrams) delivered per unit of time (hours), C_s is the concentration of the saturated solution, K is the membrane permeability, A is the semipermeable membrane surface area, h is the thickness of the membrane, and π is the osmotic pressure. The rate of drug release is constant (zero-order) until the excess undissolved drug is depleted (when the concentration of the drug is no longer saturated). The non-zero order release sets in when the saturated concentration, C_s, becomes unsaturated.
7. $dM/dt = (2.2\ cm^2/.025\ cm)(0.686 \times 10^{-3}\ cm^2/hr)(\ 330\ mg/cm^3) = 19.92\ mg/h$
8. $dM/dt = [(5.0 \times 10^{-2}\ cm^2/h)(200\ mg/cm^3)(3\ cm^2)]/0.025\ cm = 1200\ mg/h$
9. $k_{r0} = R_{in} = R_{out} = k_{el}\ Cp\ V_D$
   $k_{r0} = 0.2\ h^{-1}\ (10\ mg/l)(15\ l) = 30\ mg/h$
10. (a) A drug given 10 mg 4 times daily would be equivalent to 40 mg/24 h (i.e., 1.67 mg/h).
    (b) 0.5 ml/h should deliver 1.67 mg of the drug. Consequently, the concentration should be 3.34 mg/ml.
11. $K_{el} = 0.693/3.5 = 0.198\ hr^{-1}$; $V_D = 10\ l$; $Cp = 20\ mg/l$;
    $k_{r0} = 0.198\ h^{-1}\ (20\ mg/l)(10\ l) = 39.6\ mg/h$
    The total drug needed is $39.6\ mg/h \times 12\ h = 475.2\ mg$.

CASES

CASE 1

A diabetic patient was prescribed Glucotrol ® XL for the control of his blood glucose. He was curious about the product and asked the following questions to the dispensing pharmacist. Place yourself as the pharmacist and try to answer the patient's questions.

1. What is Glucotrol ® XL?
2. What is the mechanism of drug release from the system?
3. How does Glucotrol ® XL work?
4. What is the dosage recommendation?
5. What is the purpose of each of the inert ingredients?
6. What are the other products in the similar group?

Hint: Now as well as in real practice, you are encouraged to consult the patient package insert (PPI) and other sources to answer these questions.

Answers:

1. Glucotrol ® XL is a registered trademark for glipizide GITS marketed by Pfizer. Glipizide GITS (Gastrointestinal Therapeutic System) is formulated as a once-a-day controlled release biconvex tablet for oral use and is designed to deliver 2.5, 5, or 10 mg of glipizide. Inert ingredients in the 2.5, 5, and 10 mg formulations are polyethylene oxide, hydroxy-

propyl methylcellulose, magnesium stearate, sodium chloride, red ferric oxide, cellulose acetate, polyethylene glycol, opadry blue (OY-LS- 20921) (2.5 mg tablets), opadry white (YS-2-7063) (5 mg and 10 mg tablets), and black ink (S-1-81 06).

2 The Glucotrol ® XL (XL stands for extended release) tablet is similar in appearance to a conventional tablet. It consists, however, of an osmotically active drug core surrounded by a semipermeable membrane. The core is divided into two layers: an active layer containing the drug and a push layer containing pharmacologically inert (but osmotically active) cornponents. The membrane surrounding the tablet is permeable to water but not to drug or osmotic excipients. As water from the gastrointestinal tract enters the tablet, pressure increases in the osmotic layer and pushes against the drug layer, resulting in the release of drug through a small, laser-drilled orifice in the membrane on the drug side of the tablet. This extended release tablet is designed to provide a controlled rate of delivery of glipizide into the gastrointestinal lumen, which is independent of pH or gastrointestinal motility. The function of this extended release tablet depends on the existence of an osmotic gradient between the contents of the bi-layer core and fluid in the gastrointestinal tract. Drug delivery is essentially constant as long as the osmotic gradient remains constant and then gradually falls to zero. The biologically inert components of the tablet remain intact during gastrointestinal transit and are eliminated in the feces as an insoluble shell.

3. Glipizide appears to lower blood glucose acutely by stimulating the release of insulin from the pancreas, an effect dependent upon functioning beta cells in the pancreatic islets. Extrapancreatic effects also may play a part in the mechanism of action of oral sulfonylurea hypoglycemic drugs. Two extrapancreatic effects shown to be important in the action of glipizide are an increase in insulin sensitivity and a decrease in hepatic glucose production.
4. There is no fixed dosage regimen for the management of diabetes mellitus with Glucotrol XL Extended Release Tablet or any other hypoglycemic agent. The recommended starting dose of Glucotrol XL is 5 mg per day, taken with breakfast. The recommended dose for geriatric patients is also 5 mg per day. Dosage adjustment should be based on laboratory measures of glycemic control.
5. Please consult text.
6. First generation agents: Tolubutamide (Orinase®), Chlorpropamide (Diabinese®), Tolazamide (Tolinase®), Acetohexamide (Dymelor®). Second generation agents: Glyburide (Diabeta®), Glpizied (Glucotrol®), Glimepiride (Amaryl®).

CASE 2

A patient was prescribed Ditropan XL® for the control of her urinary incontinence. She was curious about the product and asked the dispensing pharmacist the following questions. Place yourself as the pharmacist and try to answer the patient's questions.

Hints: Now as well as in real practice set up, you are encouraged to consult patient package insert (PPI) and other sources to answer these questions

1. What is Ditropan XL®?
2. What is the mechanism of drug release from the system?
3. How does Ditropan XL® work?
4. What is the dosage recommendation?
5. What is the purpose of each of the inert ingredients?
6. What are the advantages of Ditropan XL ® (10 mg, qd) over Oxubytynin (5 mg tid)?
7. What are the other products in the similar group?

Answers:

1. Ditropan XL® is an extended release dosage form containing the antispasmodic anticholinergic agent oxybutynin chloride as the active ingredient. It is marketed by Alza as an extended release tablet containing 5, 10, or 15 mg of oxybutynin chloride USP, formulated as a once-a-day controlled release tablet for oral administration. Ditropan XL® also contains the following inert ingredients: cellulose acetate, hydroxypropyl methylcellulose, lactose, magnesium stearate, polyethylene glycol, polyethylene oxide, synthetic iron oxides, titanium dioxide, polysorbate 80, sodium chloride, and butylated hydroxytoluene.
2. Ditropan XL® uses osmotic pressure (Figure 11.7) to deliver oxybutynin chloride at a controlled rate over approximately 24 h. The system, which resembles a conventional tablet in appearance, comprises an osmotically active bilayer core surrounded by a semipermeable membrane. The bilayer core is composed of a drug layer containing the drug and excipients and a push layer containing osmotically active components. There is a precision-laser-drilled orifice in the semipermeable membrane on the drug-layer side of the tablet. In an aqueous environment, such as the gastrointestinal tract, water permeates the membrane into the tablet core, causing the drug to go into suspension and the push layer to expand. This expansion pushes the suspended drug out through the orifice. The semipermeable membrane controls the rate at which water permeates into the tablet core, which in turn controls the rate of drug delivery. The controlled rate of drug delivery into the gastrointestinal lumen is thus independent of pH or gastrointestinal motility. The function of Ditropan XL® depends on the existence of an osmotic gradient between the contents of the bilayer core and the fluid in the gastrointestinal tract. Since the osmotic gradient remains constant, drug delivery remains essentially constant. The biologically inert components of the tablet remain intact during gastrointestinal transit and are eliminated in the feces as an insoluble shell.
3. Oxybutynin chloride exerts a direct antispasmodic effect on bladder smooth muscle. In patients with conditions characterized by involuntary bladder contractions, cystomatric studies have demonstrated that oxybutynin increases bladder (vesical) capacity, diminishes the frequency of uninhibited contractions of the detrusor muscle, and delays the initial desire to void. Oxybutynin thus decreases urgency and the frequency of both incontinent episodes and voluntary urination.
4. The recommended starting dose of Ditropan XL® is 5 mg once daily. The dosage may be adjusted in 5-mg increments to achieve a balance of efficacy and tolerability (up to a maximum of 30 mg/day). In general, dosage adjustment may proceed at approximately weekly intervals.

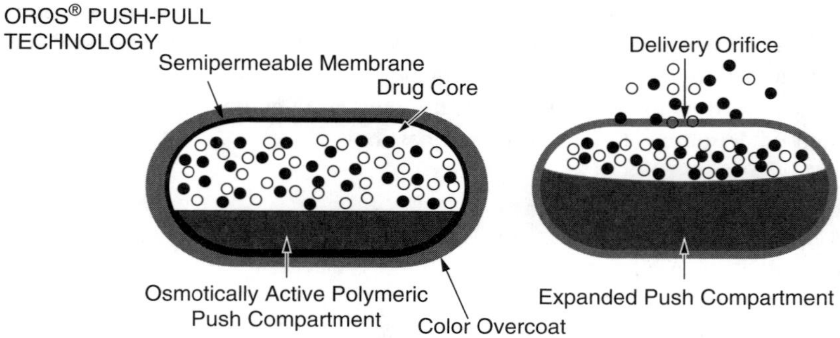

FIGURE 11.7 OROS® Push-Pull technology provides once-a-day Ditropan XL® as advertized in *Pharmacy Today,* April 2001.

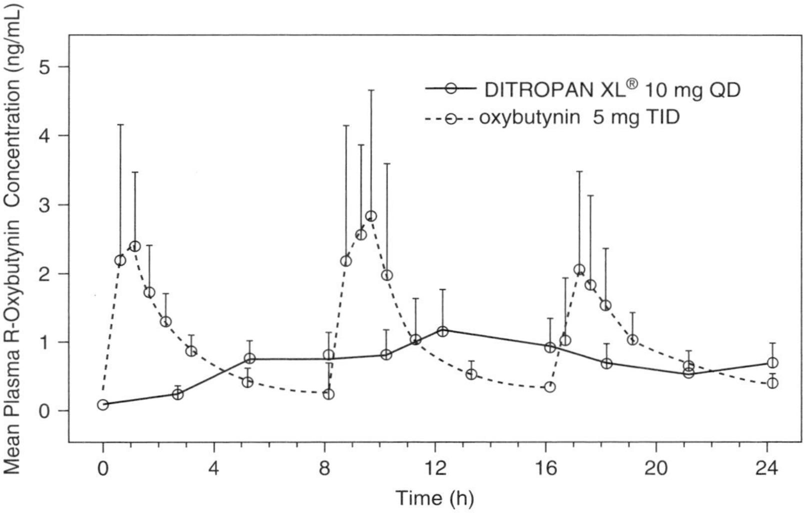

FIGURE 11.8 Mean R-oxybutynin plasma concentrations following a single dose of Ditropan XL® 10 mg and oxybutynin 5 mg administered every 8 h (n = 23 for each treatment), as reported in literature insert.

5. Please see text.
6. Advantages: Follow the plasma concentration–time profile above (Figure 11.8). It is obvious that Ditropan XL® maintains a constant profile without much peak and trough over an extended period of time with a single once-a-day dose as opposed to 5 mg of oxybutynin administered every 8 h.
7. Tolterodine (Detrol®)

CASE 3

You have been offered a summer intern position in a biopharmaceutical company. Your rotation for the first 6 weeks is in the drug delivery department. Two new drugs are about to be developed into oral controlled-delivery systems. Your first assignment is to evaluate the preformulation data (shown in the table below) available for the two drugs and choose the candidate most likely to be successful as an oral controlled drug delivery system.

Preformulation Data	Drug X	Drug Y
Molecular weight	180	640
pKa	9	5
Oral bioavailability (from solution)	0.90	0.50
Absorption rate constant (K_a) h^{-1}	0.3	0.1
Clearance (liters/hour)	80	50
Apparent volume of distribution (liters)	470	110
Log octanol/water partition coefficient	3.2	0.15

1. Which drug will you consider as the right candidate for the controlled release formulation and why?
2. Calculate the half-life of each drug from the clearance and the volume of distribution.
3. On the basis of the oil/water partition coefficient, which of the two drugs will penetrate cell membranes?

Answers:

1. Drug X: Consider the ideal characteristics of a compound to be selected for controlled release formulation. Also, consider the influence of the pKa of each drug and the pH of the gastrointestinal tract on drug ionization and absorption. A molecular weight approaching 600 normally creates a concern with respect to absorption in the gastrointestinal tract.
2. Clearance = volume of distrbution × elimination rate constant

$$\text{For drug X, K} = (80\ \text{L/h})/470\ \text{l} = 0.170\ \text{h}^{-1}$$

$$\text{Therefore } t_{1/2} = 0.693/0.170\ \text{h}^{-1}$$

$$= 4.071\ \text{h}$$

$$\text{For drug Y, K} = (50\ \text{l/h})/110\ \text{l} = 0.455\ \text{h}^{-1}$$

$$\text{Therefore } t_{1/2} = 0.693/0.455\ \text{h}^{-1}$$

$$= 1.525\ \text{h}$$

3. Drug X

CASE 4

The following article has been published: "Reinforcing Concepts by Studying Experts: An Integrated Approach to the Teaching of Pharmaceutics" (Akala, et. al. Am. J. Pharm. Educ., 65, 110S, 2001). The article describes a technique that can be used to reinforce the theory learned by the students).

This technique has been developed recently to facilitate the teaching and learning of pharmaceutics. There are many research papers that bring various concepts of pharmaceutics (pharmaceutical mathematics, physical pharmacy, pharmaceutical technology, biopharmaceutics and pharmacokinetics) in the solution of a particular pharmaceutical problem. Such papers not only give students the opportunity to see the interrelationship among the concepts taught in pharmaceutics courses (i.e., the single thread that runs through pharmaceutics), but also expose students to the applicability of the concepts in the real world of pharmaceutical research to develop a formulation presentable to the ultimate consumer: the patient. The technique is to be applied in Case 4.

This is a library exercise to be done with the help of the professor. Each student should retrieve and study the following article from the library: Kartz, B., et al., Controlled-release drug delivery systems in cardiovascular medicine, Am. Heart J., 129, 359–368, 1995. Students should analyze the paper using the questions below to develop lines of thought.

1. The two main components of controlled release drug delivery have been defined by the authors as prolongation of drug delivery and predictability of drug-release kinetics. What is the practical significance of these two components in terms of adequate therapy for patients?
2. What are the attributes, according to the authors, of drugs that are potential candidates for formulation into controlled release products?
3. Comment on the mechanisms of controlled release of drugs discussed by the authors.
4. What are the similarities between oral controlled drug delivery systems and transdermal controlled release systems mentioned by the authors?

5. Do the transdermal controlled delivery systems have any advantage over the oral controlled delivery systems?
6. What are the important criteria for the design of transdermal controlled delivery systems?
7. Name the three classes of calcium channel blockers mentioned by the authors.
8. Briefly discuss the types of controlled release formulations available for the drugs in the three classes you have mentioned with emphasis on the mechanisms of drug release.
9. Mention some of the problems that have been encountered with controlled release drug delivery systems according to the authors.
10. The professor can select similar research articles from the literature and join the students in the analysis of the paper.

REFERENCES

1. Akala, E. O., Elekwachi, O., Chase, V., Johnson, H., Lazarre, M., and Scott, K., Organic redox-initiated polymerization process for the fabrication of hydrogels for colon-specific drug delivery, *Drug Dev. Ind. Pharm.,* 29, 375–386, 2003.
2. Akala, E. O., Kopeckova, P., and Kopecek, J., Novel pH sensitive hydrogels with adjustable kinetics of swelling. *Biomaterials,* 19, 1037, 1998.
3. Amidon, G., Choe, S., and Know, T., *Biopharmaceutics, Version 1.0,* College of Pharmacy, University of Michigan, Ann Arbor. PCCAL International, 1997.
4. Ansel H. C., Allen, L. V., and Popovich, N. G., *Pharmaceutical Dosage Forms and Drug Delivery Systems,* 7th ed., Lippincott Wiiliams & Wilkins, Philadelphia, 1999, p. 233.
5. Baker, R. W., and Lonsdale, H. K., in *Controlled Release Mechanisms and Rates,* Tanquary, A. C. and Lacey, R. E., Eds., Plenum Press, New York, 1974, p. 15.
6. Barry, B. W., Development of a drug as a dosage form, in *Target Therapy and Drug Delivery: International Conference on Drug Delivery System Development,* Goldberg, A. S., Ed., London, March 9–11, 1981, p. 1.
7. Bruck, S.D., Pharmacological Basis of Controlled Drug Delivery, in *Controlled Drug Delivery,* Vol. 1, *Basic Concepts,* Bruck, S. D., Ed.), CRC Press, Boca Raton, FL, 1999, p. 9.
8. Cardinal, J. R., Drug Release from Matrix Devices, in *Recent Advances in Drug Delivery Systems,* Anderson, J. M. and Kim, S. W. Eds., Plenum Press, New York, 1984, p. 229.
9. Chandrasekaran, S. K., and Shaw, J. E., Controlled Transdermal Delivery, in *Controlled Release of Bioactive Materials,* R. W. Baker, Ed., Academic Press, New York, 1980, p. 99.
10. Cowsar, D. R., Introduction to Controlled Release, in *Controlled Release of Biologically Active Agents,* Traquary, A. C. and Lacey, R. E., Eds., Plenum Press, New York, 1974, p. 1.
11. Crommelin, D. J. A., Formulation of biotech products, including biopharmaceutical considerations, in *Pharmaceutical Biotecnology,* Cromelin, D. J. A. and Sindelar, R. D., Eds., Hardwood Academic Publishers, UK, 1997, p. 71.
12. Florence, A. T., and Attwood, D., *Physicochemical Principles of Pharmacy,* 2nd ed., Chapman and Hall/Routledge, New York, 1988, p. 318.
13. Gibaldi, M., *Biopharmaceutics and Clinical Pharmacokinetics,* 4th ed., Lea & Febiger, Philidelphia, 1991 p. 40.
14. Gibaldi, M., and Perrier, D., *Pharmacokinetics,* 2nd ed., Marcel Dekker, New York, 1982, p. 189.
15. Groning, R., Weyel, S., Akala, E., Minkow, E., and Lambow, N., Computer-controlled release of oxprenolol from capsules using gas producing cells and electronic circuits, *Pharmazie,* 54(7), 510, 1999.
16. Gupta, P. K., and Robinson, J. R., Oral controlled release delivery, in *Treatise on Controlled Drug Delivery,* Kydonieus, A., Ed., Marcel Dekker, New York, 1992, p. 255.
17. Heilman, K., *Therapeutic Systems Rate-Controlled Drug Delivery: Concepts and Development,* 2nd rev. ed., Georg Thieme Verlag/Thieme-Straton, New York, 1984, p. 51.
18. Higuchi, T., Mechanism of Sustained-Action Medication, *J. Pharm Sci.,* 52, 1145, 1963.

19. Hui, H.-W., Robinson, J. S., and Lee, V.H.L., Design and fabrication of oral controlled release drug delivery systems, in *Controlled Drug Delivery, Fundamentals and Applications,* 2nd ed., Robinson, J. R. and Lee, V. H. L., Eds., 1987, p. 373.
20. Jantzen, G. M., and Robinson, J. R., Sustained and controlled release drug delivery systems, in *Modern Pharmaceutics,* 3rd ed., Banker, G. S. and Rhodes, C. T., Eds., Marcel Dekker, New York, 1996, p. 575.
21. Jost, W., *Diffusion in Solids, Liquids and Gases,* Academic Press, New York, 1960, p. 37.
22. Katz, B., Rosenberg, A., and Frishman, W. H., Controlled-release drug delivery systems in cardiovascular medicine, *Am. Heart J.,* 129(2), 359, 1995.
23. Langer, R., and Peppas, N. A., Chemical and Physical Structure of Polymers as Carriers for Controlled Release of Bioactive Agents: A Review. *JMS Rev. Macromol. Chem. Phys.,* C23(1), 61, 1983.
24. Lippoid, B. C., Controlled Release Products: Approaches of Pharmaceutical Technology, in *Oral Controlled Release Products: Therapeutic and Biopharmaceutic Assessment,* Gundert-Remy, U. and Moller, H., Eds., Wissenschaftliche Verlagsgesellschaft, Stuttgart, 1990, p. 39.
25. Mathiowitz, E., Kretz, M. R., and Brannon-Peppas, L., Microencapsulation, in *Encyclopedia of Controlled Drug Delivery,* Vol. 2, Mathiowitz, E., Ed., John Wiley & Sons, New York, 1999, p. 493.
26. Paul, D. R., Polymers for Controlled Release Technology, in *Controlled Release Polymeric Formulations,* Paul, D. R., and Harris, F. W., Eds., A.C.S. Symp. Ser. 33, Washington, DC, 1976, p. 1.
27. Peppas, N. A., Mathematical modelling of diffusion processes in drug delivery polymeric systems, in *Controlled Drug Bioavailability,* Vol. I, *Drug Product Design and Performance,* Smolen, V. F. and Ball, L. A., Eds., Wiley, New York, 1984, p. 206.
28. Ritschel, W. A., and Kearns, G. L., *Handbook of Basic Pharmacokinetics including Clinical Applications,* 5th ed., American Pharmaceutical Association, Washington, DC, 1999, p. 124.
29. Robinson, J. R., Controlled drug delivery: past, present, and future, in *Controlled Drug Delivery: Challenges and Strategies,* Park, K., Ed., American Chemical Society, Washington, DC, 1997, p. 1.
30. Robinson, J. R., Preface, in *Sustained and Controlled Release Drug Delivery Systems,* Robinson, J. R., Ed., Marcel Dekker, New York, 1978, p. iv.
31. Robinson, J. R., and Lee, T. W.-Y., Controlled-release drug-delivery systems, in *Remington: The Science and Practice of Pharmacy,* 20th ed., Gennaro, A. R., Ed., Lippincott Williams & Wilkins, Baltimore, 2000, p. 903.
32. Roseman, T. J., and Higucchi, W. I., Release of Medroxyprogesterone Acetate from a Silicone Polymer, *J. Pharm. Sci.,* 59, 353, 1970.
33. Shargel, L., and Yu, A., *Applied Biopharmaceutics and Pharmacokinetics,* 4th ed., Appleton & Lange, Stamford, CT, 1999, p. 169.
34. Williamson, J. S., Wyandt, C. M., What pharmacists need to know about breast cancer, *Drug Topics,* 146, 84–91, 2002
35. Wood, D. A., Polymeric Materials Used in Drug Delivery Systems, in *Materials Used in Pharmaceutical Formulation,* Florence, A. T., Ed., Blackwell Scientific Publishers, London, 1984, p. 71.

12 Oral Liquid Dosage Forms: Solutions, Elixirs, Syrups, Suspensions, and Emulsions

*William M. Kolling and Tapash K. Ghosh**

CONTENTS

* No official support or endorsement of this article by the FDA is intended or should be inferred.

0-4152-8863-0/05/$0.00+$1.50
© 2005 by CRC Press LLC

Learning Objectives

After studying this chapter the student will have thorough knowledge of:

- Definitions of different types of oral liquid dosage forms
- Advantages and drawbacks of oral liquids
- Formulation components of oral liquids
- Formulation considerations of oral liquids
- Manufacturing considerations of oral liquids
- Stability considerations of oral liquids

I. INTRODUCTION

Three traditional liquid dosage forms are used to administer pharmacological agents via the oral route. They are solutions, suspensions, and emulsions. Solutions used for the delivery of oral medications commonly fall into two subcategories, based on the solution components: elixirs and syrups. Many patients believe it is easier to swallow a volume of solution than a tablet or a capsule.

From a historical perspective, pharmacologic agents were frequently obtained from plants by using extraction techniques into potable solvents such as water and alcohol. The active agent was now in solution, and the patient ingested the required volume to receive the necessary amount of medication. These types of dosage forms are as old as pharmacy and the healing arts. As the pharmacologic agents became more potent and our understanding of the stability of those agents increased, the dosage forms also increased in sophistication. Today the Food and Drug Administration (FDA) and the United States Pharmacopeia (USP) set standards to guide pharmacists and the industry in the production of stable, efficacious dosage forms.

II. DEFINITIONS OF COMMONLY USED LIQUID DOSAGE FORMS

Every profession has its own language and pharmacy is no exception. Students may not be aware of it, but they are undergoing a socialization process. Their use of older more common words becomes more refined, and some words have completely new meanings. New words are also expanding their vocabulary. Therefore some pharmaceutical terms used throughout this chapter are now defined.

A. Solution

The USP states that "oral solutions are liquid preparations, intended for oral administration, that contain one or more substances with or without flavoring, sweetening, or coloring agents dissolved in water or cosolvent-water mixtures." A solution is a homogeneous, one-phase system, or product that has two or more components. An example is Syrup, NF 19. This product consists of water, the solvent, and sucrose, the solute. It is homogeneous; that is, wherever you take a sample from within a bottle of syrup, the concentration of sucrose would be the same. The solutes are present in a monomolecular dispersion; each molecule of sugar is surrounded by thousands of water molecules. Statistically, at the molecular level, you would not expect to find two sucrose molecules next to each other; rather they would be separated by many water molecules.

Another way of looking at this dosage form, as explained by Aulton (3), is to consider the solvent to be the phase in which the monomolecular dispersion occurs; the solutes are the components that are dispersed as individual molecules (un-ionized, e.g., sucrose) or ionic salts (individual ions, e.g., K^+ and penicillin V^-). Our definition above also included the phrase "one-phase system." That means that wherever dosage form is sampled, only a liquid can be obtained. There are no solids or immisible liquids present within the container.

The strengths of commercial liquids are usually expressed as milligrams/milliliter or milligrams/5 ml. Percentage expressions are frequently used that represent percent weight-in-volume. A 3% solution of hydrogen peroxide has 3 g of H_2O_2 in 100 ml of product. There are many oral solutions on the market. Some examples are propranolol hydrochloride 80 mg/ml, morphine sulfate 10 mg/5 ml, and fluoxetine hydrochloride 20 mg/5 ml.

B. Elixir

An elixir is a type of solution. Therefore, it is a homogeneous, one-phase product. Note that an elixir has three or more components. Two of the components are water and alcohol. Be aware that when the single word alcohol is used alone, without modifiers, in pharmacy it refers to ethyl alcohol only. An elixir is a solution since all of the components are present in one phase. Traditionally, alcohol has been used in products to help solubilize the active drug. In some products the alcohol is present for its own dubious pharmacologic action. Further discussion of this point can be found in the *Handbook of Nonprescription Drugs* (1). Because of the potential for adverse effects associated with the ingestion of alcohol, many manufacturers try to limit the amount of alcohol in their elixirs. Patients have experienced disulfiram-type reactions from the small amounts of alcohol in some products. All patients should be advised about the potential side effects of alcohol if it is present in a product they are taking. The term elixir is an old one in the history of pharmacy. Some products that no longer contain alcohol in their formulation continue to use the word elixir as part of their marketed name. An example is Tylenol Children's Elixir, an alcohol-free product. The point to remember is that the use of the word elixir does not necessarily mean the product contains alcohol. As with all dosage forms, the practitioner must be familiar with the product or read the list of ingredients prior to dispensing.

C. Syrup

A syrup is another type of solution. Like a solution or elixir, it is a homogeneous, one-phase product. Syrups can be medicated or nonmedicated. Medicated syrups contain three or more components. Water and the sweetening agent are two of the components. While sucrose has traditionally been used as the primary sweetener, other agents are being used as sweeteners today, including sorbitol, fructose, and aspartame. Most syrups contain a high proportion of sucrose, usually 60 to 80% (weight-in-volume). Concentrated sugar solutions are inherently stable and quite resistant to microbial growth. The most commonly used syrup is syrup NF, also known as simple syrup, which is prepared by dissolving 85 g of sucrose in enough purified water to make 100 ml of the syrup. In some extemporaneous preparation set ups, simple syrups are prepared and set aside in bulk. Medicinal agents are dispensed in the syrup according to the prescription.

Syrups are generally prepared by four general methods: (a) with the aid of heat, (b) by agitation, (c) by addition of sugar to a medicated liquid, and (d) by percolation of either the source of medicinal agent or sugar. In the interest of saving time, most syrups are prepared with the aid of heat. However, caution should be exercised against using excessive heat. Under excessive heat, sucrose, a disaccharide, may be hydrolyzed into the monosacharides, glucose (dextrose), and fructose (levulose). This process is known as *inversion of sugar*, and the combination of the two monosaccharides is called *invert sugar*. During inversion, the normally colorless syrup darkens owing to the effect of heat on the fructose portion of the invert sugar. When the syrup is even more overheated, it becomes amber colored and the product is called caramelized syrup.

Historically, syrups were sweet, viscous products that were used to help mask the taste of various drugs. However, during inversion and caramelization, the sweetness of the syrup is enhanced and the product becomes more susceptible to microbial growth. Therefore, syrups should never be sterilized by autoclaving. Today, we still use syrups to help mask the taste of certain agents, and the use of artificial sweeteners, flavors, colors, and viscosity enhancers can result in products that resemble the traditional syrup. Examples of medicated syrups include cetirizine hydrochloride 5 mg/5 ml and promethazine hydrochloride 6.25 mg/5 ml. Nonmedicated syrups (vehicles) include cherry syrup, cocoa syrup, orange syrup, raspberry syrup, syrup NF and Ora-Sweet and Ora-Sweet-SF.

D. Suspension

A suspension is a dispersion of insoluble drug particles (the disperse phase) in a liquid, usually water (the dispersion medium). This is at least a two-phase system since there is at least one kind of solid particle dispersed in a continuous fluid medium. Not all drugs are soluble in water at the concentration required to give a convenient volume for the patient to ingest. An example is the common antibiotic amoxicillin. A typical dose of amoxicillin for a 15 kg child is 400 mg. The solubility of amoxicillin trihydrate in water is approximately 4 mg/ml, as listed in the Merck Index. A child will not take 100 ml of an antibiotic solution two times a day, no matter what flavor it is, so the product is formulated as a suspension. A powder composed of fine particles contains the drug, buffers, colors, flavors, preservatives, stabilizers, sweeteners, and suspending agents. The powder mixture is packaged in an appropriate-size plastic container. When the product needs to be dispensed, the pharmacist adds the required volume of water to the container and shakes the resulting product to disperse the components. The resulting product is a suspension with a concentration of 400 mg/5 ml. Now the 400 mg dose can be administered in only 5 ml. Suspensions are useful for administering less soluble drugs in convenient volumes. Many antibiotics are available as oral suspensions, including cefaclor 125 mg/5 ml, cefixime 100 mg/5 ml, and erythromycin ethylsuccinate 200 mg/5 ml and 400 mg/5 ml. Most antacids are suspension formulations, including Mylanta, Riopan, and Maalox.

E. Emulsion

An emulsion is a two-phase system with at least three components. It is composed of an oil and water and an appropriate emulsifying agent. If water droplets are dispersed throughout a continuous oil phase, then it is a water-in-oil emulsion. If oil droplets are dispersed throughout a continuous water phase, it is an oil-in-water emulsion. Probably the most common emulsions are salad dressings. An Italian dressing is an oil-in-water emulsion. You can experiment in the laboratory, or at home, and observe that if you mix water and olive oil together, you may see some droplets of oil and water dispersed in each other. The system is quite unstable with respect to time, as the oil and water want to separate back to their starting positions, water on the bottom of the bottle and oil floating on top.

To form a stable emulsion from water and an oil requires an emulsifying agent. Emulsions for oral use are not that common in today's pharmacy practice. An example of an oral emulsion is an oil-in-water formulation of castor oil in water. One advantage of this product is that the emulsion may help to mask the taste and change the mouth-feel of the liquid. Pharmacists frequently see that organoleptic properties can play a major role in medication compliance. Although pharmacists do not

TABLE 12.1
Representative Oral Liquid Products

Description	Concentration of Active	Indication
Theophyline oral solution	80 mg/15 ml	Bronchial asthma
Fluoxetine HCl oral solution	20 mg/5 ml	Depression and obsessive compulsive disorder (OCD)
Haloperiodol oral solution	2 mg/ml	Neuropsychiatric conditions
Sodium fluoride oral solution	0.5 mg/ml	Prophylaxis for dental caries
Cimetidine HCl oral solution	300 mg/5 ml	Peptic ulcer
Docusate sodium oral solution	10 mg/ml	Laxative
Diphenhydramine syrup	12.5 mg/5 ml	Allergy
Albuterol sulfate syrup	2 mg/5 ml	Bronchodilator
Acetaminophen suspension	160 mg/ml	Fever reducer
Ibuprofen suspension	100 mg/5 ml	Fever reducer
Digoxin elixir	50 μg/ml	Congestive heart failure

see many oral emulsions in practice, emulsions are frequently used in topical skin care products and in some parenteral formulations. Some representative oral liquid formulations are listed in Table 12.1.

III. BIOPHARMACEUTICS OF SOLUTION, SUSPENSION, AND EMULSION DOSAGE FORMS

Before a medicinal agent that is present in a dosage form can exert a pharmacologic effect at the cellular level, the active molecule must be in solution. Recall that a solution is a monomolecular dispersion, and the active molecule is surrounded by solvent molecules, most frequently water. This point cannot be over-emphasized. For all dosage forms, before the drug can exert an effect, it must be able to interact with some type of cellular structure. For that to occur, the drug must be in solution. There are potent compounds that cannot be used as drugs because they do not have the requisite solubility in water.

This section contains a brief review of some general aspects of bioavailability for solutions, suspensions, and emulsions and explores some advantages and disadvantages of these liquids compared with other types of dosage forms.

A. Solutions

When a drug is dissolved in a liquid dosage form (solution, syrup, or elixir), the pharmacologic agent is present as a monomolecular dispersion. Once the dosage form is administered, depending on the pK_a of the compound, the drug may or may not precipitate in the stomach. If the drug precipitates in the stomach, then it will redissolve in the gastrointestinal tract when the pH rises. If the drug remains in solution in the stomach, then it will most likely precipitate in the gastrointestinal tract and slowly redissolve and be absorbed as a consequence of LeChatelier's principle. In either case, starting with the drug already in solution, rather than in a solid dosage form such as a tablet or capsule, avoids the processes of disintegration and deaggregation. In general, when drugs are administered as solutions, syrups, and elixirs, the time for the drug to reach the maximum concentration, t_{max}, in the plasma is shorter than for the same drug administered as a tablet or capsule. There can be exceptions to this guideline depending on the formulation of the solid dosage form and the rate-limiting step in the drug absorption process.

The subject of precipitation and subsequent dissolution is by no means simple because changes in an agent's solubility can result from the chemical interaction of the drug with other substances in the stomach or gastrointestinal tract. One classic example is the interaction of the tetracycline class of antibiotics with calcium ions, resulting in the formation of insoluble complexes that cannot be absorbed by the body. Physiologically, a delayed gastric emptying time might result in accelerated degradation of the active drug by stomach acid. Conversely, some patients have a deficit of hydrochloric acid in their stomachs and must be supplemented with a dilute HCl solution to help absorb certain compounds. As with many subject areas in pharmacy, and the diligent practitioner will be aware of the many exceptions to the general rule.

B. Suspensions

Drugs with limited solubility, or large dose requirements, are often formulated in a suspension dosage form. Absorption of the drug requires that the insoluble agent go into solution somewhere in the stomach or gastrointestinal tract. Unlike a solution, where the agent is present at a concentration below its solubility limit, in a suspension the drug is present in the dosage form at a concentration above its solubility limit. Since the drug will have some solubility (otherwise it would not be active), LeChatelier's principle guarantees that for each molecule in solution that is absorbed by the body, another molecule from the solid particle will dissolve in the gastrointestinal tract to keep the local concentration in the intestinal body fluid saturated. This process continues until all of the drug has dissolved and been absorbed. Because of this process, t_{max} is usually longer for a drug in suspension than a drug in solution. Other factors can play a role in the absorption process, such as physiological conditions and the physico-chemical properties of the drug, but this is a conservative general rule.

C. Emulsions

Oral emulsions are no longer used to deliver pharmacologic agents, other than the oil itself. The emulsions in use are oil-in-water preparations designed to mask the feel and taste of the oil. The oils may be used as stool softeners, laxatives, or nutritional support agents. The first two uses do not lead to absorption of the oil. Generally, the absorption, or bioavailability, of nutritional support agents is not monitored, unless there is some underlying pathophysiology of the small intestinal that might interfere with normal food and drug absorption.

IV. ADVANTAGES AND DRAWBACKS OF ORAL LIQUIDS

If the chosen pharmacologic agent is available in several dosage forms, the choice of a suitable one will depend on several factors including the age of the patient, the patient's physical condition, the patient's physiologic status, the desired speed of onset, and potential storage requirements. As expected, liquid formulations have several advantages and disadvantages associated with their use when compared to other dosage forms.

A. Advantages

1. The active agent is homogeneously dispersed throughout the product.
2. The active agent is in solution and does not need to undergo dissolution; therefore, the therapeutic response is generally faster than if a tablet or capsule dosage form is used for treatment.
3. The dose of the active agent is easily and conveniently adjusted by measuring a different volume.
4. Solutions may be swallowed by patients who have difficulty taking tablets or capsules, as might be the case with pediatric or geriatric patients.
5. Drugs such as potassium chloride that may cause ulceration to the mucosa in a tablet formulation avoid this side effect when present in solution.

B. Disadvantages

1. The active ingredients, when present in solution, are usually more susceptible to chemical degradation, particularly hydrolysis, than when they are in a solid dosage form.
2. As a consequence of item 1, the solution product has a shorter shelf life than the solid formulation.
3. Some pharmacologic agents taste or smell bad enough in solution that the patient has difficulty taking the medication.
4. Liquid dosage forms are heavier and take up more shelf space than corresponding solid dosage forms. If the container breaks, the product is irretrievably lost.
5. Liquid dosage forms may require special storage facilities in very cold or very hot conditions. One example of this involves taking medicine, such as an antibiotic suspension, on a lengthy trip. In one case the drug might need to be kept refrigerated, and in another case the patient may need to protect the drug from freezing.
6. The delivery of the dose depends upon the patient, or care-giver, measuring the proper volume. This can be a significant issue for vision-impaired patients, patients with arthritis, or patients unable to read the numbers on an oral dosing syringe or medicine cup.
7. Solutions are often susceptible to microorganisms, and therefore preservatives are frequently incorporated into the formulation. Some patients may be allergic to certain preservatives.

It would be incorrect to tally up the pluses and minuses of the solution dosage form and make a determination regarding its utility. As with so many other issues in pharmacy, the practitioner needs to understand the advantages and drawbacks of the dosage form and consider the needs of the patient when arriving at a final product selection.

V. ROUTES OF ADMINISTRATION FOR ORAL LIQUID DOSAGE FORMS

The title of this section may remind one of the question, "What color is an orange?" However, oral solutions, syrups, elixirs, suspensions, and emulsions are frequently given to patients by means other than the oral route. This section discusses several alternative routes of administration for oral solutions.

Manufacturers formulate these products for delivery to the patient via the oral route. As devices and procedures for patient care have scientifically and technically advanced, routes of administration are available where the use of liquid dosage forms has a significant advantage over trying to crush tablets or empty capsules and put the powders down a tube.

Depending on the practice site, a pharmacist may encounter patients with various tubes inserted into different locations within the gastrointestinal tract. Solutions are frequently given through nasogastric tubes, gastric tubes, or tubes placed in sequentially further locations within the intestinal tract. In each case, the pharmacist must be aware that the shortened residence time may affect the bioavailability of the drug. In general, drugs in solution will be more fully absorbed than solid dosage forms in patients whose intestinal tracts have been operated on, or otherwise compromised, and might be missing some segment. Patients with parts of their intestinal tract resected need careful monitoring to ascertain that proper absorption is occurring, irrespective of dosage form used for treatment.

Valproic acid solution (Depakene® syrup) has been administered into the lower colon via a rectal tube in patients with certain types of acute seizures. Paraldehyde, which is rarely administered orally because of its odor, has been administered via a rectal tube after extemporaneous mixing with olive oil. Paraldehyde must be mixed with a compatible oil. The oil must solubilize the paraldehyde and be absorbed for the drug to act. Therefore, the use of mineral oil is contraindicated when trying to deliver paraldehyde since mineral oil is not absorbed to any appreciable extent and any pharmacologic activity will be significantly delayed or nonexistent.

VI. FORMULATION COMPONENTS OF SOLUTIONS, ELIXIRS, SYRUPS, AND SUSPENSIONS

A. Solvents

1. Water

Water is the most commonly used solvent in pharmaceutical preparations. In routine use it lacks toxicity, is compatible with bodily fluids, and can dissolve most compounds that are used as pharmacologic agents. The USP specifies the type of water that pharmacists, and the pharmaceutical industry, must use in the preparation of products. It is important to remind the student, and sometimes the practitioner, that drinking water, bottled or from the municipal tap, is not covered by a compendial monograph and therefore cannot be used in the preparation of solutions, elixirs, syrups, or suspensions. Bottled distilled water cannot be used for the same reason. The compendial requirement for water used these preparations is Purified Water, USP. Purified water must meet requirements for total organic carbon and conductivity. Water used for compounding oral solutions and the reconstitution of oral suspensions must meet official standards.

Several years ago there was a controversy regarding the extemporaneous compounding of captopril for oral solution. Pharmacists from different parts of the country reported their experiences using regular tap water for making the solution. Interestingly, one municipality had a high concentration of metal ions in the water that resulted in accelerated degradation of the captopril and loss of activity. This is just one example why Purified Water, USP should be used in the pharmacy to prepare solutions and suspensions. Depending on where you practice, the local tap water supply may vary significantly in quality. It is not uncommon in many locations to see wide variations in chlorine content, undissolved solids, and metal ions, which can result in color changes and perceptible taste and odor changes. To guarantee that the products dispensed will be stable, Purified Water, USP must be used.

The term *solubility* has been used several times in this chapter, and this is a convenient place to list the descriptive terms used by the USP, and often found in the Merck Index, when describing the solubility of a compound (Table 12.2). The drawback of this table is that the data cannot be used to make the most concentrated solution. Only the range within which you can expect the compound to dissolve is provided.

Chapter 3 covered the topic of solubility, and we will emphasize here that pharmacologic agents are soluble because they are able to interact with a solvent through two principle mechanisms. The first mechanism applies to nonionic compounds that can be solubilized in water through

TABLE 12.2
Solubility: Relative Terms

Descriptive Term	Parts of Solvent Required for 1 Part of Solute
Very soluble	Less than 1
Freely soluble	1–10
Soluble	10–30
Sparingly soluble	30–100
Slightly soluble	100–1,000
Very slightly soluble	1000–10,000
Practically insoluble or insoluble	10,000 and over

dipole–dipole bonding, of which hydrogen bonding (H-bonding) is the most important type. This type of interaction is responsible for the solubility of sucrose in water. The second mechanism applies to the many drugs that are produced as salts. They interact with water through ion–dipole bonding. This type of interaction is responsible for the solubility of the ions of diphenhydramine hydrochloride in water.

2. Alcohol (Ethyl Alcohol and Ethanol)

The other common solvent used in oral liquid dosage forms is alcohol. In pharmacy, the term *alcohol* refers to ethyl alcohol (ethanol) only. Alcohol USP contains ethanol, C_2H_5OH, not less than 92.3% and not more than 93.8%, by weight, which corresponds to not less than 94.9% and not more than 96.0%, by volume. Ethanol is miscible with water, glycerin, propylene glycol, and polyethylene glycol 400. Alcohol is second to water in terms of pharmaceutical utility as a solvent. Water–alcohol mixtures can be very effective in solubilizing drugs.

The FDA has expressed regarding the amount of alcohol in products. For over-the-counter medications, the FDA has proposed that manufacturers limit the amount of alcohol in their products based on the proposed age of the typical user. For patients under 6 years of age, the recommended alcohol limit is 0.5%, for patients 6 to 12 years of age, the recommended limit is 5%, and for children older than 12 years and adults, the recommended limit is 10%. Alcohol remains an important formulation ingredient because some compounds cannot be effectively solubilized by other agents. When alcohol is present in formulations, the label usually lists the percentage of alcohol volume-in-volume in the formulation. A further discussion of alcohol in formulations can be found in Reference 13.

3. Glycerin

Glycerin USP is a clear, colorless, viscous liquid. It is miscible with water, alcohol, propylene glycol, and polyethylene glycol 400. It is a triol alcohol without the central nervous system depressant activity of ethanol. It has humectant and preservative properties. It is used in both internal and external preparations. The solubilizing properties of glycerin are comparable to alcohol, but the increased viscosity imparted to the final product may be an undesired outcome of the use of this solvent.

4. Propylene Glycol

Propylene glycol USP is a clear, colorless, viscous liquid. It too is miscible with water, alcohol, and polyethylene glycol 400. It is a diol, has no central nervous system activity, and like glycerin has humectant and preservative properties. It is used in both internal and external products. It is being used more often in modern formulations, possibly replacing glycerin.

5. Polyethylene Glycol 400

Polyethylene glycol 400 NF is a polymer composed of ethylene oxide and water. There are various molecular weights of the polyethylene glycols, and physically they range from a liquid to a waxy solid. Polyethylene glycol 400 is a clear, colorless, viscous liquid. Like glycerin and propylene glycol, polyethylene glycol 400 is miscible with water and alcohol. Polyethylene glycol 400 is a liquid at room temperature, and it is the most common polyethylene glycol used in drug product formulations. The chemical is used in both internal and external formulations.

Products are formulated in solution dosage forms through the use of two or more of the above solvents. These are referred to as co-solvent systems. If the product requires dilution prior to dispensing, the pharmacist must use a similar co-solvent mixture as the one used to formulate the solution. Consider the following product, which is a solution. The formula is provided below.

Drug X: 5 mg/ml
Polyethylene glycol 400: 60%
Propylene glycol: 10%
Ethanol: 19%
Water: qs

Approximately 90% of this solution is nonaqueous. The solubility of drug X in water is about 0.05 mg/ml (50 mcg/ml). If a physician wanted a product dispensed that had a drug X concentration of 2.5 mg/ml, how would you accomplish that task? If you did not know the solubility of drug X in each of the formulation components, your most prudent choice would be to dilute the product using the same mixture of polyethylene glycol 400, propylene glycol, and ethanol as used in the original formulation. If you tried using water alone to dilute the product, the final concentration of each component, except water, would be halved, and it is unlikely that the drug X would remain in solution. Further, you would have no idea how the stability of the product is affected by the addition of water to the formulation. So the answer to this problem is to make a stock solution of the three nonaqueous solvents and water and use that solution to dilute the product by a factor of two.

Here is another example: Digoxin pediatric elixir is formulated in a co-solvent system at a concentration of 50 mcg/ml. The solubility of digoxin in water is approximately 80 mcg/ml. The formula of the product is given below:

Digoxin: 50 mcg/ml
Ethanol: 10%
Water: qs

If the physician wanted a solution with a concentration of 25 mcg/ml, how would you make it? The solubility of digoxin in the solution is not a problem. Digoxin is soluble to the extent of 80 mcg/ml. If you reduce the concentration to 25 mcg/ml, the digoxin will still be soluble in water. You still need to be concerned with product stability. A useful reference to consult for stability information is the book by K. A. Connors (6), in which the monograph on digoxin has information that the proposed dilution is stable and soluble. You could first make a 10% ethanol solution in water and use that for the dilution, or based on the information in the monograph you could use water alone. In that case, the final concentration of alcohol in the formula would be 5%. So you could successfully make the dilution as requested by the physician. Whenever you need to manipulate a product through dilution or the addition of another solute, you must ascertain how the intervention will affect product stability.

B. Preservatives

Preservatives are chemical compounds that are added to formulations to protect them from microbial contamination. All the compounds used as preservatives have a concentration dependent activity. That means that if you dilute a commercial product with a diluent without a preservative you may be lowering the concentration of the preservative to a level below which it cannot exert its biologic effect. This is not a problem for a dose that will be used immediately, but if the product is intended to have a significant shelf life, then it might be left unprotected by the dilution. Preservatives must be compatible with the other ingredients in the product, and manufacturers test their products for chemical compatibility. Students or pharmacists interested in compounding should check other references for preservative suitability. Common preservatives for products intended for administration via the oral route include benzoic acid (sodium and potassium salts); sorbic acid (potassium salt); and the parabens, esters of para-hydroxybenzoic acid, butylparaben, ethylparaben, methylparaben, and propylparaben. Benzoic acid and sorbic acid are only effective as preservatives in their un-ionized state. Raising the pH above 5 will lead to the loss of antimicrobial preservative activity. This is another important example of the caution pharmacists must take when they manipulate a commercial product.

C. Antioxidants

Antioxidants are chemical compounds that inhibit oxidation. Oxidation can occur through the gain of oxygen, loss of hydrogen, production of an electron, or via free radical mechanisms. Oxidation of drugs can occur through several different steps, and chemicals have been found to interfere with the various mechanisms. The about dilution of antimicrobial preservatives holds for antioxidants also. Antioxidants for products intended for administration via the oral route include ascorbic acid, potassium and sodium metabisulfite, ascorbyl palmitate, butylated hydroxyanisole (BHA), butylated hydroxytoluene (BHT), and alpha-tocopherol. Chelating agents also interfere with oxidative processes by binding metal ions and include disodium edetate and edetic acid (EDTA).

D. Buffers

Many pharmacologic agents are either weak acids or weak bases. Their solubility in solution depends on their ionic state. To maintain the drugs in the solution state, buffers are sometimes used to control the pH and therefore keep the ionic state constant. In general, low buffer capacities are used to limit the exposure of the patient to large amounts of potentially detrimental salts. Low buffer capacities also help ensure that the local tissue pH is not adversely affected by the administration of the product. A further discussion of buffers can be found in Reference 10. Characteristics of some common excepients are listed in Table 12.3.

TABLE 12.3
Some Common Excepients for Oral Liquid Dosage Forms

Name	Usual Concentration	Water Solubility
	Preservatives	
Benzoic acid (Na and K salts)	0.1–0.3%	0.33–50%
Sorbic acid (K salt)	0.05–0.2%	0.1–22%
p-Hydroxybenzoic acid esters		
Methylparaben	0.05–0.25%	0.25%
Propylparaben	0.02–0.04%	0.04%
	Antioxidants	
Ascorbic acid	0.05–3%	33%
Sodium bisulfite	0.1%	25%
Sodium metabisulfite	0.02–1%	50%
Ascorbyl palmitate	0.01–0.2%	
Butylated hydroxyanisole (BHA)	0.005–0.01%	—
Butylated hydroxytoluene (BHT)	0.01%	—
Alpha-tocopherol	0.01–0.1%	—
	Chelating Agent	
Disodium edetate	0.1%	Very soluble

E. Flavors, Colors, and Sweetening Agents

Most oral liquid dosage forms are flavored with synthetic flavors or naturally occurring products such as vanilla, raspberry, orange oil, anise oil, or lemon oil. Many other flavors are used to make products more palatable. As with the other components, the flavoring agent must be compatible with the formulation ingredients and not adversely affect product stability. Some practitioners regularly alter the flavor of products to enhance the acceptability of the drug to their patients. Pharmacists should not do this routinely. Manufacturers are required to demonstrate product stability to the FDA, and pharmacists should not circumvent the extensive studies carried out by the companies. When changing the flavor is necessary for the successful treatment of the patient, care should be taken to limit the amount of product altered and to note a conservative expiration date on the product dispensed to the patient. The extemporaneous flavoring of products needs to proceed cautiously until compatibility and stability information become more readily available to the practitioner.

Colors are used to help improve the patient's acceptance of the product. Frequently, the color goes along with the flavor, as in a red cherry-flavored product. As with all ingredients, the coloring agent used must not adversely affect the formulation's stability. The initials FD&C in the list of ingredients of various common products, in addition to medicines, stand for food, drugs, and cosmetics. The FDA certifies the dyes for use for humans before they can be used in products.

Sweeteners are used to help mask the taste of the drug in a solution. The sweetener may also contribute to an increase in viscosity of the solution or the suspension. Sucrose has commonly been used, and Syrup NF is official in the current USP. Sorbitol and mannitol have also been used to impart sweetness, but they are less sweet than sucrose on a per weight basis. Aspartame (L-aspartic acid–L-phenylalanine methyl ester) is being incorporated into products as a sweetener. The FDA recommends that patients diagnosed with phenylketonuria limit their intake of aspartame. This is one example of why pharmacists must be aware of their patient's concurrent conditions and choose products accordingly.

The everyday common use of sucrose as a sweetening agent in oral liquids should not mislead you about its potential for problems in certain patient populations. Pharmacists must always think about the effects of the whole product, not only the active ingredient. Sucrose, the drug, and other components of the formulation can contribute significantly to the osmotic load of the product. The administration of hyperosmotic solutions to pediatric, geriatric, and sometimes adult patients can lead to diarrhea or gastrointestinal upset.

VII. FORMULATION CONSIDERATIONS

The following factors should be taken into consideration in order to formulate a liquid preparation that is potent, stable, and esthetically appealing.

A. Solubility

The solubility of the drug under consideration should be determined in a solvent that is similar to the one intended for use in the final product. Also, the formulation should be designed so that the solute(s) should remain in solution even at temperatures as low as 4°C. Water is always the first choice of solvent for most oral liquid preparations. However, many drugs do not have the required intrinsic solubility to be formulated at the required dosage strength. Therefore, some manipulation may become necessary to keep the required amount of drug in a given volume of solvent. Techniques include: (a) incorporation of additional solvent(s) known as co-solvents, (b) making the salt or ester of the drug, (c) changing the pH of the formulation to make the drug more ionizable and water soluble, and (d) adding surfactants above their critical micelle concentration to effect solubilization.

B. Preservation

Oral liquid products should be adequately preserved, as contamination may be incorporated into the formulation from raw materials, processing containers and equipment, manufacturing environment, operators, packaging materials, and the user. Choosing the right preservative that is compatible with the other formulation ingredients and effective at the formulation pH is extremely important. Often instead of using a single preservative, a blend of preservatives is recommended for the best result.

C. Stability

Both physical stability (color, odor, viscosity, clarity, taste, etc.) and chemical stability (degradation) of the formulation are important. Therefore care should be taken so that none of these characteristics are altered during the shelf life of the product.

All chemical compounds are susceptible to degradation. Practitioners are usually concerned with two factors: how fast the drug degrades and the consequences of degradation. All drugs in solution degrade faster the same drug in a solid formulation such as tablets or capsules. That is true even for optimal liquid formulations. The reason is that drugs in a monomolecular dispersion are readily available to undergo chemical reactions compared with the solid state. This is an important fact and should be a reminder that the pharmacist needs to use conservative expiration dates for extemporaneously prepared liquid products. Also, any manipulation of a commercially available liquid product, such as dilution or the addition of a compound to the solution, must be dated with stability issues in mind.

Several factors affect the degradation reaction rate, and pharmacists need to be aware of these, even if they do not have the stability data to interpret the influence of a particular factor. The pH of the solution can have a significant effect on the rate of drug degradation. The pH can play a major role in the solubility of the drug and other components of the formulation. Changing the pH can prematurely degrade the drug or lead to precipitation. Oral liquids frequently employ a mixture of solvents in the formulation. If the product needs to be diluted prior to dispensing, the pharmacist should try to use the same composition of solvents as in the manufacturer's formulation. Varying the solvent mixture from that used by the manufacturer can lead to accelerated degradation. Last, it is important to recall that exposing the product to light can have deleterious effects on its integrity. Ultraviolet, visible, and infrared radiation can all accelerate degradation.

The practitioner will find significant information on extemporaneous product stability in sources such as the *American Journal of Health System Pharmacy, Hospital Pharmacy,* and the *International Journal of Pharmaceutical Compounding.* The practitioner must remember that the pharmaceutical industry spends significant amounts of time and money determining stability profiles for their products, and pharmacists unaware of the issues regarding stability are not acting in the best interests of their patients.

The topics of chemical kinetics and drug stability have been reviewed by several authors and it will serve our purpose to briefly examine several points they have made regarding the stability of active components in determining their suitability in dosage forms.

1. Chemical degradation of the active drug may lead to suboptimal blood levels of the agent when using the appropriate dose. Therefore, interpretation of the clinical effect of the drug is confounded by lack of knowledge of the true dose administered to the patient.
2. Although the amount of drug lost by degradation is small, the degraded products may be toxic. This has been reported for penicillin and its derivatives. The degradation products can lead to severe allergic reactions depending on the amount of allergens present.

3. A decrease in bioavailability can occur that is not the result of degradation, but rather of a physical instability such as adsorption of the drug in solution to some component of the storage system (e.g., the cap or bottle). Many compounds have some degree of surface activity and may be lost through adsorption.
4. There may be obvious changes in the physical appearance of the dosage form. Patients do not like to see their medication turning colors or precipitating in the bottle. The patients' lack of confidence in their medication, or pharmacist, may interfere with the treatment plan.

D. Organoleptic Properties

Some standards are available for the selection of the right flavor and color for a given formulation. For example, if the active ingredient is sweet, it should be flavored in fruit or vanilla flavor and may be colored red. If the active ingredient tastes sour, it may have citrus flavor and orange color.

VIII. MANUFACTURING CONSIDERATIONS

The following factors should be taken into consideration for manufacturing an oral liquid preparation.

A. Selection of Raw Materials

All of the raw materials, including the active agents, any excipients, and the solvents should confirm to well-defined specifications in order to assure batch to batch uniformity.

B. Equipment and Operator

All equipment must be thoroughly cleaned and sanitized before use per the preset standard operating procedure, and proper records should be maintained for such cleanliness. A major source of microbial contamination is the processing equipment and operator. Head covers, gloves, and face masks should be worn during manufacturing.

C. Containers, Labeling, and Storage

Containers, usually glass or plastic bottles for oral liquids, hold the formulation and are in direct contact with the product. All containers must be clean before they are filled. For stability concerns, the container must not physically or chemically interact with the product so as to alter the strength, quality, or purity of the product beyond the official requirements. The manufacturer addresses all these concerns prior to the selection of a suitable container. The product is eventually shipped to pharmacies for dispensing. Frequently the shipping container is larger than the volume needed by the patient. This necessitates repackaging into proper containers for the patient by the pharmacist. Today, we usually repackage liquids into plastic, light-resistant bottles. The manufacturer's recommendation for packaging and dispensing should always be followed, since there may be products that should be dispensed in glass bottles. In all cases the caps should be closed tightly to avoid the loss, or premature degradation, of the drug or any of the components. The USP should be consulted regarding appropriate containers for the dispensing of drug products.

The original manufacturer's container carries an expiration date for the product in a particular container. When pharmacists repackage a product, they are exempt from the expiration date labeling requirement. For that reason, it is imperative that the pharmacist take the necessary precautions to preserve the strength, quality, and purity of repackaged drugs. It is the pharmacist's responsibility to place a suitable beyond-use date on the label that accompanies the product when it is dispensed to the patient. The USP states that "in the absence of stability data or information to the contrary, such date should not exceed 1) 25% of the remaining time between the date of repackaging and the expiration date on the original manufacturer's bulk container, or 2) a 6-month period from the date the drug is repackaged, whichever is earlier." Frequently patients will ask if they may use some medication they have at home that was dispensed for another person or for an earlier occurrence of the same condition. Upon further questioning they may remember that the product is two or three years old. For the safety of the patient and for treatment efficacy, the patient must not use old medication.

If the product is repackaged and held in the pharmacy prior to dispensing, the product should be stored in a humidity-controlled environment and at the temperature specified by the manufacturer. When the product is dispensed to the patient or care-giver, they must store the product as recommended by the manufacturer. Generally, it is best to store the product in a dark place, away from direct sunlight. The traditional bathroom medicine cabinet is usually a poor choice because of the changes in humidity, possible frequent exposure to light, and variations in temperature that occur on a routine basis.

A final note is necessary about the storage of pharmaceuticals. Drugs and all potentially toxic compounds should be stored in such a manner so that people unable to recognize the danger (children and mentally impaired individuals) do not have access to the products.

D. Quality Control/Quality Assurance

Like any other dosage form, oral liquid dosage forms have specifications for drug substance and drug products. To ascertain batch to batch uniformity and to ensure stability of the products over the recommended shelf life, manufacturers follow those specifications. An established and validated stability-indicating assay method is the key for quality control/quality assurance. Some of the parameters that are routinely monitored for liquid dosage forms are content uniformity, viscosity, pH, color, and odor. Most of the oral liquids also contain a preservative. The efficacy of the preservative over the shelf life also should be monitored. For suspensions and emulsions, the effect of storage on flow properties and particle size also should be monitored.

IX. PREPARATION AND MANUFACTURING OF LIQUID DOSAGE FORMS

Most solutions are prepared by simply dissolving the solutes (actives and inactives) in the solvent or solvent mixture. On an industrial scale, solutions are prepared in large mixing vessels with ports for mechanical stirrers. The vessels are generally thermostatically controlled to maintain a certain temperature if desired. The order of addition of components is fixed through product development and scale-up exercises.

X. CONCLUSION

The pharmacist has traditionally been the intermediary between the physician and the patient when medication is involved. Oral liquid dosage forms probably have a maximum frequency of extemporaneous dispensing. Therefore an in-depth knowledge of the drug product and formulation is essential to further the patient's well being.

HOMEWORK

1. What are the three traditional oral liquid dosage forms?
   (a) Solutions, emulsions, and syrups
   (b) Syrups, elixirs, and suspensions
   (c) Solutions, suspensions, and emulsions
   (d) Elixirs, emulsions, and suspensions
   (e) None of the above
2. Elixirs and syrups are two examples of suspensions.
   (a) True
   (b) False
3. Which of the following phrases describe the solution dosage form?
   (a) A homogeneous system.
   (b) The solute is in a monomolecular dispersion.
   (c) The product contains at least two components.
   (d) All of the above.
   (e) None of the above.
4. Which of the following phrases describe the elixir dosage form?
   (a) A homogeneous one-phase system.
   (b) The solute is ethyl alcohol.
   (c) The product contains isopropyl alcohol as a preservative.
   (d) All of the above.
   (e) None of the above.
5. Which of the following phrases describe the suspension dosage form?
   (a) Prior to shaking, it is a homogeneous two-phase system.
   (b) Following shaking, it is a homogeneous one-phase system.
   (c) The drug is present at a concentration above its solubility limit.
   (d) The drug is present at a low concentration but at high pH.
   (e) After shaking, it is a homogeneous system with at least two phases.
6. Medicated syrups have three, or more, components. Which of the following could be present in a medicated syrup?
   (a) A pharmacologically active chemical
   (b) Sucrose
   (c) Water
   (d) Artificial sweeteners and flavoring agents
   (e) All of the above
7. What are the three components present in all pharmaceutical emulsions?
   (a) Water, olive oil, and an oil soluble drug
   (b) Water, mineral oil, and an emulsifying agent
   (c) Water, an oil, and an emulsifying agent
   (d) Water, flavoring agents, and alcohol
   (e) None of the above
8. In general, all pharmacologically active agents must be in solution before they can exert their effect.
   (a) True
   (b) False

9. By definition, all products labeled with the word elixir contain ethyl alcohol.
   (a) True
   (b) False
10. The term t_{max} refers to:
    (a) The minimum amount of time the patient can wait before drinking an acidic beverage, such as orange juice or Coke.
    (b) The maximum dose of a drug that can be taken safely.
    (c) The amount of time required for the drug to be 100% renally eliminated.
    (d) The amount of time required for the drug to reach its maximum plasma concentration.
    (e) The amount of time required for LeChatelier's principle to assure complete drug absorption from a suspension.
11. One advantage of the solution type of dosage form over an enteric coated tablet is that the solution avoids potential mucosa ulceration.
    (a) True
    (b) False
12. In general, solution dosage forms have a longer shelf life than the same drug formulated as a capsule.
    (a) True
    (b) False
13. Which of the following routes of administration cannot be used for safe delivery of oral liquid solutions?
    (a) Via a nasogastric tube
    (b) Via a gastric tube
    (c) Rectally
    (d) Intranasally
14. Tap water can be used for the preparation of solutions and suspensions if:
    (a) The pH is within the required range of 5 to 7.
    (b) The metal ions in solution are present at a concentration of less than 0.01 M.
    (c) No chlorine is present in the water.
    (d) No fluoride ion is present in the water.
    (e) Incorrect. Tap water should never be used for pharmaceutical preparations.
15. The interaction of water with drugs containing nonionizable functional groups occurs primarily through ion–dipole interactions.
    (a) True
    (b) False
16. When the word *alcohol* is used in pharmacy, it refers specifically to:
    (a) Nothing. The particular alcohol must be specified for safety reasons.
    (b) Isopropyl alcohol, which is also called rubbing alcohol.
    (c) An ethanol–water mixture, 50% of each component, by volume.
    (d) Ethyl alcohol.
    (e) Methyl alcohol.
17. The FDA does not make any recommendations to manufacturers regarding the allowable concentration of alcohol in over-the-counter medications.
    (a) True
    (b) False
18. Which of the following solvents can be mixed with water, in any proportion?
    (a) Ethanol
    (b) Glycerin
    (c) Propylene glycol
    (d) Polyethylene glycol 400
    (e) All of the above

19. Which of the following solvents can be ingested orally with a formulation?
    (a) Ethanol
    (b) Glycerin
    (c) Propylene glycol
    (d) Polyethylene glycol 400
    (e) All of the above
20. Patients diagnosed with which of the following conditions should limit their intake of phenylalanine?
    (a) Diabetes
    (b) Hypercholesterolemia
    (c) The flu
    (d) Phenylketonuria
    (e) Measles

Answers: 1. (c); 2. (b); 3. (d); 4. (a); 5. (c) and (e); 6. (e); 7. (c); 8. (a); 9. (a); 10. (d); 11. (a); 12. (b); 13. (d); 14. (e); 15. (b); 16. (d); 17. (b); 18. (e); 19. (e); 20. (d).

CASES

CASE I

A pharmacist noted during dispensing that an oral liquid preparation stored in an amber-colored bottle appeared cloudy on the shelf. What parameters should he check before he dispenses that preparation?

Nature of the preparation: If it is a suspension or emulsion it should be inherently cloudy. However, if it is a solution, it should be clear.

Expiration date: If it has already expired, it should not be dispensed.

Formulation: If the preparation is a solution and within its shelf life, then the possible reason for cloudiness is microbial growth. Therefore he needs to look at the formulation to check if the right amount and choice of preservatives have been added to the formulation or not. Cloudiness may also appear for several other reasons, for example, leaching of unwanted substances from containers and closures, change of pH leading to precipitation, chemical reactions leading to degradation. In these cases a thorough investigation is necessary, and the manufacturing company should be consulted and informed about the situation.

CASE II

During formulation development, a scientist noticed some crystals deposited at the bottom of the bottle of an oral liquid solution preparation after 3 days. What may be the possible reason for this phenomena?

The phenomena may be attributed to different reasons. However, the most plausible one is that at the given concentration of the drug, the pH of the formulation was initially such that the entire amount was in solution. However, during the standing period, the pH might have shifted to a point at which the amount of drug exceeded the solubility limit in that solvent. That might have caused deposition of the excess drug at the bottom.

CASE III

In an extemporaneous dispensing facility, simple syrup NF is kept prepared for a few days as a vehicle in which to dispense other medications. The day after the syrup was prepared, a pharmacist noticed that the syrup was brownish instead of being colorless. Should he dispense the medication in this syrup?

No, because possibly caramelization has occurred in the syrup. Caramelized syrup is sweeter than normal syrup and more susceptible to microbial growth. Therefore, that syrup should be discarded and new syrup may be prepared paying attention not to over-heat the syrup during preparation.

REFERENCES

1. Allen, L. V., et al., Eds., *The Handbook of Nonprescription Drugs,* 12th ed., American Pharmaceutical Association, Washington, DC, 2000.
2. Ansel, H. C., Allen, L. V., and Popovich, N. G., *Pharmaceutical Dosage Forms and Drug Delivery Systems,* 7th ed., Lippincott Williams & Wilkins, Philadelphia, PA, 1999.
3. Aulton, M. E., *Pharmaceutics: The Science of Dosage Form Design,* Churchill Livingstone, London, 1988.
4. Banker, G. S., and Rhodes, C. T., Eds., *Modern Pharmaceutics,* 4th ed., Marcel Dekker, New York, 2002.
5. Budavari, S., et al., Ed., *The Merck Index,* 12th ed., Merck and Co., Whitehouse Station, NJ, 1996.
6. Connors, K. A., Amidon, G. L., and Stella, V. J., *Chemical Stability of Pharmaceuticals: A Handbook for Pharmacists,* 2nd ed., Wiley, New York, 1986.
7. *Drug Facts and Comparisons,* Wolters Kluwer, St. Louis, MO, March 2004.
8. Gennaro, A. R., Ed., *Remington: The Science and Practice of Pharmacy,* 20th ed., Lippincott Williams & Wilkins, Philadelphia, PA, 2000.
9. Khan, M. A., and Reddy, I. K., *Pharmaceutical and Clinical Calculations,* 2nd ed., Technomic, Lancaster, PA, 2000.
10. Martin, A. N., and Bustamante, P., *Physical Pharmacy,* 4th ed., Lea & Febiger, Philadelphia, PA, 1993.
11. Schoenwald, R. D., *Pharmacokinetic Principles of Dosing Adjustments,* Technomic, Lancaster, PA, 2001.
12. Shargel, L., and Yu, A. B. C., *Applied Biopharmaceutics & Pharmacokinetics,* 4th ed., Appleton & Lange, Stamford, CT, 1999.
13. Thompson, J. E., *A Practical Guide to Contemporary Pharmacy Practice,* Williams & Wilkins, Baltimore, MD, 1998.
14. *United States Pharmacopeia,* 26th rev., United States Pharmacopeial Convention, Rockville, MD, 2002.

13 Parenteral Routes of Delivery

Yon Rojanasakul and Carl J. Malanga

CONTENTS

0-4152-8863-0/05/$0.00+$1.50
© 2005 by CRC Press LLC

Learning Objectives

After completion of this chapter, the student will have a thorough knowledge of:

- The fundamental concepts and terminology associated with parenteral dosage forms and their administration
- The advantages and limitations of parenteral routes of drug delivery
- Different routes and sites of parenteral drug delivery
- The characteristics of various parenteral products and their components
- The importance of sterilization and methods employed to sterilize parenteral products
- The requirements for the storage and handling of parenteral products
- Parenteral drug compatibility and stability issues
- The calculations involved in routine preparation and administration of parenteral products

I. INTRODUCTION

Injection and infusion are the major clinical procedures used for the administration of parenteral products. Because parenteral administration circumvents the body's protective barriers, for example, the skin and mucous membranes, great care must be taken to avoid any possible contaminations or health hazards associated with the administration. All injectable parenterals are sterile products, and exceptionally high standards are promulgated for their preparation, quality control, and administration. Pharmacists are responsible for preparing and dispensing safe, high quality products as well as counseling patients regarding their drug therapy. Therefore, pharmacists must have a functional knowledge of the principles of parenteral drug administration, that is, their preparation and use, to ensure patient safety and positive patient outcomes. Discussed in this chapter are key aspects of parenteral drug administration and sterile product preparation. While intraocular injections are included under injectable parenteral products, ophthalmic preparations for topical administration are also sterile products but are discussed separately in chapter 16.

II. DEFINITIONS AND USES

Parenteral means a sterile preparation of drugs for injection through one or more layers of the skin or mucous membrane. Parenteral products are prepared scrupulously by methods designed to ensure that they meet regulatory requirements for sterility, pyrogens (microbial byproducts such as endotoxins and exotoxins), particulate matter, and other contaminants. The *United States Pharmacopeia* describes five classes of sterile preparations for parenteral use: solutions, suspensions, emulsions, and dry solids or liquid concentrates that are dissolved or dispersed in sterile vehicles prior to administration. These products can further be classified as small volume parenterals or large volume

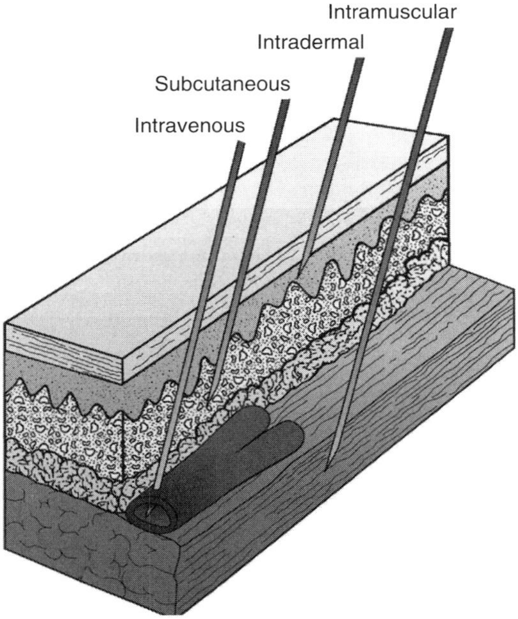

FIGURE 13.1 Routes of parenteral administration.

parenterals. Large volume parenterals are single-dose injections that contain more than 100 ml, whereas small volume parenterals are either single or multi-dose products that contain 100 ml or less.

Injection and infusion are the predominant methods of parenteral administration. Other methods of parenteral administration such as dialysis and irrigation are sometimes used. Administration by injection can be via different means depending on the intended therapeutic purpose. These means include injection into a vein (intravenous, i.v.), into a muscle mass (intramuscular), into subcutaneous tissue, within the superficial layers of the skin (intradermal), and into the cerebrospinal fluid (intrathecal). Some of the common routes of parenteral administration are depicted in Figure 13.1. An infusion involves the i.v. administration of a large volume parenteral fluid given over a prolonged period of time, usually ranging from 15 min to several hours. Dialysis and irrigation involve the use of specially formulated large volume parenteral fluids to remove biologically harmful entities from physiological fluids through an interposing semipermeable membrane (dialysis) or to flush, bathe, or moisten body cavities and wounds (irrigation).

III. ADVANTAGES AND LIMITATIONS OF PARENTERAL ADMINISTRATION

A. Advantages

There are several advantages associated with the parenteral administration of drugs:

1. It offers an alternative for patients who are unable to take medications orally, for example, owing to unconsciousness or vomiting. Fluids, electrolytes, and nutrients may also be given simultaneously to these patients, that is, by infusion.
2. Some drugs that are therapeutically inactive when administered orally, for example, owing to enzymatic degradation, poor absorption, or hepatic first-pass metabolism, can be given parenterally to avoid such problems.

3. Drugs that cause irritation or other local reactions in the gut or are unbearably unpalatable or nauseating can be administered parenterally.
4. Drugs given parenterally often have a faster onset of action and are generally more effective than those given orally.
5. Parenteral administration may sometimes be used to produce a local effect. The local anesthetic used by the dentist is injected near the trunk of a nerve to relieve pain in the immediate area.
6. Depots of drugs in long-acting delivery systems given by injection or implantation may offer prolonged or convenient therapy in certain circumstances. Examples of parenteral products and their indications are shown in Table 13.1.

B. Limitations

Like other methods of drug delivery, delivering drugs by parenteral administration is not without disadvantages. These include:

1. This method is inconvenient and requires proper techniques and trained personnel for administration.
2. Because drugs are injected directly into the tissue, there may be pain or discomfort associated with the administration. Thus, patient compliance may be a problem.
3. Once administered, a parenteral product cannot be removed. Thus, problems with overdose or adverse effects, if they occur, are difficult or impossible to reverse.
4. Parenteral products are more difficult and costly to produce than other dosage forms because they must conform to strict regulatory requirements for sterility, pyrogenicity, and particulate matter.

IV. TYPES OF PARENTERAL ADMINISTRATION

There are several routes of parenteral administration. A pharmacist must know each of these routes as well as their characteristics to be able to select specific products suitable for the intended use. Selected principles of administration by these parenteral routes are presented below and are summarized in Table 13.2.

A. Subcutaneous

A subcutaneous or hypodermic injection is intended for administration of small amounts of medication, typically 1 ml or less. Greater amounts may cause painful pressure and thus should be avoided unless they are administered very slowly. The injection of a drug is made into the loose tissue beneath the skin between the dermis and muscle. The skin may be pinched up during the injection to avoid deeper penetration of the needle into the muscle. Typical administration of drugs via the subcutaneous route includes insulin and pain medication injectables.

Drug absorption from a subcutaneous injection is usually slow owing to limited blood flow in the region. Consequently, the onset of action of subcutaneously administered drugs is slower than those given intravenously or intramuscularly. The duration of action is more prolonged but also depends on the drug's formulation. For example, drugs given subcutaneously that are formulated as suspensions have slower rates of absorption and act longer than those formulated as solutions. Insulin injection is a solution that upon subcutaneous injection has an onset of action of 1 h and a duration of action of 6 h, whereas insulin zinc suspension has an onset of 2 h and a duration of 24 h.

TABLE 13.1
Examples of Parenteral Drugs by Category

Category	Type	Commercial Products
		Antibiotics
Penicillins	Natural	Penicillin G
	Penicillinase-resistant	Nafcillin (Unipen®, Wyeth)
	Aminopenicillins	Ampicillin, ampicillin/sulbactam (Unasyn®-Pfizer)
	Broad-spectrum	Ticarcillin (Timentin®-GlxSmKl), piperacillin (Pipracil®-Lederle)
Cephalosporins	First generation	Cefazolin (Kefzol®-Lilly), cephadine (Velosef®-Bristol-Meyers Squibb), cephapirin (Cefadyl®-Apothecon)
	Second generation	Cefoxitin (Mefoxin®-Merck), cefamandole (Mandol®-Lilly), cefotetan (Cefotan®-AstraZeneca), cefuroxime (Kefurox®-Lilly)
	Third generation	Cefotaxime (Claforan®-Aventis), cefoperazone (Cefobid®-Pfizer), ceftaxidime (Fortaz®-GlaxoSmithKline), ceftizoxime (Cefizox®-Fujisawa), cCeftriaxone (Rocephin®-Roche)
Aminoglycosides		Gentamicin, amikacin (Amikin®-Geneva), kanamycin (Kantrex®-Geneva), tobramycin (Nebcin®-Lilly)
Tetracyclines		Doxycycline (Vibramycin®-Pfizer), minocycline (Minocin®-Lederle)
Macrolides (Erythromycins)		Azithromycin (Zithromax®-Pfizer), erythromycin (Erythrocin®-Abbott)
		Other Anti-Infectives
Antifungals		Amphotericin B (Fungizone®-Apothecon), itraconazole (Sporanox®-Ortho), fluconazole (Diflucan®-Roerig)
Antivirals		Acyclovir (Zovirax®-GlaxoWellcome), ganciclovir (Cytovene®-Roche), aidovudine (AZT; Retrovir®-GlxSmKl)
Antiprotozoals		Metronidazole (Flagyl®-Searle), pentamidine (Pentam®-Lyphomed)
Combination products		Trimethoprim/sulfamethoxazole (various; Bactrim®-Roche)
		Cardiovascular Drugs
Drugs acting through adrenergic receptors	α,β-Adrenergic agonists	Epinephrine (various; Adrenalin®-ParkeDavis), norepinephrine (Levophed®-Sanofi Winthrop)
	α,β-Adrenergic blockers	Labetalol (various; Normodyne®-Key)
	α-Adrenergic agonists	Metaraminol (Aramine®-Merck), methoxamine (Vasoxy®-GlaxoWellcome)
	α-Adrenergic blockers	Phentolamine (Regitine®-Ciba)
	β-Adrenergic agonists	Isoproterenol (various; Isuprel®-Sanofi Winthrop), dobutamine (Dolbutrex®-Lilly)
	β-Adrenergic blockers	Propanolol (Inderal®-WyethAyerst), esmolol (Brevibloc®-Baxter), metoprolol (Lopressor®-Novartis)
	Cardiac glycosides	Digoxin (various; Lanoxin®-GlaxoWellcome), antidote: digoxin immune fab (Digibind®-GlxSmKl)
Drugs affecting cardiac strength and rhythm	Inotropic agents	Inamrinone (Abbott Hosp), milrinone (Primacor®-SanofiWinthrop)
	Antiarrhythmics	Quinidine gluconate, procainamide, lidocaine (Abbott; Xylocaine®-Astra), phenytoin (various; Dilantin®-ParkeDavis), bretylium (various), verapamil (Isoptin®-Knoll), atropine (various)
Antihypertensives		Methyldopate (Aldomet®-Merck), hydralazine, nitroprusside, diazoxide (Hyperstat®-Schering), enalapril (Vasotec®-Merck)

TABLE 13.1 (continued)
Examples of Parenteral Drugs by Category

Category	Type	Commercial Products
		CNS Drugs
Analgesics	Narcotics	Morphine, hydromorphone (various; Dilaudid®-Knoll), meperidine (various; Demerol®-Abbott), antagonist: naloxone (various; Narcan®-Dupont)
	Mixed narcotic agonist-antagonists	Butorphanol (Stadol®-Mead Johnson), nalbuphine (various; Nubain®-Dupont), pentazocine (Talwin®-Sanofi Winthrop)
Sedatives and hypnotics	Barbiturates	Phenobarbital (various; Luminal®-Sanofi Winthrop), amobarbital (Amytal®-Lilly), pentobarbital (Nembutal®-Abbott), thiopental (Pentothal®-Abbott)
Sedative/anxiolytics	Benzodiazepines	Diazepam (various; Valium®-Roche), chlordiazepoxide (Librium®-Roche), lorazepam (various; Ativan®-WyethAyerst), midazolam (Versed®-Roche), antagonist: flumazenil (Romazicon®-Roche)
Antiemetic/adjuncts to anesthesia/analgesia	Miscellaneous	Droperidol (various; Inapsine®-Taylor), promethazine (various; Phenergan®-WyethAyerst)
Anticonvulsants	Hydantoins	Phenytoin (various; Dilantin®-ParkeDavis)
		Cytokines and Hematopoietic Factors
Interleukins	IL-2	Aldesleukin (Proleukin®-Chiron)
	IL-11	Oprelvekin (Neumega®-Genetics Institute)
Interferons	INF-α	INF-α2a (Roferon A®-Roche), INF-α2b (Intron A®-Schering), INF-αn3 (AlferonN®-Interferon Sci), INF-αcon-1 (Infergen®-Amgen)
	INF-β	INF-β1a (Avonex®-Biogen), INF-β1b (Betaseron®-Berlex)
	INF-γ	INF-γ1b (Actimmune®-InterMune)
Hematopoietic factors	G-CSF	Filgrastim (Neupogen®-Amgen)
	GM-CSF	Sargramostim (Leukine®-Immunex)
	Erythropoetin	Epoetin α (EPO;Epogen®-Amgen; Procrit®-Ortho)
		Hematologic Agents
Anticoagulants		Heparin (various); antidote: protamine SO_4 (various)
Thrombolytic agents		Streptokinase (Streptase®-Astra), anistreplase (Eminase®-Roberts), alteplase (Activase®-Genentech), urokinase (Abbokinase®-Abbott), reteplase (Retevase®-Boehringer Mannheim)
Hemostatics		Aminocaproic acid (various, Amicar®-Immunex), factor VIII (AHF, various, Kogenate®-Bayer), factor IX (Mononine®-Aventis), desmopressin (various, DDAVP®-Aventis)
		Hormones
Corticosteroids		Hydrocortisone (various, Solu-Cortef®-Upjohn), dexamethasone (various, Decadron®-MSD), methylprednisolone (various, Solu-Medrol®-Upjohn)
Estrogens		Estradiol valerate in oil (various, Delestrogen®-Bristol-Myers Squibb)
Growth hormones		Somatotropin (various), aomatrem (Protropin®-Genentech)
Insulin		Insulin (Humulin®-Lilly, Novolin®-NovoNordisk); Antidote: glucagon rDNA (Emergency Kit-Lilly)
Pituitary agents		Vasopressin (various, Pitressin®-Monarch), oxytocin (various, Pitocin®-ParkeDavis), gonadorelin (Factrel®-Wyeth Ayerst)

TABLE 13.1 (continued)
Examples of Parenteral Drugs by Category

Category	Type	Commercial Products
		Others
Antiemetics		Metoclopramide (various, Reglan®-Robins), diphenhydramine (various, Benadryl®-ParkeDavis), prochlorperazine (various, Compazine®-Smith-Kline Beecham)
Diuretics	Loop diuretics	Furosemide (various, Lasix®-Hoechst Marion Roussel), bumetanide (various, Bumex®-Roche), ethacrynate (Edecrin®-Merck)
	Thiazide diuretics	Chlorothiazide (Sodium Diuril®-Merck)
		Mannitol (various, Osmitrol®-Baxter)
	Osmotic diuretics	Immunoglobulin IV (various, Sandoglobulin®-Novartis)
Immune modulators	Immunostimulants	Azthioprine (vVarious, Imuran®-GlaxoWellcome), cyclosporine (Sandimmune®-Sandoz), lymphocyte immune globulin (Atgam®-Pharmacia)
	Immunosuppressants	

TABLE 13.2
Major Parenteral Routes and Their Administration

	Injection Site	Systemic Absorption	Injection Volume	Syringe Size	Needle Size and Length	Examples
Subcutaneous	Subcutaneous fat tissue beneath the skin	Low	<1 ml	1 ml	25 to 30 gauge, 1/2 to 5/8 in length	Insulin, histamine phosphate
Intradermal	Just below the skin between the epidermis and dermis	Low	0.02 to 0.5 ml	1 ml	25 to 28 gauge, 3/8 to 5/8 in length	Tuberculin, histoplasmin
Intramuscular	Muscle mass	Variable	Adults: 2 to 5 ml Children: <1 ml	1 to 5 ml	20 to 22 gauge, 1/2 to 1½ length	Antibiotics, toxoids
i.v.	Veins	Direct	Adult: <3 l/day Children: less	1 to 60 ml	20 to 22 gauge, 1/2 to 1½ length	Large volume parenteral fluids, lidocaine, digoxin

B. Intradermal

A small amount of drug solution, typically less than 0.2 ml, is injected between the layers of the skin. Drug absorption from an intradermal site is slow but is useful when local action or limited absorption from the injection site is required. Intradermal injection is primarily used for sensitivity and immunological testing. Examples of intradermal injections include tuberculin skin test agent, diphtheria toxin, procaine, and various allergens.

C. Intramuscular

An intramuscular injection is made deep into the layers of muscle. Drug absorption following an intramuscular injection is rapid for an aqueous solution but slow for suspensions and oily solutions. The onset of intramuscularly administered drugs is slower than those given intravenously. The maximum blood concentration of an aqueous solution of penicillin following an intramuscular injection is about 30 min, while its duration of action is about 4 h. A much more prolonged action of penicillin, that is, over 48 h, can be achieved if formulated in an oily vehicle.

Common intramuscular injection sites include the deltoid (arm), gluteus maximus (buttocks), and vastus lateralis (top of leg). These sites are relatively free of major nerves and blood vessels. The volume administered by intramuscular injection is limited by the mass of the injected muscle. Because of its larger size, a large volume of up to 5 ml can be administered into the gluteal muscle, while not more than 2 ml is recommended for the smaller deltoid muscle. For children, proportionally less volume is used according to age and muscle size. For small children, the vastus lateralis is the recommended muscle for intramuscular injection since it is the largest muscle in children under 3 years of age and it is free of major nerves and vessels. It is important to note that the maximum volume for intramuscular injection in children less than 3 years of age is 1 ml.

D. Intravenous

The primary use of an i.v. injection is to provide a rapid and immediate drug action. An i.v. injection is made directly into a vein. Volume of injection is less of a problem with i.v. therapy. However, there are certain restrictions associated with the administration, for example, the maximum volume is 3 l per day for adults and less for children. Because no absorption process is involved, an exact dose or blood concentration of the drug can be obtained with an accuracy and rapidity not possible by any other methods of administration. The immediate response from an i.v. injection is essential in emergencies. This route of administration is also used for fluid, electrolyte, and nutrient supplement. In cases of shock and severe bleeding or dehydration, the only way of rapidly increasing the blood volume and restoring the electrolyte–water balance is by the i.v. administration of parenteral fluids. Comatose patients can be sustained by the i.v. administration of parenteral nutrition fluids.

i.v. administration is subdivided into three categories: (a) intravenous bolus or i.v. push, (b) intermittent infusion, and c) continuous infusion.

1. Intravenous Bolus

The drug solution (1 to 2 ml) is injected directly into a vein via a syringe and needle or i.v. catheter tubing that goes into a vein. This is done over a short period of time (seconds to a few minutes) and can be repeated at intervals. This method of delivery is most commonly used in emergency situations when immediate action of the drug is required. Because the drug is given in a bolus over a short period of time, problems associated with drug toxicity may occur and must therefore be carefully monitored. Problems associated with drug irritation are also more commonly observed with this type of administration since there is no dilution of the drug product prior to the injection, as is the case in other types of i.v. administration.

2. Intermittent Infusion

The drug is diluted in an intermediate volume of parenteral fluid (25 to 100 ml) and is infused over a moderate period of time (15 to 60 min) at spaced intervals, for example, every 6 h. This method of administration provides a safer form of drug administration than the i.v. bolus but is less convenient. Compared to the continuous infusion method, this method provides a less consistent drug plasma level and may not be used to provide adequate amounts of fluids and electrolytes. Intermittent infusion may be administered through a secondary i.v. set or *piggyback*. In this method, a drug solution is administered intermittently through an established primary i.v. system (Figure 13.2). The advantage of the piggyback method is that it allows administration of a second drug solution without having to make another venipuncture. Because the drug from the intermittent infusion container comes to contact with the primary i.v. solution below the piggyback injection port, it is important that the secondary drug solution and the primary solution are compatible.

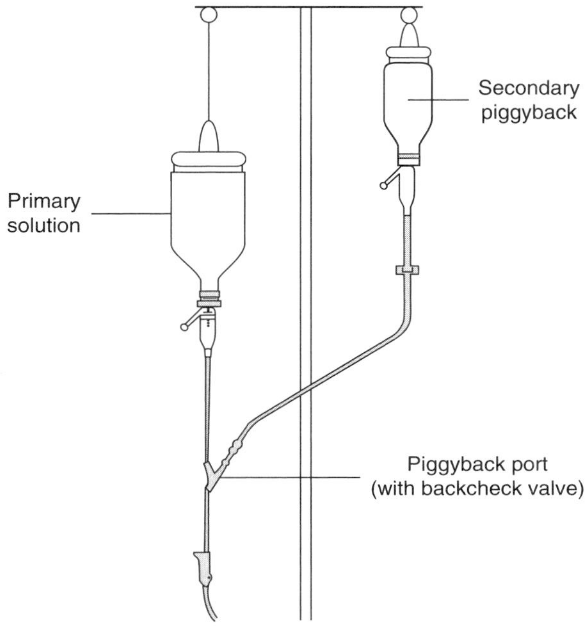

FIGURE 13.2 Primary intravenous system with piggyback.

3. Continuous Infusion

The drug is added to a large volume parenteral fluid (up to 1000 ml), and the solution is then slowly and continuously administered into a vein. This method allows fluid and drug therapy to be administered simultaneously. It provides an excellent control of drug plasma levels over a prolonged period of time. Drug plasma levels can be easily controlled by adjusting the infusion flow rate, and constant drug plasma levels can be achieved by fixing the continuous flow rate. Problems associated with drug toxicity and irritation are also minimized with this method of administration. Potential disadvantages of the continuous infusion method include: (a) It requires greater monitoring because it runs continuously; (b) certain unstable drugs cannot be administered by this means because of the extended run times; (c) it cannot be used in fluid-restricted patients because of the large volume of administration; and (d) it cannot be used to admix poorly soluble drugs dissolved in water-miscible solvents or complex hydroalcoholic solutions.

E. Others

There are many other types of injections that are intended for specific purposes. An *intrathecal* injection is intended to provide a high therapeutic concentration of a drug in the cerebrospinal fluid. Most drugs do not reach cerebrospinal fluid when administered by other injection methods and must therefore be injected intrathecally or directly into the cerebrospinal fluid. Drugs administered intrathecally include antineoplastics, antibiotics, antiinflammatory, and diagnostic agents. An *intra-articular* injection is made into the synovial cavity of a joint, usually to obtain a local therapeutic effect, for example, an intraarticular injection of a corticosteroid to achieve an antiinflammatory action in an arthritic joint. An *intracardiac* injection is made directly into the heart chamber. A common procedure is the injection of epinephrine solution to stimulate heart muscle as an emergency measure in cardiac arrest. An *intraperitoneal* injection is made into the peritoneal cavity, that is, the abdominal cavity holding the viscera. This route of administration is more commonly used in animal experimentation because the peritoneal cavity is large and provides a large absorbing area from which the drug is rapidly absorbed into the circulation. An *intraarterial* injection is made directly into an artery that has been surgically isolated within a deep seated part of the body prior to injection. This route of administration is used if it is necessary to deliver a high concentration of drug, for example, antineoplastic agents, to a diseased organ with minimal distribution to other systemic locations. An *intraocular* injection is made directly into the eye, for example, into the vitreous humor, in cases where topical application of drugs, such as antivirals or antibiotics for ocular infections, does not provide adequate drug permeation to the affected sites within the eye.

V. PRODUCT CHARACTERISTICS AND COMPONENTS

Various forms of parenteral products are available for use. These include solutions, suspensions, emulsions, concentrated liquids, and lyophilized solids. Lyophilized solids include drugs that are not stable in liquid media and are prepared by a freeze-drying process. These solids are reconstituted with an appropriate vehicle, for example, sterile water for injection, before administration to form solutions or suspensions. Among the various forms of drugs, the solution is the fastest acting form and is therefore often used to provide an immediate drug action. The suspension form is formulated when a delayed action is desired or to improve drug stability. Drugs that are oily or insoluble in an aqueous medium are often formulated as emulsions. All parenteral products must be free of microorganisms, pyrogens, and irritation. With large volume parenterals, the pH and tonicity of the fluids should be physiologically compatible with the body fluids. With small volume parenterals, the fluids need not be isotonic with the body fluids since the large volume of blood and other body fluids rapidly dilutes them. The pH of small volume parenterals can also vary appreciably from the physiological pH because the blood buffering system rapidly adjusts and maintains the blood at normal pH.

Parenteral products are made up of multiple components with individualized amounts of each. These components and their characteristics must be considered when formulating or mixing the products. The active ingredients in parenteral products must be of exceptionally high quality. Purity specifications must be indicated and should include requirements for potency and purity of the ingredients, level of specific trace contaminants, physicochemical characteristics, and freedom from microbial and pyrogenic contamination. In addition to the active ingredient, most parenteral products contain several other added substances. A number of these substances and their function are listed in Table 13.3.

A. Vehicles

Water is the most commonly used vehicle for parenteral preparations. There are several types of water for parenteral preparations:

- *Water for Injection, USP* is pyrogens-free, purified water and contains no more than 1 mg of trace elements in 100 ml. It is prepared by distillation or reverse osmosis and is intended to be used within 24 h after its collection. The USP monograph does not require this water to be sterile when used as a vehicle in the manufacture of a bulk product. However, the final product requires terminal sterilization.
- *Sterile water for injection, USP* is water for injection that has been sterilized. It is stored in single dose containers no larger than 1 l. It may contain slightly more than 1 mg of trace elements in 100 ml because the glass container may leach small quantities of its content during the sterilization process. This water is commonly used as a vehicle or diluent for dry-powder and liquid-concentrate injectable products.
- *Bacteriostatic water for injection, USP* is sterile water for injection that contains one or more antimicrobial agents. It is packaged in glass vials or prefilled syringes containing no more than 30 ml. It is mainly used as a vehicle for the preparation of small volume

TABLE 13.3
Pharmaceutical Additives Used in Parenteral Products

Function	Agent	Concentration (%)
Preservatives	Benzalkonium chloride	0.01
	Benzyl alcohol	2.0
	Chlorobutanol	0.5
	Chlorocresol	0.1–0.3
	Cresol	0.3–0.5
	Methyl paraben	0.18
	Propyl paraben	0.02
	Phenol	0.5
	Phenylmercuric nitrate	0.002
	Thimerosal	0.01
Antioxidants	Ascorbic acid	0.02–0.1
	Butyl hydroxyanisole	0.005–0.02
	Butyl hydroxytoluene	0.005–0.02
	Sodium bisulfite	0.1–0.15
	Sodium formaldehyde sulfoxylate	0.1–0.15
	Thiourea	0.005
	Tocopherol	0.05–0.075
Chelating agents	Ethylenediamine tetraacetic acid salts	0.01–0.075
Buffering agents	Acetic acid and a salt	1–2
	Citric acid and a salt	1–3
	Acid salts of phosphoric acid	0.8–2
Tonicity agents	Dextrose	5
	Sodium chloride	0.9

injectables. Its use in large volume injectables is limited because of the potential toxicity associated with the antimicrobial agents. This water is contradicted for neonates and for the administration of intraspinal or epidural drugs. In addition, certain drugs are incompatible with antimicrobial agents and must be prepared or reconstituted with sterile water for injection. Because of the presence of antimicrobial agents, however, the bacteriostatic water for injection is popularly used as a vehicle in multiple-dose vials.

- *Sodium chloride injection, USP* is a sterile isotonic solution of sodium chloride (0.9%) in water for injection. It contains no antimicrobial agents and is often used as a vehicle in preparing parenteral solutions and suspensions.
- *Bacteriostatic sodium chloride injection, USP* is a sterile sodium chloride injection that contains antimicrobial agents. Like the bacteriostatic water for injection, this water may not be packaged in a container that hold more than 30 ml. Because of the presence of antimicrobial agents, care must be exercised to ensure the compatibility of the added substances.

1. Water-Miscible and Nonaqueous Vehicles

While water is most frequently used in preparing parenteral products, its use may be precluded in formulations that contain water-insoluble drugs or drugs that are susceptible to hydrolysis. In such cases, water-miscible solvents or nonaqueous vehicles may be used. Water-miscible solvents include propylene glycol, glycerin, ethyl alcohol, and polyethylene glycol. When these solvents are used as vehicles, the preparations should not be diluted with water or precipitation may occur. For example, phenytoin sodium injection contains phenytoin that is solubilized in a water-miscible solvent at a pH of 12. If this injection is added to a large volume i.v. solution, precipitation of the drug will occur. Nonaqueous vehicles such as fixed vegetable oils may sometimes be used to dissolve water-insoluble drugs. Fixed oils commonly used in parenteral preparations include corn oil, sesame oil, cottonseed oil, and peanut oil. Mineral oil cannot be used as a parenteral vehicle because it is not absorbed from the body tissues. Fixed oils may be used to form an oily suspension of water-soluble drugs. Such a preparation provides a prolonged release of drugs from the site of administration. As some individuals are allergic to specific vegetable oils, the specific oil used in the product must be stated on the label.

B. Buffering Agents

Most drugs and pharmaceutical additives are either weak acids or bases, and are therefore subjected to partial ionization under a given pH. Buffering agents are used to adjust and maintain the pH of solutions in order to increase stability, solubility, absorption, and activity of specific ingredients in the system. Buffers commonly used in parenteral systems include phosphates, citrates, acetates, and glutamates. The acceptable pH range of parenteral products is 3 to 10.5. Extreme pHs may cause complications, especially when administered locally in large volumes. For example, pH above 9 can cause tissue necrosis, while pH less than 3 can induce severe pain.

C. Preservatives

Preservatives are substances added to dosage forms to protect them from microbial contamination. They are used primarily in multiple-dose containers to inhibit the growth of microorganisms that may be introduced inadvertently during or subsequent to the preparation. Preservatives must be compatible with other ingredients in the formula. Commonly used preservatives for parenteral products include benzyl alcohol, phenol, chlorobutanol, phenylmercuric nitrate, thimerosal, benzethonium chloride, benzalkonium chloride, and parabens.

D. Antioxidants and Chelating Agents

Antioxidants are substances that prevent or inhibit oxidation. They are added to dosage forms to protect components of the dosage form, which are subject to chemical degradation by oxidation. Antioxidants can be classified as true antioxidants (oxygen radical scavengers) and reducing agents. Examples of the former are α-tocopherol, butylated hydroxyanisole, and butylated hydroxytoluene, while those in the latter category include ascorbic acid and sodium bisulfite. Chelating agents are compounds that can form complexes with metal ions, and in so doing inactivate the catalytic activity of the metal ions in the oxidation process. Examples of chelating agents are ethylenediamine tetraacetic acid (EDTA) and disodium edetate.

E. Pharmaceutical Examples

Rx *Midazolam HCl Injection* (Versed®, Roche)
- Midazolam hydrochloride: 1 mg
- Benzyl alcohol: 1 %
- Disodium edetate: 0.01 %
- Sodium chloride: 0.8 %
- Sterile water for injection, USP *q.s.:* 1 ml

Midazolam hydrochloride is a short-acting benzodiazepine central nervous system depressant available as a sterile, nonpyrogenic parenteral dosage form for i.v. or intramuscular injection. The preparation also contains benzyl alcohol (as a preservative), disodium edetate (a chelating agent), and sodium chloride (a tonicity agent) in sterile water for injection. The pH of the solution is adjusted to approximately 3 with hydrochloric acid and, if necessary, sodium hydroxide.

Rx *Methyldopate HCl Injection* (Aldomet®, Merck)
Methyldopate hydrochloride: 250 mg
Methyl paraben: 7.5 mg
Propyl paraben: 1 mg
Sodium bisulfite: 16 mg
Monothioglycerol: 10 mg
Disodium edetate: 2.5 mg
Citric acid anhydrous: 25 mg
Sterile water for injection, USP *q.s.*: 5 ml

Methyldopate hydrochloride is an antihypertensive agent for i.v. use. Methyl and propyl parabens are added as preservatives, while sodium bisulfite and monothioglycerol are added as antioxidants. Disodium edetate serves as a chelating agent and citric acid as a buffering agent.

VI. DRUG PREPARATION AND CONTAINERS

A. Drug Preparation

1. Small-Scale Preparations

Small-scale preparations and i.v. admixtures are commonly prepared by pharmacists. Many pharmacists, particularly those working in hospitals or home health care practices, routinely handle and prepare IV admixtures and small-scale parenterals. These pharmacists require special knowledge and training and are responsible for following strict standards of practice when preparing or mixing parenteral products.

In a pharmacy setting, parenteral drugs are prepared under a laminar flow hood. The laminar flow hood is designed to prevent airborne contaminants from entering the sterile parenteral preparations and to protect the user from toxic products, for example, chemotherapy, prepared aseptically. Air enters the back of the hood and is circulated through a high-efficiency particulate air filter before it is directed out into the work area in uniform, parallel streams. The direction of flow may be vertical (Figure 13.3) or horizontal (Figure 13.4). The selection of airflow direction may depend on the type of operational movements in the work area. If most of the movements are vertical, the selected system should blow air horizontally. If most of the operational movements are horizontal, the selected flow should be vertical. To protect the user in preparing anticancer chemotherapy, ultraviolet lamps are placed in the air ducts and work area. Operators wear sterile lint-free gowns, caps, gloves, face masks, and shoe covers.

2. Large-Scale Preparations

When many parenteral products are to be prepared, the production process becomes much more complex and is usually done by the pharmaceutical industry. An overview of the manufacturing process is presented by the flow diagram in Figure 13.5. In general, industrial pharmacists are responsible for the manufacturing and quality control of the products. Many problems may be encountered in maintaining the control of the process and environment, in assuring the identity and purity of ingredients, in measuring and handling the large quantities of ingredients, and in coordinating the personnel and processes involved. Several in-process testing steps are usually performed to ensure of the quality and sterility of the bulk and finished products.

B. Containers

After the product is prepared, it must be stored in an appropriate container. The most important requirement for a container is to maintain the integrity of the product as a sterile, pyrogen-free,

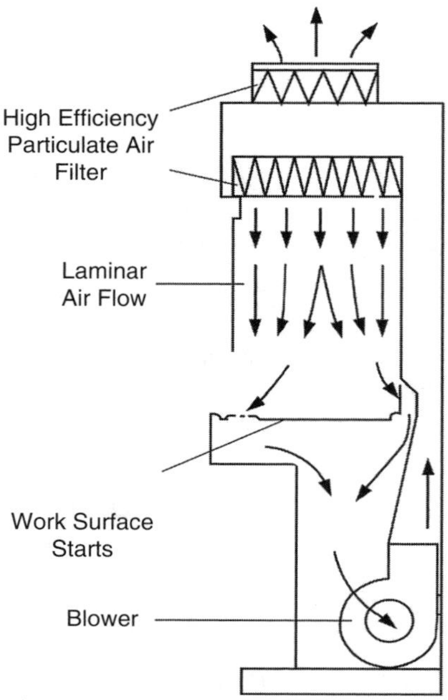

FIGURE 13.3 Vertical laminar flow hood (side view).

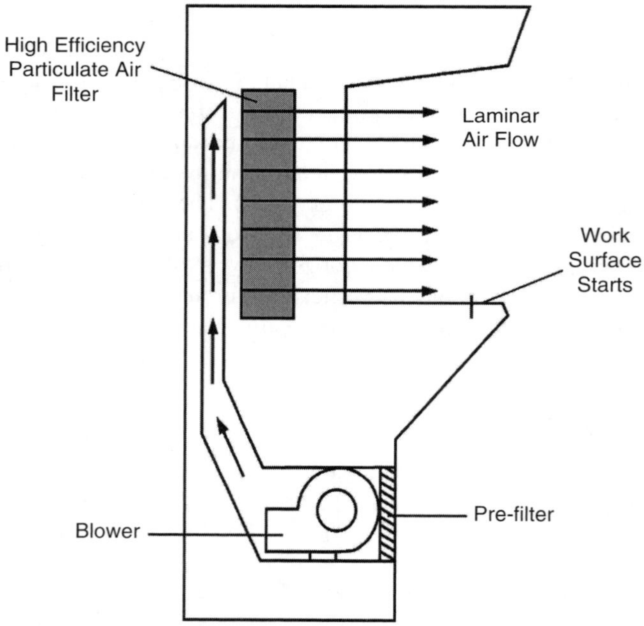

FIGURE 13.4 Horizontal laminar flow hood (side view).

high purity preparation until it is used. In addition, it should be easy to sterilize and transport and not interact with the product. The two major types of containers for parenteral products are the single-dose container and the multiple-dose container. A *single-dose container* is a hermetic container that holds a sufficient quantity of drug for a single-dose administration. Examples of single-dose containers include prefilled syringes, cartridges, and fusion-sealed ampules. When opened,

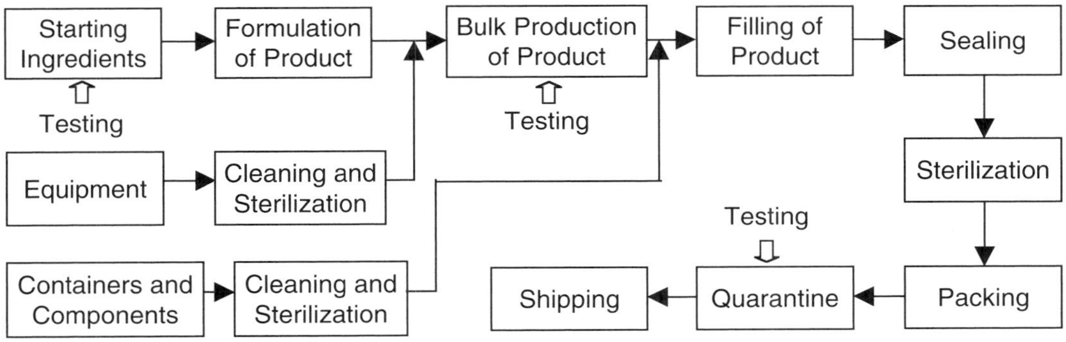

FIGURE 13.5 The manufacturing process of parenteral products.

these containers cannot be resealed with assurance that sterility has been maintained. A *multiple-dose container* is a hermetic container that allows withdrawal of successive portions of the contents without changing the strength, quality, or purity of the remaining portion. This container is normally affixed with a rubber closure to permit the penetration of a needle, i.e., to withdraw a portion of the product or to introduce a vehicle to a dry-powder preparation. Upon withdrawal of the needle, the closure reseals and protects the contents from airborne contamination. All multiple-dose parenteral products are required to contain added antimicrobial agents. In addition, all multiple-dose containers are limited to no more than 30 ml.

Most containers for parenteral products are made of glass. Glass is relatively nonreactive, provides protection against vapor, and withstands steam sterilization. Its major disadvantage is fragility, which may be problematic during transport and handling. There are four types of glass for pharmaceutical use: I, II, III, and NP. *Type NP* is for nonparenteral products. *Type I* is the best quality and can be used for any parenteral. It is a borosilicate glass that possesses low leachability. It can be repeatedly autoclaved and is satisfactory for unbuffered solutions and water. *Types II and III* are soda-lime glass, which is affected by autoclaving and must therefore be sterilized by dry heat. Both are less chemically resistant than Type I, and they do leach. Type II has been treated with an acidic gas at high temperature to increase its resistance to water and acids. In general, Type II glass is used in single-dose containers and for products that have a pH of less than 7. Type III glass is only used for sterile dry products.

Containers for i.v. fluids may also be made of plastic, which can be either a flexible plastic material such as polyvinyl chloride or a semirigid plastic material such as polyethylene or polyolefin. i.v. fluids in glass containers are packaged under vacuum. During administration of the fluid, the vacuum is dissipated through a mechanism that permits air to enter the container, which allows fluid to leave the container. Flexible plastic containers do not require air introduction in order to function. The flexible wall simply collapses as the fluid flows out of the container. Additional advantages of plastic over glass containers include their light weight, flexibility, and sturdiness.

However, plastics also present disadvantages: (a) Adsorption of drugs onto the wall of the container occurs more frequently in plastics than in glass; (b) some plastics leach their contents, for example, plasticizers, into the product and some permit permeation of vapor through the wall; (c) many plastics cannot be heat sterilized because they will soften or melt, while some are not clear and therefore make it difficult to inspect the product contents. Many plastics have been developed to overcome these problems. Some that are used as packaging materials for parenteral products are as follows: *Polyethylene* (polyolefin) is inert, translucent, and relatively impermeable to vapor or oxygen. It does not leach because it does not need a plasticizer. *Polycarbonate* is clear and heat resistant but is breakable and permeable to water vapor. *Polypropylene* is translucent and heat resistant. It has low permeability to vapor and gases but can become brittle if stored at low temperature. *Polyvinyl chloride* is transparent and inexpensive but is brittle and permeable to vapor and gases. It often leaches plasticizer and particulate matter into the product content. *Polymethylpentane* is clear, strong, and heat resistant but is more expensive.

VII. ADMINISTRATION DEVICES

A. Needles and Syringes

Most needles and syringes are available commercially as presterilized, disposable items in ready-to-use packaging. A needle consists of a cannula fused to a female hub fitting designed to accept attachment to the tip of a syringe or to the needle adapter on an i.v. set. Needles are described in terms of cannula length, for example, ∫ or 1 in, and cannula outside diameter, for example, 16G or 24G, where G is a gauge value that increases as the diameter decreases. The thin wall needle has a wider lumen diameter than the regular wall needle. The thin wall needle allows a better flow of fluid and is commonly used for administration of viscous fluids or suspensions. The sharpened end of the needle comes in three bevels denoting the angle of cut: long, short, and ultrashort. The ultrashort is mainly used for intradermal injection.

A syringe consists of a snugly fitting plunger in a graduated barrel. Syringes have either a Luer locking or Luer-slip tip, which allows attachment to a needle. The locking tip has treads onto which a needle is twisted to prevent detachment. The slip tip syringe is attached to a needle by pressure and stays in place by friction. Therefore, it can slip off during use if the fluid expulsion pressure is too high or if the patient moves rapidly during administration.

VIII. ADMINISTRATION SETS

i.v. administration sets deliver i.v. fluids and regulate flow. They are disposable, sterile, and pyrogen-free. Various types of administration sets are available, but they share common basic components. They usually consist of a plastic spike to pierce the closure seal on the i.v. container, a drip chamber to trap air and permit adjustment of flow rate, and a polyvinyl chloride tubing that ends in a rubber injection port. At the tip of the port is a rigid needle or catheter adapter. The injection port is self-sealing; therefore additional medication can be conveniently added to the i.v. system through this port. Regulation of fluid flow is achieved through an adjustable clamp that is attached to the tubing. Glass containers that lack air tubes require air-inlet filters designed as part of the administration set.

A. Venipuncture Devices

Venipuncture devices are the means of connecting i.v. administration sets to the patient's vein. Three types of venipuncture devices are available: a winged infusion set, an over-the-needle catheter, and a through-the-needle catheter. The *winged infusion set*, sometimes called a butterfly needle, has a steel needle with plastic wings that are attached to the cannula hub for easy manipulation during insertion. After the vein has been entered, the wings lie flat against the skin and provide a means for securing the needle and tubing. The winged infusion set is often used for short-term therapy or for a one-time i.v. push injection, as in chemotherapy. The rigid needle catheter associated with the winged infusion set may produce infiltration more frequently than a flexible catheter would. For long-term therapy, the *over-the-needle catheter* is preferred. A flexible, plastic catheter that is left in the vein, this device is more comfortable for the patient and is less likely to puncture the vein than the winged infusion set. Some over-the-needle catheters also have wings for easy anchoring and holding during insertion. The *through-the-needle catheter* is used when the i.v. is expected to remain in place for several days. It consists of a long catheter (8 to 12 in) that stays in the vein after insertion and a short introducer needle (1 to 2 in) that is pulled back out of the skin after insertion. This longer catheter decreases the risk of infiltration and is mainly used in long arm veins.

TABLE 13.4
Recommended Methods of Sterilization

Item	Examples	Dry-Heat Oven	Autoclave	Gaseous	Filtration
Thermostable solutions	Morphine, vasopressin	—	121°C, 30 min	—	—
Thermostable powders	Sulfamerazine, talc, zinc oxide	150–170°C, 1–2 h	—	—	—
Thermolabile solutions	Emetine, physostigmine	—	—	—	Include a bacteriostatic agent in solution
Thermolabile powders	Penicillin	—	—	Ethylene oxide	—
Oily liquids	Desoxycorticosterone, dimercaprol	150–170°C, 1 h	—	—	—
Glassware	Ampule, bottle	170°C, 2 h	121°C, 20 min	—	—
Rubber	Stoppers, tubing	—	121°C, 30 min	—	—
Metals	Apparatus	—	121°C, 30 min	—	—

IX. METHODS OF STERILIZATION

Sterilization is a process of killing or removing microorganisms. All parenteral products as well as the equipment, glassware, tubing, and gowns used in the preparation of parenteral products must be sterile. Sterilization can be accomplished in various ways, including physical, chemical, and mechanical methods. Frequently, more than one method is used before the total processing of a sterile product is completed. Examples of recommended methods of sterilization for specific drugs and materials used in the preparation of sterile products are given in Table 13.4. The aseptic technique is the use of procedures and precautions that prevent contamination by microorganisms. In some cases, parenteral products can be prepared by mixing sterile components of the products using aseptic techniques and then packaging them in sterile containers. There are several methods of sterilization for parenteral products. Their characteristics and uses are described below.

A. Dry Heat

Dry heat is the simplest and most economical method of sterilization. Compared with steam sterilization, however, this method requires higher temperatures and longer exposure times to achieve the same microbial-killing efficiency (see below). A major problem associated with dry-heat sterilization is nonuniform distribution of temperature. A fan is usually installed in the oven to overcome this problem, but unless the fan is properly baffled it may blow around powders that are to be sterilized. Furthermore, dry heat cannot be used with materials that are heat sensitive. It is mainly used for sterilization of glass and metal objects. It can also be used to sterilize thermostable powders and fatty substances.

B. Steam Under Pressure

Steam sterilization under pressure is carried out in an autoclave, which is an airtight, jacketed chamber designed to maintain a high pressure of saturated hot steam. Because the autoclave permits the attainment of high moist-heat temperatures and because heat exchange by steam is more rapid than by dry heat, this method of sterilization is more efficient than the dry-heat method. Steam sterilization is the most reliable and satisfactory method for the sterilization of aqueous solutions, glassware, and rubber articles. It is not suitable for sterilizing solutions of drugs that are thermolabile or are liable to degrade in the presence of moisture. Oily or fatty substances and powders cannot be sterilized by this means.

C. Filtration

Unlike other methods of sterilization, filtration works by removing microorganisms rather than killing or destroying them. This method also provides a means of removing particulate matter, which cannot be accomplished by other sterilization methods. Sterilization by filtration is most popularly used in pharmacy owing to its simplicity, reliability, and convenience. This increased popularity is attributed to the availability of sterile disposable filters that come in sizes suitable for normal procedures encountered in hospital practice. Most currently available filters are made of inert synthetic polymers such as cellulose nitrate and polycarbonate. They are manufactured to have uniform and controllable pore sizes. To remove microorganisms, the filter must have a pore size of approximately 0.2 μm. Larger pore sizes (e.g., 5 μm) are used to remove particulate matter.

D. Gaseous Sterilization

Gaseous sterilization is accomplished by exposure to a gas that kills microorganisms. The most commonly used gas for sterilization is ethylene oxide. Ethylene oxide is a cyclic ether that has flammable and explosive properties, especially when confined and mixed with oxygen. It is therefore often used in combination with inert gases such as carbon dioxide to avoid hazard. The effectiveness of ethylene oxide in the sterilization process depends on relative humidity, gas concentration, temperature, exposure time, and the extent of contamination. The normal working ranges for the relative humidity and ethylene oxide concentration are 30 to 60% and 500 to 1000 mg/l, respectively. At these ranges, the times required for complete sterilization are 2 to 5 h at 55°C. In practice, an exposure time of 6 h or more is normally used to provide a safety margin. After sterilization, the residual gas is removed by a vacuum, followed by an air wash with filtered sterile air. Ethylene oxide sterilization is effective in sterilizing a wide variety of materials including surgical instruments and gloves, plastic syringes, disposable needles, tubing sets, and dialysis units. In addition, it is often used to sterilize thermolabile powdered drugs such as penicillins. However, some drugs, for example, thiamine, riboflavin, and streptomycin, lose potency when treated with ethylene oxide.

E. Radiation Sterilization

Radiation sterilization is accomplished by exposure to ultraviolet (UV) light or high-energy ionizing radiation such as gamma rays. Because of its low energy, UV radiation has very little penetrating power, that is, it does not penetrate most substances, and it acts primarily on surfaces that are directly radiated. Consequently, this method of sterilization is not effective in sterilizing drugs, foods, and fabrics. Despite this limitation, UV radiation is extremely useful in reducing the number of airborne microorganisms. In hospitals, UV radiation is used to control the spread of infection during or after surgical procedures. In production areas of parenteral products, the use of UV radiation and the aseptic technique is considered to be a good manufacturing practice. As UV radiation has a limited range, the circulating air must pass close to the radiation source, normally located in the air duct and immediate vicinity of the sterile area, to be effective. When properly operated, the UV radiation technique can reduce the level of airborne bacterial contamination in a room by 90% within 30 min.

Ionizing radiation such as beta rays, gamma rays, X-rays, and accelerated electron beams, can be used to kill microorganisms. While these radiation techniques have rarely been used by pharmacists, they are routinely used by industrial firms in the sterilization of medical devices and certain parenteral drugs. Thermolabile drugs such as penicillin, streptomycin, thiamine, and riboflavin have been effectively sterilized by ionizing radiation. Pharmaceutical products are more resistant to degradation by ionizing radiation if they are in a powdered form than in a liquid form. Insulin and heparin solutions are inactivated by the radiation. Because radiation sterilization depends on the dose of irradiation and exposure time, it is necessary that these conditions be properly controlled to ensure the stability and potency of the exposed drugs.

X. STORAGE AND HANDLING

The proper storage and handling of parenteral products is crucial to their stability and sterility. The products should be stored in a dry, clean, ventilated area with even temperature and limited access. The proper storage temperature is dependent on specific products and should be strictly followed. In most cases, parenteral products are stored under refrigeration. Prior to use, they are removed from the refrigerator and allowed to come to room temperature.

i.v. admixtures should be administered as soon after preparation as possible. If this cannot be done, they may be stored in the refrigerator, but not for more than 24 h after the initial preparation or mixing. i.v. admixtures in flexible plastic bags are sometimes frozen soon after preparation and are thawed when needed. Thawing should be performed gradually. Rapid thawing in a microwave oven, which is routinely done in some pharmacies and hospitals, may lead to drug instability. Unless the process has been properly validated for a specific product, this practice should be avoided. Many parenteral products, notably biological products, are unstable upon repeated freezing and thawing. These products must be stored under refrigeration. Aside from temperature and thawing, protection from direct sunlight is a crucial consideration for the stability of many products. Most parenteral products are shipped from the manufacturer in some type of outer protective container, carton, or overwrap. These products should remain in their protective covering until the time of use. In fact certain light sensitive parenteral products such as sodium nitroprusside injection, used for hypertensive emergencies by i.v. infusion, must be protected from light during administration by wrapping the infusion bottle with aluminum foil to protect the solution from photodegradation.

XI. DRUG COMPATIBILITY AND STABILITY

The preparation of parenteral drugs is accompanied by the risks of incompatibility and instability. These problems occur as a result of an undesirable reaction between the drug and other components including the container. *Incompatibility* is normally associated with physical changes that are often visible, for example, a color change, haze, turbidity, precipitate, or gas formation. Incompatibility problems occur most frequently with i.v. admixtures and often involve precipitate formation, such as that seen when diazepam is added to dextrose 5% in water. *Instability* normally involves chemical degradation of the drug that may not be visible. Hydrolysis, oxidation, and photolysis are major chemical reactions responsible for the instability of parenteral drugs. An important factor contributing to these instabilities is the degree of protonation of the drug. Most drugs are either weak acids or bases and thus are susceptible to pH changes. To maintain a desired pH, buffers are normally added to the drug. The concept of stability is best exemplified by penicillin. This drug is most stable in a slightly acidic environment (pH ~6.5) but deteriorates if added to a very acidic or alkaline medium. Factors that affect drug stability are listed in Table 13.5.

XII. CALCULATIONS FOR PARENTERAL PRODUCTS AND ADMINISTRATION

A. Concentration Expression

The concentration or amount of drug in a parenteral dosage form is usually expressed in metric units such as milligrams and milliliters. However, units such as milliequivalents (mEq) and millimoles (mmol) are also used, especially to express the amount of electrolytes in parenteral solutions. The relationship between milliequivalent and milligram expressions of a substance is: milliequivalent = milligram/equivalent weight. The equivalent weight is the atomic weight divided by the valance or charge of the substance.

TABLE 13.5
Factors Affecting Drug Stability

Factor	Effect	Example
Temperature	High temperatures usually accelerate chemical instability of drugs.	Cephalothin is only stable for 6 h at room temperature, but up to 48 h when refrigerated.
Time	The length of time the drug is in solution affects stability.	Ampicillin is only stable for 4 h when added to dextrose injection.
Light	Some drugs are light sensitive and degrade more rapidly when exposed to light.	Exposure of levarterenol to light results in faster degradation of the drug.
pH	Many drugs undergo acid and/or base catalysis in solutions.	Penicillin degrades more rapidly in an acidic or alkali medium.
Dilution	Some drugs are unstable when presented at high concentrations in solutions.	Only limited amounts of heparin and hydrocortisone are stable in amphotericin solution.
Additive	The greater the number of drugs and additives contained in the admixture, the greater the chance of one or more of the ingredients becoming unstable.	Multiple additives in large volume parenteral fluids.
Container	The composition of the container may affect the stability of the drug.	The potency of insulin is reduced by at least 20% when stored in a plastic container.

Example 1

A prescription order calls for 25 mEq K^+ to be added to an large volume parenteral solution. If the source of K^+ is from KCl, how many milligrams of KCl are needed?

The atomic weight of KCl is 74.5. The valance of K^+ is 1.
Therefore, the equivalent weight of KCl is 74.5/1 = 74.5
The amount of KCl needed is $74.5 \times 25 = 1117.5$ mg

Example 2

If 1.5 mEq Ca^{++} are ordered, how many mg of $CaCl_2.2H_2O$ are needed?

The atomic weight of $CaCl_2.2H_2O$ is 147. The valance of Ca^{++} is 2.
Therefore, the equivalent weight of $CaCl_2.2H_2O$ is 147/2 = 73.5
The amount of $CaCl_2.2H_2O$ needed is $73.5 \times 1.5 = 110.25$ mg

Note that the dihydrate form of $CaCl_2$ is more stable than the anhydrous form and is therefore more frequently used. A significant error would have been made if the equivalent weight of anhydrous $CaCl_2$ (111/2 = 55.5) were used in the calculation.

B. Tonicity and Osmolality

A solution that is isotonic has the same tonicity as plasma or body fluids. Tonicity is often indicated by *osmolality*, a term that represents the number of solute particles per kilogram of solvent and is expressed as milliosmoles (mOsm). The osmolality of body fluids is approximately 310 mOsm/l. Most parenteral solutions are isotonic or iso-osmotic. Examples of these solutions include 0.9% sodium chloride, 5% dextrose, and lactated Ringer's. Some parenteral fluids are hypotonic or hypertonic; that is, the osmolality is less than 240 mOsm/l or greater than 340 mOsm/l, respectively. If body cells are exposed to hypotonic solutions, water will be drawn into the cells and cause them to swell or burst. In contrast, the cells will shrink if exposed to hypertonic solutions. Clinically, hypotonic solutions are used to dilute excess serum electrolytes, as in hyperglycemia. An example of a hypotonic i.v. solution is 0.45% sodium chloride. Patients receiving hypotonic solutions should be monitored closely, as these solutions may cause hemolysis. Hypertonic solutions are used to correct electrolyte imbalances, as in loss from excess vomiting and diarrhea. Patients receiving hypertonic solutions should be monitored to prevent body fluid overload, especially if the solutions are extremely concentrated or are being given at a rapid rate. Examples of parenteral hypertonic solutions include 3% and 5% sodium chloride, 20% and 50% dextrose in water, and 5% dextrose in lactated Ringer's.

A common practical problem related to solution tonicity is the calculation of a specific solute quantity that must be added to a drug solution to make it isotonic. This can be done by several methods, including the sodium chloride equivalent, freezing point depression, and other related methods. The sodium chloride equivalent method is perhaps the simplest and most frequently used method for calculating the amount of sodium chloride necessary to prepare an isotonic solution. This method is based on the fact that a 0.9% concentration of sodium chloride in water gives an isotonic solution. The *sodium chloride equivalent* is defined as the amount of sodium chloride that will produce the same osmotic effect as one unit of a drug. For example, the sodium chloride equivalent of aminophylline is 0.17. Thus, 1 g of aminophylline in water would give the same osmolality as 0.17 g of sodium chloride in water.

Example 3

How much sodium chloride is required to make the following prescription isotonic?

Rx

Aminophylline: 1.5 %
Sterile water, *qs* ad: 10 ml
Make isotonic with sodium chloride

1. Calculate the number of grams of drug in the solution: 1.5% × 10 ml = 0.15 g.
2. Determine the weight in grams of sodium chloride that is equivalent to the weight in grams of aminophylline in this solution:

$$\frac{x \text{ g NaCl}}{0.15 \text{ g Drug}} = \frac{0.17 \text{ g NaCl}}{1 \text{ g Drug}}; \quad x = 0.0255 \text{ g NaCl}$$

3. Calculate the number of grams of sodium chloride needed to make the drug solution isotonic if no other solutes were present: 0.9% × 10 mL = 0.09 g NaCl.
4. Subtract the weight of sodium chloride that is equivalent to the weight of the drug from the weight of sodium chloride that would be needed to make the solution isotonic: 0.09 g – 0.0255 g = 0.0645 g NaCl needed.
5. Therefore, to prepare the prescribed solution would require 0.15 g aminophylline, 0.0645 g NaCl, and sufficient sterile water to make 10 ml.

C. Intravenous Admixtures

The preparation of intravenous admixtures involves the addition of one or more drugs to large volume sterile fluids such as sodium chloride injection, dextrose injection, and total parenteral nutrition fluid. Ttotal parenteral nutrition, or hyperalimentation, is the feeding by i.v. infusion of all basic nutrients needed by the patient. Among the components generally included in total parenteral nutrition fluids are dextrose, electrolytes, amino acids, vitamins, and trace elements.

The administration of medicated fluids requires extreme care and meticulous observation so that exactly the prescribed amount of medication in the fluid is administered to the patient.

Example 4

A medication order for a child weighing 44 lb calls for polymyxin B sulfate to be administered by the i.v. drip method in a dosage of 7,500 units/kg of body weight in 500 ml of 5% dextrose injection. Using a vial containing 500,000 units of polymyxin B and sodium chloride injection as the solvent, explain how you would obtain the polymyxin B needed in preparing the infusion.

$$44 \text{ lb} = 44/2.2 = 20 \text{ kg}$$

$$20 \text{ kg} \times 7{,}500 \text{ units/kg} = 150{,}000 \text{ units}$$

Dissolve contents of vial (500,000 units) in 10 ml of sodium chloride injection.
Add 3 ml of reconstituted solution to 500 ml of dextrose injection.

Example 5

A medication order for a patient weighing 154 lb calls for 0.25 mg of amphotericin B per kilogram of body weight to be added to 500 ml of dextrose injection. If the amphotericin B is to be obtained from a reconstituted injection that contains 50 mg/10 ml, how many milliliters should be added to the dextrose injection?

$$154 \text{ lb} = 154/2.2 = 70 \text{ kg}$$

$$70 \text{ kg} \times 0.25 \text{ mg/kg} = 17.5 \text{ mg}$$

$$17.5 \text{ mg} \times \frac{10 \text{ ml}}{50 \text{ mg}} = 3.5 \text{ ml}$$

D. Rate of Flow of Intravenous Fluids

Unlike small volume injections, which can be injected directly into the body by a hand-held syringe and needle, large-volume parenteral fluids are continuously infused into the patient's body over a prolonged period of time. These parenteral fluids are hung at the patient's bedside and allowed to drip slowly into a vein by gravity or through the use of electric or battery-operated volumetric infusion pumps. These pumps can be calibrated to deliver 0.1, 1, 10, 100, and other numbers of milliliters per minute, depending on the drug and patient's requirements.

The physician specifies the rate of flow of intravenous fluids in milliliters/minute, drops/minute, amount of drug (milligrams/hour), or as the duration of time needed to administer the total infusion volume. The pharmacist may be asked to make or to check the calculations involved in converting the desired total time interval into a flow rate of drops per minute.

Example 6

The physician's order reads "1000 cc D5W IV in 24 hr." How many drops per minute should the i.v. infusion run if the i.v. administration set delivers 60 drops/ml?

$$\frac{1000\ \text{ml}}{24\ \text{h}} \times \frac{1\ \text{h}}{60\ \text{min}} \times \frac{60\ \text{drops}}{1\ \text{ml}} = 42\ \text{drops/min}$$

Example 7

An intravenous infusion contains 10 ml of a 1:5000 solution of isoproterenol hydrochloride and 500 ml of dextrose injection. At what flow rate (millileters/minute) should the infusion be administered to provide 5 μg of isoproterenol per minute, and what time interval will be necessary for the administration of the entire infusion?

10 ml of a 1:5000 solution contains 2 mg of isoproterenol

$$10\ \text{ml} \times \frac{1\ \text{g (or 1000 mg)}}{5000\ \text{ml}} = 2\ \text{mg}$$

2 mg or 2000 μg of isoproterenol are contained in a volume of 510 ml

$$\frac{5\ \mu\text{g}}{1\ \text{min}} \times \frac{510\ \text{ml}}{2000\ \mu\text{g}} = 1.275\ \text{ml/min}$$

$$510\ \text{ml} \times \frac{1\ \text{min}}{1.275\ \text{ml}} = 400\ \text{min}$$

Example 8

If 10 mg of a drug is added to a 500-ml--volume parenteral fluid, what should be the flow rate (milliliters/hour) to deliver 1 mg of drug per hour? If the infusion set delivers 15 drops/milliliter, what should be the rate of flow in drops per minute?

$$\frac{1\ \text{mg}}{1\ \text{h}} \times \frac{500\ \text{ml}}{10\ \text{mg}} = 50\ \text{ml/h}$$

$$\frac{15\ \text{drops}}{1\text{ml}} \times \frac{50\ \text{ml}}{1\ \text{h}} \times \frac{1\ \text{h}}{60\ \text{min}} = 12.5\ \text{drops/min}$$

XIII. CONCLUSION

The parenteral route of administration has increasingly been used to administer drugs in various health care settings, including the hospital, outpatient, and home care settings. With its increased use, pharmacists have assumed increased responsibility in preparing and dispensing parenteral drug products as well as in providing consultation to other health professionals and patients regarding the drug therapy. A competent pharmacist should therefore have a functional knowledge of parenteral drug administration and preparation of sterile products needed for their administration. This knowledge is necessary both for the safety of the patient and to ensure quality patient outcomes.

TUTORIAL

1. All parenteral products must meet the following requirements:
   (a) Sterility
   (b) Absence of pyrogens
   (c) pH neutrality
   (d) (a), (b), and (c)
   (e) (a) and (b)
2. Parenteral products must be free of all forms of microorganisms, live or dead.
   (a) True
   (b) False
3. Examples of pyrogens include:
   (a) Bacterial endotoxins
   (b) Bacterial exotoxins
   (c) Bacterial enzymes
   (d) (a), (b), and (c)
   (e) (a) and (b)
4. An IV administration is needed when:
   (a) An immediate action is required.
   (b) An oral administration is ineffective.
   (c) A prolonged action is required.
   (d) (a), (b), and (c)
   (e) (a) and (b)
5. Which of the following statements is not true regarding parenteral administration?
   (a) Once the drug is administered it cannot be retrieved.
   (b) Systemic drug absorption occurs more rapidly than from oral administration.
   (c) It has greater potential health hazard than oral administration.
   (d) All parenteral products must be isotonic.
   (e) Dry-powder parenteral products must be reconstituted with a suitable sterile vehicle before administration.
6. For small children under the age of three, the ______________ is recommended as a site for intramuscular injection:
   (a) Deltoid
   (b) Gluteus maximus

(c) Vastus lateralis
(d) Ventrogluteal
(e) Intramuscular is not recommended

7. The method of choice for tuberculin skin testing is:
(a) Intramuscular
(b) Intradermal
(c) i.v.
(d) Intrathecal
(e) Subcutaneous

8. The size range of intramuscular syringes and needles is:
(a) 1 ml, 25 to 30 gauge
(b) 1 to 5 ml, 20 to 22 gauge
(c) 1 to 60 ml, 20 to 22 gauge
(d) 1 to 5 ml, 30 gauge
(e) 1 ml, 25 to 28 gauge

9. The maximum injection volume for an intradermal administration is:
(a) 0.5 ml
(b) 1 ml
(c) 2 ml
(d) 5 ml
(e) 3 l

10. The primary reasons that buffers are used in parenteral products are:
(a) To stabilize the solution against pH changes
(b) To make the pH more comfortable for injection
(c) To increase the dissociation constant of the drug
(d) All of the above
(e) None of the above

11. Which of the following statements is true?
(a) Dry heat is more effective in killing microorganisms than steam.
(b) Steam sterilization is effective in sterilizing thermolabile drugs.
(c) UV radiation has poor penetrating power and is ineffective in sterilizing most parenteral products.
(d) Ethylene oxide is an inert gas that works best under low humidity conditions.
(e) Ionizing radiation, owing to its potent microbial-killing activity, is popularly used in pharmacy.

12. Which of the following statements is true?
(a) Sterilization by filtration prevents thermal stress on the product.
(b) During aseptic filtration, the solution is passed through a sterile 2-μm filter.
(c) Filtration cannot be used to sterilize parenteral suspensions.
(d) (a), (b), and (c)
(e) (a) and (c)

13. All forms of heat and radiation sterilization methods are intended to eliminate what from the final product?
(a) Bacteria
(b) Viruses
(c) Pyrogens
(d) Viable microorganisms
(e) None of the above

14. The method of choice for sterilizing thermostable aqueous solutions is:
(a) Dry heat
(b) Autoclaving

   - (c) Filtration
   - (d) Ethylene oxide
   - (e) Radiation
15. Total parenteral nutrition products:
   - (a) Are often administered by a pharmacist
   - (b) Can be administered intramuscularly
   - (c) Are often used in comatose patients
   - (d) Should never be combined with other parenteral products
   - (e) Have little potential for patient injury

Answers: 1. (e); 2. (b); 3. (e); 4. (e); 5. (d); 6. (c); 7. (b); 8. (b); 9. (a); 10. (a); 11. (c); 12. (e); 13. (d); 14. (b); 15. (c)

HOMEWORK

1. Discuss the advantages and disadvantages of parenteral drug administration vs. oral drug administration.
2. What are the different types of parenteral drug administration? Which type provides the fastest onset of drug action systemically?
3. What is total parenteral nutrition, and what is its intended use?
4. Why do some parenteral products come in the form of freeze-dried solids?
5. Describe the piggyback method of drug administration and its intended use.
6. Name the different aqueous solutions and nonaqueous solvents that are used as vehicles for parenteral products?
7. Distinguish between water for injection and bacteriostatic water for injection.
8. What is the acceptable pH range for parenteral products?
9. What are preservatives? Name three preservatives commonly used in parenteral products.
10. Name the different types of glass. Which type is best for use with parenteral products?
11. Compare and contrast the different methods of sterilization. Name two methods that can be used to sterilize heat-sensitive drugs.
12. A prescription requires a 500-ml solution of potassium chloride to be made, so that it will contain 400 mEq of K^+. How many grams of KCl (MW 74.5) are needed? (29.8 g)
13. How much sodium chloride is required to make the following prescription isotonic, given that the sodium chloride equivalent of ephedrine sulfate is 0.23? (*Answer:* 0.132 g)

    Rx

    Ephedrine sulfate: 2%
    Sterile water qs ad: 30 ml
    Make isotonic solution with sodium chloride

14. A pharmacist receives a medication order of 300,000 units of Penicillin G potassium to be added to 500 ml of dextrose injection. The directions on the 1,000,000-unit package state that if 1.6 ml of solvent is added, the constituted solution will measure 2 ml. How many milliliters of the constituted solution must be withdrawn and added to the dextrose injection? (*Answer:* 0.6 ml)

15. An intravenous infusion for a patient weighing 132 lb calls for 7.5 mg of kanamycin sulfate per kilogram of body weight to be added to 250 ml of dextrose injection. How many milliliters of a kanamycin sulfate injection containing 500 mg per 2 ml should be used in preparing the infusion? (*Answer:* 1.8 ml)
16. In preparing an intravenous solution of lidocaine in dextrose injection, a pharmacist added a concentrated solution of lidocaine (1 g/5 ml) to 250 ml of dextrose injection. What was the final concentration of lidocaine (milligrams/milliliters)? (*Answer:* 3.92 mg/ml)
17. Calculate the i.v. flow rate for a liter of normal saline to be infused in 6 h. The infusion set is calibrated for a drop factor of 15 drops/ml. (*Answer:* 42 drops/min)
18. A 1-l bag of intravenous solution contains 2.5 million units of ampicillin. How many units of the drug will have been infused after 6 h with the flow rate of 1.2 ml/min? (*Answer:* 1.08 million units)
19. A patient is to receive 2 μg/kg/min of nitroglycerin from a solution containing 100 mg of the drug in 500 ml of dextrose injection. If the patient weighs 154 lb and the infusion set delivers 60 drops/ml, (a) how many milligrams of nitroglycerin would be delivered per hour and (b) how many drops per minute would be delivered? (*Answer:* 8.4 mg, 42 drops/min)
20. The drug alfentanil hydrochloride is administered by infusion at the rate of 2.2 μg/kg/min for inducing anesthesia. If a total of 0.55 mg of the drug is to be administered to a 175-lb patient, how long should be the duration of the infusion? (*Answer:* 3.14 min)

CASES

CASE I: PEDIATRIC IRON TOXICITY AND I.V. DESFERAL

SCENARIO 1

A frantic mother calls her local pharmacist to report that her 3-year-old child's lips and teeth are stained green, and she believes that her child has swallowed her iron tablets that she had lying on her kitchen counter prior to transferring them from the blister pack to her weekly tablet organizer. Her phone had rung and she went into the bedroom to talk with her mother because her older child had the CD player blasting. The mother tells the pharmacist that she had counted out seven tablets, one daily for a week, and that all seven tablets are missing and that she was not gone from the kitchen for more than 30 minutes. The child must have eaten them, thinking they were candy, within the past half hour. The iron tablets were ferrous sulfate 325 mg. The child weighs 33 lb.

1. What is the first thing the pharmacist should do in response to this call?
   (a) Determine which iron-containing tablet the child ingested.
   (b) Calculate the dose of elemental iron ingested.
   (c) Tell the mother to come into the pharmacy to get a bottle of syrup of ipecac.
   (d) Look up the phone number of the nearest poison control center for the mother, give her the number, and strongly recommend that she call the poison center immediately for instructions.
2. If ferrous sulfate tablets contain 20% elemental iron, what is the maximum amount of elemental iron the child could have ingested?
   (a) 32.5 mg

(b) 65 mg
(c) 130 mg
(d) 390 mg
(e) 455 mg

3. If the mother was instructed to give her child syrup of ipecac at home and then bring the child as quickly as possible to the hospital emergency room, what is the usual dose of ipecac syrup for her child? (While most pharmacists have committed to memory the standard doses of ipecac syrup to be used as an emergency emetic in home poisonings, if you do not know the dose, in what reference would you quickly look it up?)
Reference: ________________________
(a) 2.5 ml
(b) 5 ml
(c) 15 ml
(d) 30 ml

Scenario 2

You are a hospital pharmacist on duty the day the mother brought her 3-year-old to the emergency room following the accidental ingestion of ferrous sulfate tablets. Upon medical examination of the patient, the physician orders KUB x-rays and a serum iron level and then decides to treat the child with chelation therapy to lower the child's elevated serum iron levels.

4. If the child weighs 33 lb. and did ingest seven 325-mg ferrous sulfate tablets, how many milligrams per kilogram of elemental iron did the child ingest?
(a) 4.3 mg/kg
(b) 8.6 mg/kg
(c) 26 mg/kg
(d) 30.3 mg/kg
(e) 260 mg/kg
5. What is the drug of choice for use as chelation therapy in iron toxicity?
(a) Ethylenediamine tetraacetic acid
(b) Deferoxamine mesylate
(c) Dimercaprol
(d) Cholestyramine
(e) Sodium polystyrene sulfonate

The physician orders the chelating agent under its tradename, Desferal®. The dose ordered is 15 mg/kg/h for 6 h by slow i.v. infusion. Desferal comes in 5-ml vials containing 500 mg of a freely soluble lyophilized powder.

6. How should this parenteral solution for i.v. infusion be prepared according to acceptable compounding practices and the manufacturers directions?
(a) Aseptically add 5 ml bacteriostatic water for injection to dissolve the Desferal powder in the vial then transfer 5 ml of the solution to 95 ml of dextrose injection in a large volume parenteral bag.
(b) Aseptically add 5 ml of sterile saline to dissolve the lyophilized powder and then add the contents to 995 ml of sterile water for injection in a large volume parenteral bag.
(c) Aseptically add the smallest volume (2 ml) of sterile water for injection to dissolve the lyophilized powder and then add the total volume of Desferal solution aseptically to 100 to 150 ml of dextrose injection in a large volume parenteral bag.

(d) Add the smallest volume (2 ml) of alcohol, USP to the vial to dissolve the Desferal powder and then add the alcohol solution to 500 ml of bacteriostatic water of injection in a large volume parenteral bag.

7. Based upon the physician's prescription order, what is the total dose of Desferal that must be infused into this child?
   (a) 15 mg
   (b) 225 mg
   (c) 495 mg
   (d) 1350 mg
8. How many vials of Desferal must be reconstituted to provide the total dose of Desferal ordered for this child?
   (a) One
   (b) Two
   (c) Three
   (d) Four

Answers: 1. (d); 2. (e); 3. *USPDI* or any pharm/tox reference, (c); 4. (d); 5. (b); 6. (c); 7. (d); 8. (c)

CASE II. ACUTE ALCOHOL WITHDRAWAL AND I.V. ALCOHOL DRIP

Scenario 1

You are a community pharmacist in a small town and receive a phone call from a close friend, Barney, who is a newly hired deputy sheriff at the local jail. He is concerned because a local street person, named Cecil, who was picked up and has been jailed for over 24 hours on charges of harassment associated with drunk and disorderly conduct was behaving very strangely in his cell. Cecil was alternately lying on his cot sweating profusely, then pacing around his cell, breathing rapidly, shaking, vomiting blood, and screaming that he feels like he is going to die. Your friend the deputy asks your advice. He wants to know what he should do, since he has never seen anything like this before.

1. What should you tell your friend Barney?
   (a) Tell your friend to just leave him alone; he will get over it, because street people do that sort of thing all the time to get attention.
   (b) Tell Barney to give him two or three aspirin and a glass a water and check on him in a hour or so.
   (c) Tell Barney to order an alcohol blood level from the local hospital's clinical lab.
   (d) Tell Barney to call the county ambulance service through 911 to get Cecil to the hospital emergency room as soon as possible.
2. Based on your knowledge of pharmacotherapeutics and chemical dependency learned in pharmacy school, what is your most educated guess as to Cecil's medical diagnosis?
   (a) Acute alcohol intoxication
   (b) Chronic alcoholism and acute alcohol abstinence syndrome, with the possibility of impending delerium tremens
   (c) Delerium tremens
   (d) Fetal alcohol syndrome

Scenario 2

It is eventually determined that Cecil is probably in the second or third stage of alcohol withdrawal and is suffering from recurrent gastrointestinal bleeding resulting from alcohol-induced gastritis.

He is being prepared for surgery, but the emergency room physician wants to control the alcohol withdrawal syndrome first and orders an i.v. ethanol drip from the pharmacy. You are the hospital pharmacist on duty and receive the prescription order to prepare the intravenous alcohol solution. Upon consultation with the anesthesiologist, the emergency room physician orders 1 l of a 20% ethanol solution in dextrose injection. The dose is 6 ml of ethanol per hour, which is well within the recommended dosage range of ethanol (2.5 to 10 ml/h) to prevent withdrawal. Upon inquiry you discover that i.v. ethanol is only available commercially as a 10% solution in dextrose injection. Answer the following questions about the preparation of the parenteral ethanol solution to be sent to the emergency room.

3. What do you do about the request for 20% ethanol i.v. upon discovering that only 10% is available commercially?
   (a) Just substitute the 10% ethanol and tell the emergency room nurse to give half the dose.
   (b) Substitute 10% ethanol and tell the emergency room nurse to set up two i.v.s, one in each arm, and double the drip rate.
   (c) Substitute 10% ethanol and increase the dose by recommending a shot of whiskey every hour.
   (d) Do the appropriate calculations and prepare to compound a 20% ethanol solution for i.v. drip.
4. You remember from your physical pharmacy education that there is a unique characteristic surrounding the admixture of ethanol and water. What physical chemical property of alcohol–water admixtures should you take into account in diluting the alcohol, USP with dextrose injection for i.v. infusion?
   (a) When a given volume of alcohol, USP and water are mixed, the resultant final volume is less than the sum of the two added volumes; therefore you do not QS to a final volume to make a diluted concentration of alcohol.
   (b) When alcohol, USP is mixed with water, the admixture produces an endothermic reaction that results in an icy cold diluted solution that must be warmed to room temperature before use.
   (c) When diluting alcohol, USP to prepare a diluted alcohol solution you must always add the alcohol to the water and never water to alcohol to avoid spattering owing to the boiling of small volumes of water poured into alcohol from the generation of an exothermic reaction.
   (d) A small amount of methanol must be added to prevent the foaming that occurs when water and alcohol are mixed.
5. What is the concentration of ethanol in alcohol, USP?
   (a) 100%
   (b) 95%
   (c) 75%
   (d) 50%
6. What volumes of alcohol, USP and dextrose injection must be admixed to prepare 1 l of a 20% ethanol solution for i.v. infusion as prescribed?
   (a) 200 ml alcohol, USP and 800 ml dextrose injection
   (b) 210 ml alcohol, USP and 790 ml dextrose injection
   (c) 267 ml alcohol, USP and 733 ml dextrose injection
   (d) 400 ml alcohol, USP and 600 ml dextrose injection
7. What method of sterilization would you employ to prepare the sterile i.v. ethanol infusion described above?

   - (a) Autoclave
   - (b) Dry heat
   - (c) Ethylene oxide
   - (d) Filtration through a 0.22 micron millipore filter
8. After compounding and sterilizing the 20% ethanol parenteral solution and providing an i.v. infusion kit calibrated to deliver 20 drops per milliliter, what infusion rate would you recommend to deliver the prescribed dose of 6 ml of ethanol per hour from the 20% ethanol solution?
   - (a) 1 drop/min
   - (b) 5 drops/min
   - (c) 10 drops/min
   - (d) 20 drops/min

Answers: 1. (d); 2. (b); 3. (d); 4. (a); 5. (b); 6. (b); 7. (d); 8. (c)

CASE III: SMALL VOLUME NEONATAL INJECTION AND DEXAMETHASONE IN ENDOTRACHEAL EXTUBATION

SCENARIO 1

You are a pharmacist assigned to the neonatal intensive care unit and have been called in on a neonatal intensive care unit physician/nurses conference to discuss corticosteroid antiinflammatory therapy for neonates who have had routine endotracheal intubation/ventilation procedures. A number of special problems involving pharmaceutical intervention surround these procedures. One problem that surrounds the extubation process is that the endotracheal ventilation tube causes an upper airway inflammation, the swelling of which may close off the airway lumen once the tube is removed. Another problem that many preterm neonates experience is bronchopulmonary dysplasia, which occurs in the lower respiratory airways and alveoli, often as a sequela of respiratory distress syndrome, common in many preterm neonates and requires surfactant administration along with O_2 therapy and long-term intermittent mandatory ventilation. i.v. corticosteroids are used for their antiinflammatory properties both to reduce swelling prior to endotracheal extubation and to facilitate weaning from the ventilator in neonates with bronchopulmonary dysplasia. The physician wishes to use dexamethasone injection administered i.v. push in a dose of 0.25 mg/kg. Upon checking the pharmaceutical availability of dexamethasone, you find that there are three forms available: dexamethasone base, dexamethasone acetate, and dexamethasone sodium phosphate. Two of those forms are available parenterally, but only one is indicated for i.v. administration. N.B. also, to avoid dosing errors, a standing hospital pharmacy rule is that nurses must be provided with parenteral solutions of a concentration that do not require an injectable volume of less than 0.1 ml; that is, 0.1 ml is the lowest permitted injectable volume. At the same time, i.v. push injected volumes in neonates should be minimized and less than 1 ml to avoid hypervolemia.

1. Which form of dexamethasone would most likely not be used parenterally?
   - (a) Dexamethasone base
   - (b) Dexamethasone acetate
   - (c) Dexamethasone sodium phosphate
2. Which is the only form of dexamethasone indicated for i.v. administration?
   - (a) Dexamethasone base
   - (b) Dexamethasone acetate
   - (c) Dexamethasone sodium phosphate

Scenario 2

A 32-week gestational age preterm neonate weighing 1000 g has been attached to a ventilator for 2 days, and the physician deems that the neonate's lungs are sufficiently developed to permit spontaneous breathing and the patient should be extubated. He orders dexamethasone at a dose of 0.25 mg/kg i.v. push 4 h prior to extubation and then q 8 h for three doses. When you check the pharmacy stock, you find out that the i.v. dexamethasone salt is available in concentrations of 10 mg/ml in 1-ml and 10-ml vials, 20 mg/ml in 5-ml vials, and 24 mg/ml in 5-ml and 10-ml vials.

3. Based on the neonatologist's prescription order, what volume of the currently available commercial parenteral solutions would you advise the neonatal intensive care unit nurse to administer to provide the correct dose for this neonatal patient?
   (a) 0.025 ml of the 10 mg/ml solution
   (b) 0.125 ml of the 20 mg/ml solution
   (c) 0.25 ml of the 10 mg/ml solution
   (d) 0.1 ml of the 24 mg/ml solution
   (e) None of the above
4. Which of the following dispensing scenarios would you choose to administer the doses of dexamethasone ordered by the neonatologist?
   (a) Send a 1-ml vial of dexamethasone 10 mg/ml and instruct the nurse to withdraw 0.025 ml, which would contain the prescribed dose of 0.25 mg 4 h prior to extubation; repeat for the subsequent three doses q 8 h.
   (b) Send a 5-ml vial of the 20 mg/ml dexamethasone solution and instruct the nurse to withdraw 0.125 ml, which would contain the prescribed dose of 0.25 mg 4 h prior to extubation; repeat for the subsequent three doses q 8 h.
   (c) Make a 1:10 dilution of the 10 mg/ml dexamethasone by withdrawing 1 ml from the 10-ml vial, diluting it in 10 ml of sterile water for injection and instruct the nurse to administer 0.25 ml 4 h prior to extubation; repeat q 8 h for three doses, then discard the unused portion.
   (d) Make a 1:10 dilution of the 10 mg/ml dexamethasone solution by aseptically withdrawing 0.5 ml from the 1-ml vial, diluting it in 4.5 ml of sterile water for injection and instruct the nurse to administer 0.25 ml i.v. push 4 h prior to extubation; repeat q 8 h for three doses after extubation.
   (e) None of the above.

Scenario 3

A 26-week gestational age neonate weighing 700 g at birth and suffering from respiratory distress syndrome has been intubated on a ventilator for about 2 months. The patient now weighs about 1000 g and shows signs of spontaneous breathing, and the neonatologist orders that the patient be extubated. The neonate no doubt has developed bronchopulmonary dysplasia and will require longer-term therapy with antiinflammatory corticosteroids to help wean the 35-week infant from the ventilator following extubation. Long-term corticosteroid therapy predisposes the patient to hypothalamic-pituitary-adrenal suppression, and the neonatologist emphasizes that the dosage regimen with dexamethasone that he is prescribing must be carefully followed to ensure a period of tapering down the dexamethasone dose during the time the neonate is off the ventilator. The dose of dexamethasone he prescribed is as follows: 0.25 mg/kg 4 h prior to extubation, followed by 0.25 mg/kg q 8 h for 3 days; then 0.15 mg/kg q 12 h for 3 days; then 0.125 mg/kg q 12 h for 1 day; then 0.22 mg/kg qd for 1 day; then finally 0.20 mg/kg qd for 1 day. Upon consultation with the neonatologist regarding the required dilution of the commercially available dexamethasone 10 mg/ml i.v. solution, you are informed that the physician does not want any bacteriostatic preservatives added to the diluent to prepare the final dilution of the dexamethasone. The commercially available dexamethasone i.v. solution therefore must be diluted in sterile water for injection.

5. The above information has guided your decision to prepare the dexamethasone dilution in a multidose vial in a volume to be used in ________, with instructions to discard the unused portion after that time.
   (a) 1 day
   (b) 5 days
   (c) 9 days
   (d) 12 days
6. How many total days of dexamethasone therapy, including the taper period, have been prescribed?
   (a) 3
   (b) 6
   (c) 9
   (d) 12
7. What volume of a 1:10 dilution of the 10 mg/ml dexamethasone i.v. solution will be injected at each administration on day 4 of the protocol?
   (a) 0.01 ml
   (b) 0.10 ml
   (c) 0.125 ml
   (d) 0.15 ml
   (e) 0.25 ml
8. What is the total dose of dexamethasone administered on day 2 vs. day 7 of the protocol?
   (a) Day 2 = 1.0 mg; day 7 = 0.125 mg
   (b) Day 2 = 0.75 mg; day 7 = 0.25 mg
   (c) Day 2 = 0.75 mg; day 7 = 0.125 mg
   (d) Day 2 = 0.5 mg; day 7 = 0.5 mg
   (e) Day 2 = 0.25 mg; day 7 = 0.125 mg

Answers: 1. (a); 2. (c); 3. (e); 4. (d); 5. (a); 6. (c); 7. (d); 8. (b)

REFERENCES

1. Ansel, H.C., Popovich, N.G., Allen, L.V., Jr., *Pharmaceutical Dosage Forms and Drug Delivery Systems,* Lea & Febiger, Malvern, PA, 1995.
2. Gennaro, A.R., *Remington: The Science and Practice of Pharmacy,* Mack, Easton, PA, 1995.
3. Khan, M.A. and Reddy, I.K., *Pharmaceutical and Clinical Calculations,* Technomic, Lancaster, PA, 1997.
4. Leff, R.D. and Roberts, R.J., *Practical Aspects of Intravenous Drug Administration,* ASHP, Bethesda, MD, 1992.
5. Pickar, G.D., *Dosage Calculations,* Delmar, Albany, NY, 1999.
6. Stoklosa, M.J. and Ansel, H.C., *Pharmaceutical Calculations,* Williams & Wilkins, Media, PA, 1996.
7. Terry, J. *Intravenous Therapy: Clinical Principles and Practice. Philadelphia,* W.B. Saunders, Philadelphia, 1995.
8. Thompson, J.E., *A Practical Guide to Contemporary Pharmacy Practice,* Williams & Wilkins, Baltimore, MD, 1998.
9. Turco, S., *Sterile Dosage Forms,* Lea & Febiger, Philadelphia, 1994.

Module IV

14 Transdermal and Topical Drug Delivery Systems

*Bhaskara R. Jasti, William Abraham, and Tapash K. Ghosh**

CONTENTS

* No official support or endorsement of this article by the FDA is intended or should be inferred.

0-4152-8863-0/05/$0.00+$1.50
© 2005 by CRC Press LLC

Learning Objectives

After studying this chapter, the student will have thorough knowledge of:

- Transdermal and topical drug delivery systems
- Different parameters that determine the permeability of drugs across skin
- Different types of transdermal delivery systems
- Different modes of permeation enhancement
- Various topical delivery systems
- Differences between transdermal and topical delivery systems

I. TRANSDERMAL DELIVERY SYSTEMS

A. Introduction

Over the past 3 decades, developing controlled drug delivery systems has become increasingly important in the pharmaceutical industry. This is in response to the general belief that a substantial number of active drugs exist for different therapeutic areas, but their uses are limited owing to their side effects. The pharmacological response, both the desired therapeutic effect and the undesired adverse effect, to a drug is dependent on the concentration of the drug at the site of action, which in turn depends upon the dosage form and the extent of absorption of the drug. Thus, developers of advanced drug delivery systems seek multiple benefits such as improved efficacy through sustained delivery, improved efficacy through better bioavailability, less toxicity and reduced side-effects, and more patient-friendly dosing. Also, advanced delivery systems extend the proprietary status of drugs facing patentexpiration, thereby providing increased profitability.

For these reasons, a broad range of drug delivery technologies are being developed. Transdermal drug delivery technology has emerged at the forefront of these new technologies starting in the

TABLE 14.1
Examples of Transdermal Systems Currently on the Market

Brand Name	Sponsor	Active	System Design	Indication
Alora	Watson	Estradiol	Adhesive matrix	Hormone replacement
Androderm	Glaxo SmithKline	Testosterone	Reservoir/membrane	Hypogonadism
Androgel	Unimed	Testosterone	Gel	Hypogonadism
Catapres-TTS	Boehringer Ingelheim	Clonidine	Reservoir/membrane	Hypertension
Climara	Scherig AG/Berlex	Estradiol	Adhesive matrix	Hormone replacement
Combipatch	Novartis	Estradiol/norethindrone aceate	Adhesive matrix	Hormone replacement
Duragesic	J & J, Janssen	Fentanyl	Reservoir/membrane	Analgesic
Esclim	Women First Healthcare, Inc.	Estradiol	Adhesive matrix	Hormone replacement
Estraderm	Novartis	Estradiol	Reservoir/membrane	Hormone replacement
Habitrol	Novartis	Nicotine	Adhesive matrix	Smoking
Minitran	3M	Nitroglycerin	Adhesive matrix	Angina
Nicoderm	Glaxo Smith Kline	Nicotine	Reservoir/membrane	Smoking
Nicotrol	Pharmacia and Upjohn	Nicotine	Adhesive matrix	Smoking
Nitrodur	Key Pharma/ Schering	Nitroglycerin	Adhesive matrix	Angina
Orthoevra	R.W. Johnson	Norelgestremin/ethiyl estradiol	Adhesive matrix	Contraception
Oxytrol	Watson	Oxybutynin	Adhesive matrix	Urinary incontinence
Prostep	American Home Products	Nicotine	Gel matrix	Smoking
Testim	Auxillium	Testosterone	Gel	Hypogonadism
Testoderm-TTS (nonscrotal)	Alza	Testosterone	Reservoir/membrane	Hypogonadism
Testoderm (Scrotal)	Alza	Testosterone	Adhesive matrix	Hypogonadism
Transderm-Nitro	Novartis	Nitroglycerin	Reservoir	Angina
Transderm-Scop	Novartis	Scopolamine	Reservoir	Motion sickness
Vivelle	Novartis	Estradiol	Adhesive matrix	Hormone replacement

1980s, owing to its proven commercial success and ready patient acceptance. Transdermal drug delivery development was mostly carried out in 1980 to 1995 as indicated by the number of patents awarded and products that entered interstate commerce. The total sales from advanced drug delivery sales is expected to reach $14 billion, with transdermal products accounting for $1 to 1.5 billion by 2001. Strong sales of advanced drug delivery systems are anticipated over the next decade, as an unprecedented number of drugs losing patent protection and the management of drug product life cycles is ever increasing. Also, the increase in biotechnology-derived drug compounds has necessitated the use of drug delivery technology in the early stages of drug development.

Any type of formulation that allows a drug substance to transit from the outside of the skin through the layers of the skin to underlying tissues and finally into the systemic circulation to exert a pharmacological action is defined as a transdermal system. Examples of transdermal systems currently on the market are listed in Table 14.1.

B. Medical Rationale

1. Bypasses First-Pass Metabolism

Upon oral administration, two competing processes, absorption and degradation, determine the fate of the ingested drug substance. The degradation of drug could be hydrolytic owing to the fairly

TABLE 14.2
Advantages of Transdermal Delivery

Avoids the unpredictability associated with gastrointestinal absorption and produces more uniform plasma levels.
No hepatic first pass effect (bypasses first pass metabolism) and an improvement in bioavailability.
Noninvasive, thus no pain associated with administration.
Desired plasma levels can be maintained over longer duration.
Increases patient compliance.
Reduces frequency of administration.
Reduces side effects.

hostile acidic pH of the stomach or enzymatic owing to the enzymes present within the cell-wall or inside the cells when the drug is absorbed into the cells. The drug absorbed across the mucosa of the different parts of the gastrointestinal tract from the stomach to the colon passes into the hepatic portal vein, which carries the blood directly to the liver, an organ where major biotransformation processes take place (first pass effect). Some well-known drugs such as aspirin, nitroglycerin, verapamil, desipramine, metoprolol, and others undergo extensive metabolism in the liver. Additionally, interaction with food, drinks, or other drugs in the stomach may prevent the drug from permeating through the mucosal lining of gastrointestinal tract. Transdermal delivery avoids the erratic drug absorption through the gastrointestinal tract, as it bypasses the hepatic clearance by delivering the drug directly into the systemic circulation.

2. Noninvasive

The second most utilized routes for drug administration are the parenteral routes such as intravenous, intramuscular, and subcutaneous injection. However, these are invasive and not readily accepted by patients. Transdermal application is noninvasive and is readily accepted.

3. Ideal for Drugs with Short Biological Half-Life

Some biological molecules have a very short half-lives ($t_{1/2}$) (e.g., $t_{1/2}$ for nitroglycerin is 3 min), whereas others have very long half-lives (e.g., $t_{1/2}$ for digitoxin is 7 to 9 days). For oral and parenteral drugs with very short biological half-lives and high body clearance rates, frequent dosing is required to elicit and maintain a pharmacological effect. A highly frequent dosage schedule may not generally suit patients. A transdermal system can provide a steady drug input across skin, thereby maintaining the desired steady drug concentration in blood over a scheduled period of time. The desired blood drug concentration can be obtained by designing the transdermal delivery system with appropriate delivery profiles. This is in contrast to oral and intravenous bolus injection administration, where the duration of action (the phase in which the plasma concentrations are within the therapeutic window) is relatively short. As a result, transdermal delivery systems improve patient compliance by reducing the frequency of administration. We currently have transdermal delivery systems available in the market that can deliver a drug over a period of 7 days (Catapress TTS). If such a system results in an adverse reaction or overdose, the patient can easily remove the delivery system, terminating the adverse effect. The advantages of transdermal delivery are summarized in Table 14.2.

C. The Barrier Property of Skin

1. Structure of Epidermis

The epidermis makes up the outermost layers of skin and is 150 to 200 μm thick. In contrast to gastrointestinal mucosa, which is made to efficiently absorb different molecules into the body, the

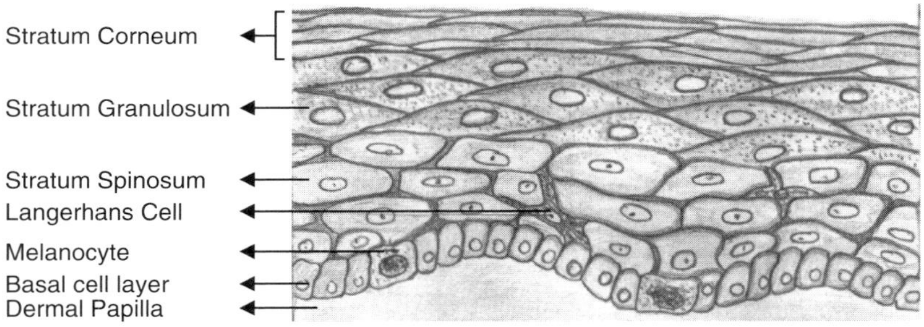

FIGURE 14.1 Structure of the epidermis.

primary function of skin is to protect against the entry of foreign bodies into the body. The epidermis is a stratified squamous epithelium and is not well suited for drug absorption. There are four morphologically distinct regions of the epidermis: the basal layer, the spiny layer, the stratum granulosum, and the uppermost region, the stratum corneum. These layers provide a formidable barrier to the passage of drug molecules as shown in Figure 14.1. The epidermis undergoes programmed death. During this process, the epidermal cells undergo extensive differentiation starting from the highly proliferative basal layer until they form the stratum corneum, consisting of highly cornified (dead) cells embedded in a continuous matrix of lipid membranous sheets. These extracellular membranes are unique in their composition and are composed of ceramides, cholesterol, and free fatty acids. The nonpolar lipids are saturated and the membranous structures formed by these lipids are less permeable than a typical plasma membrane. Phospholipids, a common component of most biological membranes, are absence in the stratum corneum. The unique architecture of the stratum corneum, often compared to a wall made of bricks and mortar (the bricks representing the corneocytes and the mortar representing the extracellular lipids), presents a tortuous pathway around the corneocytes and across the extracellular membranes, thus providing a formidable barrier to the diffusion of drug molecules. The extracellular lipid route occupies ~1% of the stratum corneum's productive diffusional area (in the plane of the skin, as shown in Figure 14.1).

2. Pathways of Absorption

The different pathways by which drugs can penetrate through skin before they reach the systemic circulation are shown in Figure 14.2. A drug can penetrate skin by transcellular (across the cells), intercellular (between the cells), or transappendeageal (through hair follicles, sebaceous glands,

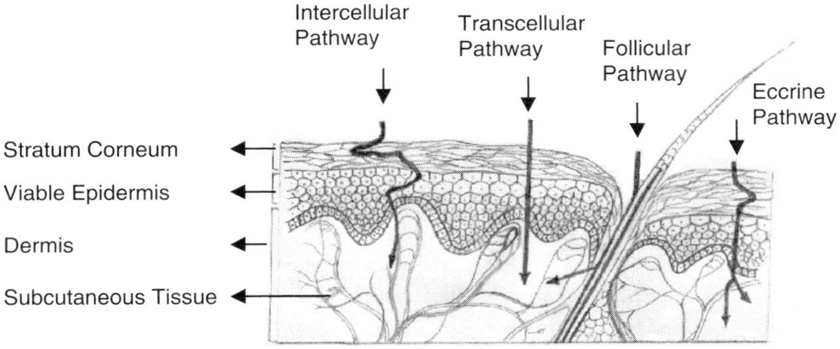

FIGURE 14.2 Pathways of transport through skin.

etc.) pathways. The pathway by which a majority of drug transport occurs is the transcellular pathway. Since the extracellular space is concentrated with lipids, the lipid solubility of a molecule is very important for its transdermal delivery. The intercellular pathway contributes an insignificant amount to the total transport of any molecule. However, peptides and other polar molecules are transported mainly by transappendeageal pathways, especially under iontophoretic conditions.

D. Drug Selection Criteria Based on Pharmacokinetic Parameters and Physicochemical Properties of the Drug

In addition to the medical rationale described earlier, the following selection criteria need to be considered in evaluating the suitability of a transdermal route for the delivery of a drug substance:

- Pharmacokinetic information
- Skin permeability
- Physicochemical properties of the drug
- Skin toxicity
- Aesthetic properties

1. Pharmacokinetic Consideration

Pharmacokinetic information about the drug is a critical factor in deciding its suitability for delivery by the transdermal route. Pharmacokinetics provides information on the disposition of a drug in our body, whereas pharmacodynamics provides information on its pharmacological effect. Pharmacokinetic parameters such as effective plasma level, necessary for maintenance of steady state plasma levels, will determine whether a transdermal delivery can be developed or not. For example, insulin secretion in the human body is pulsatile, as it is controlled by a feedback mechanism that is determined by blood glucose levels. Maintenance of constant levels of insulin in diabetics is not desirable, as individuals may become hypoglycemic and eventually pass out.

One can estimate the skin input rate of a drug required from its transdermal system based on pharmacokinetic parameters such as volume of distribution (V_d), total body clearance (CL_t), and steady state or therapeutic plasma concentration (C_{Pss}). Under steady state conditions, the drug input rate from its transdermal system is expected to equal its output rate, determined by total body clearance multiplied by the therapeutic plasma concentration. This relationship can be expressed using the following mass balance equation.

$$\text{Input rate} = \text{dosing rate} \times \text{bioavailable fraction } (F)$$

$$\text{Output rate} = \text{total body clearance} \times \text{steady state plasma concentration at steady state,}$$

$$\text{Input rate} = \text{output rate}$$

or

$$\text{F} \times \text{dosing rate} = CL_t \times C_{pss} \tag{14.1}$$

where CL_t is total body clearance, and C_{pss} is average target plasma concentration. Since the epidermis is metabolically inert, $F = 1$ for most drug compounds. Recall that total body clearance is the product of volume of distribution and total elimination rate (k_e)

$$CL_t = k_e \times V_d \tag{14.2}$$

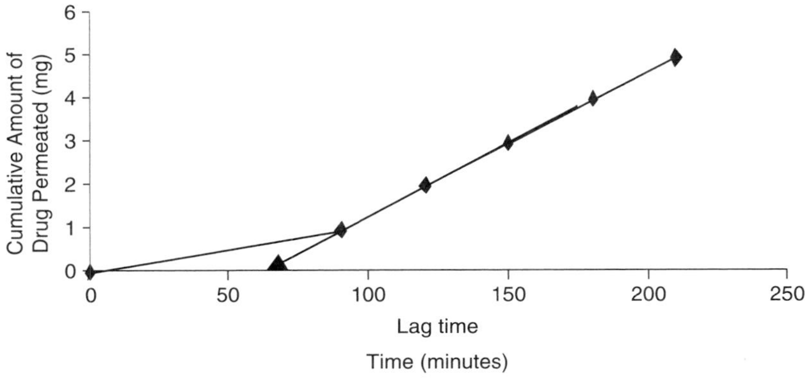

FIGURE 14.3 Percutaneous drug penetration.

Thus, the required flux from a transdermal patch can be calculated by normalizing the dosing rate (Equation 14.1) for the surface area (A, cm²):

$$\text{Flux, } J_{ss} = \frac{CL_t \times C_{pss}}{A} \tag{14.3}$$

2. Skin Permeability

The skin permeability of a drug candidate is a key criterion in assessing its delivery by the transdermal route. Skin permeation from a delivery system can be maximized by maximizing the drug's thermodynamic potential at the skin surface by maintaining the drug as a saturated solution. The amount of drug (M_t) passing through a unit area of skin in time t from a saturated solution is summarized as:

$$M_t = (K \times C_s)\left[\frac{D_{sc} \times t}{h_{sc}} - \frac{h_{sc}}{6} - \frac{2 \times h_{sc}}{\Pi^2} \times \sum_{i=1}^{\infty}\left\{\frac{-1^n}{n^2}\right\} \times \exp\left\{-D_{sc} \times n^2 \times \Pi^2 \times \frac{t}{h_{sc}}\right\}\right] \tag{14.4}$$

in which C_s is the saturated concentration of the drug and K is the skin–vehicle partition coefficient. The product KC_s gives the saturation concentration of the drug within the skin surface. In this equation, D_{sc} and h_{sc} are the effective diffusion coefficient and effective thickness of the stratum corneum, respectively. An illustrative curve based on Equation 14.4, where the cumulative amount of drug permeated is plotted as a function of time as shown in Figure 14.3.

The curve shows that no drug penetrates the skin immediately upon application of the drug delivery system on the skin. The flux gradually increases after the first molecules pass through the skin and reaches a steady level. Thus, the permeation of drug across skin as $t \Rightarrow \infty$ is described as

$$M_t = \frac{K \times D_{sc} \times C_s}{h_{sc}}\left\{t - \frac{h_{sc}}{6D_{sc}}\right\} \tag{14.5}$$

Equation 14.5 describes a situation in which the amount of drug penetrating per unit time is constant, that is, the steady state condition. So, as the system approaches steady state, the drug flux, J_{ss}, is obtained by differentiating Equation 14.5 with respect to t, as shown below:

$$J_{ss} = \frac{\partial M_t}{\partial_t} = \frac{K \times D_{sc} \times C_s}{h_{sc}} \tag{14.6}$$

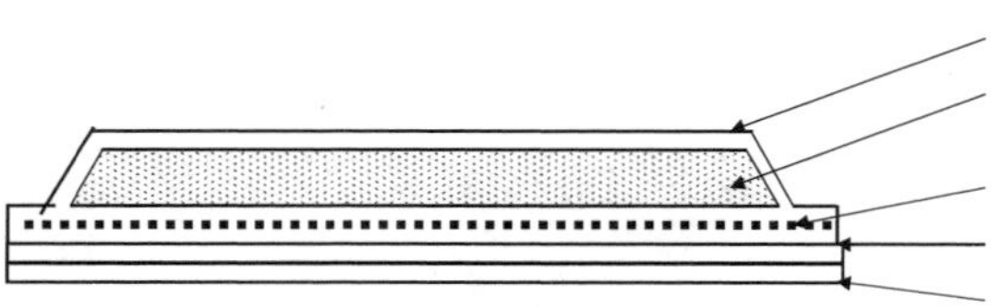

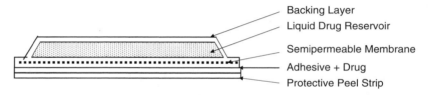

FIGURE 14.4 Reservoir type design.

As shown in Figure 14.4, below, no drug permeates immediately upon application to the skin, but slowly reaches steady state; the time interval between these two points is the lag time (t_L). Lag time is determined by extrapolating the steady state line to the time axis and is the value of t at $M = 0$. When Equation 14.6 is integrated and simplified to $M = 0$, lag time can be related to thickness of the skin and the effective diffusion coefficient.

$$t_L = \frac{h_{sc}^2}{6D_{sc}} \tag{14.7}$$

Thus, by measuring the amount of drug diffusing across the epidermis in vitro and plotting against time (similar to Figure 14.3), one could determine the lag time for a given drug substance from a given formulation from the intercept on the time axis. Also, Equation 14.7 illustrates that the lag time is inversely related to the drug diffusivity, D_{sc}. While assessing the delivery of drug by a transdermal route the lag time needs to be a small fraction of the projected time of application. A drug with low diffusivity and therefore long lag time is not a good candidate for transdermal delivery. A rule of thumb in selecting a drug candidate is that the lag time needs to be less than 20% of the intended duration of the system.

a. Permeability Coefficient

The thickness in Equation 14.7 is the length of the diffusion pathway across the stratum corneum or the barrier membrane and cannot be physically measured. So it is not feasible to calculate the diffusion coefficient of a drug substance across skin. Thus, for a complex membrane such as the skin, derived values for D_{sc}, K, and h_{sc} depend on assumptions made about the membrane. However, one could combine the diffusion coefficient, membrane thickness, and the partition coefficient into a single parameter, the permeability coefficient, which could be experimentally measured for different drug substances. Thus, Equation 14.6 could be rewritten in terms of permeability coefficient as

$$J_{ss} = P \times C_s \times A \tag{14.8}$$

and P is defined as

$$P = \frac{KD}{h_{sc}} \tag{14.9}$$

where P is the mass transfer coefficient or permeability coefficient. Equation 14.8 describes the flux across skin and is determined by the slope of the cumulative plot shown in Figure 14.4, while Equation 14.9 is used to determine the permeability coefficient from *in vitro* transport study data, using the values for flux, the concentration of drug used in transport studies, and the area of skin across which transport was measured. The permeability coefficient is dependent on the chemical structure of the drug, the physicochemical nature of the medium of application, and the skin. When the delivery medium and skin are fixed, then the permeability coefficient is dependent on the differences in chemical structure and therefore the physicochemical properties of the drug and are a useful predictor of the suitability of the drug as a transdermal candidate.

The larger the value of P in Equation 14.8, the greater the flux of the compound. The permeability coefficients for human skin range from $\sim 10^{-6}$ to $\sim 10^{-2}$ cm.h^{-1}. If we use a reference drug concentration of 1 mg/ml and the maximum permeability coefficient of 1×10^{-2} cm.h^{-1}, an estimated 10 μg of drug passes through the skin per hour per square centimeter. This amounts to ~0.24 mg.cm^{-2}.day^{-1}. If we assume a maximum patch size of 100 cm^2, this amounts to a maximum of 24 mg of drug delivered across skin. Nitroglycerin is at the high end of the permeability scale (10^{-2} cm.h^{-1}) and represents the maximum amount of a drug delivered by the transdermal route. At the lower end of the permeability scale (10^{-6} cm.h^{-1}), the maximum amount of drug delivered in a 24-h period would be negligible.

3. Physicochemical Factors of Drug Molecules Affecting Skin Permeability

The following physicochemical factors play critical roles in the permeability of molecules across skin.

- Solubility
- Crystallinity
- Molecular weight
- Polarity

a. *Solubility*

As shown in Equation 14.9, the concentration of the drug in the delivery medium is a critical factor in determining its permeability. The limiting permeation rate is obtained when permeation is from a saturated solution. Thus, one must know the solubility of the drug in the medium used in the assessment of permeability coefficient. Several factors have a bearing on solubility, such as polarity and melting point of the drug molecule as described below.

b. *Crystallinity and Melting Point*

The concentration of the drug in any medium is related to heat of fusion and melting point. According to ideal solution theory, the solubility of a drug is related to two important thermodynamic parameters, heat of fusion and melting point as described by the equation:

$$\ln a = \Delta H_m \left(T - T_m\right) / RT_m T \tag{14.10}$$

where a is the thermodynamic activity and is defined as $a = x\gamma$, x is the mole fraction, and γ is the activity coefficient; ΔH_m is the heat of fusion, T_m is the melting point, and R is the universal gas constant. For crystalline compounds, ΔH_m is of the order of 5000 Kcal/mole, and Equation 14.10 predicts that high-melting, hard crystalline materials with large enthalpies of fusion are less soluble than soft, low-melting compounds. As a first approximation, the deliverable amount of crystalline compounds goes down by a factor of 20 when the melting point is 200°C and a milligrams per day dose of such compounds by the transdermal route becomes difficult to attain. Hydrophobic molecules generally have a low degree of crystalinity and accordingly tend to be more soluble. However, owing to the very small net negative free energy of solution of hydrophobic molecules in water, these molecules have low aqueous solubility. They are highly soluble in the lipophilic phase of skin and hence have large permeability coefficients that tend to compensate for their low solubility in water. Molecules containing multiple hydrogen bonding centers or strong dipoles are generally high melting. These molecules dissolve in water owing to their strong interaction with water but do not partition into the lipoidal phase of the stratum corneum. Consequently, their transport across skin from even saturated solutions is minimal.

c. Molecular Weight

Molecular weight of the drug molecule is another important property that affects its ability to diffuse across skin. The diffusion coefficient is a measure of diffusional resistance offered by the membrane to any molecule and is related to the molecular size as

$$D = D_0 \bullet \exp^{-\beta \bullet MV} \tag{14.11}$$

where D_0 is the diffusivity at zero void volume, β is the measure of the average free-volume available for diffusion, (30 cm^3/mol for lipid membrane), and MV is the molecular volume, which can be substituted with molecular weight. By substituting D from Equation 14.9 and rearranging Equation 14.11, the permeability coefficient can be related to molecular volume as

$$\log \frac{P}{K} = \log \frac{D_0}{h_{sc}} - \beta \bullet MV \tag{14.12}$$

For small molecules, molecular volume can be approximated to molecular weight to get

$$\log \frac{P}{K} = \log \frac{D_o}{h_{sc}} - \beta \bullet MW \tag{14.13}$$

in which is a measure of the reciprocal average free-volume available for diffusion, (~30 cm^3/mol for lipid membranes). D is related to molecular volume, which in turn depends upon the molecular weight and the permeability, and consequently on the flux.

d. Polarity

The polarity of a drug molecule affects its skin permeability by impacting the partition coefficient in Equation 14.6. The octanol–water partition coefficient ($K_{o/w}$) is often used instead of the skin/vehicle partition coefficient. Based on the diffusion of a number of molecules, an empirical relationship between permeability and $K_{o/w}$ has been developed. This empirical relationship holds well in the range of log $K_{o/w}$ −2.3 and +2. For compounds with log $K_{o/w}$ between −2.3 and +2.0, the permeability coefficients can be empirically determined as log P = log $K_{o/w}$ − 3.698. For compounds with log $K_{o/w}$ <−2.301, a lower permeability limit of 1 × 10–6 cm/h was assigned, and for compounds with log $K_{o/w}$ >2.0, an upper limit of permeability coefficient of 1 × 10–2 cm/h was assigned.

4. Skin Toxicity

A drug could have all the physicochemical properties that are essential to be a good transdermal candidate but elicit a powerful toxic response. This toxicity may be a simple skin irritation that can be treated with nonsteroidal antiinflammatory agents (NSAID's) or a complex immunological reaction that would be difficult to treat. Such immune responses may result from drugs that are well tolerated when given orally. This unwelcome toxicity of transdermal route of administration has been attributed to the presence of histamine-releasing mast cells, which in turn trigger the immunological toxicity. A typical example of this is the triprolidine (antihistamine) transdermal delivery system, which was found to elicit a powerful sensitization response in clinical testing, and consequently the transdermal system development was withdrawn.

TABLE 14.3
Ideal Properties of a Drug Candidate for Transdermal Drug Delivery

Properties	Comments
Dose	Should be low (<20 mg/day)
Half-life	10 h or less
Molecular weight	<400
Melting point	<200°C
Partition coefficient	Log P (octanol–water) between –1.0 and 4
Skin permeability coefficient	>0.5 × 10^{-3} cm/h
Skin reactions	Nonirritating and nonsensitizing
Oral bioavailability	Low
Therapeutic index	Low

5. Aesthetic Properties

A drug candidate may have very favorable physicochemical and pharmacokinetic properties but may need a larger patch owing to a large dose (>100 cm^2); or the patch adhesive may ooze out during wear; or the adhesive might be too strong and difficult to peel after use, rendering it unacceptable to patients. The size requirement for the patch quite often depends on the target patient population. Any one of these aesthetic requirements could prevent a transdermal system from reaching the market. Ideal properties of a drug candidate for transdermal drug delivery are summarized in Table 14.3.

E. Designs of Passive Transdermal Systems

Several transdermal delivery systems have reached the market in the U.S. Often, they are referred to as *matrix system*, *reservoir system*, or *drug in adhesive system*. These terms are based on the drug release mechanisms from such systems and can be misleading. An easier way to classify these systems is based on the components and the fabrication of the systems. Transdermal systems presently on the market can be classified into the following four types:

- *Type I:* Semisolid amorphous ointment, cream lotion, or viscous dispersions applied directly on the skin (Nitrobid, Progestagel, and Estragel). *Nitrobid is an ointment containing 2% nitroglycerin in a lanolin and white petrolatum base. Each inch, as squeezed from the tube, contains approximately 15 mg of nitroglycerin.*
- *Type II:* Reservoir (liquid form, fill and seal laminate structure; Transderm-Nitro and Estraderm). Transderm-Nitro is a transdermal system that delivers 0.1 to 0.6 mg/h of nitroglycerin. The required dose can be delivered by choosing systems of different surface areas. Transderm-Nitro is shown in the following schematic (Figure 14.4). In this example, the system is composed of four layers proceeding from the top layer. These layers are (a) a tan backing layer that is impermeable; (b) a drug reservoir containing nitroglycerin adsorbed on lactose, colloidal silicon dioxide, and silicone medical fluid; (c) an ethylene vinyl acetate copolymer membrane that is permeable to nitroglycerin; and (d) a layer of hypoallergenic silicone adhesive. Prior to use, the protective peel strip is removed from the adhesive surface. The drug is mixed in a liquid form that is filled and sealed to be put on the skin. Transdermal scientists routinely refer this as pillow design owing to the thickness of the system.

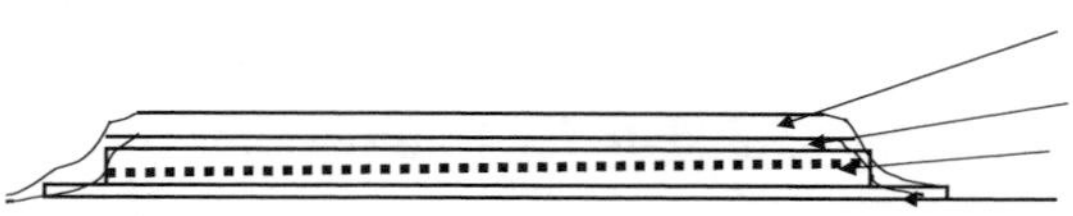

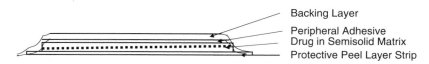

FIGURE 14.5 Matrix type design.

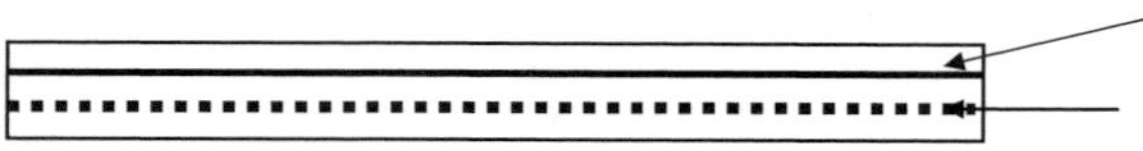

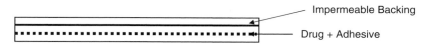

FIGURE 14.6 Drug in adhesive type design.

- *Type III:* Matrix (peripheral adhesive laminate structure, Nitro-Disc, NitroDur I). *In this design, the adhesive layer is separated from the drug-containing semisolid (as opposed to liquid in the reservoir type) matrix layer by a barrier membrane. The adhesive layer exposed on the periphery of the transdermal system keeps it in place (Figure 14.5)*
- *Type IV:* Drug in adhesive (solid state laminate structure, Nitro Dur, Transderm Scop, Catapres, Deponit, Nicotrol). *In this design, the drug is uniformly dispersed in the adhesive. An example is NitroDur from Key Pharmaceuticals. This design yields a thin transdermal system with controlled release of nitroglycerin. The delivered dose can be adjusted by selecting systems of different surface areas. This system consists of an impermeable backing layer and an adhesive that contains nitroglycerin in acrylic-based polymer adhesives with a resinous cross-linking agent to provide a continuous source of active ingredient. The adhesive keeps the system in place during the wear period. This system is the most advanced design and provides a great degree of comfort and aesthetic appeal (Figure 14.6).*

The majority of the passive transdermal drug delivery systems are prepared by one of the following methods:

- *Form-fill-seal method:* This method is used to create reservoir patch designs in which a drug-containing gel or solution is metered onto the backing membrane; a laminated web consisting of a membrane, adhesive, and release liner is added, and the backing is sealed to the membrane around the periphery of the metered gel or solution to form a reservoir compartment.
- *Solvent cast method:* In this method, the adhesive polymer solution is coated over a backing polyethylene/polyester membrane (at a predetermined wet thickness). This coated adhesive film is dried by passing through a variable temperature oven to drive the solvent off. At this stage, a release liner is laid on top of the cured adhesive and laminated. The complete process is carried out under GMP environment. Patches of fixed surface area are die cut and pouched for clinical trials.
- *Extrusion method:* A drug-containing elastomer is extruded or molded between a suitable backing and a release liner to a predetermined cross sectional shape, then sliced to the desired thickness.

Ideal properties of a transdermal drug delivery system are listed in Table 14.4.

F. Product Development

The product development steps involved in transdermal drug delivery are somewhat unique compared to traditional dosage forms. Other than selection of the drug candidate and excipients,

TABLE 14.4
Ideal Properties of a Transdermal Drug Delivery System

Properties	Comments
Shelf life	Up to 2 years
Patch size	$<40\ cm^2$
Dose frequency	Once a day to once a week
Aesthetic appeal	Clear, tan, or white color
Packaging	Easy removal of release liner and minimum number of steps required to apply
Skin adhesion	No fall-off during dosing and no cold-flow
Skin reactions	Nonirritating and nonsensitizing
Release	Consistent pharmacokinetic and pharmacodynamic profiles over time

selection of the patch design is an integral component of the product development cycle. Because of the uniqueness of this dosage form, the following questions need to be answered to define the final product:

- Target therapeutic concentration
- Dose to be systemically delivered
- Maximum patch size acceptable by the target population
- Preferred site of application
- Preferred application period (daily, biweekly, weekly, etc.)

Once the preferred final product description has been established, an evaluation of the drug candidate begins. Because of the limitation of loading dose in a patch, and a practical patch size, not all drugs can be candidates for transdermal drug delivery. The following physicochemical and pharmacological parameters of a drug molecule are useful in assessing suitability and predicting the maximum attainable delivery via the transdermal route:

- Chemical structure and molecular weight
- Melting/boiling point
- PKa (s)
- Oil–water partition coefficient
- *In vitro* skin permeation rate
- Aqueous and polymer solubility
- Solid sate and solution stability
- Known chemical incompatibilities
- Known decomposition pathways
- Pharmacokinetic parameters (ADME, half-life, clearance, etc.)
- Toxiclogical profile

The ultimate decision regarding which of the four transdermal product design alternatives to select should be based on a thorough evaluation of the physicochemical and pharmacological characteristics of the drug candidate, the preferred product description, and consideration of the specific strengths and weakness of each design. Often a paper feasibility assessment compiling all the above information is made before proceeding with further steps.

The product development of a transdermal formulation generally includes the following stages:

- Selection of drug candidate
- Selection of the appropriate physical form (e.g., acid, base, or salt)

- Selection of the desired design (e.g., reservoir, matrix, laminate, etc.)
- Preparation of prototype formulations and testing of their physicochemical properties (tack, shear, peel adhesion, skin adhesion, etc.)
- Evaluation of *in vitro* permeation
- Development of analytical methods to quantitate drug in the formulation, skin layers, release medium, and blood (if applicable)
- Evaluation of potential for systemic adverse events (e.g., carcinogenicity, teratogenicity, mutagenicity, etc.)
- Evaluation of skin toxicity (irritation, sensitization, etc.) in animals and humans
- Microbial and preservative testing, if necessary
- Phase I, II, and III human clinical trials
- Scale-up activities including development of specifications
- Postapproval market surveillance

Figure 14.7 shows an overall scheme of the various stages involved in the product development of a transdermal dosage form. Though the figure was developed mostly with the development of topical dosage forms in mind (discussed later), owing to the similarity of the two dosage forms, most of the features are also applicable for transdermal dosage forms. The figure shows that even after the proprietary product has been launched, the competition from the other brand and generic manufacturers necessitates the production of line extensions or prescription to over-the-counter switches. Any changes in the original formulation during these stages will again require the manufacturer to go through the product development process. However, the Food and Drug Administration (FDA) has published guidance (SUPAC) to make these postapproval changes simpler.

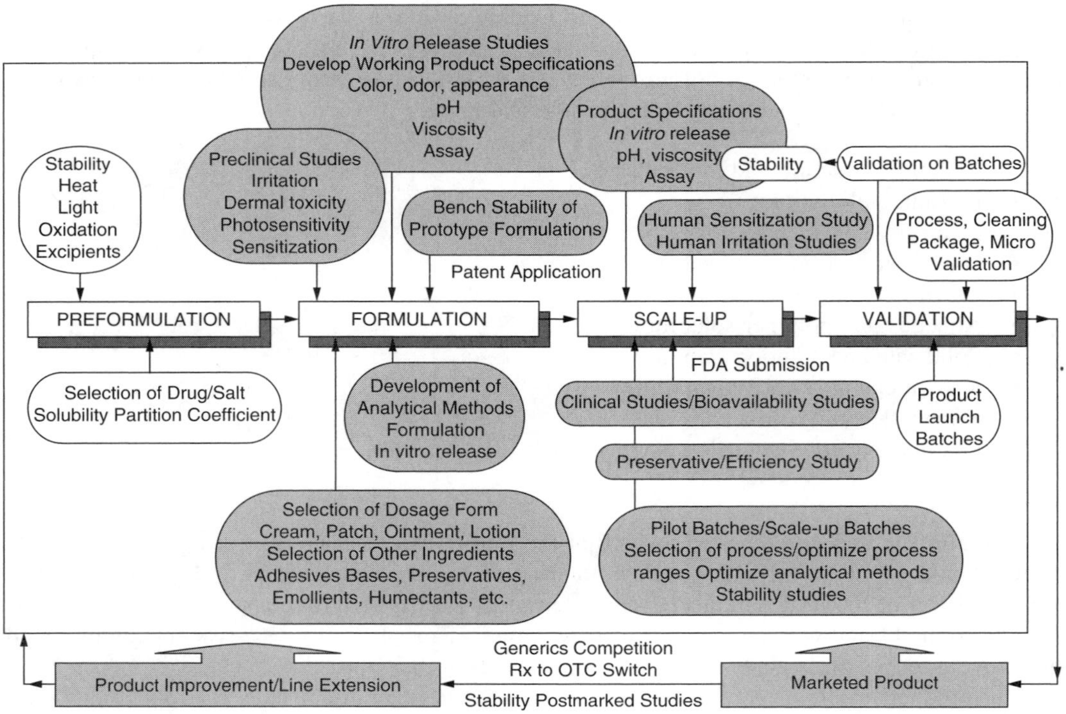

FIGURE 14.7 Product development scheme of transdermal/topical dosage forms.

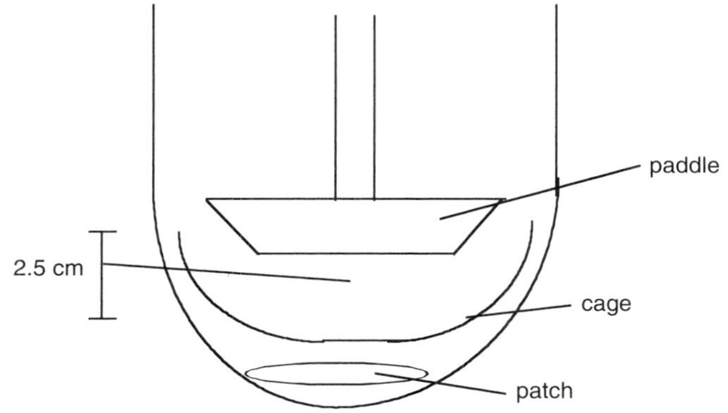

FIGURE 14.8 Schematic of USP dissolution apparatus.

G. Quality Control

1. Stability of the Assembled System

After the transdermal systems are fabricated, the systems are monitored for stability of the drug release profile as a function of equilibration time. Any phase separation or partitioning of drug or excipients during equilibration could significantly affect the drug release profile if the initial system did not reach equilibrium during the fabrication process. The assembled systems are usually evaluated for potency, content uniformity, purity, residual solvent, release profiles, *in vitro* skin flux, tack, shear, peel adhesion, skin adhesion, etc. as a function of aging.

2. Release Profile

For transdermal drug delivery systems, drug release studies are generally performed using a USP type 5 dissolution apparatus with 100-ml cells as shown in Figure 14.8. The paddles are adjusted to a height of 2.5 cm from the bottom of the release cage. In each cell a 1.267-cm^2 patch is fixed to the bottom, with the adhesive side up. Then a cage is placed on top of the patch. An aqueous solution that provides sink conditions for the release of drug is used as the release medium. The paddles are rotated at 49.0 ± 0.3 rpm, and the bath temperature is set at 32.5°C. Aliquots of 1.5 ml are withdrawn at intervals of 8 and 16 h and replenished with an equal amount of fresh buffer. The drug content in the samples is analyzed by assay procedures provided in compendia. The release profiles of prototypes upon storage should be within 10% of that obtained immediately after fabrication. Any deviation could indicate potential stability problems.

3. *In Vitro* Assessment of Transdermal Delivery Systems

Skin permeation of a drug from transdermal delivery systems is evaluated by Franz diffusion cells (several versions are commercially available) with a defined diffusional area, which can accommodate receiver volume of 3 to 7.5 ml as shown in Figure 14.9. Franz diffusion cells have two half-cells: a donor chamber in which the formulation (if liquid or kept empty if it is a patch) is placed and a receiver chamber that is filled with the release medium and monitored for drug transport. The release medium is generally an aqueous or hydroalcoholic solution in which the drug has sufficient solubility to maintain sink conditions for transport. A suitable size of heat-separated human epidermis is die cut and mounted between the two halves of the Franz cell, with the stratum corneum side facing the transdermal drug delivery system. Aliquots of predetermined

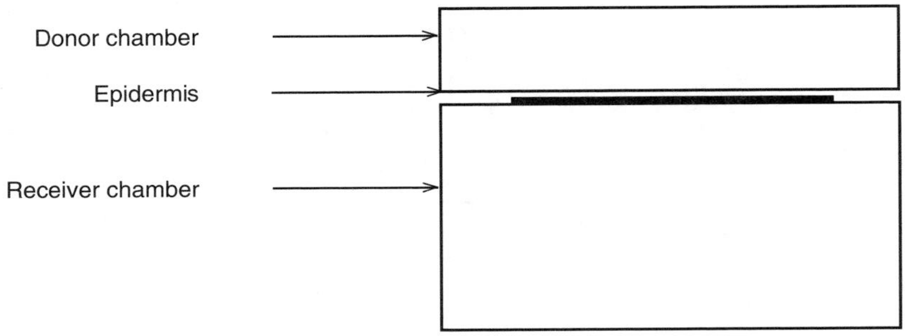

FIGURE 14.9 Schematic of Franz diffusion cell.

volume are withdrawn from the receiver chamber at periodic intervals, and the receiver medium is replenished during transport studies. All samples are analyzed for drug content using the specified analytical method. The amount of drug accumulated in the receiver fluid is measured as a function of time and plotted. Steady state flux is calculated as the slope of the linear region of this cumulative plot.

4. *In Vivo* Assessment of Transdermal Delivery Systems

Transdermal drug delivery has similarities to both oral extended release and parenteral dosage forms. Transdermal drug delivery resembles oral extended-release dosing because drug absorption occurs over an extended period of time, whereas it is similar to infusion dosing because delivery tends to achieve a steady state within a single dosing interval. Currently, the FDA has a guidance, "Bioavailability and Bioequivalence Studies for Orally Administered Drug Products: General Considerations," issued in October 2000, that is also generally applicable to nonorally administered drug products where reliance on systemic exposure measures is suitable to document BA and BE (e.g., transdermal delivery systems and certain rectal and nasal drug products).

A transdermal product is generally labeled to provide information on the amount of drug that is delivered to the body (systemic circulation). This information is generally obtained from (a) a pharmacokintic study comparing the bioavailability of the drug from the transdermal system to that from intravenous administration, and (b) an assay of the transdermal patches for the remaining drug content (after intended usage). Although information from both studies are important to account for mass balance, the bioavailability study provides more accurate information regarding *in vivo* release and a subsequent pharmacokinetic profile for comparison with traditional routes of delivery (oral and i.v.) for the same compound. The unique pattern of drug delivery from the transdermal system is generally characterized by single and multiple dose pharmacokintic studies. Information related to accumulation of the drug over long-term application is generally obtained from the multiple dose studies followed until steady state. The pharmacokintics of all relevant molecules (drug and metabolites) are determined and preferably compared to the pharmacokintic following administration of the drug from the traditional formulations. However, considering the uniqueness of this complex dosage form, some additional parameters also need to be looked into for this dosage form. They include but are not limited to patch adhesion (tack, shear, etc.), effect of anatomical site of application (as permeability may vary from one site to another), and changing environmental or weather conditions.

At this time, there are no *in vitro* methods that can be substituted for *in vivo* bioequivalence requirements that apply to transdermal drug delivery products. Therefore, a clinical study must be used to evaluate bioequivalence. A generic transdermal drug delivery system is currently required to demonstrate pharmacokinetic equivalence to the reference listed product under a single-patch application.

H. Modes of Permeation Enhancement

Skin permeation is governed by the diffusion equation shown earlier:

$$J = \frac{KD_{sc}C_s}{h_{sc}} \tag{14.6}$$

According to this equation, transdermal flux can be enhanced by (a) increasing the drug solubility in skin (KC_S), (b) increasing the diffusivity of the drug across skin (D_{SC}), or (c) by decreasing the diffusion path length (h_{SC}). This could be accomplished by either chemical or physical means. Thus, we have systems using chemical enhancement as well as physical enhancement.

1. Chemical Enhancement

Adjuvants are commonly used in transdermal formulations for a variety of reasons. Some of these adjuvants or chemical excipients when included in transdermal formulations can enhance the permeability of drugs through a number of mechanisms such as (a) increasing the solubility of drug in the vehicle and hence increasing the total delivered dose from a single application (but this does not necessarily mean the rate of absorption is enhanced); (b) increasing drug solubility in the stratum corneum, that is, facilitating partitioning of drug from the vehicle into the skin; (c) reducing the diffusional barrier of the stratum corneum by either perturbing the intercellular lipid domains or to a lesser degree perturbing the intracellular keratin networks or amplifying transport via the appendages (sweat glands, follicles); and (d) promoting drug partitioning at the stratum corneum–viable tissue interface.

The following classes of chemicals have been studied for permeation enhancement of several drug models: sulfoxides, alcohols, polyols, alkanes, fatty acids, esters, amines and amides, terpenes, surfactants, cyclodextrins, etc. Among these, alcohols, polyols, fatty acids, esters, and mineral oil are used in commercial systems.

The following criteria should be met for an excipient to be used as a penetration enhancer: (a) The chemical excipient should not have inherent pharmacological activity. (b) The permeation enhancement should be specific; (c) reversible; (d) have good chemical and physical stability; (e) be compatible with the rest of the components of the drug delivery system; and (f) should be devoid of toxic, allergic, and irritation potential and should be classified as a generally recognized as safe (GRAS) material by the FDA.

2. Physical Enhancement

Advances made in biotechnology have lead to the development of many therapeutic peptides and proteins. More than 100 such biotechnology based pharmaceuticals are currently in clinical trials. These peptides and proteins are extremely potent and highly specific in their therapeutic activity but are difficult to administer clinically. They are administered by parenteral routes. These potent peptides are intrinsically extremely short acting and therefore require repeated injections. Therapeutic applications and commercialization of such peptides rely on the successful development of viable delivery systems to improve their stability and systemic bioavailability. The difficulties associated with the efficacious delivery of these compounds have forced pharmaceutical companies to evaluate alternate drug delivery modes, and physically enhanced transdermal delivery is at the forefront of such efforts. The major advantage of the transdermal route for systemic delivery of peptides and proteins is the low proteolytic activity of skin compared with oral and other non-parenteral routes. However, the large molecular weight and the polar nature of the therapeutic peptides have precluded them as candidates for passive transdermal delivery. While passive transdermal delivery relies on the thermodynamic potential of the drug molecule in the patch to drive

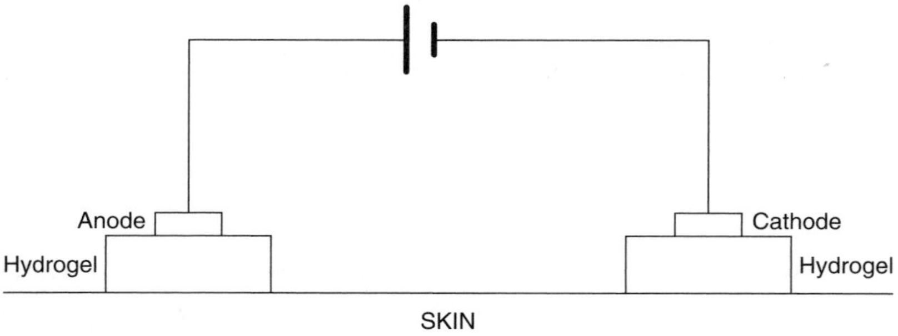

FIGURE 14.10 Schematic of an iontophoretic delivery device.

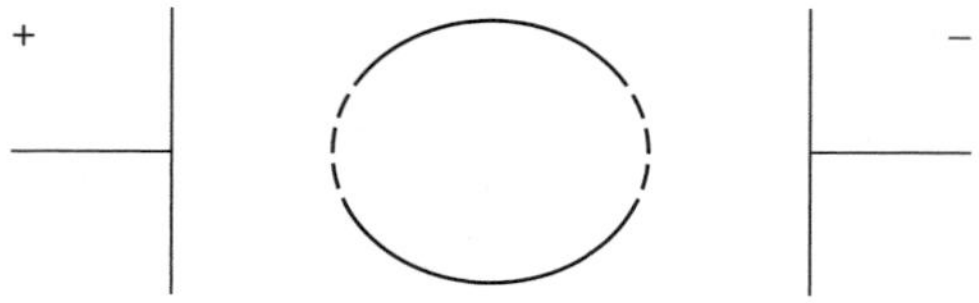

FIGURE 14.11 Schematic showing the formation of pores under electric field.

the compound across skin and is dependent on the molecular size of the compound as described earlier, active transdermal delivery relies on electrical or ultrasound energy to drive these large compounds across skin.

a. *Iontophoresis*

The use of an electric field to enhance the transport of charged and uncharged molecules across skin (iontophoresis) has been explored for transdermal delivery of peptides. Transport can be enhanced by electrostatic repulsion and by electro-osmotic flow. The electric driving force can be in the form of either constant current or constant field (voltage) across the two electrodes. The iontophoretic device consists of a power supply, electrodes, a drug reservoir, and an adhesive to hold the system in place (Figure 14.10). Iontophoresis has the advantage of being programmable since the electric potential can be turned on or off. A constant or a pulsatile delivery profile can be achieved, and the later is an especially attractive alternative to multiple injections of potent peptides that need to be delivered on demand.

b. *Electroporation*

Two decades ago researchers discovered that applying an electric field to a living cell for short periods of time led to the formation of transient pores in the plasma membrane, resulting in increased permeability. When the electric field is discontinued, the pores close in ~1 to 30 minutes without significantly damaging the exposed cells. This phenomenon of transient pore formation under electric field is known as electroporation and is being used to incorporate DNA fragments into cells. As shown in Figure 14.11, the pores are located primarily on the surfaces of cells nearest to the electrodes. This mechanism is currently being investigated in the clinic for the treatment of accessible solid tumors and gene delivery. This is also being explored to deliver peptides and proteins across skin.

c. *Sonophoresis*

Sonophoresis is the ultrasound cousin of iontophoresis. Sonophoresis is the use of sound energy, especially ultrasound, to enhance drug transport across skin. Ultrasound frequencies (from KHz to MHz), when applied to the skin, have been shown capable of increasing skin permeability. In some cases, spectacular results have been reported (e.g., delivery of insulin and larger proteins). Mechanistically, it is unclear exactly how ultrasound elicits its effects, although not surprisingly cavitation

TABLE 14.5
Representative Topical Dermatological Products Used in Different Diseases

Disease	Agents
Acne	Azelaic acid benzoyl peroxide (gel, cream, ointment, etc.)
	Clindamycin (gel, solution, lotion, etc.)
	Erythromycin (gel, solution, ointment, etc.)
	Erythromycin/benzoyl peroxide cream
	Sulfacetamide lotion
	Sulfur/sulfacetamide lotion
	Corticosteroids, retinoids
Rosacea	Metronidazole (gel, cream, and lotion)
Psoriasis	Coal tar, corticosteroids, retinoids
Eczema/dermatitis/pruritis	Corticosteroids, retinoids, tacrolimus, pimecrolimus, antihistamines, emollient creams and ointments, menthol, camphor, coal tar, etc.
Fungal infections	
Tinea corpis	Terbinafine
Tinea pedis	Terbinafine
Onychomycosis	Terbinafine
Cadidiasis	Ketoconazole
Tinea vesicolor	Ketoconazole, itraconazole, fluconazole
Bacterial infections	Bacitracin, neomycin sulfate, polymixin
Herpes viral infections	Acyclovir
Photodamage (wrinkles)	Retinoids
Hirsutism (excessive facial hair)	Eflornithine
Androgenic alopecia	Minoxidil solution 2%

is believed to play a role. Unlike iontophoresis, however, sonophoresis acts on the skin barrier and not on the drug, and the degree of control possible has yet to be firmly established. Also unclear are the long-term effects of ultrasound on the skin (since the impact of the energy is difficult to confine exclusively to just the stratum corneum), and the complexity of how to evolve the present state of the art into a practical drug delivery system.

II. TOPICAL DRUG DELIVERY SYSTEMS

A. Introduction

Topical delivery systems represent formulations that are intended to deliver drugs locally rather than systemically to treat local pathophysiologic conditions. Topical delivery systems are formulated for dermatological, ophthalmic, otic, vaginal, anorectal, or oral use. This chapter will focus on products for the skin. A topical product is applied to the skin as a thin film, with or without rubbing action, such that no substantial mass remains on the skin following the application. Dermatological products represent a large market within the pharmaceutical and cosmetic industry. Some of the topical dermatological products used are listed in Table 14.5.

The major classes of agents that are delivered topically and act by chemical action include corticosteroids, antifungals, acne products, antibiotics, emollients, antiseptics, local anesthetics, and antineoplastic agents, in that order. Topical agents are used as protectives, adsorbents, emollients, and cleansing agents and act primarily through physical action. Astringents, irritants, rubefacients, and keratolytic agents are also topical agents that have chemical reactivity. The following is a brief explanation of each of these categories with examples (Table 14.6).

TABLE 14.6
Commonly Used Categories and Uses of Topical Preparations

Name of Active Ingredient	Uses
Boric acid, chlorohexidine, triclosan, hydrogen peroxide	Antiseptics and germicides
Aluminum acetate, calcium hydroxide, calamine, alcohol, alums	Astringent
Zinc oxide, allantoin, dimethicone, benzoin, dexpanthenol,	Protective
Mineral oil	Emollient
Capsiacin, methyl salicylate, menthol, camphor	Counter irritant
Salicylic acid, urea	Keratolytic agent

- *Protectives:* Shield exposed skin surface and other membranes from harmful stimuli.
- *Absorbents:*Absorb moisture from skin and local wounds and thereby maintain dry conditions to discourage bacterial growth.
- *Demulcents:*Can alleviate irritation of mucous membranes or abraded tissues.
- *Emollients:*Fat or oily substances used to increase the moisture content of skin and other membranes and render them soft shiny and pliable.
- *Astringents:*Arrest blood hemorrhage by coagulating blood. These agents help wounds and cuts heal quickly.
- *Counter Irritants:*Used to promote a secondary irritation that helps to counter an initial irritation.
- *Rubefacients:* Increase the skin temperature by increasing the circulation at the surface.
- *Caustics:*Destroy skin at the applied site (corrosive). They are useful in the treatment of warts, keratoses, and hyperplastic tissues.
- *Keratolytics:* Cause desquamation (peeling) of skin. These agents are useful in the treatment of eczema, acne, etc.

B. Delivery Systems Employed for Topical Administration

The dosage forms available for the delivery of topical agents include ointments, pastes, creams, lotions, gels, and powders. Depending upon the site of application and therapeutic need, each topical dosage form offers unique characteristics. Even though the details of some of the dosage forms are covered in other chapters, brief descriptions are provided below:

- *Ointments:* Ointments are semisolid preparations intended for application on skin or other mucosal surfaces. Generally ointments that contain medications are used for local treatment, whereas ointments that do not contain any medicaments are applied for their protective action. Ointments can also be used for transdermal delivery of medicinal agents, but as the technology has matured, they are less frequently used for that purpose, as controlling delivery rate from transdermal patches can be more precise.
- *Creams:* Creams are semisolid dosage forms containing one or more drug substances dissolved or dispersed in a suitable base. Creams are fluid compared to other semisolid dosage forms such as ointments and pastes, as the bases used in creams are generally oil-in-water emulsions. Occasionally, water-in-oil emulsions are also employed in the formulation of creams such as cold cream.
- *Pastes:* As per USP, pastes are defined as semisolid dosage forms that contain one or more drug substances intended for topical application. They differ from ointments in their consistency, as they contain larger amount of solids and consequently are thicker and stiffer. They can be made either of fatty bases such as petrolatum and hydrophilic petrolatum or aqueous gels such as celluloses. Pastes are well adsorbed into skin and

absorb watery solutions so that they can be used around oozing lesions. Pastes can be easily removed from skin, which is an important consideration when they are applied on traumatized skin.

- *Lotions:* Lotions are not defined in USP separately from solutions and suspensions, but the British Pharmacopeia defines them as either liquid or semisolid preparations that contain one or more active ingredients in an appropriate vehicle. Lotions are intended to be applied to intact (unbroken) skin, gently without excessive friction. Solid particles incorporated in lotions should be in a finely divided state.
- *Gels:* Gels are semisolid systems containing either a suspension of small inorganic particles or large organic molecules interpenetrated by a liquid. Gels can exist as a single phase system in which large organic molecules are dispersed in a liquid without any clear boundaries (e.g., carbomer or tragacanth in water) or as a two-phase system in which small particles are suspended in a liquid (e.g., aluminum hydroxide gel).
- *Powders:* Powders intended for topical action may or may not contain active ingredients. Dusting powders such as simple talcum powder are used to prevent skin irritation and chafing. Medicated powders are used to treat diaper rash and athlete's foot. Powders, whether used mostly for topical use, are prepared similarly as described in Chapter 10.
- *Liposomes:* Liposomes are vesicles consisting of amphiphatic lipids arranged in one or more concentric bilayers. The drug is usually entrapped in the vesicle structure. Unlike a traditional emulsion, liposomes are thermodynamically unstable. The amount of drug entrapped within a liposome and its encapsulation efficiency are functions of the partition coefficient of the drug between aqueous compartments, the lipid bilayers, and the solubility of the drug in each phase.
- *Aerosols:* Aerosols contain the formulation concentrate (with the active agent) and a propellant in a pressurized container. The active agent is dispensed from the container through a valve primarily as a spray or foam. Several topical products, including first aid products, burn medications, topical antiseptics, and foot-care antifungals, are available as aerosols.

Topical dosage forms should have some desirable characteristics. Owing to the various types of topical delivery systems, it is somewhat difficult to generalize the desirable attributes of topical delivery systems because they depend on the nature of the disease and the condition being treated. The general properties of an ideal topical skin product are listed in Table 14.7.

C. Product Development

Figure 14.7 illustrates the overall scheme of the various stages involved in product development of a topical dosage form. The development of a topical formulation generally includes the following stages:

- Selection of drug candidate
- Selection of the appropriate physical form (e.g., acid, base, or salt)
- Selection of the desired dosage form (e.g., cream, lotion, ointment, solution, sprays)
- Preparation of prototype formulations and testing of their physicochemical stability, micrbiological stability, and cosmetic qualities
- Evaluation of *in vitro* and *in vivo* skin permeation
- Development of analytical methods to quantitate drug in the formulation, skin layers, release medium, and blood (if applicable)
- Evaluation of potential for systemic adverse events (e.g., carcinogenicity, teratogenicity, mutagenicity, etc.)
- Evaluation of skin toxicity (irritation, sensitization, etc.) in animals and humans
- Microbial and preservative testing, if necessary

TABLE 14.7
Properties of an Ideal Topical Skin Product

Characteristics	Description
Physicochemical stability	No or little reversible change in properties such as color, odor, appearance, pH, and viscosity over the shelf life of the product.
Chemical stability	Loss of potency with time should be within specified limits over the shelf life of the product.
Microbiological stability	The product (including container-closure) should be within microbiological limits over the shelf life of the product.
Aesthetic	The product (including container-closure) should be aesthetically appealing to the user, compatible with skin, and easy to apply and remove from the skin.
Skin toxicity	The product should not elicit any irritation or sensitization to the skin.
User friendly	The product should be user and tamperproof.
Manufacturing/scale-up	The formulation and manufacturing should be simple and easy to scale-up from laboratory to large scale production.

- Phase I, II, and III human clinical trials
- Scale-up activities including development of specifications
- Postapproval market surveillance

D. Preparation of Topical Dosage Forms

Topical dosage forms generally contain the following ingredients:

- Base or body of the dosage form
- Medicinal agent (not always the case)
- Preservative

However, the composition may vary slightly from one dosage form to another. As ointments and creams are two major categories of topical dosage forms, the general methods of preparation of only these two dosage forms will be described in the following section.

1. Preparation of Ointments

The objective of the ointment preparation and manufacturing process is to uniformly distribute the medicinal agent in an ointment base. The type of preparation or manufacturing method depends upon the type of base and the quantity of preparation. There are five different types of ointment bases:

- *Ointment bases:* An ointment base forms the body of any ointment. Ointment bases are widely categorized into four classes: (a) hydrocarbon bases, (b) absorption bases, (c) water-removable bases, and (d) water-soluble bases.
- *Hydrocarbon bases:* Hydrocarbon bases are also called oleaginous bases and contain petrolatum and/or modified petrolatum waxes or paraffin oil to lower viscosity. They do not contain water, and only small amounts of water can be incorporated into them. Hydrocarbon bases are greasy, since they are not readily miscible with water, and stay on the skin surface longer. These bases are generally used for emollient and occlusive action.
- *Absorption bases:* These are hydrophilic anhydrous materials that form water-in-oil emulsions or hydrous bases that are already water-in-oil emulsions that can absorb

additional quantities of water. The term *absorption* in these bases refers to their ability to absorb water. Absorption bases are less greasy than hydrocarbon bases but offer emollient and occlusive properties.

- *Water-removable or water-washable bases:* These are oil-in-water emulsions that can be washed with water. These bases resemble creams and represent the majority of dermatological preparations. Since these agents can be diluted with water or aqueous solvents, when applied to wounds, they have the ability to mix with serous fluids and thus deliver drugs to wounds. Water -removable bases contain three components: an oil phase, a water phase, and an emulsifier. A complete classification is provided in Chapter 6.
- *Water-soluble bases:* All the components are soluble in water and do not contain any water-insoluble components (no oil-phase or emulsifying agents). These bases can be washed with water and thus are not greasy. The majority of water-soluble bases contain polyethylene glycol of one different molecular weight or a mixture of different molecular weights. Occasionally gelling agents such as cellulose derivatives, sodium alginate, or silicates are also included to enhance stability. The firmness of water-soluble ointment bases can be improved by including stearyl alcohol.

Once the appropriate base is selected, ointments are prepared by one of the following processes:

- *Incorporation:* In incorporation the different components are mixed together by various means. Preparation of small quantities of ointments, such as extemporaneous preparation in a pharmacy, can be performed by mixing the different components using a spatula on an ointment slab. One to several ounces of ointment can be prepared by this method. On an industrial scale up to 1500 kg of ointment can be produced in a single batch. To mix such large quantities, commercial mixers such as Hobart mixers and Pony mixers are employed.
- *Levigation:* Levigation is the incorporation of a small quantity of powder medication into a small amount of base to get a concentrate, followed by dilution with more base to get the required amount of ointment.
- *Fusion:* In the fusion method, all the components that can be melted are combined and melted together, and components that cannot be melted are added to the molten mixture while cooling and congealing. The heat-labile and -volatile components are mixed toward the end of the ointment preparation when the temperature of the melt is below their degradation temperature. Ointments containing beeswax, paraffin, stearyl alcohol, and high molecular weight polyethylene glycols do not blend into solid state, and they need to be melted to mix uniformly. While preparing such ointments, generally the one with the lower melting point is melted first and the rest of the components are added in the order of increasing melting points. In doing so, the first melted component exerts solvent action on the subsequent components, making them melt at a lower temperature.

Topical preparations such as ointments are not designed to be sterile, but microbial species of pseudomonas and staphylococcus may be present or grow in them, especially in water-removable and water-soluble bases. To prevent the growth of such organisms, ointments are either filter sterilized or chemical preservatives are used, when necessary. The selection of a preservative may depend upon the ointment base, the physico-chemical properties of the medicinal agent, the site of application, etc. Commonly used chemical preservatives in ointments include benzoic acid, p-hydroxybenzoates, quarternary ammonium compounds, sorbic acid, and other compounds.

2. Preparation of Creams

Creams are usually prepared by the fusion method, in which all the oil-type components are melted first, and the heat-stable water (aqueous) dissolving components are separately heated to the same temperature. Then the aqueous solution is slowly added to the oil phase by mixing. The heat-labile compounds are added toward the end when the temperature is low.

E. Stability Testing of Topical Formulations

1. Preformulation Stage

In this stage, systematic studies are conducted to select the most suitable form of the drug substance (i.e., free drug or the best salt form). The following studies are generally performed for that purpose:

- Physicochemical characterization (i.e., solubility, crystalinity, thermal properties, pH solubility profiles)
- Stability characterization (e.g., stress testing to evaluate the effects of temperature, light, humidity, oxygen)
- Stability-indicating analytical method development to quantify, detect, or identify active ingredient, impurities, and degradation products
- Compatibility with excipients (e.g., phase separation, solubility in different phases, effect of shear, particle size, pH, viscosity, etc.)

2. Product Development Stage

Based on preformulation data, a number of formulations can be considered at this stage. Accelerated stability tests are conducted on these potential formulations for screening purposes. Both physical and chemical characteristics are monitored closely. Chemical stability may be evaluated by storing the formulations under various conditions of stress, such as high temperature and high humidity, light, and freeze-thaw cycling and assaying for potency at several intervals. Once a suitable formulation is selected, the next stage is the selection of an appropriate container-and-closure system for the formulation. Again, short-term accelerated stability studies at elevated temperatures and humidity are performed to evaluate any deleterious effect of the container-and-closure system on the formulation. The formulation as a whole including the container-and-closure system is now finally selected and undergoes Phase 3 clinical trials for final FDA approval for marketing.

3. Post-NDA Stage

After the new product is approved, stability testing should continue for as long as the company manufactures and markets the product.

F. Quality Control

1. *In Vitro* Release Test

In vitro release is one of several standard methods that can be used to characterize performance characteristics of a finished topical dosage form, that is, semisolids such as creams, gels, and ointments. It is also important to assure batch-to-batch reproducibility of the delivery systems. Important changes in the characteristics of a drug product formula or the thermodynamic properties of the drug it contains should show up as a difference in drug release. Release is theoretically proportional to the square root of time (/t) when the formulation in question is in control of the release process because the release is from a receding boundary.

Several methods have been proposed, extensively discussed, and reviewed for this purpose. However, the *in vitro* release method using an open chamber diffusion cell system such as a Franz cell system, usually fitted with a synthetic membrane, is most widely used. The test product is placed on the upper side of the membrane in the open donor chamber of the diffusion cell, and a sampling fluid is placed on the other side of the membrane in a receptor cell. Diffusion of drug from the topical product to and across the membrane is monitored by assay of sequentially collected samples of the receptor fluid. The *in vitro* release methodology should be appropriately validated. Sample collection can be automated.

2. *In Vivo* Studies

For systemic drugs, the concentration of drug or active metabolite in blood or urine are quantitated as a measure of bioavailability or bioequivalence. The problem in the topical dermatological area is that systemic absorption (penetration) is very low and currently there are no recognized surrogate measures available for clinical efficacy studies that demonstrate bioequivalence between a generic and the reference (innovator's) product. The sponsor (company) is still required to conduct a human pharmacokinetic study on a fairly small number of patients to show the safety of the product. Depending in the indication (disease) for the product, the sponsor conducts pharmacokinetic studies under maximum usage conditions (i.e., applying the formulation covering maximum area of skin or lesions contemplated in clinical practice). The design of *in vivo* bioequivalence studies for semisolid dosage forms varies depending on the pharmacological activity of the drug and dosage form. With the exception of comparative skin blanching studies as in topical corticosteroids (FDA, *Topical Dermatological Corticosteroids: In Vivo Bioequivalence*, June 2, 1995), the only means a U.S. generic company has to demonstrate bioequivalence of a topical, dermatological product to an innovator's product is through clinical trials. However, clinical efficacy trials are less sensitive, time consuming, and expensive. Several other *in vitro* techniques are under experimental development at this point to be used for demonstrating bioequivlence for topical products. They are dermatopharmacokinetics (skin stripping) measurement of transepidermal water loss, *in vitro* skin permeation, confocal laser scanning microscopy, and microdialysis.

F. Conclusion

Often the terms *transdermal* and *topical* delivery systems are discussed simultaneously. Students should recognize that the goal of a transdermal system is to deliver the drugs across the skin into the systemic circulation, whereas the goal of a topical system is to administer the drugs for a local or regional effect. A competent pharmacist should have adequate knowledge to address patients' questions related to transdermal and topical products and a basic understanding of the possible methods of making these delivery systems and the pros and cons of each.

TUTORIAL

1. Define and distinguish transdermal and topical dermal delivery systems.

Quite often in the literature, transdermal delivery is used generally to describe any application to the skin. Thus, it is important to distinguish the correct terminology, even though in both cases, drug delivery systems applied to skin and individual variations in skin are difficult to control.

Topical	Transdermal
Topical dermal delivery can be defined as application of a system to skin directly at the effected area to deliver the drug in dermal and the underlying tissue at the site of application.	Transdermal delivery can be defined as application of a system to skin at a practically accessible site such as behind ear lobes, buttocks, and abdomen, with an intent of treating a systemic disease, for which drug is required to achieve therapeutic levels in plasma.
The formulations commonly used to deliver drugs by this route are ointments, creams, lotions, gels, sprays, medicated plasters, and powders.	The dosage form commonly used to deliver drugs by this route is a patch. However, gels and ointments are also used occasionally.
High concentrations of drugs can be reached locally by administering through this route.	Local concentrations may not reach high levels.
Only small amounts of drug are absorbed into the systemic circulation, avoiding toxic effects.	Drug absorption is high, resulting in therapeutic plasma levels.
Applied dose (amount applied per unit area) varies with applying individual. Absorption may change with application technique such as the pressure applied, etc.	Currently available transdermal systems are very practical, and applied dose is constant and is usually defined by the manufacturer.
Overall, system design is easy, but application is complex and difficult to control.	System design is difficult, whereas application is simple and uniform.

2. The transdermal route of administration offers many advantages. If so, why can we not develop patches for every drug? How can we estimate patch size?

The above questions may be answered with the following example:

Background Information

The mass of drug delivered across skin, M, (mass/unit area/time) is based on estimated permeability coefficient (P) and can be calculated as: $M = P_{\text{estimate}} \times C_s$ where C_s is the saturation solubility.

The required flux (J_{SS}) or input rate can be calculated from the pharmacokinetic properties of the drug as shown in pharmacokinetic considerations as follows:

$$J_{ss}(\text{mass/time}) = CL \times C_{pss}$$

to obtain the transdermal dose.

From these calculated values, one can estimate the required patch size using

$$\text{Area of the patch,}\ \ A(cm^2) = \frac{J_{ss}}{M} \tag{14.8}$$

Example

You are a product development scientist in charge of extending patents' lives through developing novel drug delivery systems in a multinational company. You have been asked to determine the feasibility of extending the patent life of an anticonvulsive drug, primidone, by devoloping a 10-cm^2 transdermal patch. Following the administration of 750 mg/day tablets of primidne, drug absorption was complete, yielding a therapeutic concentration of 10 μg/ml. The total body clearance and elimination half-life of primidone were determined to be 0.78 ml/Kg/min and 4 h, respectively. The permeability coefficient of primidone is 5×10^{-3} $cm.h^{-1}$ and the saturation solubility is 1 mg/ml.

Since the drug is absorbed completely, F = 1. The required output rate of primidone can be calculated using Equation 14.3.

$$CL = 0.78 \text{ ml/Kg/min} \times 70 \text{ kg (normal body weight)} = 54.6 \text{ ml/min} = 3276 \text{ ml/h}$$

$$\text{Output rate} = \frac{3276 \text{ ml/h} \times 10 \text{ µg/ml}}{10 \text{ cm}^2} = 3276 \text{ µg/cm}^2\text{/h}$$

Thus, an input rate, transdermal flux in this case, of 3.3 mg/cm²/h is required from the transdermal patch for primidone.

The mass of drug that can be delivered across skin is $M = P_{\text{estimate}} \times C_s = 5 \times 10^{-3} \times 1 =$ 5 µg/cm²/h. Using Equation 14.13, the area of the patch required to deliver therapeutic plasma levels of primidone is

$$A\left(\text{cm}^2\right) = \frac{J_{ss}}{M} = \frac{3300 \text{ µg/h}}{5 \text{ µg/cm}^2\text{/h}} = 660 \text{ cm}^2$$

Thus, a large portion of the body must to covered with a transdermal patch to achieve therapeutic concentrations; thus, transdermal delivery from a 10 cm² patch is not feasible for primidone.

REFERENCES

1. Chien, Y. W., *Novel Drug Delivery Systems,* 2nd ed., Marcel Dekker, New York, 1992, ch. 7.
2. Dermal and Transdermal Drug Delivery. Second International Symposium of the International Asoociation for Pharmaceutical Technology, Frankfurt, 1991.
3. Flynn, G. L. and Stewart, B., Percutaneous drug penetration: choosing candidates for transdermal development, *Drug Development Research,* 13, 169–185, 1988.
4. Ghosh, T. K., Pfister, W. R., and Yum, S.I., Eds., *Transdermal and Topical Drug Delivery Systems,* Interpharm Press, Buffalo Grove, NY, 1997.
5. Goldsmith, L. A., Ed., *Biochemistry and Physiology of Skin,* Oxford University Press, New York, 1983. (This is a two volume book that has several chapters on structure of skin: everything you would want to know about skin and more.)
6. Guy, R. H., Ed., *Advanced Drug Delivery Reviews,* Elsevier Sciences, Oxford, UK, Vol. 9, 1992.
7. Guy, R. H. and Hadgraft, J., Rate control in transdermal drug delivery? *International Journal of Pharmaceutics,* 83, R1–R6, 1992.
8. Kydonius, A. F. and Berner, B., Eds., *Transdermal Delivery of Drugs,* CRC Press, Boca Raton, FL, 1987.
9. Michaels, S., Chandraswkaran, S. K., and Shaw, J. E., Drug permeation through human skin: theory and in vitro experimental method, *Amercan Institute of Chemical Engineering Journal,* 21, 985–996, 1975.
10. Potts, R. O. and Guy, R. H., Predicting skin permeability, *Pharmaceutical Research,* 9(5), 663–669, 1992.
11. Shah, V. P. and Maibach, H. I., Eds., *Topical Drug Bioavailability, Bioequivalence, and Penetration,* Plenum Press, New York, 1993.
12. U.S. Department of Health and Human Services Food and Drug Administration Center for Drug Evaluation and Research, SUPAC-SS: Nonsterile Semisolid Dosage Forms; Scale-Up and Postapproval Changes: Chemistry, Manufacturing, and Controls; *In Vitro* Release Testing and *In Vivo* Bioequivalence Documentation (http://www.fda.gov/cder/guidance/index.htm) (posted 6/16/97)

HOMEWORK

1. Upon application of a transdermal delivery system, the primary pathway contributing to drug absorption into systemic circulation is:
   (a) Trancellular pathway.
   (b) Intercellular pathway.
   (c) Transappendegeal pathway.
   (d) All the above contribute equally.
2. You are developing two transdermal patches for the delivery of estradiol: one to be worn for a day and the other to be worn for a week. Which of the following is acceptable from a lag time point of view?
   (a) Lag time of 3 h for the 1-day patch and lag time of 20 h for the 1-week patch
   (b) Lag time of 10 h for both patches
   (c) Lag time of 3 h for both patches
   (d) (a) and (c)
   (e) (a) and (b)
3. Transdermal flux for a drug compound was determined to be 0.5 $\mu g.cm^{-2}.h^{-1}$. The target dose for this drug is 480 $\mu g.day^{-1}$. The estimated area for the transdermal patch is:
   (a) 10 cm^2
   (b) 20 cm^2
   (c) 30 cm^2
   (d) 40 cm^2
   (e) 50 cm^2
4. Drug A has a melting point of 100°C and a molecular weight of 100. Drug B has a melting point of 200°C and a molecular weight of 200. Which of the following statements accurately describe the suitability of these compounds for transdermal delivery?
   (a) Drug A is better suited for transdermal delivery than drug B.
   (b) Drug B is better suited than drug A.
   (c) Both compounds are good candidates for transdermal delivery.
   (d) Neither compound is suitable for transdermal delivery.
5. The advantages of transdermal drug delivery include:
   (a) Bypassing liver metabolism
   (b) Feasibility that every drug can be formulated into a transdermal delivery device
   (c) Less fluctuations in plasma drug levels in the therapeutic window
   (d) (a), (b), and (c)
   (e) (a) and (c)
6. Not all drugs can be delivered by the transdermal route owing to:
   (a) Practical limitations on patch size
   (b) Some drugs having long lag times
   (c) Skin forming a formidable barrier to their permeability
   (d) (a), (b), and (c)
   (e) (a) and (b)
7. Crystalline drugs have low permeability because they have higher melting points.
   (a) True
   (b) False

8. Drug X has lower water solubility and a higher partition coefficient than drug Y.
   (a) Drug X is better suited for transdermal delivery than drug Y.
   (b) Drug Y is better suited than drug X.
   (c) Both drugs are equally good candidates for transdermal delivery.
9. Ketamine gel applied to the joints for reducing pain in knee joints can be classified as:
   (a) A topical delivery system
   (b) A transdermal delivery system
   (c) Both
   (d) Neither
10. Iontophoresis, chemical enhancers, sonophoresis, and electroporation are the techniques employed to enhance transport across skin.
    (a) True
    (b) False
11. The following are semisolid topical preparations:
    (a) Ointments
    (b) Creams
    (c) Lotions
    (d) All the above
12. The most striking difference between creams and ointments is:
    (a) Creams are thicker than ointments.
    (b) Ointments are thicker than creams.
    (c) Creams are emulsions, whereas ointments are suspensions.
13. The solid content is highest in:
    (a) Ointments
    (b) Creams
    (c) Lotions
    (d) Solutions
14. The presence of petrolatum-like bases renders them:
    (a) Occlusive
    (b) Greasy
    (c) Water washable
    (d) (a) and (b)
    (e) All the above
15. The majority of creams are oil-in-water type emulsions.
    (a) True
    (b) False
16. During preparation of ointments by fusion, the fatty/waxy materials are:
    (a) Added in the order of increasing melting points
    (b) Added in the order of decreasing melting points
    (c) Added randomly

Answers: 1.-(b); 2. (d); 3. (d); 4. (a); 5. (e); 6. (d); 7. (a); 8. (a); 9. (a); 10. (a); 11. (d); 12. (c); 13. (a); 14. (d); 15. (a); 16. (a)

CASES

CASE I

The objective of this case study is to enable students to go through an initial assessment of a class of compounds in a particular therapeutic area using the principles learned in this chapter and come up with a medical rationale for selecting one or more drug candidates for administration by the transdermal route.

Therapy of epilepsy falls under two categories: rapid relief and chronic therapy.

1. Many patients, such as those having a series of seizures, need medication administered rapidly by a nonoral route. Would you administer the drug by the transdermal route? Explain.
2. Chronic oral therapy is difficult for epileptic patients who have seizures that interfere with cognitive function or who simply cannot comply with a complex, multiple dosing regimen. The following table lists commonly used anti-epileptic drugs along with their pharmacokinetic information. Select one or more drug candidates for administration by the transdermal route, using the rationale and the selection criteria that you learned in this chapter.

Drug	Daily Dose (mg)	Effective Plasma Concentration (μg/ml)	Clearance (ml/kg/min)	Half-life (h)
Carbamazepine	600–1200	4–12	0.58 ± 0.12	5–26
Clonazepam	3.5–14	0.12–0.08	0.92 ± 0.25	18–50
Ethosuximide	750–2000	40–100	0.26 ± 0.05	40–60
Phenobarbital	120–250	15–40	0.09 ± 0.04	53–140
Phenytoin	300–400	10–20	0.20 ± 0.03	8–59
Primidone	750–1500	10–14	0.78 ± 0.62	4–22
Vigabatin	1000–3000	15	1.69–1.83	—

CASE II

A local semisolid generic company is trying to develop topical delivery systems for the following indications. The manager of its marketing department developed a strategy wherein he sent three employees to local pharmacies under disguise as patients for consultation with the pharmacists.

Ms. Smith came to Pharmacy A with the following complaints: Her 5-year-old daughter has head lice, and her family friend remembered using a malathion product. She forgot the nature of the product and wanted to know from the pharmacist whether malathion lotion, cream, or ointment is appropriate for her daughter and if all three products are available, which one is better suited for her needs.

Mr. Flower came to Pharmacy B with a complaint of pain. His other medications include insulin injections for insulin, and he reported that tricyclic antidepressants did not relieve his pain from diabetic neuropathy. His doctor prescribed ketamine gel made from pluronic lecithin organogel microemulsions. Mr. Flower inquired why there is no lotion for ketamine.

Mrs. Jackson, 48 years old, came into Pharmacy C and said that she heard on the radio about the availability of topical estrogen therapy to cover her skin wrinkles (aging skin). She requested a consult from the pharmacist and asked what kind of topical dosage form is better for the estrogen therapy.

Imagine you are an on-call pharmacist at the desk in each of the three pharmacies when the decoys approached and provide brief explanations. This exercise is provided to differentiate the different topical dosage forms, so ignore their therapeutic potential in providing the explanations.

ANSWERS TO CASE STUDIES

CASE I

1. Patients having a series of seizures need medication administered rapidly by a nonoral route. The medication will probably be administered by i.v. route and not by the transdermal route. The transdermal route is not suited for rapid delivery owing to the inherently long **lag time** associated with the low skin permeability.
2. Students need to calculate the **target input rate** for each of the drug candidates shown in the Table. For example, the total body clearance for Carbamazepine is 0.58 $ml.kg^{-1}.min^{-1}$. Assuming an average body weight of 70 kg for an adult, $CL_T = 0.58 \times 70 = 40.6$ $ml.min^{-1}$ or 2436 $ml.h^{-1}$. C_{SS} = 4 to 12 $\mu g.ml^{-1}$. Target input rate = $CL_T \times C_{SS} = 2436 \times 4 = 9.744$ $mg.h^{-1}$, but the C_{SS} has a range of 4 to 12 $\mu g.ml^{-1}$. Thus, target input rate = 9.7 – 29.2 $mg.h^{-1}$ (232.8 to 698.4 mg/day). Keeping in mind the practical limitation of the maximum amount feasible to be delivered from a patch (≈10 mg/day, compare all the marketed patches) and maximum patch size (≈50 cm^2), the required target input rate for this drug is unachievable.

Upon repeating this exercise for other drugs listed in the table and considering the given parameters, one would end up selecting **clonazepam** as the only candidate for transdermal delivery among the list of anti-epileptics shown here.

CASE II

Malathion is a yellow to brownish liquid with a characteristic odor and is an effective insecticide, especially in mice. It inhibits acetylcholinesterase and is rapidly broken down by hepatic enzymes when ingested by humans and is thus safe for human use. Malathion is only soluble in water but is miscible in alcohols and hydrocarbons. Thus, a lotion of malathion is the simplest formulation one could develop wherein the insecticide can be dissolved in alcohol and a suitable fragrance can be added to mask the odor. When such a lotion is applied to the scalp, the alcohol evaporates, leaving a coat of malathion, which can be washed off with a shower. Since it is miscible with hydrocarbons, an ointment can also be easily developed. However, such an ointment would be very greasy and difficult to keep on the head for long periods. Moreover, after wear, cleaning is difficult for ointments made from hydrocarbon bases. An oil-in-water type emulsion based cream is also feasible for malathion, but preparing a large-scale emulsion is more complex and expensive than the simple lotion and ointments.

Ketamine is an N-methyl-D-aspartate receptor antagonist and elicits its pharmacological activity by blocking intracellular events that inhibit hyperexcitability. Owing to the intolerable side effects associated with systemic administration, local application may be preferable. For pain relieving effect, ketamine needs to be applied up to 4 hours. Gels have a more rigid nature (solid consistency) than liquids, and the application site can be easily controlled. Since PLO gel is a micro-emulsion, it can be easily applied.

Even though the role of hormones on aging is not clearly understood, there is a growing belief that estrogen supplements may improve skin texture. This is becoming popular with women who are at the perimenopausal stage (50 years). For this indication, quite often an applicator is essential to use. In such cases the best dosage form is oil-in-water emulsion-based creams, as they are easy to apply and nongreasy. A lotion formulation for such use may not be recommended, as they are too liquid to stay long enough for any beneficial effect. On the contrary ointment may be difficult to apply because of its thicker consistency.

15 Rectal and Vaginal Routes of Drug Delivery

Nikhil R. Shitut, Sumeet K. Rastogi, Somnath Singh, Feirong Kang, and Jagdish Singh

CONTENTS

0-4152-8863-0/05/$0.00+$1.50
© 2005 by CRC Press LLC

Learning Objectives

On studying of this chapter, the student will have thorough knowledge about:

- Advantages and limitations of suppositories
- Mechanism of drug absorption from rectal and vaginal dosage forms
- Various types of rectal and vaginal dosage forms
- Rectal and vaginal product characteristics and types of bases
- Preparation, storage, and handling of suppositories
- How to do dose displacement calculations for suppositories

I. INTRODUCTION

The rectum and vagina have a rich blood and lymph supply, and drugs can cross through their mucosa like the other lipid membranes. The vaginal and lower rectal tissues are supplied with perineal venous plexus, which flows into the pudentum vein and finally into the vena cava, which circumnavigates the liver in first pass. In contrast to this, the gastrointestinal blood circulation drains into the portal vein and passes directly to the liver before returning into the general circulation. Thus, rectal and vaginal routes of drug administration can be of particular importance for drugs such as progesterone and estradiol, which are poorly bioavailable upon oral administration because they are extensively metabolized by the liver. Currently these routes are becoming more important for the delivery of peptides and proteins. The superior hemorrhoidal vein drains from the upper portion of rectum and enters into portal system. Hence, drug administered only at the lower portion of the rectum will be able to bypass the hepatic portal system. Therefore, it is the responsibility of pharmacists to counsel patients on the proper use of rectal suppositories. This chapter discusses the preparation and safe use of various dosage forms for rectal and vaginal administration.

A. Definitions and Uses

Various types of dosage forms have been used for rectal and vaginal administration, for example suppository, tablet, capsule, ointment, cream, aerosol foam, sponge, powder, jelly, and gel. Suppositories are an easily fusible, solid dosage form to be inserted into body orifices, such as the

TABLE 15.1
Examples of Rectal Suppositories

Product	Commercial Products	Effect	Category
Bisacodyl suppository, USP	Dulcolax Suppositories (Boehringer-Ingelheim)	Local	Cathartic
Chlorpromazine, USP	Thorazine Suppositories (Smith Kline & French)	Systemic	Antiemetic, suppository, tranquilizer
Cyclomethycaine Sulfate Suppository, USP	Surfacaine Suppositories (Lilly)	Local	Local anesthetic
Dibucaine Suppository, USP	Nupercainal Suppositories (Madison)	Local	Local anesthetic
Dimenhydrinate, USP	Dramamine Supposicones (Searle)	Systemic	Anti-emetic, suppository, antihistaminic, antivertigo
Ergotamine Tartarate and Caffeine Suppository, USP	Cafergot Suppositories (Sandoz)	Systemic	Antiadrenergic
Glycerin Suppository, USP	Various	Local	Cathartic
Oxymorphone HCl Suppository, USP	Numorphan Suppositories (Endo)	Systemic	Narcotic analgesic
Prochlorperazine Suppository, USP	Compazine Suppositories (Smith Kline & French)	Systemic	Anti-emetic

TABLE 15.2
Examples of Vaginal Tablets and Suppositories

Product	Commercial Products	Effect	Category
Candidicin vaginal tablets, USP	Vanobid Tablets (Merrill-National)	Local	Antifungal
Diethylstilbestrol Suppository, USP	Various	Local	Estrogen
Clotrimazole Vaginal Tablets, USP	Gyne-Lotrimin (Schering)	Local	Antifungal

rectum or vagina, where they melt to exert local or systemic effect. Rectal suppositories for adults are tapered at one or both ends and weigh 2 g if cocoa butter has been used as a base. Infant rectal suppositories weigh about 1 g. Vaginal suppositories or pessaries are globular or oviform in shape and weigh about 5 g. Urethral suppository or bougies are slender, pencil-shaped suppositories intended for insertion into the male or female urethra. Some rectal and vaginal suppositories are shown in Table 15.1 and Table 15.2, respectively. The following are formulations of some commercially available products.

1. Dulcolax® Suppositories

Each suppository contains:

Bisacodyl: 5 mg (children) or 10 mg (adult)
Nonmedicinal ingredients: hard fat (hydrogenated vegetable oil)

Dulcolax suppositories contain bisacodyl as active ingredient, which is a contact laxative. It acts directly on the colonic mucosa to increase peristalsis throughout the large intestine.

2. Thorazine® Suppositories

Each suppository contains:

Chlorpromazine: 25 mg or 100 mg
Inactive ingredients: glycerin, glyceryl monopalmitate, glyceryl monostearate, hydrogenated coconut oil fatty acids, and hydrogenated palm kernel oil fatty acids

Chlorpromazine is the first tranquilizer of the phenothiazines group of compounds. It is effective in the management of manifestations of psychotic disorders, nausea, and vomiting.

3. Gyne-Lotrimin® Vaginal Tablets

Each tablet contains:

Clotrimazole: 100 mg or 200 mg.
Inactive ingredients: benzyl alcohol, cetearyl alcohol, cetyl esters wax, octyldodecanol, polysorbate 60, purified water, sorbitan monostearate

Clotrimazole is a broad spectrum antibiotic. It is indicated for the treatment of vaginal fungal infection.

B. Advantages and Limitations

1. Advantages of Rectal Administration

1. Safe and painless for administration and possible to remove.
2. Drugs liable to degradation in gastrointestinal tract can be administered.
3. First pass elimination of high clearance drugs is partially avoided owing to bypassing the liver.
4. Even a large dose can be administered.
5. The duration of action can be controlled by using a suitable base.
6. Drugs can be administered rectally in the long-term care of geriatric and terminally ill patients.
7. It is a useful way to administer medication to children who are unwilling or unable to tolerate the drug by the oral route.
8. Administration of a rectal suppository or capsule is a simple procedure that can be undertaken even by unskilled healthcare personnel and the patient.
9. It is useful for nauseous/vomiting patients.

2. Limitations

1. Patient acceptability and compliance is poor, especially for long-term therapy.
2. Upward movement of suppository from local site can increase first-pass metabolism.
3. Suppositories can leak.
4. Insertion of suppository may be problematic.
5. Generally, drug absorption from suppositories is slow in comparison to oral or intravenous administration.

II. PHYSIOLOGY AND DRUG ABSORPTION

In order to understand the mechanism of drug absorption from the rectum and vagina it is important to know the anatomy and physiology of the region (Figure 15.1).

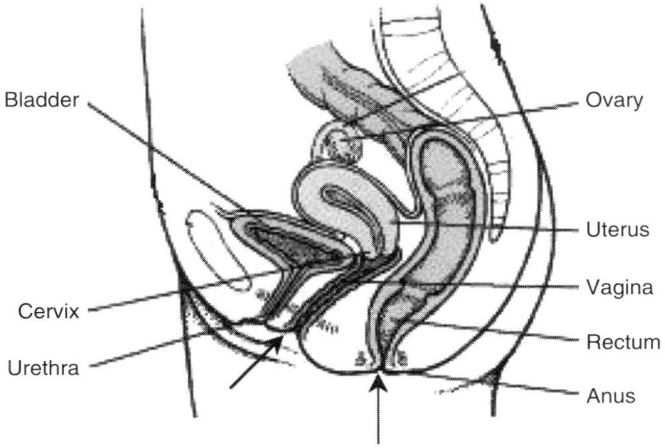

FIGURE 15.1 Anatomy of the anorectal and vaginal region with sites of insertion (↑) of different dosage forms.

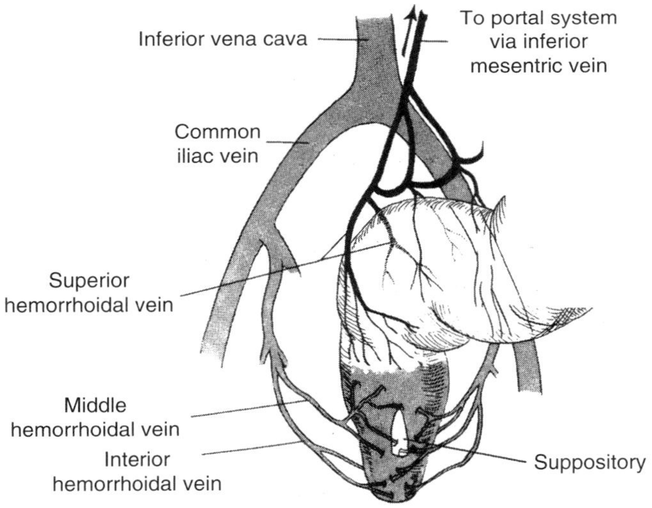

FIGURE 15.2 Major blood vessels of the rectum and proper position of the suppository.

A. The Rectum

The rectum constitutes the last part of the large intestine, whose main function is to absorb water and electrolytes from chyme and to serve as a vault for fecal contents. The large intestine is capable of secretion if it is irritated. Sections of the rectum are neutral in pH and have no buffer capacity. Anorectal dynamics may alter with age, trauma, hemorrhoids, rectal prolapse, and other disease states. The rectum's main supply of blood is delivered by the rectal artery and is drained by the superior, middle, and inferior hemorrhoidal veins (Figure 15.2). Dilation of these veins causes internal and external hemorrhoids. The fate of a substance absorbed form the rectum depends upon its position in the rectum. In general, the chance of a drug being delivered to the portal circulation and thus directly to the liver for first pass metabolism increases as the absorption site distance from the anus increases. The inferior and middle hemorrhoidal veins empty into the iliac vein and the vena cava (and thus feed directly into the general circulation), while the superior hemorrhoidal vein empties into the portal vein as seen in Figure 15.2. Absorption from the rectum is also accomplished by the lymphatics, which are located all along the gastrointestinal tract.

B. The Vagina

The vagina is a thin-walled, fibromuscular tube about 7.5 cm long. It is located behind the bladder and urethra and in front of the rectum and anal canal. The uterus is above the vagina and the vestibule below. The female urethra is intimately related to the lower two thirds of the vagina and opens into the vestibule immediately in front of the vaginal orifice. Blood is supplied to the vagina by the vaginal artery and is returned to the circulation by way of the vaginal plexus to the internal iliac veins. Lymph vessels also drain the vagina. The outermost layer of cells in the vagina is protected by a complex mixture of proteins and polysaccharides called mucus. Most of this fluid comes from the cervix and is usually thick and glutinous. The viscosity of this cervical fluid is controlled by hormone levels. The vaginal epithelium is very sensitive to estrogen and progesterone. Estrogen decreases the fluid viscosity and progesterone increases the viscosity of cervical fluid. The pH of the adult vagina is acidic because of the conversion of glycogen to lactic acid. The glycogen is derived from epithelium, which stores and secretes it when necessary. The glycogen serves as a nutrient to the ovules and the sperms in the vagina. The pH of the vagina is usually between 3.5 and 4.2 during childbearing age. The pH of the vagina is affected by hormone levels, age, and individual lifestyle.

C. Mechanism of Absorption

The mechanism of absorption from the rectum, vagina, and urethra is similar to other mucus membranes such as the stomach and intestine. Absorption takes place by passive diffusion of un-ionized molecules following apparent first-order kinetics. Effects of pH and partition coefficients are in accordance with the pH partition hypothesis. However, differences exist in the amount and the content of fluid in rectum and vagina. Rectal fluid has a pH slightly below 7, and it has no buffer capacity. There are no enzymes or electrolytes in rectal or vaginal fluid. Vaginal fluid has an acidic pH owing to lactic acid. The surface areas of the rectum and vagina are less than that of the small intestine.

D. Factors Affecting Absorption

Drug solubility, concentration, and particle size; presence of mucus; and surface area affect bioavailability from the rectum and vagina. Solutions and suspensions administered rectally act more quickly than tablets, capsules, or suppositories administered by the same route. A slowly liquefying base may be desirable for localized action. Implanted vaginal devices are meant to release the drug for a period of weeks or months. Retention enemas are more rapid and efficient than suppositories, and in an evacuated rectum, blood levels of drug may be similar to intravenous administration after 1 hour. However, several biopharmaceutical and patient factors influence this outcome. An uncooperative patient could expel the product. Fecal matter could also affect the rate and amount of the drug absorbed. A cleansing enema may help absorption by cleansing the fecal matter from the site of absorption. Suppositories tend to migrate after insertion. Hence, even if initial placement is correct, the suppository may move upward. Thus, its content may be carried directly to the liver after absorption.

III. TYPES OF RECTAL AND VAGINAL DOSAGE FORMS

The various types of rectal and vaginal dosage forms, depending upon the drug incorporated and specific use, are given below:

1. Suppositories
2. Tablets and capsules
3. Ointments, creams, and aerosol foams
4. Gels and jellies
5. Contraceptive sponges
6. Intrauterine devices
7. Powders
8. Solutions

A. Suppositories

The word *suppository* comes from the Latin word *supponere,* meaning "to place under," derived from *sub* (under) and *ponere* (to place). They are the most commonly used rectal and vaginal solid dosage forms. They are inserted into the body orifices such as the rectum and vagina, where they melt, soften, or dissolve and exert localized or systemic effects. Suppositories of different sizes and shapes can be prepared so that they can be inserted easily into the rectum or vagina. In order to release the drug over a certain period of time the suppositories must be retained in the rectum, vagina, or urethra. Vaginal suppositories are called *pessaries,* are usually globular, oviform, or cone shaped.

B. Tablets and Capsules

Vaginal tablets are more widely used than vaginal suppositories, owing to better stability and ease in handling. Vaginal tablets, also called *vaginal inserts,* are ovoid in shape and packaged with a plastic device for convenient placement of the tablet in the vagina. The vaginal tablets contain drugs similar in function to suppositories. Bioadhesive tablets are the new form of rectal tablets, which are mainly composed of a polymer, which hydrates and swells to a gel-like consistency. The drug is released at a controlled rate from the tablet matrix. Rate of drug release from these tablets depends upon several factors such as composition of the matrix, drug solubility, and the partition coefficient of the drug.

Some vaginal inserts may be capsules containing a drug in gelatin shell. Capsules are also used rectally in pediatrics to administer medication to children who are unwilling or unable to tolerate a drug orally. The insertion of the capsule into the rectum is facilitated by first lightly wetting the capsule with water.

C. Ointments, Creams, and Aerosol Foams

Rectal ointments are primarily used for local action. Ointments are inserted into the rectum with the help of an insertion and delivery tip. After placement of the rectal tip on the ointment tube, the tip is slowly and carefully inserted into the anus. The ointment tube is depressed to release the medication through the rectal tip and into the anal canal. After use the rectal tip should be thoroughly cleaned and the ointment tube cap must be replaced. While selecting an ointment for rectal use, close attention should be given to the product formulation base. Water-soluble bases are easier and less messy to clean from applicator tips than oleaginous bases. The commonly used rectal ointments are benzocaine ointment (USP), dibucaine ointment (USP), and pramoxine ointment (USP).

Vaginal ointments and creams contain drugs such as anti-infectives, estrogenic hormone substrates, and contraceptive agents. Contraceptive creams and foams contain spermicidal agents such as nonoxylol-9 and octoxynol and are used alone or in combination with a cervical diaphragm or special inserter-applicator. Contraceptive creams are used just prior to intercourse.

Foam is a dispersion of a gas in a medicated liquid resulting in a light, frothy mass. Aerosol foams are commercially available containing estrogenic substances and contraceptive agents. Foams are used intravaginally in a similar manner to creams. The aerosol contains an inserter device, which is filled with foam, and contents are placed in the vagina through activation of the plunger.

D. Gels and Jellies

Jellies are a class of gels in which the structural coherent matrix contains a high proportion of liquid, which is usually water. Jellies are used intravaginally in a similar manner to creams. Spermicidal agents are primarily incorporated into jellies. There are a number of contraceptive jellies available on the market today, for example, Koromex Jelly and Gynol II Extra Strength Contraceptive Jelly. These jellies mainly incorporate drugs such as benzalkonium chloride, nonoxynol 9, and octoxynol 9. A number of antiseptic, antifungal, anesthetic, and nonmedicated lubricant jellies are available, primarily for use by physicians in their rectal, urethral, and vaginal examination procedures. One important additive in jellies is an antimicrobial preservative, because they are susceptible to contamination owing to high viscosity and water content. The tubes of jellies should be tightly closed when not in use, as they have a tendency to lose water to the air.

E. Contraceptive Sponges

A vaginal contraceptive sponge was introduced in 1983 and is the largest-selling, over-the-counter female contraceptive in the U.S. It consists of a resilient, hydrophilic, polyurethane foam sponge, impregnated during manufacture with nonoxynol-9, a spermicidal agent. Sperms are absorbed into the sponge, where the spermicidal agent exerts its action. The contraceptive sponge is manufactured in a slightly concave, rounded form designed to fit snugly in the upper vagina. Before insertion of the sponge into the vagina, it should be moistened to activate the spermicidal agent. It provides contraception for a 24-hour period and is designed to remain in place at least 6 hours following intercourse.

F. Intrauterine Devices

Intrauterine devices (IUDs) are small flexible devices made of metal or plastic that prevent pregnancy when inserted into a woman's uterus. The most widely used IUDs are copper-bearing IUDs. Inert (unmedicated) and progestin-releasing (levonorgestrel or progesterone) IUDs are less widely available. IUDs are a safe and effective method of reversible, long-term contraception for most women. The Progestasert System (Alza Corporation) (Figure 15.3) releases an average of 60 μg

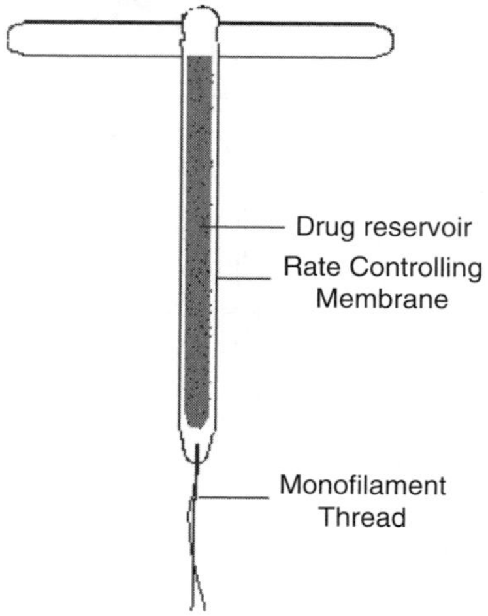

FIGURE 15.3 Diagrammatic illustration of an intrauterine device.

of progesterone per day for one year after insertion. The continuous release of progesterone into the uterine cavity provides a local rather than a systemic action. The device might act in two ways. Progesterone-induced inhibition of sperm survival and alteration of the uterine milieu so as to prevent nidation. The uterine device also contains a smaller dose of progesterone, a much smaller amount than given by other doses. An alternative method of improving the acceptability of inert IUDs is to solve the pain and bleeding problems of some of the uncomfortable but effective (that is, resulting in a low pregnancy rate) devices. To alleviate these problems, an antifibronolytic drug (e.g., aminocaproic acid, tranexamic acid, etc.) can be incorporated into the device. A device with such a drug is effective as a topical treatment for excessive menstrual bleeding. Antifibronolytic drugs are particularly effective if released during the first few months after insertion, when excessive menstrual bleeding is particularly high. A therapeutically useful device, therefore, releases the drug at an initially high rate that declines to a negligible value after three months.

G. Powders

Powders are used to prepare solutions for vaginal douches for the irritative cleansing of the vagina. Powder for constituting douches may be available in unit or bulk form. A unit package contains the required amount of powder to prepare the specific volume of douche solution. In bulk packaging the user simply adds the prescribed amount of powder to the appropriate volume of warm water and stirs until dissolved. The components of the douche powder are:

1. Antiseptic (e.g., boric acid or sodium borate)
2. Astringents (e.g., potassium alum, ammonium alum, zinc sulfate)
3. Antimicrobials (e.g., oxyquinoline sulfate, povidone-iodine)
4. Quaternary ammonium compounds (e.g., benzethanium chloride)
5. Detergents (e.g., sodium lauryl sulfate)
6. Oxidizing agents (e.g., sodium perborate)
7. Salts (e.g., sodium citrate, sodium chloride)
8. Aromatics (e.g., menthol, thymol, eucalyptol, methyl salicylate, phenol)

H. Solutions

1. Vaginal Douches

These are prepared from liquid concentrates by diluting with the prescribed amount of warm water. They have same components as powders.

2. Retention Enemas

Many solutions are administered rectally for local effects (e.g., hydrocortisone) or for systemic absorption (e.g., aminophylline). In the case of aminophylline, the rectal route of administration minimizes the undesirable gastrointestinal reactions associated with oral therapy. Corticosteroids are administered as retention enemas or continuous drip as adjunctive treatment of some patients with ulcerative colitis. Oil retention enemas serve to lubricate the rectum and lower bowel and soften the stool. Salad oil and liquid petrolatum are commonly used at a temperature of 91°F (32.8°C).

3. Evacuation Enemas

Evacuation enemas may be given for the following purposes:

1. To remove feces when an individual is constipated or impacted.
2. To remove feces and cleanse the rectum in preparation for an examination.

3. To remove feces prior to a surgical procedure to prevent contamination of the surgical area.
4. To administer drugs or anesthetic agents.

They act by stimulation of bowel activity through irritation of the lower bowel and by distention with the volume of fluid instilled. When the enema is administered, the individual should be lying on the left side, which places the sigmoid colon (lower portion of the bowel) below the rectum and facilitates infusion of fluid. The length of time it takes to administer an enema depends on the amount of fluid to be infused.

Precautions and requirements for enemas are:

1. The rectal tube used for infusion of the enema solution should be smooth and flexible to decrease the possibility of damage to the mucous membrane that lines the rectum.
2. Tap water is commonly used for adults but should not be used for infants because of the danger of electrolyte imbalance.
3. The colon absorbs water, and repeated tap water enemas can cause cardiovascular overload and electrolyte imbalance.
4. Repeated saline enemas can cause increased absorption of fluid and electrolytes into the bloodstream, resulting in overload.

IV. PRODUCT CHARACTERISTICS AND COMPONENTS

In order to thoroughly understand rectal and vaginal product characteristics, it is essential to have a working knowledge of the factors affecting drug absorption from the anorectal region. The pH of the rectal region is 7 to 8, and there is no sufficient buffering capacity associated with the rectal fluids. The lipid water solubility of the drug is an important factor while formulating the base. The suppository base employed has a marked influence on the release of active constituents incorporated into it.

The ideal suppository base should have following properties:

1. Nontoxic and nonirritating to the mucous membrane
2. Compatible with a variety of drugs
3. Melting or dissolving in rectal fluids
4. Stable on storage, should not bind or otherwise interfere with the release and absorption of drug substances

The bioavailability of a drug from the suppository dosage form is dependent on the physicochemical properties of the drug as well as the composition of the base. Relative solubility of the drug in the vehicle used to formulate the base is an important measure. Lipid soluble drugs present in low concentration in a cocoa butter base tend to diffuse into the rectal fluids. Drugs that are only slightly soluble in the lipid base partition readily into the rectal fluid. Water-soluble bases assume a rapid dissolution rate, and the rate-limiting step in absorption is transport of the drug through rectal mucosa.

Figure 15.4 shows a schematic of the steps involved in the absorption of drug from the rectum in a fatty base suppository. The following hypothesis could be suggested: An interface is produced

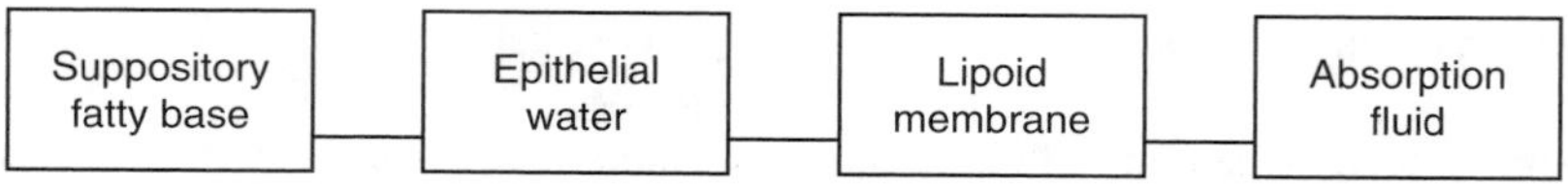

FIGURE 15.4 Four-compartment model for absorption of drug from fatty base suppositories.

TABLE 15.3
Rate of Drug Release from Different Suppository Bases

Drug vs. Base Characteristics	Approximate Drug Release Rate from Dosage Form
Oil-soluble drug with oily base	Slow release; poor escaping tendency
Water-soluble drug with oily base	Rapid release
Oil-soluble drug with water miscible base	Moderate release rate
Water-miscible drug with water miscible base	Moderate release; based on diffusion; all water soluble

between the molten fatty base and the water layer present all along the epithelial layer of the membrane. Therefore, a four-compartment model may be suggested to explain the absorption of the drug.

Table 15.3 provides a general summary of the relationship of drug release to the drug and the suppository base.

A. Suppository Bases

Suppository bases are classified on the basis of their physical characteristics into three main categories:

1. Fatty or oleaginous bases
2. Water-miscible or water-soluble bases
3. Miscellaneous bases

1. Fatty or Oleaginous Bases

These are **the** most frequently used suppository bases, for example, theobroma oil or cocoa butter along with other hydrogenated fatty acids of vegetable oils such as palm kernel oil and cottonseed oil. Fat based compounds containing compounds of glycerin with high molecular weight fatty acids, such as palmitic acid and stearic acid or various combinations of these ingredients, are used to achieve the desired melting point and hardness to withstand shipping. Suppository bases are also prepared with emulsified fatty materials or with an emulsifying agent to present prompt emulsification when a suppository base makes contact with aqueous body fluids. Cocoa butter NF is a triglyceride of oleopalmitostearin and oleodistearin. Cocoa butter melts between 30 and 36°C; hence it is an ideal suppository base, melting just below body temperature and yet maintaining solidity at room temperature. However, owing to its triglyceride content, cocoa butter exhibits marked polymorphism. Consequently, quick chilling of melted base (cocoa butter) results in a metastable crystalline form (α crystals) with a melting point lower than normal for cocoa butter. A crystal in metastable condition reverts very slowly to the stable form (β crystals), having a melting point higher than room temperature. This transition will not take place quickly enough to ensure a defect free finished product after the suppositories have been removed from the mold. Substances such as phenol and chloral hydrate tend to lower the melting point of the cocoa butter when incorporated with it. Certain hardening agents (bees wax, cetyl esters wax) may be used if the melting point of the suppository base has been lowered by any ingredients of the formulation.

2. Water-Miscible and Water-Soluble Bases

a. Water-Miscible Bases

These bases are primarily composed of glycerinated gelatin. Glycerinated gelatin bases are most commonly used in preparation of vaginal suppositories, where the prolonged localized action is desired. The glycerinated gelatin base is slower to soften and mix with the physiologic fluids than is cocoa butter and therefore provides a more prolonged release. The hygroscopic nature of the gelatin explains this slow softening of the base within the rectum. The hygroscopic effect may even cause irritation to the rectal mucosa owing to dehydration, so water is often incorporated in the formula. Such hygroscopic bases need to be protected from atmospheric moisture in order to maintain their shape and consistency.

The following is the formula for a glycerinated vaginal suppository:

Granular gelatin	20%
Glycerin	70%
Aqueous drug solution	10%

Urethral suppositories may be prepared from a glycerinated base of a somewhat different formula:

Gelatin	60%
Glycerin	20%
Aqueous drug solution	20%

Urethral suppositories of glycerinated gelatin are more easily inserted than suppositories made with a cocoa butter base, owing to the brittleness of cocoa butter and its rapid softening at body temperature.

b. *Water-Soluble Bases*

Polyethylene glycols are polymers of ethylene oxide and water, which can be prepared to various chain lengths to get desired molecular weights and physical states. Polyethylene glycols with average molecular weights of 200, 400, and 600 are clear colorless liquids. Those with average molecular weights of greater than 1000 are wax-like, white solids with the hardness increasing with an increase in molecular weight. Various combinations of polyethylene glycols may be used to prepare a suppository base of desired consistency and characteristics. In the following examples, base 1 exhibits a low melting point, whereas base 2 a high melting point.

Base 1: Polyethylene glycol 1000: 96%, polyethylene glycol 4000: 4%
Base 2: Polyethylene glycol 1000: 75%, polyethylene glycol 4000: 25%

Polyethylene glycol suppositories do not melt at body temperature but dissolve slowly in body fluids, enabling slower release of the drug on insertion into the rectum. Thus, the base does not have to melt at body temperature. This imparts a distinct advantage to these suppositories, as they need not be refrigerated and do not melt at room temperature or warm weather conditions. Their solid nature also permits them to be inserted slowly into the body without the fear that they might melt on the fingertips, thereby ensuring greater patient compliance. Because they do not melt at body temperature, but mix with mucous secretions upon their dissolution, polyethylene glycol–based suppositories do not leak from the orifice, as do many cocoa butter suppositories. Polybase™ is a preblended suppository base that is a white solid consisting of a homogeneous mixture of polyethylene glycols and polysorbate 80. It is a water-soluble base that is stable at room temperature, has a specific gravity of 1.177 at 25°C with an average molecular weight of 3440, and does not require mold lubrication.

c. *Miscellaneous Bases*

Other bases may be mixtures of lipid and water-soluble bases, for example, polyoxyl 40 stearate, which is a mixture of monostearate and distearate esters of mixed polyoxyethylene diols and the free glycols. Recently, hydrogels defined as macromolecular networks that swell but do not dissolve in water have been recommended as bases for rectal and vaginal drug delivery. The swelling of hydrogels is attributed to the presence of hydrophilic functional groups attached to the polymeric network. Crosslinks between adjacent macromolecules result in aqueous insolubility of the hydrogels. The use of a hydrogel matrix for drug delivery involves the dispersal of the drug into the matrix, followed by drying of the system and concomitant immobilization of the drug. When the hydrogel delivery is placed in the rectum or vagina, the hydrogel swells in an aqueous environment, enabling the drug to diffuse out of the macromolecular network. Rate and extent of drug release from these hydrogel matrices depend on the rate of water migration into the matrix and the rate of drug diffusion out of the swollen matrix. Although the hydrogel-based drug delivery systems are yet to appear in suppository or insert form commercially, research efforts in this direction are increasing because of their potential for controlled drug delivery, bioadhesion, retention at the site of administration, and biocompatibility.

V. PREPARATION OF SUPPOSITORIES

Suppositories are prepared by three methods: molding, compression, and hand rolling and shaping.

A. Molding

The most commonly used method for producing suppositories on both a small and large scale is the molding process. The steps in molding include (a) melting the base, (b) incorporating required medicaments, (c) pouring the melt into moulds, and (d) allowing the melt to cool and congeal into suppositories and removing the formed suppositories from the mold.

The molds in common use today are made from stainless steel, aluminum, brass, and plastic. Molds are of various capacities, depending upon the scale of production. A community pharmacy would have molds capable of producing 6 to 12 suppositories. The mold shown in Figure 15.5 is a manually operated mold with a capacity of 50 suppositories. Molds are opened longitudinally for cleaning before and after preparation of suppositories. Care must be taken while cleaning molds, as scratches may affect the desired smoothness of the final product.

1. Lubrication of the Mold

This is an important step in the preparation of suppositories. Lubrication is required for clean and easy removal of suppositories. Cocoa butter and polyethylene glycol bases seldom require lubrication, as they contract on cooling within the mold to separate from the inner surfaces and allow

FIGURE 15.5 A partially opened suppository mold with a capacity of 50 suppositories.

their easy removal. Lubrication is necessary when glycerinated gelatin suppositories are prepared. A thin coating of mineral oil is applied to the molding surface, which provides sufficient lubrication.

2. Calibration of the Mold

It is necessary to account for different densities of different bases. For instance, the weight of suppositories prepared from cocoa butter differs from that of polyethylene glycol for a particular mould. Furthermore, any added medicinal agent alters the density and hence the weight of the resulting suppository. Empty suppositories are prepared from base material alone. The suppositories are weighed and their volume is determined by melting. Thus the density is computed for that particular base. In the case of extemporaneous preparations the pharmacist needs to determine the amount of the base required to be incorporated along with the medicaments, so that each suppository provides the required amount of the drug. The volume of the base required is determined by subtracting the volume of the medicaments from the total volume of the mold. Suppository bases are solids at room temperature, so the density of the material is used to calculate the volume of the base required; for example, if 12 ml of cocoa butter is required to fill a mold, and if the volume of the medicaments is 3.4 ml, then the volume of the cocoa butter required is 8.6 ml. The weight of the base required is then obtained by multiplying the density of cocoa butter (0.86 g/ml) with 8.6 ml, which is 7.4 g.

3. Preparing and Pouring the Melt

The weighed base is melted over a water bath by using the least possible heat. A porcelain casserole is generally used for melting the base, as the melt can be conveniently poured into the cavities of the mold from this utensil. Medicaments are usually incorporated into a portion of the melted base by mixing on a porcelain or glass tile with a spatula. This material is then added with stirring to the remaining base, which is cooled almost to its congealing point. Any volatile or thermolabile substances should be added to the base with thorough stirring at this point. This material is now poured carefully and continuously into the chilled mold. If any undissolved solid materials with a tendency to settle are present, the material should be poured with constant stirring. The chilled mold avoids settling of the solids within the mold and also aids in congealing. If the melt is not at the congealing point when poured into the cavities, the solids may settle within each cavity of the mold; this will cause inaccurate distribution of the active ingredient in the bulk of the suppository. Filling of the suppository cavity should be continuous to avoid layering, which may lead to a product that is easily broken when handled.

To ensure complete filling of the mold upon congealing, the melt is poured above the level of the mold over each opening. The excessive material prevents the formation of recesses at the ends of the suppositories. After solidification the excess material is scrapped off evenly with a spatula. The mold is usually kept in the freezer to promote hardening of the suppositories. When the suppositories are hard, the mold is removed from the freezer, the sections of the mold are separated, and the suppositories are removed, with pressure being exerted at the ends.

Industrial preparation of suppositories on a large scale involves steps similar to those discussed above. However, it also involves automation for each of the basic processes such as pouring the melt into molds, refrigeration, and removal of suppositories from the mold. The automatic molding machine is one in which all filling, ejecting, and mold-cleaning operations are fully automated (e.g., Sarong SpA equipment). The output of a typical rotary machine ranges from 3500 to 6000 suppositories an hour. The essential parts of a rotary molding machine are shown in Figure 15.6.

In this machine, the chrome-plated brass molds are installed radially in the cooling turntable. The prepared mass is fed into the filling hopper, where it is continuously mixed and maintained at constant temperature. The suppository mold is lubricated by brushing or spraying and then filled to slight excess. After the mass solidifies in the molds, the excess material is collected for reuse.

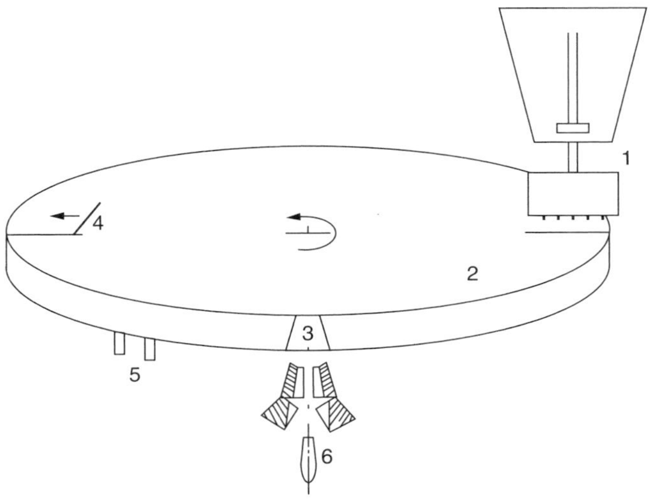

FIGURE 15.6 Schematic of an automated rotary machine for large scale production of suppositories showing (1) feeding device and filling hopper, (2) rotating cooling turntable, (3) suppository ejection station, (4) scraping device, (5) refrigerant inlet-outlet, and (6) suppository.

The cooling cycle is adjusted, as required by the individual suppository mass, by adjusting the speed of the rotary cooling turntable. The solidified suppositories are then taken to the ejecting station, where the mold is opened and the suppositories are pushed out by steel rods. The empty mold then proceeds for the repetition of the cycle.

The inner surface of the molds must be kept clean in order to prevent improper closure during production. Incomplete closure of the molds results in overweight suppositories with mold marks. For ensuring smooth and uniform final product, air jets are installed in the machine. They routinely clean the molds by blowing out any particles sticking to the inside of the mold at the end of each cycle.

B. Compression

An alternative method of preparing suppositories is by forcing the mixture of the suppository base and the medicaments into special molds using suppository making machines. The suppository base and the medicaments are combined by thorough mixing, and the heat produced by the friction of the process causes the base to soften into a paste. Mortar and pestle are used on a small scale, and on a large scale mechanically operated kneading mixers and warmed vessels are employed.

The suppository mass is filled into a cylinder, which is then closed, and the mass is forced out the other end by applying pressure mechanically or by turning a wheel. After the die is filled completely, the movable end plate at the back of the die is removed and pressure is applied to the mass in the cylinder to collect the formed suppositories.

This method is especially useful for making suppositories containing drugs that are thermolabile and for those containing substances that are insoluble in the base. There is no chance of settling of insoluble matter during preparation as in the case of preparation by melting. The limitation of this method is that it requires a special suppository machine.

C. Hand Rolling and Shaping

This is a historic part of the art of the pharmacist and hence this method will be discussed concisely. A cocoa butter base is employed in this method. A plastic-like mass is formed from the base of grated cocoa butter and other ingredients by severe trituration. The mass is rolled into a ball with the palms of the hands. Since the suppository base melts near body temperature, before handling it, hands need to be cooled below body temperature by dipping them in ice water. The mass is shaped into a cylinder with the help of the spatula. The cylinder is cut into appropriate lengths with a knife, each length equal to one suppository. The final shaping of the suppository is done with the fingertips with suppositories rolled in a glassine powder paper to prevent the warmth of the palm melting the base. Alternatively, the cylindrical base can be pushed into a mould to give the suppositories a final uniform shape.

VI. STORAGE AND HANDLING

Glycero-gelatin suppositories are usually packaged in well-closed screw-capped glass containers, preferably at a temperature below 35°C. Cocoa butter–based suppositories are usually packaged by wrapping them individually into partitioned boxes to prevent contact and adhesion. Suppositories containing light-sensitive drugs are generally wrapped in an opaque material such as a metallic foil. Most commercially available suppositories are individually wrapped in either foil or polyvinyl chloride (PVC)-polyethylene. Some suppositories utilize strip-packaging, with individual suppositories being separated by tearing along perforations located between suppositories.

Since suppositories are sensitive to heat, they should be stored in a cool place, but they should not be frozen. Cocoa butter suppositories must be stored below 30°C, preferably in a refrigerator (2 to 8°C). Glycerinated gelatin suppositories should be stored below 35°F and can be stored at controlled room temperature (20 to 25°C). Polyethylene glycol–based suppositories can be stored at usual room temperatures.

Suppositories are adversely affected by humidity. High humidity causes them to become spongy, whereas an extreme dry environment results in the loss of moisture from suppositories, which makes them brittle.

VII. DRUG INCOMPATIBILITIES AND STABILITY

The active ingredients must be compatible with the suppository base. A wide range of ingredients are incompatible with polyethylene glycol bases, including benzocaine, iodochlorhydroxyquin, sulfonamides, ichthammol, aspirin, silver salts, and tannic acid. Other materials, such as sodium barbital, salicylic acid, and camphor, tend to crystallize out of polyethylene glycol bases.

The completed products are generally considered dry or nonaqueous and thus provide a stable dosage form as long as they are protected from moisture and heat. The USP description of stability consideration for suppositories includes observations for excessive softening and evidence of oil stains on packaging materials. The major indication of instability in suppositories is excessive softening. Sometimes suppositories may dry out, harden, or shrivel. Storage in a refrigerator is usually recommended. The pharmacist may have to check for evidence of instability by removing the wrappings of individual suppositories.

Finished suppositories are routinely examined for their appearance. Several tests are performed on the finished suppositories to ensure their quality. The melting range test determines the melting time of the suppository when immersed in a 37°C water bath. The apparatus used in this test is a USP tablet disintegration apparatus. The softening time test measures the liquefaction time of suppositories in an apparatus that simulates *in vivo* conditions. A cellophane tube is tied to both ends of a condenser. Water (37°C) is circulated through the condenser at a rate that causes the lower half of the cellophane tube to collapse while the upper half opens. The suppository is dropped into the tube and sits at the level shown in Figure 15.7. The time it takes the suppository to melt completely in the tube is measured. The breaking test is designed to measure the fragility of suppositories. Suppositories with different shapes have different breaking points. The desired

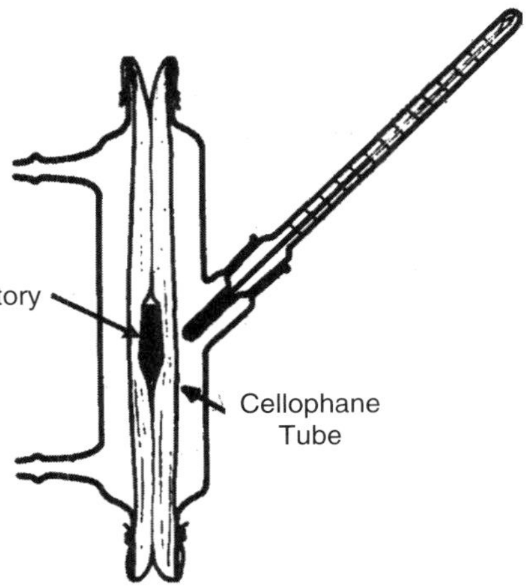

FIGURE 15.7 Apparatus for measuring the liquefaction time of rectal suppositories.

breaking point of differently shaped suppositories is set as the level at which the suppository withstands the break forces induced by all kinds of handling such as production, packaging, shipping, and patient in-use handling. The dissolution test determines the *in vitro* release rate of drug from suppositories.

VIII. DENSITY (DOSE DISPLACEMENT) CALCULATIONS

To determine the weight of the individual suppositories, it is important to know the density of the incorporated materials. It is usually assumed that the volume occupied by the active drug is not significant and need not to be considered if its quantity is less than 100 mg. This is generally based on a 2 g suppository weight. Therefore if a suppository weighs less than 2 g or if the quantity of the active drug is greater than 100 mg, the powder volume should be considered.

To determine the weights of the ingredients needed, the pharmacists should know the density factors of various bases and drugs. The density factors of cocoa butter are known. If the density factor of a base is not known, it can be calculated as the ratio of the blank weight of the base and cocoa butter.

Three methods are used to calculate the quantities of base that the active medication will occupy and the quantities of ingredients required.

A. Method 1: Dosage Replacement Factor Method

$$f = \frac{100(W - A)}{(A)(Y)} + 1$$

where f is the dosage replacement factor, W is the weight of the pure base suppositories, and A is the weight of suppositories with $Y\%$ of the active ingredient.

Cocoa butter is arbitrarily assigned a value of 1 as the standard base. If the dosage replacement factor, f, of the incorporated drug is known, the weight of a suppository with $Y\%$ of the active ingredient, A, can be calculated by the above equation. The dosage replacement factors for some drugs are listed in Table 15.4.

TABLE 15.4
Dosage Replacement Factors of Some Medicaments

Medicament	Dosage Replacement Factor
Balsam Peru	0.83
Bismuth subgallate	0.37
Bismuth subnitrate	0.33
Boric acid	0.67
Camphor	1.49
Castor oil	1.0
Chloral hydrate	0.67
Ichthammol	0.91
Phenobarbital	0.81
Phenol	0.9
Procaine HCl	0.8
Quinine HCl	0.83
Resorcin	0.71
Silver protein, mild	0.61
Theophylline sodium acetate	0.6
White/yellow wax	1.0
Zinc oxide	0.15–0.25

B. Method 2: Density Factor Method

1. Determine the average blank weight, A, per mold, using the suppository base of interest.
2. Weigh the quantity of suppository base necessary for 10 suppositories.
3. Weigh 1.0 g of medication. The weight of medication per suppository, B, is then equal to 1 g/10 suppositories = 0.1 g/suppository.
4. Melt the suppository base, incorporate the medication, mix, pour into mold, cool, trim, and remove from the mold.
5. Weigh the 10 suppositories and determine the average weight (C).
6. Determine the density factor as follows:

$$\text{Density factor} = \frac{B}{A - C + B}$$

   where A is the average weight of the blank suppository, B is the weight of medication per suppository, and C is the average weight of the medicated suppository.
7. Take the weight of the medication required for each suppository and divide it by the density factor of the medication (Table 15.5) to find the replacement value of the suppository base.
8. Subtract this quantity from the blank suppository weight.
9. Multiply by the number of suppositories required to obtain the quantity of suppository base required for the prescription.
10. Multiply the weight of drug per suppository by the number of suppositories required to obtain the quantity of active drug required for the prescription.

C. Method 3: Occupied Volume Method

1. Determine the average weight per mold (blank) using the suppository base of interest.
2. Weigh the quantity of suppository base necessary for 10 suppositories.
3. Divide the density of the active drug by the density of the suppository base to obtain a ratio.

TABLE 15.5
Density Factors of Some Medicaments for Cocoa Butter Suppositories

Medicament	Density Factor
Alum	1.7
Aminophylline	1.1
Aspirin	1.3
Balsam Peruvian	1.1
Barbital	1.2
Belladonna extract	1.3
Benzoic acid	1.5
Bismuth carbonate	4.5
Bismuth salicylate	4.5
Bismuth subgallate	2.7
Bismuth subnitrate	6.0
Boric acid	1.5
Castor oil	1.0
Chloral hydrate	1.3
Cocaine HCl	1.3
Digitalis leaf	1.6
Glycerin	1.6
Ichthammol	1.1
Iodoform	4.0
Menthol	0.7
Morphine HCl	1.6
Opium	1.4
Paraffin	1.0
Phenobarbital	1.2
Phenol	0.9
Potassium bromide	2.2
Potassium iodide	4.5
Procaine	1.2
Quinine HCl	1.2
Resorcinol	1.4
Sodium bromide	2.3
Spermaceti	1.0
Sulfathiazole	1.6
Tannic acid	1.6
White wax	1.0
Witch haze fluid extract	1.1
Zinc oxide	4.0
Zinc sulfate	2.8

4. Divide the total weight of active drug required for the total number of suppositories by the ratio obtained in step 3 (this will give the amount of suppository base displaced by the active drug).
5. Subtract the amount obtained in step 4 from the total weight of the prescription (number of suppositories multiplied by the weight of the blanks) to obtain the weight of suppository base required.
6. Multiply the weight of active drug per suppository by the number of suppositories to be prepared to obtain the quantity of active drug required.

IX. CONCLUSION

Though not very popular in the U.S., suppositories are used widely in Europe, Asia, and other parts of the world. Even in the U.S., the importance of suppositories in special circumstances should be recognized. Some of those circumstances involve unconscious patients or those not able to take medication by oral route. Under those circumstances suppositories are definitely a better alternative than parenteral routes of delivery. Also, in many instances, pharmacists may have to dispense suppositories extemporaneously. Therefore, pharmacists need a thorough knowledge of the preparation, storage, and handling of suppositories. This knowledge is necessary to counsel patients on the use and efficacy of this relatively underused dosage form.

TUTORIAL

1. Advantages of rectal administration include:
   (a) Convenient self-insertion and removal
   (b) No gastrointestinal degradation of drug
   (c) At least partially circumvents first pass elimination
   (d) All of the above
2. Suppositories can be used for both local and systemic effects.
   (a) True
   (b) False
3. __________ is a hygroscopic suppository base.
   (a) Cocoa butter
   (b) Polyethylene glycol
   (c) Glycerinated gelatin
4. Which of the following suppository vehicles exhibits polymorphism:
   (a) Glycerinated gelatin
   (b) Polyethylene glycol
   (c) Cocoa butter
   (d) Polyoxyl 40 stearate
5. Polyethylene glycol base is ideal for use with ichthammol.
   (a) True
   (b) False
6. Water is an ideal solvent for dissolving medicaments before incorporating into suppositories.
   (a) True
   (b) False
7. Suppositories can be prepared by:
   (a) Hand rolling
   (b) Compression
   (c) Molding
   (d) All of the above
8. Volume contraction results in cavity formation at the back of the suppository during cooling.
   (a) True
   (b) False

9. Glycerin suppositories are preferably stored at what temperature?
   (a) 20 to 25°C
   (b) Below 35°F
   (c) 37°C
10. Polyethylene glycol–based suppositories can be stored at room temperature.
    (a) True
    (b) False
11. The volume occupied by active drugs in a 2-g suppository need not to be considered if its quantity is less than:
    (a) 500 mg
    (b) 300 mg
    (c) 200 mg
    (d) 100 mg
12. Indication of instability in suppositories includes:
    (a) Excessive softening
    (b) Drying out
    (c) Hardening
    (d) Shriveling
    (e) All of the above
13. Factors that influence stability of suppositories are:
    (a) Heat
    (b) Humidity
    (c) Light
    (d) All of the above
    (e) None of the above
14. Layering in suppositories may lead to:
    (a) Softening
    (b) Drying out
    (c) Breaking
    (d) (a) and (b)
15. Which of the following combinations produces a rapid release of a drug?
    (a) Oil-soluble drug in an oily base
    (b) Oil-soluble drug in a water-miscible base
    (c) Water-soluble drug in a water-miscible base
    (d) Water-soluble drug in an oily base

Answers: 1. (d); 2. (a); 3. (c); 4. (c); 5. (b); 6. (b); 7. (d); 8. (a); 9. (b); 10. (a); 11. (d); 12. (e); 13. (d); 14. (c); 15. (d)

HOMEWORK

1. What is a suppository? Name the types of dosage forms used for rectal and vaginal administration.
2. What are the advantages of rectal and vaginal administration of drugs?
3. Name an advantage of suppositories over tablets.

4. What are the limitations of suppositories?
5. Describe the characteristics of an ideal suppository base. Mention different types of bases used.
6. What are the different methods for the preparation of suppositories? What is the advantage of preparing suppositories by compression?
7. Explain the method to calibrate a suppository mold.
8. Explain the density factor method to calculate the quantity of the base needed for preparing a suppository.
9. Describe the storage temperature requirements for suppositories prepared with cocoa butter, glycerinated gelatin, and polyethylene glycol base.
10. A prescription for 10 acetaminophen 300 mg suppositories, using cocoa butter as the base, needs to be filled. The average weight of the cocoa butter blank is 2.0 g, and the average weight of the medicated suppository is 1.8 g. Using the density factor determination method, calculate the amount of cocoa butter and acetaminophen needed for the prescription. (*Answer:* acetaminophen: 3.0 g; cocoa butter: 15.0 g)
11. A pharmacist wants to prepare suppositories containing 100 mg phenobarbital ($f = 0.81$) using cocoa butter as the vehicle. The weight of the blank cocoa butter suppository is 2.0 g. The resorcinol will occupy 5% of the total weight. Using the dosage replacement factor method, calculate the total weight of each medicated suppository? (*Answer:* 2.019 g)
12. A prescription for 300-mg zinc oxide suppositories needs cocoa butter as the vehicle. The density factors for zinc oxide and cocoa butter are 4.0 and 0.9, respectively. If the suppository mold holds 2.0 g of cocoa butter, what quantities of cocoa butter and zinc oxide would be required to prepare 6 suppositories? (*Answer:* zinc oxide: 1.8 g; cocoa butter: 11.59 g)

CASES

CASE I

A 28-year-old woman is complaining of mood swings, irritability, and loss of libido. She is having an irregular menstrual cycle, bloating, and breast swelling and tenderness. She has noticed impaired work ability and strained interpersonal relationships. The physician, after a thorough examination, diagnosed the condition as PMS (premenopausal syndrome) and, prescribed 125-mg progesterone suppositories.

1. A ____________ base would be an appropriate suppository base for the drug, progesterone.
   (a) Hydophilic
   (b) Lipophilic
2. The pharmacist understands that progesterone is a lipophilic drug and its slow release from the suppository is desirable. Which of the following bases the pharmacist should choose?
   (a) Cocoa butter
   (b) Polyethylene glycol
   (c) Glycero-gelatin

3. The amount of progesterone per suppository required is 125 mg. In this case, the pharmacist does not need to do density displacement factor calculations.
   (a) True
   (b) False

The pharmacist selects the base for the suppositories to be dispensed. This base is 40% polyethylene glycol 8000 and 60% polyethylene glycol 400. The average weight per suppository with the polyethylene glycol base was only 2.371 g. The density displacement factor of a drug is the number of grams of drug that will displace 1 g of the base.

4. The pharmacist needs to calculate the density displacement factor of progesterone in the chosen polyethylene glycol base. It turns out to be:
   (a) 0.67
   (b) 0.87
   (c) 0.78
   (d) 1.08
5. Polyethylene glycol–based progesterone suppositories should be stored at:
   (a) Below 0°C
   (b) Between 5 and 10°C
   (c) Room temperature
6. Prescribed progesterone suppositories will not melt at:
   (a) Between 5 and 10°C
   (b) Room temperature
   (c) Body temperature
   (d) All of the above

Answers: 1. (a); 2. (b); 3. (b); 4. (b); 5. (c); 6. (d)

CASE II

A report on a 12-year-old patient's condition has been given to a physician. The patient was suffering from nausea and vomiting caused by food given by his mother. The frequent occurrence of nausea and vomiting caused the child to be rushed to the hospital, where he was given an antidote for the toxin contained in the food. Metoclopramide was given to the child to stop the nausea and vomiting. Intravenous fluids were administered to restore the electrolyte balance and to prevent excessive fluid loss. The child is now out of danger, but the intermittent attacks of nausea and vomiting still occur owing to the irritation of the gastrointestinal tract caused by the antidote. The child also has high fever owing to the immune response produced by the poison. Mild dehydration has set in owing to the vomiting.

1. What will the physician do in response to this call?
   (a) Try to curb the vomiting by reducing gastric irritation.
   (b) Try to reduce the child's fever.
   (c) Maintain the fluid and electrolyte balance.
   (d) All of the above.
2. The physician, after thorough examination of the patient's history and previous medications given, prescribes an antiemetic (prochlorperazine) and an antipyretic (acetaminophen) to the child. What dosage form should be given?
   (a) Rectal suppositories
   (b) Oral tablets
   (c) Capsules
   (d) All of the above

The pharmacist must dispense the medication. The rectal dose for prochlorperazine is not more than 25 mg twice a day for 6 to 12 year olds. The dose for acetaminophen is 150 to 325 mg for children 6 to 12 years of age.

3. For which suppositories will the pharmacist do dose replacement calculations?
   (a) Acetaminophen
   (b) Prochlorperazine
   (c) (a) and (b)
   (d) None
4. Which type of base should the pharmacist use for acetaminophen suppositories if a rapid release of the drug from the suppository is needed? (Note: acetaminophen is water soluble.)
   (a) Cocoa butter
   (b) A water-miscible base
   (c) Both could be used

The pharmacist needs to prepare six suppositories, each containing 300 mg of acetaminophen. These are to be administered for two days to the child. Suppose the pharmacist weighs six blank cocoa butter suppositories and the average weight is 2 g. The average weight of six medicated suppositories is 1.8 g.

5. Calculate the density factor of acetaminophen using the above data.
   (a) 0.5
   (b) 0.43
   (c) 0.8
   (d) 0.6

Answers: 1. (d); 2. (a); 3. (a); 4. (a); 5. (d)

REFERENCES

1. Allen, L.V., Jr., *The Art, Science and Technology of Pharmaceutical Compounding,* American Pharmaceutical Association, Washington, DC, 1998.
2. Banker, G.S. and Chalmers, R.K., *Pharmaceutics and Pharmacy Practice,* J.B. Lippincott, Philadelphia, 1982.
3. Chein, Y.W., *Novel Drug Delivery Systemi* Marcel Dekker, New York, 1982.
4. Gennaro, A.R., *Remington: The Science and Practice of Pharmacy,* Mack, Easton, PA, 1995.
5. Singh, J. and Jayaswal, S.B., Formulation, bioavailability and pharmacokonetics of rectal administration of lorazepam suppositories and comparison with oral solution in mongrel dog. *Pharmazeutische Industrie,* 47, 664–668, 1985.
6. Stoklosa, M.J. and Ansel, H.C., *Pharmaceutical Calculations,* Williams & Wilkins, Media, PA, 1996.
7. *U.S. Pharmacopeia 24 and National Formulary 19,* U.S. Pharmacopoeial Convention Inc., Rockville, MD, 2000.

16 Ocular, Nasal, Pulmonary, and Otic Routes of Drug Delivery

Harisha Atluri, Giridhar S. Tirucherai, Clapton S. Dias, Jignesh Patel, and Ashim K. Mitra

CONTENTS

0-4152-8863-0/05/$0.00+$1.50
© 2005 by CRC Press LLC

Learning Objectives

After studying this chapter, the student will have thorough knowledge of:

- Physiological and anatomical aspects influencing ocular, nasal, pulmonary, and otic drug delivery
- Physicochemical properties that influence drug delivery through each route
- Various strategies employed to optimize drug delivery through each route
- Production and packaging of commonly used dosage forms

I. OCULAR DRUG DELIVERY

A. Introduction

A multitude of disease states affect different sites within the eye. Highly effective therapeutic agents have been developed for the treatment of various ocular pathologies. However, achieving and maintaining therapeutic drug concentrations at the target site in the eye for the desired length of time is a challenging task. Conventional drug administration through topical or systemic routes is no longer sufficient to combat ocular pathologies because of the complex ocular physiological and cellular barriers. For drugs administered through the topical route, the cornea is a major barrier for drug absorption. A topically instilled dose is also lost by protective mechanisms, such as solution drainage, lacrimation, and systemic absorption via the conjunctiva. Drugs administered systemically have poor access to the internal structures within the globe. This is partly due to the blood ocular barriers, which prevent systemically administered drug from entering into the aqueous humor and the extravascular space of the retina. Therefore effective delivery systems should be designed such that they can deliver therapeutic agents to the targeted site in a predictable manner. The process of drug development for successful ocular delivery systems demands a thorough knowledge of physicochemical properties of the drug molecule and anatomical and physiological factors affecting ocular drug disposition. A brief overview of the relative anatomy and physiology of the eye is presented followed by a discussion on the various routes of administration and design of an optimal ocular drug delivery system.

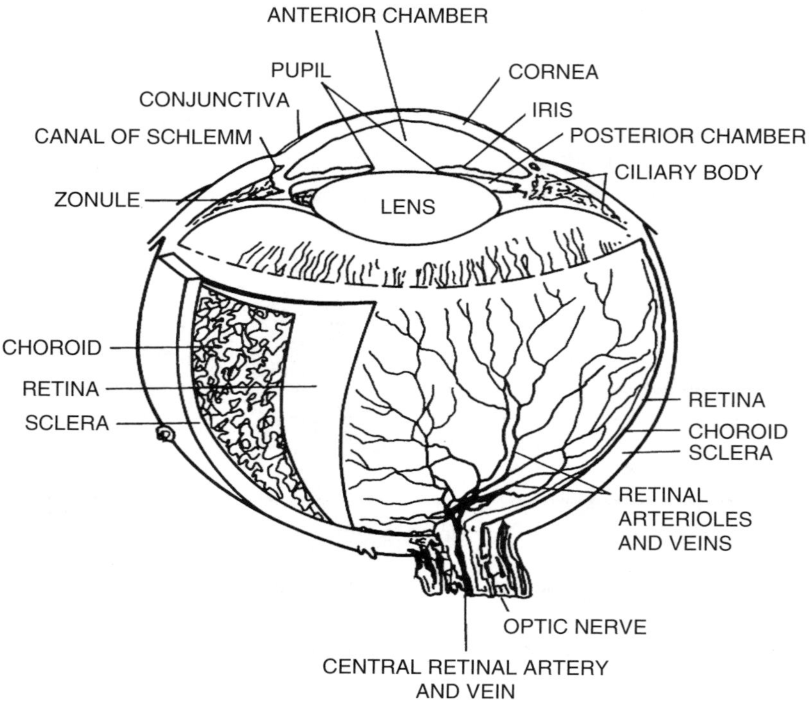

FIGURE 16.1 Anatomical structure of the eye. (Reproduced with permission from Mitra, A. K., *Ophthalmic Drug Delivery Systems,* Marcel Dekker, New York, 1993.)

B. Anatomical Features of the Eye

The eye is an isolated and highly specialized organ of photoreception. The eyelids and the orbit protect the eye. It is made three layers (Figure 16.1): the outer fibrous protective sclera, the middle vascular uvea, (including the choroid, the ciliary body, and the iris), and the inner neural retina. The eye can be divided into an anterior segment and a posterior segment by the lens, where the anterior segment includes the cornea, iris, the ciliary body, the anterior chamber, and the posterior chamber and the posterior segment includes the retina and the vitreous body.

C. Anterior Segment of the Eye

The anterior segment of the eye is composed of the cornea, the conjunctiva, the iris, the ciliary body, the anterior and posterior chambers, and the lens. The aqueous outflow pathway and the lachrymal apparatus also reside within this segment.

1. The Cornea

The anterior segment of the eye is bound anteriorly by a transparent cornea and a small portion of the sclera. The junction where the cornea and the sclera join is called the limbus. The human cornea is approximately 0.5 mm thick in the center, with the periphery being thicker than the central part. It is a transparent avascular structure that receives its nourishment and oxygen supply from aqueous humor and the tear film. The capillaries of the limbus provide oxygen and nutrition to the peripheral parts of the cornea. The tissue is composed of five layers: the epithelium, bowman's membrane, stroma, descemets membrane, and the endothelium (Figure 16.2).

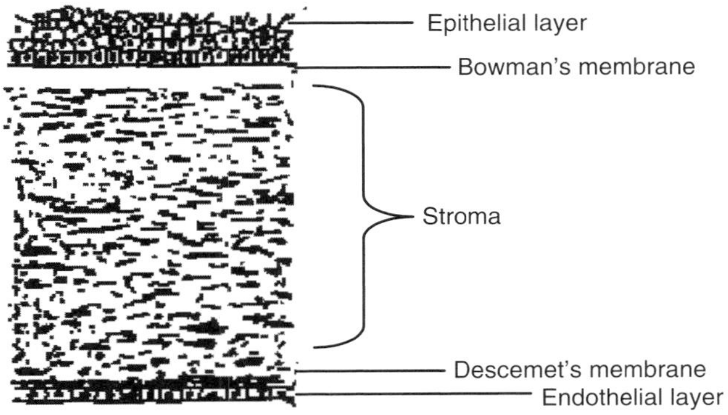

FIGURE 16.2 Cross section of cornea depicting different layers. (Reproduced with permission from Mitra, A. K., *Ophthalmic Drug Delivery Systems,* Marcel Dekker, New York, 1993.)

- *Epithelium*: The corneal epithelium acts as an outer protective barrier consisting of five to six layers of superficial squamous and basal columnar epithelial cells. Numerous microvilli protrude into the lacrimal film from these cells. The microvilli are covered by mucin and glycocalix. The superficial epithelial cell layers have tight junctional complexes between adjacent cells called zonulae occludentes. These junctional complexes prevent the passage of water-soluble drugs with low lipophilicity across the cornea. The corneal epithelium is a reasonably hydrophobic tissue, and for the drugs to cross the cornea it should have an oil–water partition coefficient greater than 1.
- *Bowman's membrane*: Bowman's membrane is an acellular thin layer made of collagen fibrils. Anteriorly it is separated from the epithelium by a basement membrane. It does not offer high diffusional resistance and as such it is not a rate limiting barrier.
- *Stroma:* This corneal layer is a highly organized structure constituting 90% of the cornea. It consists of parallel collagenous lamellae. Between the lamellae lies the modified fibroblast known as keratocytes. The stromal layer becomes a limiting barrier for permeation of lipophilic drugs. However, it offers minimal resistance to the passage of hydrophilic compounds.
- *Descemets membrane*: Descemets membrane is a thin homogeneous layer between the stroma and the endothelium.
- *Endothelium*: The corneal endothelium is a single-layered squamous epithelium on the posterior surface of the cornea. The endothelial cells do not posses tight junctions. Therefore this layer does not act as a barrier to drug molecules. Endothelial cells play a important role in regulation of corneal hydration by transferring bicarbonate and chloride into the aqueous humor.

2. Conjunctiva

The conjunctiva is a vascularized mucous membrane that covers the inner surface of the eyelids and the anterior part of the sclera. It offers less resistance to the passage of drug molecules than the cornea.

3. Iris

The iris is the most anterior portion of the uveal tract. The iris consists of the pigmented epithelial cell layer, the iridic sphincter and dilator muscles, and the stroma. The anterior surface of the iris is the stroma, which contains melanocytes, blood vessels, smooth muscle, and parasympathetic and sympathetic innervations. Individual eye color varies according to the number of melanocytes present in the stroma. Ocular pigmentation may be an important factor relative to ocular bioavailability owing to drug–melanin binding. The iris has high enzymatic activity, resulting in low ocular bioavailability. The iridic smooth muscles cause dilation and constriction of the pupil when stimulated.

4. Ciliary Body

The ciliary body is composed of the ciliary muscle and the ciliary processes. Ciliary muscle plays an important role in accommodation. The ciliary processes are the highly vascularized folds protruding into the posterior chamber. These are covered by a two-layered epithelium, the inner nonpigmented and the outer pigmented epithelium. The nonpigmented epithelial cells are attached to each other by junctional complexes, which obstruct the diffusion of large molecules into the aqueous humor. This cell layer forms the blood aqueous barrier. The nonpigmented epithelial cells secrete the aqueous humor into the posterior chamber by active mechanisms.

5. Aqueous Outflow Pathway

Aqueous humor is secreted by the ciliary processes into the posterior chamber at the rate of 2 to 2.5 μl/min. Aqueous humor flows continuously from the posterior chamber to the anterior chamber through the pupil and from there it leaves the eye through the trabecular meshwork and Schlemm's canal. The aqueous humor plays an important role in the nourishment and supply of oxygen to the avascular tissues in the eye. It also carries metabolites away from the surrounding tissues. In addition to these functions the continuous flow of aqueous humor is important for maintaining the constant volume and intraocular pressure.

6. Lens

The lens is a crystalline structure surrounded by the lens capsule. It is located behind the iris and in front of the vitreous. The lens is important for the visual function and together with the ciliary muscle it enables accommodation and protects the retina from harmful ultraviolet radiation.

7. Lacrimal Apparatus

Tear fluid is secreted by the main lacrimal gland, the acessory glands, and the meibomian glands. The normal tear flow rate is approximately 1 μl/min, and it covers the cornea and the conjunctiva. The tear fluid has a nutrition function as well as an antibacterial function. Tear fluid secretion also helps in lubricating the eye and washing away debris. It is drained through the canaliculi, lacrimal sac, nasolacrimal duct, and finally into the nasal cavity.

D. Posterior Segment of the Eye

The posterior segment of the eye is composed of the retina, vitreous, the choroid, the sclera, and the optic nerve.

1. Retina

The retina is a thin, transparent neural tissue extending anteriorly to the pars plana of the ciliary body. The retina is the innermost tissue of the three coats of the eye. It consists of the neural layer and the pigment epithelium. The pigmented epithelial cells are connected with tight junctions and form a barrier between the vascular choroid and retina.

2. Vitreous

Vitreous is a clear medium that fills the cavity between the retina and the lens. It is composed of 98 to 99.7% water. The vitreous mainly consists of dissolved collagen, hyaluronic acid, and proteoglycans. It is easily susceptible to bacterial infection, and the treatment of vitreal infections becomes challenging owing to poor bioavailability of the antibiotics. The blood retinal barrier largely prevents the entry of many drugs into the vitreous from the systemic circulation.

3. The Choroid and the Sclera

The choroid is a highly vascularized tissue between the retina and the sclera. It constitutes the posterior part of the uvea that represents the middle vascular coat of the eye. The primary function of the choroid is to nourish the outer layers of the retina. The sclera is the outermost layer, consisting of collagen bundles and elastic fibers. The sclera functions to both protect the intraocular contents and to maintain the shape of the eye. It is permeable to molecules as large as dextran; hence the transscleral route can be used to deliver drugs to the posterior segment.

E. Routes of Ocular Drug Delivery

The routes of administration for ocular therapeutics can be (Table 16.1) broadly classified as topical, periocular, intraocular, transscleral, and systemic (Figure 16.3) (3). Apart from these routes of administration, drugs can be delivered to the eye by a specialized technique called ocular iontophoresis.

1. Topical Administration

The topical route of administration is used to treat diseases that affect the anterior segment of the eye such as keratitis, conjunctivitis, and glaucoma. However, topical delivery may not provide the desired therapeutic concentration of the drug in the posterior segment of the eye to treat diseases there. Drug delivery vehicles used for topical administration include solutions, colloids, emulsions, suspensions, ointments, solid hydrophilic inserts, therapeutic contact lenses, rate-controlled release systems, and new delivery systems such as liposomes and particulates.

The advantages of topical administration are:

- Convenient mode of administration
- Noninvasive
- Easy enough for self-administration
- Fewer systemic drug effects

The disadvantages of topical administration are:

- Low ocular bioavailability
- Ineffectiveness in the treatment of posterior segment diseases

a. *Factors Affecting the Bioavailability of Topically Administered Drugs*

In developing an effective topical drug delivery system, a thorough understanding of the physiological factors affecting the bioavailability is required. The major factors determining the bioavailability of topically applied drugs are precorneal, corneal, and postcorneal factors.

i. *Precorneal Factors*

Precorneal fluid drainage: After instillation of the liquid dosage form, a significant part (80 to 90%) of it is drained into the nasolacrimal duct. This loss occurs primarily due to the

TABLE 16.1
Therapeutic Classes of Drugs Used as Ophthalmic Medications

Class	Generic Name	Trade Name	Route of Administration	Manufacturer
Mydriatics and cycloplegics (used for refraction and pupillary dilation)	Phenylephrine hydrochloride	AK-Dilate	Topical	Akorn
	Atropine sulphate	Atropine Care	Topical	Akorn
	Cyclopentolate hydrochloride	AK-Pentolate	Topical	Akorn
	Homatropine hydrobromide	—	Topical	Novartis
	Scopolamine hydrobromide	Isopto Hyoscine	Topical	Alcon
	Tropicamide	Tropicacyl	Topical	Akorn
Antibacterial agents (treatment of bacterial infections such as blepharitis, conjunctivitis, endophthalmitis, etc.)	Bacitracin	AK-Tracin	Topical/subconjunctival	Akorn
	Chloramphenicol	Chloromycetin	Topical/subconjunctival/intravitreal/intravenous	Monarch
	Ciprofloxacin hydrochloride	Ciloxan	Topical/systemic	Alcon
	Erythromycin	—	Topical/subconjunctival/intravitreal	Medical Ophthalmics
	Gentamicin sulfate	Gentak	Topical/subconjunctival/intravitreal/intravenous	Akorn
	Tobramycin sulfate	Tobrex	Topical/subconjunctival/intravitreal/intravenous	Alcon
Antifungal agents (treatment of fungal keratitis and conjunctivitis)	Amphotericin B	Fungizone	Topical/subconjunctival/intravitreal/intravenous	—
	Natamycin	Natacyn	Topical	Alcon
	Flucytosine	Ancobon	Oral/topical	—
	Miconazole nitrate	Monistat	Topical/subconjunctival/intravitreal	MacNeil
Antiviral agents (Treatment of viral keratitis and retinitis)	Trifluridine	Viroptic	Topical	Monarch
	Vidarabine monohydrate	Vira-A	Topical	King Pharmaceuticals
	Cidofovir	Vistide	Intravenous	Gilead Sciences
	Fomiversen	Vitravene	Intravitreal/intravenous	Novartis
	Foscarnet sodium	Foscavir	Intravenous	Astra Pharmaceutical
	Ganciclovir sodium	Cytovene	Intravenous/oral	Roche
	Ganciclovir	Vitrasert	intravitreal	Bausch & Lomb
Antiinflammatory agents (treatment of inflammation)	Dexamethasone sodium phosphate	AK-Dex	Topical	Akorn
	Prednisolone acetate	Pred Mild	Topical	Allergan
	Fluorometholone	Fluor-Op	Topical	Novartis
	Loteprednol etabonate	Lotemax	Topical	Bausch & Lomb
	Diclofenac	Voltaren	Topical	Novartis
	Ketorolac	Acular	Topical	Allergan
Topical anesthetics	Cocaine hydrochloride	—	Topical	—
	Proparacaine hydrochloride	AK-Taine	Topical	Akorn
	Tetracaine hydrochloride	AK-T-Caine	Topical	Akorn

TABLE 16.1 (continued)
Therapeutic Classes of Drugs Used as Ophthalmic Medications

Class	Generic Name	Trade Name	Route of Administration	Manufacturer
Agents used to treat glaucoma	Miotics			
	Carbachol	Isopto carbachol	Topical	Alcon
	Pilocarpine hydrochloride	Isopto Carpine	Topical	Alcon
	Echothiophate iodide	Phospholine Iodide	Topical	Wyeth-Ayerst
	Sympathomimetics	Epifrin	Topical	Allergan
	Epinephrine hydrochloride			
	ß-adrenergic blocking agents			
	Betaxolol hydrochloride	Betoptic-S	Topical	Alcon
	Timolol maleate	Timoptic	Topical	Merck
	Carteolol hydrochloride	Ocupress	Topical	Novartis
	Prostaglandins			
	Latanoprost	Xalatan	Topical	Pharmacia & Upjohn
Immunosuppressive agents	Cyclosporin	-----	Topical/Oral	--------
Ocuar decongestants	Naphazoline hydrochloride	AK-Con	Topical	Akorn
	Tetrahyrozoline hydrochloride	Collyrium Fresh	Topical	Bausch & Lomb

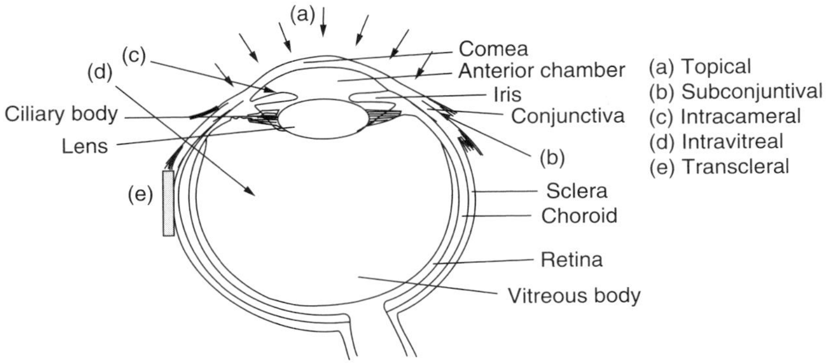

FIGURE 16.3 Schematic representation of various routes of delivery of ocular therapeutics. (From Reddy, I. K., *Ocular Therapeutics and Drug Delivery: A Multidisciplinary Approach,* Technomic, Lancaster, PA, 1996.)

tendency of the eye to maintain the precorneal fluid volume at 7 to 10 μl at all times as a protective physiological mechanism. Therefore excess fluid present in the cul-de-sac is drained into the nasolacrimal duct. Various factors that influence the drainage rate are instilled volume, viscosity, pH, tonicity, and drug type. The rate of solution drainage from the conjunctival sac is directly proportional to the instilled volume. Increasing viscosity can prolong the residence time of an instilled dose in the conjunctival sac. The physiological pH of tear fluid is ~7.4. Since tears have poor buffer capacity, instillation of an acidic or alkaline solution causes excessive tear secretion and loss of drug. To prolong drug retention at the eye surface, ophthalmic preparations must have a pH between 7.0 and 7.7. Drugs that act on the lacrimal gland can affect precorneal fluid dynamics. Agents such as epinephrine induce tear production, and local anesthetics such as tetracaine suppress tear turnover.

Drug binding to tear proteins: The protein content of tears in humans is about 0.7% of the total body protein. Therefore the drug binding to the tear proteins may result in a reduction in free drug concentrations available for pharmacological action at the target site.

Conjunctival drug absorption: The conjunctiva is a highly vascularized mucous membrane lining the inside of the eyelids and the anterior sclera. Drugs are better absorbed across conjunctiva than cornea because of its greater surface area and high permeability. However, such nonproductive conjunctival absorption is a major precorneal loss factor and may lead to side effects, since it serves as a major route of entry into systemic circulation.

Systemic drug absorption: A fraction of the topically applied dose that reaches the nasal mucosa through the nasalacrimal duct drainage may be absorbed systemically, leading to potential systemic side effects.

ii. Corneal Factors

The cornea consists of a hydrophilic stromal layer placed between a lipoidal epithelial layer and a single cell layer of endothelial cells. As a general rule corneal epithelium acts as a major barrier to the transport of hydrophilic drugs, whereas the hydrophilic stromal layer offers resistance to the passage of relatively lipophilic compounds. Various strategies such as prodrug derivatization, penetration enhancers, liposomes, and nanoparticles have been employed to deliver drugs across the cornea.

iii. Postcorneal Factors

Melanin binding: The presence of melanin pigment in the iris and ciliary body can affect the ocular bioavailabilty of the topically applied drug. Melanin imparts color to the eye. Drugs such as ephedrine and timolol can bind to the melanin with a high binding capacity, and only a small fraction of the bound drug is slowly released.

Drug metabolism: Various enzymes in the eye can metabolize the active drug, resulting in decreased ocular bioavailabilty. Drugs that are biotransformed by oxidation or reduction are less prone to metabolism than those transformed by hydrolysis because of the abundance of ocular hydrolases. The corneal epithelium and the iris ciliary body are the most metabolically active.

b. Drug Delivery Strategies

Frequent administration is required for most drugs delivered by the topical route. This is because of the rapid turnover rate of the tear film, which quickly flushes drugs out of the precorneal area, and the low permeability of corneal tissues. To overcome these disadvantages various ophthalmic drug delivery systems have been investigated. Prodrug derivatization can be employed to overcome low corneal permeability of water-soluble drugs where bioreversible derivatives of the drugs are synthesized. The drug molecule can be chemically modified to obtain suitable structural configuration and physicochemical properties, to afford maximal corneal absorption. However, a prodrug must be converted enzymatically or chemically to the parent drug *in vivo* in order to elicit its effect. The corneal epithelium has abundant choline esterases. These enzymes can be used in the delivery of more lipophilic esterified prodrugs of water-soluble compounds to the eye. Controlled release delivery systems are employed to release the drug at a constant rate to the precorneal film over a finite period of time. The benefits of these systems are diminished frequency of administration, lower toxicity, and side effects owing to systemic absorption and prolonged therapeutic action. Controlled delivery systems employed for ophthalmic therapeutics are summarized in Table 16.2.

2. Periocular Administration

This mode of administration is usually employed for the treatment of anterior segment diseases when topical administration has failed. Periocular administration may be either subconjunctival or sub-Tenon. Injection is made underneath the conjunctiva in the case of subconjunctival administration and beneath the Tenon's capsule in the case of sub-Tenon's administration. Drugs administered by this route enter the eye by diffusing through the sclera, which has high permeability to different molecules compared with the cornea. The periocular route is used for the administration of antibiotics or antivirals for the treatment of anterior segment pathologies.

3. Intraocular Administration

The treatment of many ocular diseases is hampered by poor penetration into the eye following systemic or topical administration. To circumvent the barrier, different routes of intraocular drug delivery have been investigated. Intraocular administration may be intravitreal (injections into the aqueous humor) or intracameral (injections into the vitreous). Intravitreal administration is employed for severe posterior segment diseases including endophthalmitis and retinitis. Repeated intravitreal injections may cause trauma to the eye and break down the blood retinal barrier.Intraocular sustained release ganciclovir implants have recently been developed for effective treatment of cytomegalovirus retinitis. These implants release a predetermined amount of drug at a constant rate, thereby avoiding repeated injections.

4. Transscleral Administration

Repeated long-term intravitreous delivery, which is required for treatment of chorioretinal disorders, leads to risk of local complications such as retinal detachment, endopthalmitis, and vitreous hemorrhage. Transscleral delivery may be a viable alternative to deliver drugs to the posterior segment because of the sclera's large surface area and high permeability characteristics. Protein therapeutics can be administered by this route.

5. Ocular Iontophoresis

Iontophoresis is a method of drug delivery that utilizes electric current to deliver ionized molecules to the intraocular tissues. To drive the molecules into the tissue either a cathode or an anode is used, depending on the charge of the molecule. Drugs can be delivered to the cornea or the sclera. Transcorneal iontophoretic delivery of antibiotics such as gentamicin and aprofloxacin for treatment of bacterial keratitis has yielded experimentally promising results. Transscleral iontophoresis can be used to directly deliver drugs to the vitreous for treating posterior segment diseases. It may produce discomfort and small areas of necrosis at sites of application.

F. Formulation and Manufacturing Conditions

Topical application is considered the preferred way to achieve therapeutic levels of drug in the eye. Topical formulations can be divided into the following major categories: solutions, suspensions, ointments, and others such as sprays and inserts.

TABLE 16.2
Ocular Delivery Systems

Delivery System	Description
Polymeric Gels	
Bioadhesive hydrogels	Bioadhesive polymers are macromolecular hydrocolloids and are capable of forming strong noncovalent bonds with mucin-coating biological membranes.
In situ forming gels	*In situ* forming gels are viscous liquids that upon exposure to physiological conditions shift to gel phase.
Colloidal Systems	
Liposomes	Liposomes are microscopic vesicles composed of a membrane-like lipid bilayer surrounding the aqueous compartment, which increase the absorption. Based on lipophilicity, drugs can be incorporated into either the aqueous compartment or lipid bilayer.
Nanoparticles	Nanoparticles are polymeric colloidal particles ranging in size from 10 to 1000 nm that prolong the duration of action of drugs. Drug is dissolved, entrapped, or encapsulated in the macromolecular material of nanoparticles.
Cyclodextrins	Cyclodextrins are a group of homologous cyclic oligosaccharides with a hydrophilic outer surface and a lipophilic inner cavity. They increase corneal permeability and hence improve ocular bioavailability.

1. Solutions

Solutions are the most common and preferred type of formulation for ophthalmic medications. Various adjuvants are used to maintain the stability and sterility of the formulation as well as to increase the precorneal residence time and drug penetration. In developing a successful formulation, factors such as incompatibility, antagonism, irritation potential, and toxicity should be considered.

a. Viscosity Enhancers

Viscosifiers help in prolonging the retention time of the drug in the precorneal area. They also decrease the lacrimal drainage of the drug. Water-soluble polymers are used as viscosity enhancers. Cellulose derivatives including methylcellulose, carboxy methylcellulose, and hydroxy propyl methylcellulose are generally used. The concentration range of these polymers usually varies from 0.2 to 2.5%. Administration of a highly viscous solution results in reflex tearing and blinking to restore the original viscosity. Therefore an optimum concentration of viscosity enhancers should be used. Certain polymers such as poly (acrylic acid) and hyaluronic acid, have the ability to interact with the mucus layer that covers the corneal and scleral surface. This adhesive capacity further improves the retention of the polymer solution in the preocular area.

b. Tonicity Agents

Topically instilled solutions must be isotonic with tears. Instillation of a hypotonic solution results in water flow from the aqueous layer through the cornea to the surface of the eye. Hypertonic drops may cause discomfort or irritation owing to a dehydrating effect on the corneal epithelium. This might result in reflex tears and reflex blinks, subsequently leading to drug loss. Tears and other body fluids including blood have an osmotic pressure corresponding to that of a 0.9% solution of sodium chloride. Tonicity agents such as sodium chloride, buffering salts, dextrose, and mannitol can be employed.

Problem

Calculate the amount of sodium chloride to be added to make the following solution isotonic. (Sodium chloride equivalent E is 0.23).

Rx
Ephedrine sulfate: 1 g
Sodium chloride: x g
Sterile water: enough to make 100 ml

Calculate the weight in grams of sodium chloride that is equivalent to the weight in grams of ephedrine sulfate: quantity of the drug × sodium equivalent.

$$1.0 \text{ g} \times 0.23 = 0.23 \text{ g}$$

A total of 0.9% of sodium chloride is required for isotonicity if no other salts are present.

Since ephedrine has contributed a weight of material osmotically equivalent to 0.23 g of NaCl, the amount of sodium chloride to be added is

$$0.90 - 0.23 = 0.67 \text{ g}$$

c. Buffering Agents

The physiological pH of tears is 7.4 and is governed by substances dissolved in the aqueous layer of the tears, including bicarbonate and proteins. The tear fluid has a small buffer capacity; hence the pH of tears alters rapidly with addition of greater amounts of acid or alkali. Ophthalmic formulations have to be suitably buffered for the following reasons:

- To maintain physiological pH upon topical administration. The exposure of the eye to acidic or alkaline fluids may cause damage to the epithelial cells, resulting in ocular irritation and discomfort.
- To enhance drug penetration through changes in the degree of drug ionization. The penetration of drug across the cornea can be improved by selecting the pH that favors the un-ionized form of the drug. Transcorneal flux of weak organic bases such as procaine was found to increase as the solution became more alkaline.
- To improve the stability of the product.

d. Preservatives

Ophthalmic formulations must be sterile. Most ophthalmic formulations are packed in multiple-dose containers. As a result, medication may easily become contaminated during use. A suitable antimicrobial preservative or mixture of preservatives has to be added in ophthalmic preparations. Desirable properties of effective an antimicrobial preservative are:

- Broad spectrum of activity
- Compatibility with other ingredients
- Nontoxic and nonirritant properties
- Chemical stability
- Rapid action

TABLE 16.3
Ophthalmic Preservatives

Preservative	Concentration
Quaternary ammonium compounds	
Benzalkonium chloride	0.01%
Substituted alcohols and phenols	
Chlorobutanol	0.5%
Organic mercurials	
Phenyl mercuric nitrate or acetate	0.002%
Esters of parahydroxy benzoic acid	
Methyl paraben	0.18%
Propyl paraben	0.02%

Different classes of preservatives and the concentration commonly employed are given in Table 16.3.

e. Surfactants

Surfactants are used to solubilize and disperse the drug effectively. Commonly employed surface active agents are nonionic surfactants. Some surfactants have inherent antimicrobial properties; hence proper care should be taken when adding surfactants. They may cause irritation owing to surface tension–lowering properties and may also remove the mucus layer and disrupt the tight junctional complex. Surfactants used in ophthalmic formulations are benzalkonium chloride, benzethonium chloride, polysorbate 20, and dioctyl sodium sulphosuccinate. Stabilizers such as chelating agents or antioxidants are employed in ophthalmic solutions to improve the stability.

2. Suspensions

Topical ophthalmic medications are formulated in the form of suspensions to sparingly administer soluble drugs or complexes of soluble drugs. They provide slow dissolution and prolonged release of drug. The ophthalmic suspension is generally deposited in the cul-de-sac. The rate of dissolution of suspended particles determines the rate and extent of the suspension's therapeutic activity. For optimum therapeutic effect, the rate of dissolution of particles and the rate of absorption through the cornea must be faster than the rate of loss of drug from the eye. The size of the suspended drug particle is important for patient comfort and therapeutic effectiveness. The particle size is generally less than 10 μm in diameter. Steroids are frequently administered in the form of suspensions.

3. Ophthalmic Ointments

Ointments contain one or a combination of hydrocarbons, mineral oil, lanolin, and polymers such as polyvinyl alcohol, carbopol, and methylcellulose as bases. Drugs administered as ointments have better bioavailability than drops primarily owing to reduced dilution of drug with the tears, prolonged corneal contact time, and reduced drainage. Vehicles used in ophthalmic preparations should not cause discomfort to the eye and should be compatible with other ingredients. The main disadvantage of ointments is that they cause blurring of vision and an increased incidence of contact dermatitis.

4. Ocular Inserts

An ocusert is an insoluble ophthalmic insert that releases drug at a constant and reproducible rate for a prolonged period of time. It consists of a central core or drug reservoir sandwiched between

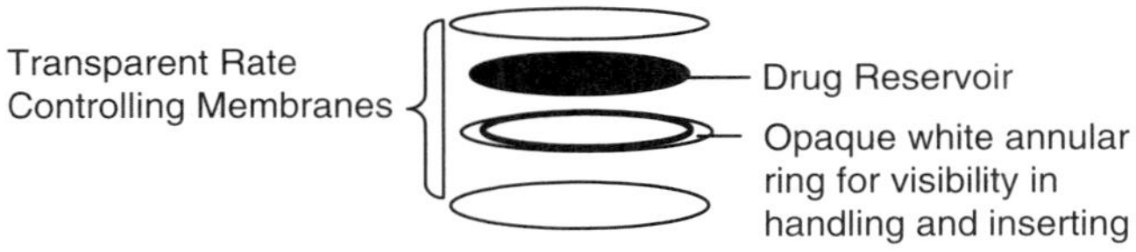

FIGURE 16.4 Schematic representation of ocusert.

two semipermeable membranes (Figure 16.4). The membrane is made up of an ethylene-vinyl acetate copolymer that controls the rate of drug release. It is placed in the cul-de-sac.

Ocuserts are commercially available for pilocarpine in two different release rates as Ocusert P-20, which delivers a dose of 20 μg/hr, and Ocusert P-40, which delivers a dose of 40 μg/hr. The release of sparingly soluble drugs from the ocusert can be calculated from Equation 16.1:

$$M = \frac{SDKC_s}{h} t \qquad (16.1)$$

where M is the accumulated of drug amount released, t is time, S is the surface area of the device, D is the diffusion coefficient of the membrane, K is the liquid–liquid partition coefficient of the ocusert membrane and the eye fluids, C_s is the aqueous solubility of the drug, and h is the membrane thickness.

G. Ophthalmic Formulations

1. GENOPTIC® (Gentamicin Sulfate Ophthalmic Solution, USP)

Each milliliter of GENOPTIC solution contains gentamicin sulfate equivalent to 3 mg (0.3%) gentamicin base; benzalkonium chloride; liquifilm® (polyvinyl alcohol) 14 mg; edetate disodium; sodium phosphate, dibasic; sodium chloride; hydrochloric acid and/or sodium hydroxide; and purified water. The above solution is buffered to a pH of 7.2 to 7.5. Gentamicin sulfate, which belongs to aminoglycoside group, is the active ingredient. Benzalkonium chloride is a preservative, and polyvinyl alcohol is a thickening agent used to prolong the contact with the eye. Edetate disodium is a sequestering agent used to limit decomposition of eye drops. Edetate disodium forms complexes with di- and tri-valent metals. Sodium phosphate, dibasic, hydrochloric acid, and/or sodium hydroxide are used to buffer the ophthalmic solution to a pH of 7.2 to 7.5. Sodium chloride is used to adjust the tonicity of eye drops.

2. GENOPTIC® (Gentamicin Sulfate Ophthalmic Ointment, USP)

Each gram of GENOPTIC ointment contains gentamicin sulfate equivalent to 3 mg (0.3%) gentamicin base in a base of white petrolatum, with methyl paraben (0.5 mg) and propyl paraben (0.1 mg). Methyl paraben and propyl paraben are preservatives.

3. FML-S® (Fluorometholone Ophthalmic Suspension, USP)

Fluorometholone ophthalmic suspension contains fluorometholone (0.1%); benzalkonium chloride (0.004%); polyvinyl alcohol (1.4%); edetate disodium; sodium chloride; sodium phosphate, monobasic; sodium phosphate, dibasic; polysorbate 80; sodium hydoxide; and purified water. Fluorometholone is the active ingredient, which is an antiinflammatory agent. Benzalkonium chloride is a preservative, and sodium phosphate, monobasic; sodium phosphate, dibasic; and sodium hydoxide are used to adjust the pH. Edetate disodium is a sequestering agent, and sodium chloride is a tonicity agent. Polyvinyl alcohol is a thickening agent, and polysorbate 80 is used as a wetting agent and to increase the dispersion of fluorometholone.

H. Production and Packaging

Factors to be considered for production of ophthalmic products are similar to those for parenterals. Ophthalmic formulations, though not introduced into internal body cavities, are in contact with tissues that are very sensitive to contamination. Therefore, similar standards to sterile products should be followed for the preparation of ophthalmic medications. Quality control tests should be conducted on incoming stock and manufacturing and finished products. Finished products should be subjected to a leaker test, a clarity test, and a sterility test. One characteristic not as important for ophthalmic medications is that the product be free from pyrogens, since pyrogens are not absorbed systemically from the eye.

Eye drops are frequently packaged in multiple-dose containers ranging from 4 to 60 ml. The proper packaging of ophthalmic medication is important for its effective usage. Important factors to be considered for ophthalmic packaging include selection of components that offer ease of administration and maintenance of sterility of the product for its entire shelf life. Ophthalmic formulations are usually dispensed in a glass dropper bottle with the dropper either inserted or packaged separately. Ophthalmic solutions used for special cases such as surgery or traumatized eye must be dispensed in single-use containers. Glass bottles are used for packing in special circumstances such as maintaining product stability and compatibility. Dual chamber containers can hold drug and diluent separately, in a single package, for stability reasons.

- *Glass packages:* Glass packages consist of following components:
- *Glass bottle:* used for packaging products that are not compatible with plastics.
- *Dropper:* A glass or plastic dropper is placed directly into the glass bottle or is packaged in a separate sterile blister packet.
- *Closures:* Made of plastic or elastomeric material.
- *Caps:* Made of aluminium or plastic with a teflon or vinyl liner to prevent leakage.
- *Plastic containers:* Plastic containers should be selected that do not contain any material that can be extracted into the formulation.
- *Droptainers:* Droptainers are plastic containers that dispense a drop of medication when inverted and gently squeezed. They consist of a plastic bottle, dispenser tip, and cap.

OCULAR REFERENCES

1. Bourlais, C. L., Ascar, L., Zia, H., Sado, P. A., Needham, T., and Leverge, R. Ophthalmic drug delivery systems: recent advances, *Progress in Retinal Eye Research,* 17, 33–58, 1998.
2. Edman, P., *Biopharmaceutics of Ocular Drug Delivery,* CRC Press, Boca Raton, FL, 1993.
3. Forrester, J. V., Dick, A. D., Mcmenamin, P., and Lee, W. R., *The Eye: Basic Sciences in Practice,* W.B. Saunders, London, 1996.
4. Gennaro, A. R., *Remington: The Science and Practice of Pharmacy,* Mack, Easton, PA, 1995.
5. Martin, A., Bustamante, P., and Chun, A. H. C., *Physical Pharmacy: Physical Chemical Principles in the Pharmaceutical Sciences,* Waverly International, Baltimore, MD, 1993.
6. Mitra, A. K., *Ophthalmic Drug Delivery Systems,* Marcel Dekker, New York, 1993.
7. Reddy, I. K., *Ocular Therapeutics and Drug Delivery: A Multidisciplinary Approach,* Technomic, Lancaster, PA, 1996.

II. NASAL DRUG DELIVERY

A. Introduction

In the past 2 decades, drug delivery through the nasal route has received considerable attention from clinicians and pharmaceutical scientists. Drugs have been administered nasally for both local and systemic action. Whereas local administration has been limited to the treatment of congestion, rhinitis, sinusitis, and related allergic conditions, nasal medications for systemic therapy have appeared in the market for indications ranging from infertility and migraine to vaccination (Table 16.4). In recent years, scientists have shown tremendous interest in exploring nasal administration for systemic and central nervous system drug delivery.

Nasal delivery is a promising alternative for the administration of drugs that are poorly absorbed via the oral route. Fast absorption and rapid onset of action are major advantages of nasal administration. The rate of absorption, peak plasma concentration and pharmacokinetic profiles are comparable to those achieved by the intravenous route of administration. Avoidance of hepatic and intestinal first pass effects is another reported benefit of nasal systemic drug delivery. In addition, ease of administration, convenience, and self-medication represent therapeutic advantages of intranasal administration that cause improved patient compliance.

B. Anatomy of the Nasal Cavity

The nose is responsible for the temperature and particulate regulation of inspired air and its passage into the respiratory tract. Figure 16.5 shows a cross-sectional view of the entire nasal lateral wall. The nasal passageways are extremely convoluted. The turbinates divide the air spaces into thin slits only a few millimeters wide, increasing the surface area of the nasal mucosa considerably. The volume of the nasal cavity is about 20 ml, and its total surface area is about 180 cm^2. In adult humans, the nasal cavities are covered by a 2- to 4-mm-thick mucosa. The nostrils are covered by skin, the anterior one-third of the nasal cavity by a squamous and transitional epithelium, the upper part of the cavity by an olfactory epithelium, and the remaining portion by a typical respiratory epithelium that contains four basic types of cells, shown in Figure 16.6.

These cells consist of nonciliated columnar cells (with microvilli on the mucosal surface), goblet cells, basal cells, and ciliated columnar cells. In addition to pseudostratified columnar cells, the olfactory epithelium also consists of specialized olfactory cells, supporting cells, and both serous and mucosal glands. The respiratory epithelium is covered by a two-component mucus layer: a bottom layer consisting of a low viscosity sol surrounding the cilia and microvilli, and on the surface, a more viscous and bioadhesive gel layer. The cilia beat continuously in the sol layer with a frequency of 1000 strokes per minute, thereby moving the surface gel layer toward the pharynx. Any inhaled dust or foreign particles in or on the mucous layer are moved posteriorly at 4 to 10 mm/min over the middle and posterior parts of the nose and at about 0.5 to 3 mm/h over the anterior part of the nose.

The blood supply to the nasal mucosa is rich, rendering this route suitable as an alternative to vascular administration. The arterial blood supply to the nasal cavity is derived from both the external and internal carotid arteries. The terminal branch of the maxillary artery supplies the sphenopalatine artery, which in turn supplies the lateral and medial wall of the nasal chamber. The anterior and posterior ethmoid branches originate from the ophthalmic artery, which is a branch of the carotid artery. These vessels supply the anterior portion of the nose. Additionally, twigs from the facial artery supply the vestibule and the anterior portion of the septum. Some vessels from the greater palatine artery pass through the incisive canal of the palate to reach the anterior part of the nose. The veins of the nasal cavity drain into the sphenopalatine foramen and then into the pterygoid plexus. Other veins accompany the ethmoid arteries and join the superior ophthalmic vein. Veins

TABLE 16.4
Nasal Products Currently on the Market

Product	Drug	Indication	Manufacturer[a]
Topical Nasal Products			
Astelin® Nasal Spray	Azelastine hydrochloride	Treatment of seasonal allergic rhinitis	Wallace Laboratories
Beconase® AQ Nasal Spray	Beclomethasone dipropionate monohydrate	Symptomatic treatment of seasonal and perennial allergic rhinitis	Glaxo Wellcome Inc.
Vancenase® AQ Nasal Spray	Beclomethasone dipropionate monohydrate	Symptomatic treatment of seasonal and perennial allergic rhinitis	Schering Plough Corp.
Rhinocort® Nasal Inhaler	Budesonide	Management of symptoms of seasonal and perennial allergic rhinitis and nonallergic perennial rhinitis	AstraZeneca
Nasalcrom® Nasal Solution	Cromolyn sodium	Symptomatic prevention and treatment of seasonal and perennial allergic rhinitis	Pharmacia
Pretz-D® Nasal Drops	Ephedrine[b]	Nasal decongestant	Parnell
Adrenalin® Chloride	Epinephrine hydrochloride[b]	Nasal decongestant	Parke Davis
Nasalide® Nasal Spray	Flunisolide	Treatment of seasonal and perennial allergic rhinitis	Dura Pharmaceuticals, Inc.
Flonase® Nasal Spray	Fluticasone propionate	Symptomatic treatment of seasonal and perennial allergic rhinitis	Glaxo Wellcome, Inc.
Atrovent® Nasal Spray	Ipratropium bromide	Symptomic relief of rhinorrhea (runny nose)	Boehringer Ingelhein Pharmaceuticals Inc.
Livostin® Nasal Spray	Levocabastine	Treatment of allergic rhinitis	Janssen Pharmaceutica, Belgium
Afrin® Nasal Spray	Oxymetazoline hydrochloride[b]	Temporay relief of nasal congestion associated with colds, hay fever, and sinusitis	Schering Plough Healthcare Products
Vicks® Sinex® Regular Decongestant Nasal Spray and Ultra Fine Mist	Phenylephrine hydrochloride[b]	Temporary relief of nasal congestion caused by colds, hay fever, upper respiratory allergies, or sinusitis	Proctor & Gamble
Benzedrex® Nasal Inhaler	Propyl hexedrine[b]	Temporary relief of nasal congestion caused by colds, hay fever, upper respiratory allergies, or sinusitis	Ascher & Company, Inc.
Nasacort® Nasal Inhaler	Triamcinolone acetonide	Treatment of seasonal and perennial allergic rhinitis	Rhone Poulenc Rorer
Otrivin® Nasal Spray and Drops	Xylometazoline hydrochloride[b]	Nasal decongestant	Novartis
Systemic Nasal Products			
Stadol NS® Nasal Spray	Butorphanol tartrate	Management of pain including migraine headache pain	Bristol Myers Squibb
Miacalcin® Nasal Spray	Salmon calcitonin	Treatment of hypercalcemia and osteoporosis	Novartis

TABLE 16.4 (continued)
Nasal Products Currently on the Market

Product	Drug	Indication	Manufacturer[a]
DDAVP® Nasal Spray	Desmopressin acetate	Diabetes insipidus	Aventis Pharmaceuticals
Migranal® nasal spray	Dihydroergotamine mesylate	Treatment of migraine	Novartis
Nitrolingual® Spray	Nitroglycerin	Prevention of angina pectoris caused by coroary artery disease	Rhone-Poulenc Rorer
Nicotrol® Inhalation	Nicotine	Smoking cessation	Pharmacia
Syntocinon® Nasal Spray	Oxytocin	Promote milk ejection in breast-feeding mothers	Novartis
Imitrex® Nasal Spray	Sumatriptan	Migraine	Glaxo Wellcome Inc.
Relenza® Powder for Inhalation	Zanamivir	Treatment of uncomplicated acute illness resulting from influenza A and B	Glaxo Wellcome Inc.

[a] Only one manufacturer has been cited, although products may be available from other manufacturers.
[b] Nonprescription products.
Reproduced with permission.

Source: From Tirucherai, G. S., Yang, C., and Mitra, A. K., *Expert Opinion in Biololgical Theory,* 1, 49–66, 2001.

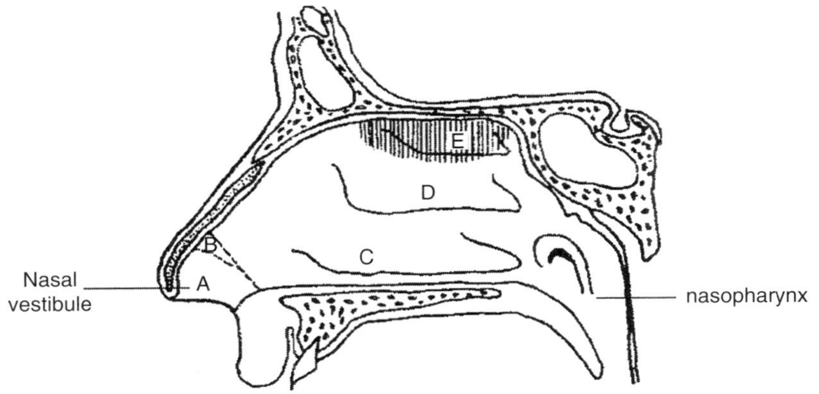

FIGURE 16.5 Schematic representation of the lateral wall of the nasal cavity. (A) Nasal vestibule, (B) internal ostium, (C) inferior concha (inferior turbinate and orifice of the nasolacrimal duct), (D) middle concha (middle turbinate and orifices of frontal sinus, anterior ethmoidal sinuses, and maxillary sinus), (E) superior concha (superturbinate and orifices of posterior ethmoidal sinuses); hatched area, olfactory region. (Reproduced with permission from Mygind, N., *Nasal Allergy,* Prodrugs in Nasal Drug Delivery, Blackwell Science, Oxford, 1979.)

of the anterior port drain into the facial vein. In the submucosa there is a proliferation of blood vessels, including sinusoids forming erectile tissue that allows for the rapid passage of drugs that can cross the epithelium into the blood stream. Thus, since the nasal mucosa is well perfused, the drug can enter the systemic circulation rapidly.

The nasal mucosa represents a complex barrier to drug absorption, which includes three components: a physical barrier composed of the mucus and epithelium, a temporal barrier controlled by the mucociliary clearance, and an enzymatic barrier acting principally on protein and peptide

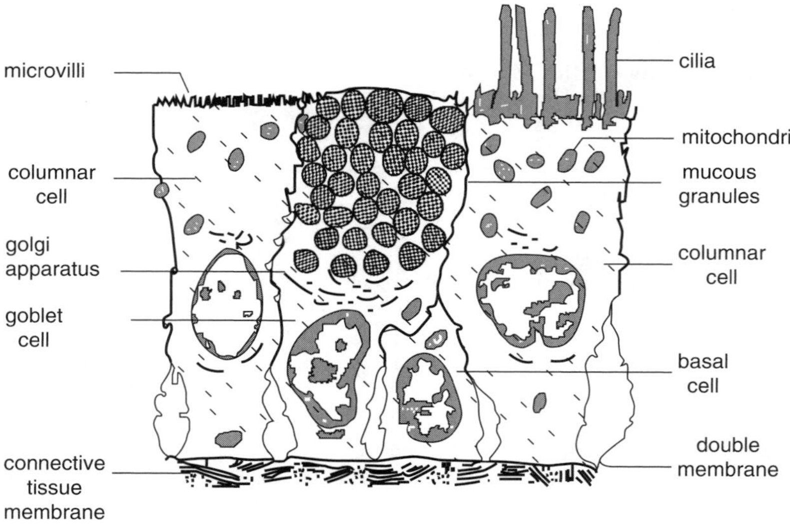

FIGURE 16.6 Schematic representation of the respiratory mucosa consisting of four major cell types. (Reproduced with permission from Mygind, N., *Nasal Allergy,* Blackwell Science, Oxford, 1979.)

drugs. The nasal epithelium the comprises the *physical barrier* consists of a lipoidal pathway and an aqueous pore pathway. Nasally administered drugs have to pass through the epithelial cell layer to reach the systemic circulation. Drug passage through this barrier can occur by the transcellular or the paracellular route.

The transcellular route involves permeation across the apical cell membrane, the intracellular space, and the basolateral membrane by passive transport (diffusion, pH partitioning) or by active transport (facilitated and carrier-mediated diffusion, specific transcellular transport mechanisms, receptor-mediated endocytosis). This route is important for the absorption of lipophilic molecules or molecules capable of specific recognition of a membrane transporter/receptor (active transport and/or receptor mediated endocytosis). The paracellular route involves passage of drug molecules by a passive process through the intercellular junctional complex of the epithelial cells. This is the main pathway for the transport of ionic penetrants and hydrophilic drugs such as oligonucleotides, oligosaccharides, and small peptides. The nasal mucosa is a relatively ineffective structural barrier because of its low membrane resistance and high permeability.

Mucociliary clearance is a protective mechanism that is highly efficient in removing inhaled and deposited particles such as allergens, toxins, bacteria, and viruses from the airway. The interaction of the epithelial cilia with the mucus is responsible for the efficiency of nasal mucociliary clearance. This clearance mechanism draws the particles toward the nasopharynx, preventing them from penetrating the nasal mucosa. This normal physiological process of the nasal mucosa constitutes a *temporal barrier* to transnasal absorption.

In addition to the permeation barrier, there also exists an *enzymatic barrier* in the nasal mucosa. Various enzymes exist in substantial quantities in the nasal epithelia, that is, aldehyde dehydrogenase, glutathione transferase, epoxide hydrolase, cytochrome P-450 dependent mono-oxygenases, and carboxylesterases.

C. Unique Features of Nasal Drug Delivery

- The richly supplied vascular nature of the nasal mucosa coupled with its low barrier to drug permeation makes the nasal route of administration attractive for macromolecular drug delivery including oligonucleotides, peptides, and proteins.

- The high obtained by smoking cigarettes is associated with rapid drug delivery and high concentrations inhaled with smoke. Nasal preparations have been developed that provide nicotine at a programmed rate to relieve withdrawal symptoms that usually accompany abrupt cessation.
- The olfactory region provides a unique advantage whereby a drug may be exposed to neurons that may facilitate its access into the cerebral spinal fluid when administered intranasally. As a result, brain-targeted intranasal delivery of drugs has become an exciting possibility.
- Induction of local and systemic immunity has been achieved by intranasal administration. Intranasal flu and pneumonia vaccines are currently being developed for administration to children.

D. Selection of Drug Candidates

The nasal route of administration is not suitable for all drugs. The potential limitations of nasally administered drugs include:

- Inadequate aqueous solubility: Typically, the entire dose is to be given in a volume of 25 to 200 μl, which necessitates high aqueous drug solubility. For drugs that are water insoluble, suspensions must be formulated.
- Potential for drug degradation by enzymes in the nasal cavity: Although the nasal route bypasses liver metabolism, the ability of the nasal mucosa to degrade drugs is substantial.
- Unsuitability for chronically administered drugs: For example, insulin for the treatment of type I diabetes may not be an appropriate drug candidate for nasal delivery because patients would need to administer this drug daily for the remainder of their lives. If a chronic drug is to be administered much less frequently, its nasal delivery may still be a viable option.
- Drug loss by rapid mucociliary clearance.

An ideal candidate for nasal drug delivery must possess the following attributes:

- Complete absence of nasal mucosal irritation
- Absence of offensive odor associated with the drug
- Adequate aqueous solubility to provide the desired dose in a typical administration volume
- Nontoxic nasal metabolites
- Adequate stability (i.e., chemical stability and enzymatic stability)

E. Absorption of Drugs from Nasal Mucosa

Various *in situ* and *in vivo* experimental models have been established to investigate the effect of physicochemical properties on nasal absorption. The most significant model is the rat nasal perfusion model. The experimental setup for the *in situ* nasal perfusion technique is shown in Figure 16.7. Rats are anesthetized by sodium pentobarbital and following a brief surgical procedure, a polyethylene tube is introduced into the nasal cavity. Drug solution with volumes ranging from 3 to 20 ml are placed in a water-jacketed beaker and maintained at a constant temperature of 37°C. This solution is circulated through the rat nasal cavity at a constant rate by a peristaltic pump. The perfusing solution passes from the nostrils through a funnel, back into the beaker. The extent of drug absorption is directly related to the amount of drug remaining in the perfusate. The concentration of drug remaining in the perfusate is analyzed over a period of time.

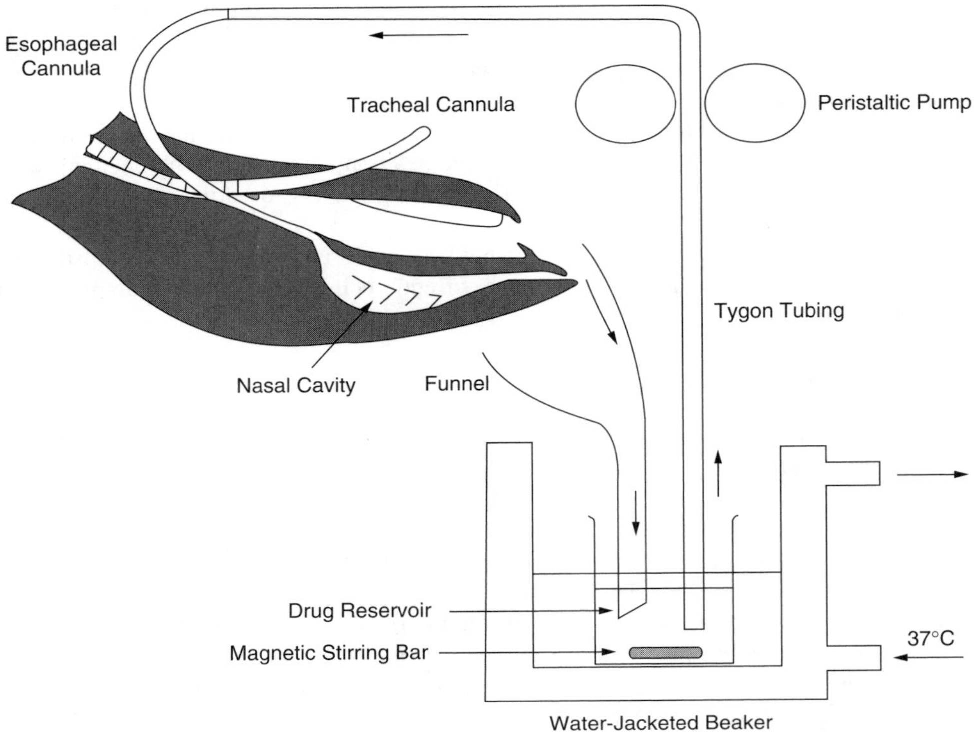

FIGURE 16.7 Schematic representation of the *in situ* nasal perfusion of the rat nasal cavity. (Adapted from Hirai, S., Yashiki, T., Matsuzawa, T., and Mima, H., *International Journal of Pharmaceutics,* 7, 317–325, 1981.)

Typically, the loss in drug concentration of the perfusate vs. time exhibits first-order kinetics, and the rate of loss can be given by Equation 16.2:

$$-dC/dt = kC \tag{16.2}$$

where C denotes the drug concentration in the perfusate, and k represents the apparent first-order rate constant of drug loss from the perfusate. The k is a rate constant for the overall loss, that is, it includes loss of the drug from the nasal mucosa caused by chemical and enzymatic degradation in addition to loss caused by absorption by the nasal mucosa. Therefore, k represents a total of three first-order rate constants according to Equation 16.3:

$$k = k_m + k_{hyd} + k_{abs} \tag{16.3}$$

where k_m represents the rate constant of metabolism of the drug by the enzymatically rich nasal mucosa, k_{hyd} is the rate constant of chemical degradation (usually hydrolysis) of the drug, and k_{abs} denotes the first-order rate constant of absorption.

The rat nasal perfusion model is useful in delineating the importance of physicochemical properties (such as pH, lipophilicity, solubility, molecular weight) of drug candidates relative to absorption. Ranking of drug candidates in terms of suitability for nasal administration is possible based on the results obtained from nasal perfusion experiments. Optimal physicochemical properties of drug molecules are essential in optimizing the rate and extent of absorption across the nasal epithelia. Physicochemical parameters influencing nasal drug absorption are discussed below.

F. Factors Affecting Nasal Drug Delivery

1. Formulation pH

The pH of the formulation as well as the membrane surface pH can affect the nasal absorption of drugs. The extent of nasal absorption is dependent on the pH of the drug solution or formulation. A higher nasal absorption is achieved at a pH lower than the pK_a at which the penetrant molecule exists primarily in its un-ionized form. With an increase in pH, the rate of absorption decreases owing to the ionization of the penetrant molecule. Nasal absorption of weak electrolytes is highly dependent on the degree of ionization.

Assuming that both the ionized and the un-ionized species of a weak electrolyte can be absorbed, and there is no loss from metabolism or chemical degradation, the overall loss of drug from the perfusate can be given by Equation 16.4:

$$-dC/dt = k_{abs}C \tag{16.4}$$

Since k_{abs} is the sum of the rates of absorption of the ionized and the un-ionized species and C is the total of the concentrations of the ionized and un-ionized species, the above equation can be written as

$$-dC/dt = k_{HA}[HA] + k_{A-}[A^-] \tag{16.5}$$

where k_{HA} denotes the first-order rate constant for absorption of the un-ionized species [HA] and k_{A-} is the first-order rate constant for absorption of the ionized species $[A^-]$.

a. *Estimation of k_{HA} and k_{A-}*

The rate constants k_{HA} and $k_{A-}$of a weakly acidic or basic drug can be determined by performing nasal perfusion studies at pH values below and above the apparent pK_a of the drug such that the drug is either predominantly un-ionized or ionized

Let us consider the case of a weakly acidic drug such as salicylic acid:

Low pH: The drug is predominantly un-ionized. The rate of absorption is given by Equation 6:

$$k_{abs} = k_{HA}[HA] \text{ or } k_{abs} = k_{HA}(H^+/H^+ + K_a) \tag{16.6}$$

where H^+ is the hydrogen ion concentration and K_a is the dissociation constant.

High pH: The drug is predominantly ionized. The rate of absorption is given by Equation 16.7:

$$k_{abs} = k_{A-}[A^-] \text{ or } k_{abs} = k_{A-}(K_a /H^+ + K_a) \tag{16.7}$$

Thus, by experimentally determining the rate constant of absorption at two pH values at which the drug is either almost ionized or almost un-ionized, the overall rate of absorption at any pH can be predicted using values of k_{HA}, k_A, and K_a (dissociation constant) in Equation 16.7.

The formulator has only limited ability to enhance nasal drug absorption by changing the pH of the formulation. To avoid nasal irritation, the pH of the formulation must be adjusted between 4.5 and 6.5. Any pH beyond either side of this range may cause cellular damage to the nasal mucosa.

2. Lipophilicity of the Drug

The rate and extent of transepithelial absorption of a drug depends largely on its lipophilicity. For a drug that permeates the nasal mucosa by Fickian diffusion, the following relationship holds:

$$P = DK/h \quad (16.8)$$

where P represents the nasal permeability coefficient of the drug, D is the diffusion coefficient, K is the membrane/aqueous partition coefficient, and h is the thickness of the nasal mucosa.

The permeability of a drug, that is, the rate at which it diffuses across a membrane, is directly proportional to the partition coefficient. The partition coefficient of a drug is an index of its lipophilicity. While low molecular weight lipophilic compounds are generally well absorbed through the nasal epithelium, hydrophilic and macromolecular compounds (e.g., peptides) are poorly absorbed. An increase in the extent of nasal absorption with an increase in lipophilicity of compounds has been observed in several studies.

However, factors other than lipophilicity, for example, molecular size, may also contribute significantly to transnasal permeation.

3. Effect of Molecular Size

The *in vivo* nasal absorption of compounds of molecular weight less than 300 is not significantly influenced by the physicochemical properties of the drug. Factors such as size of the molecule and the ability of the peptides to hydrogen bond with the penetrating medium (i.e, aqueous phase) are more important than lipophilicity and ionization state. The rate of absorption from the nasal cavity is very restrictive beyond molecular weight 300. Like the gastrointestinal tract, absorption from the nasal cavity decreases as the molecular size increases. Consequently with a drug of molecular weight less than 300, absorption via the aqueous pore pathway is favored. Nonspecific diffusion through aqueous channels between cells of the nasal mucosa impose a size restriction on permeability.

G. Formulation and Manufacturing Conditions

Several types of dosage forms are used to deliver drug formulations in the nasal cavity based on the desired absorption profile (8). Selection of a particular dosage form is dependent on the indication for which the treatment is intended, the patient population, and marketing aspects. The following is a list of the various dosage forms that are in common use:

- Nasal drops
- Solution sprays
- Powders
- Gels
- Emulsions and ointments
- Novel delivery systems (e.g. microspheres, liposomes)

1. Nasal Drops

The simplest, commonest, and most convenient form of administering drug formulations into the nose. These dosage forms are easy to manufacture at relatively low cost. However, a precise dose of the formulation cannot be delivered, making it unsuitable for prescription drugs. Microbiological and chemical stability of the product cannot be ensured. Contamination of the product by the dispenser is also a possibility.

2. Solution Sprays

Solution formulations have been administered with great precision utilizing sophisticated metered dose actuators. These formulations have almost replaced the conventional aerosol formulations that contain potentially environmentally harmful propellants. These systems can precisely deliver actuation volumes as low as 25 μl. Aspects of aerosol and solution spray technology are discussed in the section on pulmonary drug delivery.

3. Powders

Powder formulations have been developed for drugs that exhibit poor solution stability. This type of formulation is best suited for drugs acting locally in the nasal cavity. Powder administration leads to a longer contact time with the nasal mucosa. A large fraction of the administered dose is easily cleared by the nasopharynx and oropharynx, from where the powder enters the posterior part of the tongue. Administration of nasal powders therefore increases patient compliance, especially for children, if the taste and odor of the delivered drug is unpalatable. Lack of preservatives is an added advantage.

However, drug absorption from powder administration depends on the aqueous solubility of the drug substance and its absorption rate across the nasal mucosa. Powders suffer from the obvious disadvantage of causing irritation and grittiness in the nasal mucosa. This administration also has stringent morphology and size restrictions.

4. Gels

Gels are thickened viscous solutions or suspensions of drugs. The advantages of gels over other forms are:

- Localization of the formulation on the mucosa and consequent promotion of absorption
- Reduction in anterior leakage of the drug from the nasal cavity
- Diminished irritation potential of the drug or excipients owing to possible inclusion of emollients or other soothing agents

However, the use of gels in nasal administration requires a specially designed nasal adapter as an actuator.

The use of gels is limited because of their high viscosity, which results in poor spreading over the nasal mucosa. Owing to high viscosity of the gels, precompression pumps, which allow accurate dosing, have been developed.

5. Emulsions and Ointments

Drug absorption is enhanced upon rapid solubilization of the drug and prolonged residence of the formulation in the nasal cavity. Both of these criteria can be met through emulsion formulations. Charged emulsions have been shown to be more effective than neutral emulsions, presumably owing to electrostatic interactions between emulsion globules and the mucus layer. Lipid emulsions have also proven to be useful in delivering peptides across the nasal mucosa. The emulsions can be formulated such that the hydrophilic therapeutic agents are incorporated into the aqueous continuous phase. The presence of a small fraction of oil may enhance absorption owing to the hydrophobic interactions between the oil droplets and the nasal mucosa. However, the use of emulsions, containing oil as the continuous phase, may greatly reduce the amount of drug absorbed, since the oil layer at the mucosal surface is static and hence rate limiting. Major disadvantages of emulsions include poor patient acceptability and problems in delivering precise doses.

6. Novel Delivery Systems

Microspheres have been introduced with a view to prolong the residence time of the drug in the nasal cavity. Microspheres are solid polymer matrices throughout which the drug is uniformly dispersed. As opposed to solutions, suspensions, and powders, microspheres are not cleared rapidly from the nasal cavity. Microspheres with a diameter of about 50 µm are expected to deposit in the nasal cavity, whereas those under 10 µm can escape the filtering system of the nose and deposit in the lower respiratory airways. Particles larger than 200 µm have been found not to be retained in the nose after administration. Both high and low molecular weight drugs can be encapsulated into microspheres. The swelling property of microspheres is critical for enhanced drug transport across the nasal mucosa. This swelling property depends on the amount of water in the nasal cavity and the amount of spheres administered. Excess water relative to the spheres in the nose results in instantaneous release of the drug. However, if the number of microspheres far exceeds the moisture available for swelling, then a portion will remain dry and drug release will be affected. The optimal conditions for drug release require that the microspheres are just fully swollen. Microsphere formulation involves incorporation of biocompatible materials such as starch, gelatin, albumin, and dextrans. Microspheres possess a mucoadhesive property, which helps to optimize drug absorption across the nasal mucosa. Microsphere uptake is followed by the temporary dehydration of the nasal mucosa caused by moisture uptake. Enhanced drug delivery has been reported as a consequence of this reversible shrinkage of the nasal epithelial tight junctions. A novel application of microparticles is the intranasal administration of vaccines. The association of antigens with particulate carriers such as microparticles and liposomes has been well documented in the literature.

In designing an appropriate dosage form, the following requirements should be met:

- Good intracellular permeation
- Increased mucosal retention
- Potential for loading of both lipophilic and hydrophilic drugs
- Ability to combine and deliver portions of antigens

H. Characterization and Testing of Nasal Delivery Devices

The particle size of the applied or inhaled product has a major impact on the nasal deposition of the formulation. Particle size measurements can be carried out using sieving techniques, microscopy, or light scattering. The two main issues pertaining to deposition are the site of deposition and the amount of deposition. Various techniques have been used to determine the optimal parameters for both.

Several other factors such as influence of nasal air flow, administration techniques, application volumes, and cone angle also have to be considered in designing a formulation. A comprehensive treatment of this subject is provided elsewhere. Choice of the right delivery system and the correct mode of application are important in optimizing the nasal absorption rate.

NASAL REFERENCES

8. Behl, C. R., Pimplaskar, S. K., Sileno, A. P., deMeiretes, J., and Romeo, V. D., Effect of Physicochemical Properties and Other Factors on Systemic Nasal Drug Delivery, *Advanced Drug Delivery reviews.* 29: 89-116, 1998.
9. Hirai, S., Yashiki, T., Matsuzawa, T., and Mima, H., Absorption of drugs from the nasal mucosa, *International Journal of Pharmaceutics,* 7, 317–325, 1981.
10. Kublik, H. and Vidgren, M. T., Nasal delivery systems and their effect on deposition and absorption, *Advanced Drug Delivery Reviews,* 29, 157–177, 1998.
11. Mitra, R., Pezron, I., Chu, W. A, and Mitra, A. K., Lipid emulsions as vehicles for enhanced nasal delivery of insulin, *International Journal of Pharmaceutics,* 205, 127–134, 2000.
12. Mygind, N., *Nasal Allergy,* Blackwell Science, Oxford, 1979.
13. Tirucherai, G. S., Yang, C., and Mitra, A. K., Prodrugs in nasal drug delivery, *Expert Opinion Biological Theory* 1, 49–66, 2001.

III. PULMONARY DRUG DELIVERY

A. Introduction

Respiratory infections have become more prevalent in recent years. A recent survey estimates the asthma population in the U.S. to be 13 million. Chronic obstructive pulmonary disease (COPD) accounts for an additional 3 million patients. The total U.S. respiratory market is valued at more than $3 billion, with sales of metered dose inhalers and nebulizer solutions well over $2 billion. Projected growth in the number of prescriptions and revenues are expected to approach a 15% annual growth rate over the next 5 years. Need for noninvasive delivery of macromolecules such as peptides and proteins has become very acute. As an alternative to parenteral delivery, oral, transdermal, and nasal routes of delivery have not been overly successful with these agents. Owing to their large size and hydrophilicity, therapeutic amounts of most macromolecular drugs cannot readily permeate across the skin or nasal membranes without the use of penetration enhancers, such as detergents or electrical impulses. Oral delivery of proteins is more problematic because these compounds are digested before they have an opportunity to reach the blood stream. Recent studies have shown that many drug molecules are readily absorbed through the deep lung into the bloodstream without any enhancers. High systemic bioavailability renders the lung a natural target for peptides, proteins, oligonucleotides, and genes, which would benefit significantly by pulmonary delivery.

In addition, the lung is a robust organ. According to the American Conference of Governmental Industrial Hygienists, a person can inhale approximately 30 mg per day of nuisance dusts into the lung day after day for years in industrial settings without effect. Many new therapeutic agents meant for treatment of bronchial infections are available only by the oral and systemic route, which are subjected to the usual problems of low and variable bioavailability. Human lung epithelium is highly permeable and is low in metabolic activity compared with the liver and intestine. With a large surface area (approximately 1000 square feet of absorptive area) and a highly permeable membrane, the peripheral lung is an ideal drug delivery compartment for very small to very large molecules. Because this mode of delivery provides access to both pulmonary and circulatory systems, it may be used for local or systemic therapy. Relative ease of administration and the noninvasive nature of inhalation therapy render pulmonary delivery attractive to patients, especially those affected by respiratory tract infections. Medications can be delivered quickly and effectively with little disruption to the patient's lifestyle. For those who must undergo daily injections, such as patients with diabetes, noninvasive medication administration can dramatically improve their quality of life.

Physiologically the lungs offer a favorable environment for noninvasive drug delivery of macromolecules such as proteins, peptides, oligonucleotides, and genes. The primary barrier to the delivery of compounds via the lungs is the tightly packed, single-cell-thick layer known as the pulmonary, or alveolar, epithelium. Thick, ciliated, mucus-covered cells line the surface of the upper airway, but the epithelial cell layer thins out as it reaches deeper into the lungs, until reaching the tightly packed alveolar epithelium. Table 16.5 shows a comparison of other routes of delivery with the pulmonary route.

B. Anatomy and Physiology of Respiratory System

Air enters the respiratory system through either the nose or the mouth. When air enters through the nose, it is filtered, humidified, and heated to body temperature. The airways, including the lungs, may be viewed as a series of dividing passageways starting at the trachea and ending at the alveolar sac. The airways may be viewed as a pulmonary tree. The trachea is analogous to the trunk of a tree and it bifurcates to form the main bronchi. The bronchi then divide to form smaller bronchi, which lead to the individual lung lobes. The right lung has three lobes, whereas the left lung has two lobes. In each of the lobes the bronchi further divide to form the smaller airways: the bronchioles.

TABLE 16.5
Comparative Aspects of Different Routes of Drug Delivery

	Oral	i.v.	Intramuscular/ Subcutaneous (SC)	Transdermal	Pulmonary
Delivery interface to blood	Indirect; absorbed through gastrointestinal system	Direct infusion into vein	Indirect; absorbed from muscular/ subcutaneous tissue	Indirect; absorbed through relatively impermeable skin	Indirect, but drug delivered to a large, highly permeable epithelia
Delivery issues and concerns	Subject to digestive process; first-pass metabolism	Requires administration by healthcare professional	Painful injection; may require administration by healthcare professional	Highly variable, slow delivery; potential for skin reactions	Requires deep, slow inhalation of small aerosol particles
Patient convenience	High	Low	Low	Moderate	Moderate to high
Onset of action	Slow	Rapid	Moderate	Slow	Moderate to rapid
Amenable to delivery of macromolecules, peptides, proteins, oligonucleotides, and genes	No	Yes	Yes	No	Yes
Bioavailability	Low to high	Reference standard	Moderate to high	Low	Moderate to high
Dose control	Moderate	Good	Moderate	Poor	Moderate to good
Comparison to advanced pulmonary	Slower onset, subject to first-pass metabolism	Costly, inconvenient	Painful, inconvenient	Slow onset, highly variable absorption	Rapid onset, convenient, painless, small- and large-molecule delivery, low variability.

This process of branching continues and terminates with the alveolar sacs. In the classic model by Wiebel each airway divides to form two daughter airways. As a result, the number of airways is doubled in a generation. The model proposes the existence of 24 generations beginning with the trachea and terminating with the alveolar sacs.

Air entering the nose passes through the nasopharynx and air entering the mouth passes through the oropharynx before entering the airways. It then passes through the glottis and the larynx and enters the tracheobronchal tree. Airways not involved in the exchange of gases are termed *conducting*, whereas those involved in exchange are termed *respiratory*. The conducting airways extend from the trachea to the terminal bronchioles and are the principal site of airway obstruction in lung diseases such as asthma. The respiratory airways consist of the respiratory bronchioles, alveolar ducts, and alveolar sacs. The conducting airways and respiratory airways may be distinguished based on the presence or absence of alveolar pockets. Regions within the different airways may be identified based on histological appearance.

The epithelium separates the internal environment (subepithelial structures) from the external environment (airway lumen). It is a continuous sheet of cells lining the lumenal surface of the lungs. The lumenal surface is in contact with inhaled substances such as gases, particulate matter, and aerosols. Adjacent to the lumenal surface is the epithelial surface, which has tight junctions. The tight junctions limit the entry of inhaled particles. In certain pathophysiological conditions this epithelial barrier may be damaged resulting in increased entry of substances present in the airway lumen. Most of the airway epithelium cells are ciliated cells interspersed with mucus secreting cells and other secretory cells. The ciliated cells are pseudostratified and columnar in the larger airways and become cuboidal in the bronchioles. The bronchioles contain in addition to the goblet cells another secretory cell called the Clara cell. The ciliated cells, along with the mucus secreted by glands along the airways, the secretory cells, and the products of the Clara cells, form an important defense mechanism.

1. Pulmonary Blood Supply

The pulmonary circulation carries deoxygenated blood from the right ventricle to the lungs and returns oxygenated blood from the lungs to the left atrium. The pulmonary artery, which leaves the right ventricle, branches to form smaller pulmonary arteries. These arteries further subdivide to form pulmonary capillaries, which are in intimate contact with the alveolus. There are about 1000 capillaries per alveolus. The capillaries drain into postcapillary venules that unite to form veins and then larger veins. These then unite to form the pulmonary vein, which returns blood from the lungs to the left atrium. This blood is subsequently pumped by the heart to the various tissues in the body.

C. Aerosol Technology

Pulmonary delivery, like any other route of delivery, has its challenges. The major challenge is to deliver drug particles to the deeper lung tissues. Most drugs delivered via the pulmonary route are delivered in a pressurized system known as aerosol. Recently, pharmaceutical companies have developed pulmonary drug delivery systems to administer drugs mainly to the upper airways only for local therapy. Metered dose inhalers (MDIs), breath-activated dry powder inhalers (DPIs), liquid jet, and ultrasonic nebulizers are not designed to deliver drugs into the deep lung. The delivery of macromolecules using such systems is inefficient. Macromolecule delivery using inhalation therapy is dependent on the particle size of the formulation. Particle size affects the deposition site of the drug and hence is an important factor when formulating inhalation systems. Studies have established that these particles should range between 1 to 3 μm in diameter for optimal deposition efficiency. The amount of drug deposited from the device is also highly dependent on the patient's proper inhalation technique. An effective delivery device for proteins and peptides would be one that takes into account the patient's ability to inhale correctly.

Most aerosol systems in the past have allowed delivery of less than 100 μg of drug per puff. The therapeutic levels of most macromolecules require delivery of milligram doses. Systems developed for the delivery of macromolecules will require taking this factor into consideration and will have to exhibit the ability to deliver varying amounts of drug per puff. For inhalation therapy to be successful in the area of macromolecule delivery it must provide accurate and consistent doses and it must be readily absorbed into the systemic circulation. In clinical studies, insulin administered by oral inhalation effectively normalized elevated plasma glucose levels in diabetes patients without adverse effects. Bioavailability studies of the aerosolized lutenizing hormone-releasing hormone, a decapeptide, and its analogues have demonstrated that appropriate systemic levels can be achieved to treat conditions such as endometriosis and prostate cancer. The rate of drug release depends on the power with which a compressed or liquefied gas expels the contents from the container. Aerosols provide several advantages over other dosage forms. The stability of moisture sensitive and oxygen sensitive compounds can be enhanced when formulated in aerosol form. Another important advantage is the ease and accuracy of dosage administration. The drug

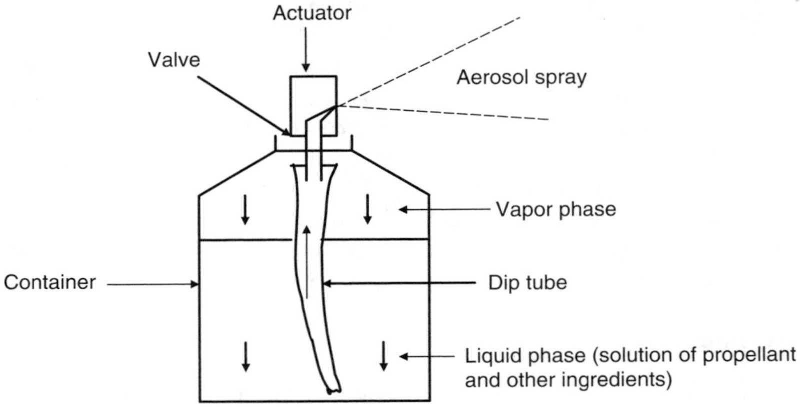

FIGURE 16.8 Typical cross-section of an aerosol device.

can be delivered to the target area in the form of a spray, stream, or stable foam. An added advantage is that sterility of the medication can be maintained even when a dose is dispensed. A typical aerosol consists of a propellant, a container, a valve and actuator and the drug solution/suspension (Figure 16.8).

1. Propellant

The propellant provides the proper pressure within the container and expels the drug solution when the valve is released. Various types of propellants are available. Chlorofluorohydrocarbons (CFCs) were widely used until they were found to deplete the ozone layer in the atmosphere. Environmentally safe alternative compounds to CFCs are now used as propellants. A single propellant may or may not have all of the desired properties. Propellants are mixed based on the vapor pressure they exert. The vapor pressure of a mixture of propellants can be calculated according to Dalton's law:

$$P = p_a + p_b \cdots \tag{16.9}$$

where P denotes the pressure exerted by the mixture, p_a is the partial vapor pressure exerted by propellant A, and p_b is the partial vapor pressure exerted by propellant B.

Another important law that applies to propellants is Raoult's law:

$$p_a = \{n_a/(n_a + n_b)\} * p_A^0 = N_A p_A^0 \ldots . \tag{16.10}$$

where p_A^0 represents vapor pressure of pure propellant A, n_a is the moles of propellant A, n_b is the moles of propellant B, and N_A denotes the mole fraction of component A.

This law is applicable only when ideal behavior is followed. When one component is present in relatively small amounts, ideal behavior is approached. Raoult's law is sufficiently accurate for most determinations and hence is widely used.

2. Containers

The most important factor to be considered when selecting the container material is the pressure it has to withstand. Pressures can range as high as 140 to 180 pounds per square inch guage (psig) at 130°F. Various materials used for containers include:

- *Metals:* Steel and aluminum are two commonly used metals for this purpose. Steel electroplated with tin on both sides is also employed. Tinplate sheets can be soldered or welded to form the containers. Welding eliminates the soldering operation and saves considerable manufacturing time and also decreases the possibility of product–container interaction. Pure stainless steel is also used but is limited to smaller containers since it is relatively expensive. Inhalation aerosols generally have steelcontainers. Aluminum is preferred for extruded (seamless) aerosols, since it has greater resistance to corrosion. Owing to its seamless nature, it is less prone to incompatibilities. However, aluminum can be corroded by pure water and ethanol. This factor needs to be considered when formulating a drug product.
- *Glass:* Glass may be used with or without plastic coatings. Glass containers may be used satisfactorily by limiting the amount and type of propellant used. Since glass does not corrode, compatibility problems are less than with metals.

3. Valves

Valves are an important component of the delivery system, as valves control the form in which the contents are delivered. Valves also regulate the amount of drug to be delivered. Different types of valves are employed for pharmaceutical preparations.

- *Continuous spray valves*: The components of this type of valve include the ferrule or mounting cup, the valve body or housing, and the stem, gasket, spring, and dip tube (Figure 16.9). The mounting cup is used to attach the valve to the container and is made up of stainless steel or aluminum. Ferrules are used with glass bottles or aluminum tubes and are usually made of aluminum or brass. The housing is generally made of nylon or Delrin and has an opening at the point of attachment of the dip tube. The stem may be made of brass, steel, or even nylon or Delrin. It has one or more orifices. Neoprene rubber is generally used for the gasket, as it is compatible with most materials. The spring holds the gasket in place. When the actuator is pressed and released, it allows the valve to revert to the closed position. The dip tube is an essential part of the device, as the formulation has to flow through it to be dispensed. It is fabricated from polymeric materials such as polyethylene and polypropylene. Viscosity and desired delivery rate can play an important role in the selection of the inner diameter of the tube.
    - *Valve Cup:* typically constructed from tinplated steel or aluminium
    - *Outer gasket:* the seal between the valve cup and the aerosol can
    - *Valve housing:* contains the valve stem, spring, and inner gasket
    - *Valve stem:* the tap through which the product flows
    - *Inner gasket:* covers the hole in the valve
    - *Valve spring:* usually stainless steel
    - *Dip tube:* allows the liquid to enter the valve
- *Metering Valves*: These valves are generally used for dispensing very potent medications. They control exactly how much drug is delivered with single actuation. Aerosols containing such valves are generally termed *metered dose inhalers*.

4. Actuator

The actuator is a specially designed button that is fitted to the valve stem. It allows for easy opening and closing of the valve. Types of actuators include spray, foam, solid stream, and special types. Actuators are available in many varieties. The essential components are a spray orifice and a location for the valve–stem connection.

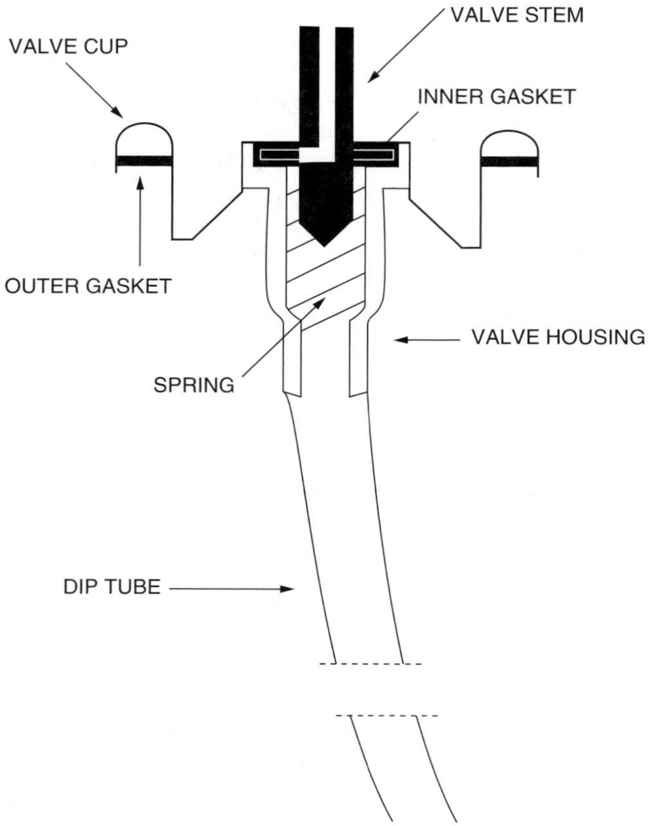

FIGURE 16.9 A typical valve assembly.

D. Formulation and Manufacturing Conditions

The drug solution is generally called a product concentrate and may contain the drug and additives such antioxidants, surfactants, and solvents. The effect of the propellants on each of these ingredients has to be studied carefully. The formulator must know the physical and chemical compatibility of the propellant mixture with the product concentrate. Formulations may be developed depending on the physical and pharmacological properties of the drug, site of application, and drug–solvent interactions.

1. Solution Systems

A solution system is used in most aerosols and it exists as an equilibrium between a vapor and a liquid phase. Depending on the solubility of the drug in the propellant, a blend is chosen. If a low vapor pressure is selected, then larger particles usually form. A lower vapor pressure may be achieved by adding less volatile solvents such as ethanol, glycerin, and propylene glycol. For inhalation products, the amount of propellant may be as high as 95%. Aerosol for inhalation or local activity in the respiratory system in the treatment of asthma may contain the following ingredients:

Ingredient	Wt %
Isoproterenol HCl	0.25
Ascorbic acid	0.10
Ethanol	35.75
Propellant	63.90

2. Water Based Systems

Water based systems are generally used to dispense the drug in the form of a spray or foam. For the drug to be expelled as a spray, an emulsion containing the propellant as the external phase may be formulated. Ethanol may be used as a cosolvent to solubilize some of the propellant in water. Surfactants may also be incorporated in such systems to produce a homogenous dispersion.

3. Suspension or Dispersion Systems

In some cases the use of a cosolvent can interfere with accurate delivery of the drug. To overcome these difficulties, a dispersion of the drug in the propellant or the propellant mixture can be formulated. Such systems may also include surfactants or suspending agents in order to decrease the rate of sedimentation of the drug particles. One drawback with dispersion systems is agglomeration. Particles form floccules, which in time form aggregates. Such agglomerated large particles appear to form cakes that may adhere to the sides of the container and clog the valves, leading to inaccurate dosage administration. The higher the moisture level, the greater is the probability that agglomeration will take place. Hence it is essential that the moisture content of the formulation be controlled. Moisture removal can be achieved by passing the propellants through a series of desiccants and drying the suspended solid material. Sometimes materials suspended in a solvent in which they are partially soluble tend to exhibit particle size growth. The particle size of the dispersed particle should be in the range of 1 to 5 microns. This range ensures good stability of the dispersion system and maximum deposition in the deep lung region.

4. Dry Powder Systems

To overcome the limitations of metered dose inhalers and other conventional lung delivery systems, scientists have developed dry powder formulations of proteins and peptides. Dry powder formulations impart stability to the formulation and have advantages such as high drug volume delivery per puff, low susceptibility to microbial growth, and applicability to both soluble and insoluble drugs. Macromolecules can be formulated as dry powders because they are more stable as solids than liquids. Compared with liquid aerosol particles, which mostly contain water (97%), dry powder aerosol particles can carry larger amounts of the drug. Liquid formulations are very dilute, leading to less drug being administered per puff and carry the risk of microbial growth. It is possible that a drug powder system requiring higher doses could decrease dosing frequency compared with nebulizers. Dry powder formulations have emerged in recent years as the system of choice for delivery of various peptides and proteins.

E. Factors Affecting Pulmonary Drug Delivery

1. Ventilatory Parameters

Large individual variation exists in the inspiration process. Considerable differences exist in the tidal volume, flow rates, and respiratory rates among individuals.

2. Aerosol Properties

Many physical and chemical factors affect aerosol efficiency and accuracy. These include particle size and shape, surface charge, density and mass of the particles generated, and physical state (i.e., a solution or a dispersed system). The hygroscopic properties of a particle are important because particles expand to six times their original size as they travel down the airways. The extent to which this expansion happens depends on the water content and the tonicity of the airways. Hygroscopic properties are more obvious in the peripheral airways where the temperature is 37°C and humidity is 44 μg/cm^3. Particle size determines the distance a particle can travel down the respiratory tract.

Larger porous particles tend to sediment under gravitational influences and become lodged in the nasopharynx, while smaller particles are influenced by Brownian motion and may be too small to be exhaled. Generally, aerosolized products for inhalation therapy have particles in the range of 1 to 5 μm.

F. Future of Pulmonary Delivery

Pulmonary drug delivery can be used as a noninvasive route of administration for a wide variety of macromolecules. Many macromolecules have the inherent disadvantage of being highly unstable in the acidic environment of the stomach and suffer from poor penetration across the gastrointestinal epithelium. Pulmonary delivery offers a viable solution to this problem. The mapping of the human genome has widened the scope of drug targeting, and it will lead to the discovery of a wide variety of macromolecules. Pulmonary delivery may become the route of choice for the delivery of such molecules. Drug administration by inhalation will gain more attention and will become the second most common route of drug administration after oral dosing. Many new agents are now under investigation for pulmonary delivery: interleukin-1 receptor (asthma therapy), heparin (blood clotting), human insulin (diabetes), −1 antitrypsin (emphysema and cystic fibrosis), interferons (multiple sclerosis and hepatitis B and C), and calcitonin and other peptides (osteoporosis).

TABLE 16.6
Pulmonary Delivery Systems in Clinical Trials or Currently on the Market

Product	Description	Manufacturer
Imavist™ (formerly known as Imagent®)	An intravenous contrast agent being codeveloped with Schering AG. An "approvable" status has been received from the FDA based on the results of two Phase 3 studies. These studies demonstrated that Imavist improved ultrasound images of the walls of the heart (endocardial border delineation). Enhanced visualization of the heart may allow better detection of cardiac abnormalities such as structural and functional defects.	Alliance Pharmaceuticals Corp.
Oxygent™	An intravascular oxygen carrier in clinical development as a temporary blood substitute to reduce or eliminate the need for donor blood transfusions during surgery.	Alliance Pharmaceuticals Corp.
LiquiVent®	An oxygen-carrying liquid drug administered directly into the lungs to open up collapsed air sacs (alveoli) and assist in transport of gases. This novel liquid ventilation product has the potential to reduce the mortality and morbidity associated with acute respiratory failure.	Alliance Pharmaceuticals Corp.
PulmoSpheres®	Microscopic hollow/porous spheres that contain a drug stabilized within the shells of the spheres. The spheres are suspended in a fluorochemical liquid for administration into the lungs. Drugs formulated in PulmoSpheres include asthma medications such as bronchodilators and anti-inflammatory agents, antibiotics, proteins, and peptides.	Alliance Pharmaceuticals Corp.
FloGel®	A thermo-reversible gel technology being developed for the reduction of adhesion formation following gynecologic surgery.	Alliance Pharmaceuticals Corp.
TORNALATE® MDI & solution for inhalation		Dura Pharmaceuticals
Unknown	Inhalable insulin for diabetes (Phase III).	Inhale Therapeutics with Pfizer
Inhalable AVONEX®	Multiple sclerosis (Phase I).	Inhale Therapeutics with Biogen
Inhalable Forteo®	Parathyroid hormone (PTH) for osteoporosis (Phase I)	

PULMONARY REFERENCES

14. Lachman, L., Lieberman, H., and Kanig, J., *The Theory and Practice of Industrial Pharmacy,* Varghese, Mumbia, 1987.
15. Patton, J., Deep lung delivery of therapeutic proteins, *Chemtech.,* 27, 34–38, 1997.
16. Weibel, E. R., *Morphometery of the Human Lung,* Springer Verlag, Berlin, 1963.

IV. OTIC DRUG DELIVERY

A. Introduction

Otic delivery systems mainly refer to otic solutions, or ear solutions, as solutions are the predominant dosage form used for this route of administration, with rare exceptions of suspensions and ointments (Table 16.7). Otic preparations are mainly delivered in very small volumes to the central canal of the ear for the removal of ear wax (cerumen) and for treating local infections and inflammation in the ear. There are very few cases in which the dermal layer of the ear is used as a systemic route of delivery owing to the physiological barriers present in the dermal layers of the ear.

Ideally, these solutions are prepared as a solid, a liquid, or rarely as a gaseous solute dissolved in a liquid solvent. These are normally unsaturated solutions. Thus, the amount of solute dissolved is much less than the solvent employed. The solution compositions are normally expressed in terms ofpercent strength. This term indicates percent weight in volume for solutions or dispersions of solid in liquid, or percent volume in volume, or percent weight in weight. Lipophilic compounds undergo slow dissolution in the required vehicle, which may be overcome by various techniques such as heat, particle size reduction, and incorporation of various solublizers or rigorous agitation during the preparation of the solution. Application of heat to aid solubilization while incorporating volatile solvents or solutes is not favored, as heat causes rapid loss of the components. Reduction in particle size causes an exponential increase in the surface area exposed to the solvent, causing rapid dissolution.

B. Anatomy of the Ear Passage

The ear has a dual sensory function. In addition to its role in hearing, it functions as the sense organ of balance, or equilibrium. It is divided into three parts: external ear, middle ear, and inner ear (Figure 16.10).

The external ear has two divisions: the flap, or modified trumpet, on the side of the head, called the *auricle*, and the tube leading from the auricle into the temporal bone, named the *external auditory meatus*. This canal is about 3 cm long and takes, in general, an inward, foreword, and downward direction, although the first portion of the tube slants upward and then curves downward. Because of this curve in the auditory canal, in adults the auricle should be pulled up and back to straighten the tube when medications are dropped into the ear. The tympanic membrane stretches across the inner end of the auditory canal, separating it from middle ear. The tympanic cavity, a tiny epithelial-lined cavity hollowed out of the temporal bone, contains three auditory ossicles: the *malletus, incus,* and *stapes*. Posteriorly the middle ear cavity is continuous with numerous mastoid air spaces in the temporal bone. The clinical importance of these middle ear openings is that they provide routes for infection to spread. Head colds, for example, especially in children, may lead to middle ear or mastoid infections via the nasopharynx–auditory tube, middle ear–mastoid path. The inner ear is also called the *labyrinth* because of its complicated shape. It consists of two main parts: a bony labyrinth and, inside this, a membranous labyrinth. The term *endolymph* is used to describe the clear and potassium-rich fluid that fills the membranous labyrinth. *Perilymph*, a fluid similar to cerebrospinal fluid, surrounds the membranous labyrinth and fills the space between this membranous tunnel and its contents, the bony walls that surround it.

TABLE 16.7
Some Marketed Otic Formulations

Drug (Brand name) (Manufacturer)	Dosage Form	Action	Dosage	Adverse Reactions	Contraindications	Patient Instructions
Acetic Acid: VoSol Otic (GlaxoSmithkline) Domeboro (Bayer)	Otic Solution: Domeboro: 2% acetic acid in aluminum acetate solution VoSol: 2% acetic acid with 3% propylene glycol diacetate	Inhibits or destroys bacteria in the ear canal	4 to 6 drops into ear canal q 2 to 3 h; or insert saturated wick for first 24 h, then continue with instillations	Ear irritation and itching Skin: uticaria Other: overgrowth of organisms	Perforated eardrum	Avoid touching ear with dropper to prevent reinfection
Boric Acid: Auro-Dri (Del Pharma, Inc.)	Otic solution: 2.75% boric acid in isopropyl alcohol	Weak bacteriostatic for ear canal; also fungistatic action	3 to 8 drops in ear canal; plug with cotton. Repeated t.i.d. or q.i.d.	Ear irritation or itching Skin: urticaria Other: overgrowth of organisms	Perforated eardrum or excoriated membranes	Avoid touching ear with dropper; cotton plug to be moistened with medication.
Carbamide-peroxide: Auro Ear Drops (Del Pharma, Inc.) Debrox (GlaxoSmithkline) Murine Ear (Abbott)	Otic Solution: 6.5% carbamide in glycerin or glycerin and propylene glycol	A ceruminolytic that emulsifies and disperses accumulated cerumen	5 to 10 drops in ear canal b.i.d. for up to 4 days. Allow solution to remain in ear canal for a few minutes; remove with warm water	None reported	Perforated eardrum	Flush the ear gently with warm water using a rubber bulb-syringe. Call doctor if pain or redness persists.

Chloramphenicol: Chloromycetin Otic (Monarch Pharma.)	Otic solution: 0.5% in vehicle	Inhibits or destroys bacteria in the ear canal	2 to 3 drops into ear canal t.i.d.	EENT: ear itching or burning Skin: pruritis, urticaria Other: overgrowth of organisms, burning, bone marrow hypoplasia, and aplastic anemia.	Perforated eardrum or hypersensitivity to any component of drug	Avoid touching ear with dropper to avoid reinfection.
Triethanolamine polypeptide oleate condensate: Cerumenex (Purdue Pharma.)	Otic Solution: 10% in 6 and 12 ml bottles with droppers	A ceruminolytic that emulsifies and disperses accumulated cerumen	Fill ear canal with solution and insert cotton plug. After 15 to 30 min, flush with warm water.	EENT: ear erythemia or itching Skin: severe eczema.	Perforated eardrums, otitis media, and otitis externa	Moisten cotton plug with medication before insertion; do not use drops more often than prescribed; discontinue if adverse reactions occur and contact physician immediately. Tightly closed container and keep it away from moisture.

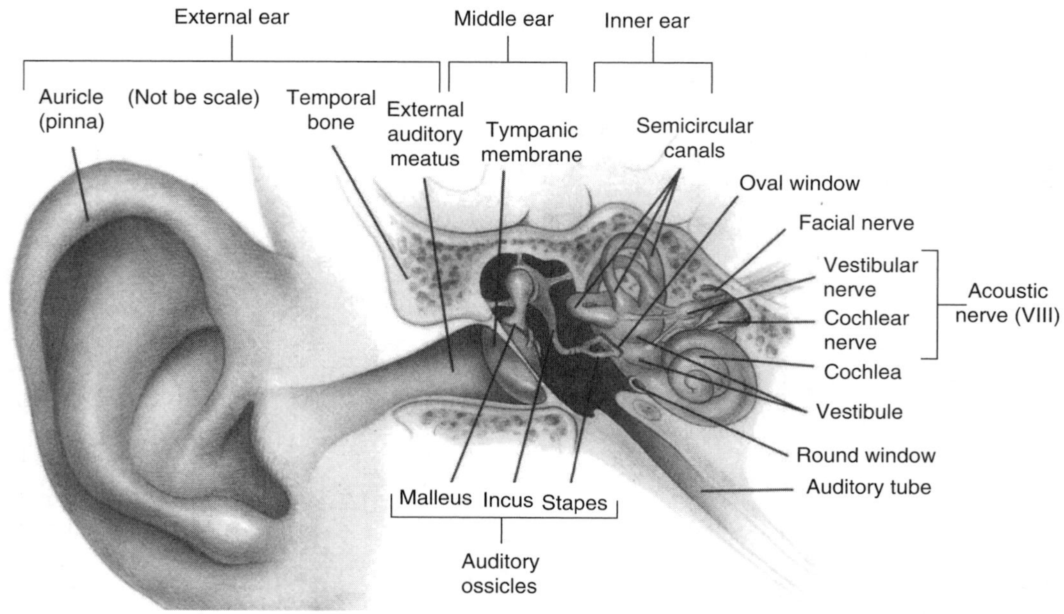

FIGURE 16.10 Schematic diagram showing anatomy of the human ear (From Thibodeau, G. A. and Patton, K. T. *Anatomy and Physiology,* 3rd ed. Mosby, St. Louis, MO.)

C. Formulation and Manufacturing Conditions

1. Sterility

Sterility is defined as the absence of living organisms from a given setting. The process of achieving total sterility is sterilization. The processes used to effect sterility in otic solutions are classified as (17):

- *Physical processes:* This involves the utilization of heat in the presence or absence of moisture or with exposure to either UV or ionizing radiation.
- *Chemical processes:* Chemical bactericides can be used in solution along with the application of heat for sterilization except for certain thermolabile solutions.
- *Mechanical processes:* The solutions are sterilized by a mechanical process of eliminating organisms by filtration.

Many compounds used in the formulation of otic solutions are affected by heat, and the amount of heat applied in the sterility process is determined by the stability issues of the incorporated active moieties. Many official pharmacopoeias around the world recommend following general methods of sterilization:

- The active ingredients are dissolved in the recommended vehicle containing prescribed antimicrobial agents. The solution is then clarified by filtration, transferred to final container, sealed, and sterilized by autoclaving.
- The active ingredients are dissolved in the recommended vehicle containing prescribed antimicrobial agents. The solution is then sterilized by filtration and transferred aseptically to sterile containers, which are then sealed to exclude microorganisms.
- The active ingredients are dissolved in the recommended vehicle containing prescribed antimicrobial agents. The solution is then clarified by filtration, transferred to the final container, sealed, and sterilized by ionizing radiation.

2. Tonicity

It was previously thought, because of discomfort during application, that otic formulations should be made isotonic since they are formulated for application as eye/ear drops. However, since otic formulations are now considered to be topical, there is no general requirement in the USP for otic preparations to be made isotonic. Generally the pain and discomfort is associated with the nature of medicament and infection, rather than tonicity.

D. Otic Dosage Forms

1. Cerumen Removing Solutions

Cerumen is a combination of the secretions of the sweat and sebaceous glands of the external auditory canal and is commonly known as earwax. These secretions, when dry, form a sticky semisolid mass that holds shed epithelial cells, hairs, dust particles, and other foreign objects making their way to the inner auditory canal. When this cerumen is allowed to accumulate, it may cause discomfort such as itching, pain, or in many cases impaired hearing, which may necessitate detailed examination by a physician. If proper care is not taken in routine cleansing of the accumulated cerumen, the impacted mass may require a more painful extraction process. Traditionally, light mineral oil, vegetable oils, and hydrogen peroxide have been used for the softening and removal of impacted cerumen. Recently, solutions of synthetic surfactants have been developed for the removal of earwax. Commonly used agents include a formulation of triethanolamine oleate in propylene glycol, which emulsifies the cerumen and aids in its removal, and carbamide peroxide in glycerin or propylene glycol, which releases oxygen on contact with cerumen, thus disrupting the integrity of cerumen and helping in its easy removal. These solutions are normally placed in the patient's ear canal and retained in place for a while, after which they are flushed with a fine stream of warm water using a rubber applicator.

2. Solutions for Ear Infections

The bacterial pathogens most commonly isolated from middle ear effusions from patients with otitis media (inflammation of the middle ear or tympanic membrane) are *Streptococcus pneumoniae*, *Haemophilus influenzae,* and *Moraxella catarrhalis*. Less commonly isolated organisms include *Streptococcus pyrogenes*, *Pseudomonas aeruginosa*, group-A streptococci, and *Staphylococcus aureus*.

The principle pathogens of otitis externa (inflammation of the external auditory canal, commonly called swimmer's ear) are *P. aeruginosa* and *S. aureus*. Gram-negative species such as *Proteus*, *Klebsiella*, and *E. coli* are also common. Organisms less commonly involved include *Enterococcus faecalis* and *Bacteroides fragilis*. The strains of *P. aeruginosa* involved in otitis externa may exhibit fewer biochemical characteristics of the species (e.g., production of pyocyanin and urease, and mucoid characteristics) than strains isolated from other sites. However, *in vitro* susceptibility tests have not been performed on these distinct strains.

It is estimated that over a million myringotomies (a surgical slit in the eardrum to insert a tube to drain the fluid out from behind the drum) with insertion of pressure equalization tubes are now performed each ear, making this the most common surgical procedure performed in the U.S. A relatively common and often frustrating complication of tympanostomy tubes is chronic suppurative otitis media (bacterial infection of the middle ear that can follow untreated acute otitis media), with a reported incidence of 3.6 to 21% in patients with tympanostomy tubes. Another even more common problem is otorrhea, or aural discharge. The irreversible tissue damage caused by the pathogens in this process make systemic therapy impractical, thus making the topical treatment the preferred therapy. In patients with perforated eardrums or tympanostomy tubes, pathogens normally found in otitis externa can gain access to the middle ear and thus become important pathogens in otitis media. In children with acute otitis media and tympanostomy tubes, the percentage incidence of in baseline infection varies with age.

These ear infections are usually treated by topical application of anti-infective drugs and topical antibiotics. The agents normally used for ear infections are chloramphenicol, colistin sulfate, neomycin, polymyxin B sulfate, and antifungal agents such as nystatin. These agents are normally formulated as solutions or suspensions containing glycerol or polyethylene glycol as the vehicle. They are made viscous to maximize the contact time between the medicinal agents and the infected tissue of the remote ear. These formulations consist of hygroscopic agents that help in dehumidifying the tissues lower the moisture content the bacteria need to survive, thus preventing the infection from spreading. For better infection management these topical preparations are to be given as an adjunct with the systemic antibiotics administered orally. Besides the antibiotics used to treat otitis externa, acetic acid in aluminum acetate solution and boric acid in isopropyl alcohol are also employed. These agents help to reacidify while the vehicles (isopropyl alcohol) help to dry the ear canal by eliminating water and hence slowing the growth of *P. aeruginosa*. Routinely, white vinegar (5% acetic acid) and boric acid 2 to 5% dissolved in either ethanol or propylene glycol is recommended for cleansing the infected ear.

3. Anti-Inflammatory and Analgesic Preparations

Solutions of dexamethasone sodium and hydrocortisone are used to treat the inflammation caused by the allergic or irritative manifestations of the ear and the pruritis, which follows the treatment of ear infections. In some case, corticosteroid ointments are also used in postinfective inflammation management. Such commercially available formulations are also employed in ophthalmic inflammations. Pain in the ear frequently accompanies ear infection and inflammation. Frequently, the pain far outweighs the severity of the infection. As the ear canal is extremely narrow, even a minor inflammation causes intense pain and discomfort. Under such circumstances, topical application of analgesics should accompany the systemic analgesics. The formulations used as analgesics contain solutions of antipyrine and a local anesthetic, for example, benzocaine in a vehicle of propylene glycol or anhydrous glycerin. Such a formulation is ideal for the treatment of acute otitis media.

OTIC REFERENCES

17. Hadgraft, J. W., Sterilization, in *Bentley's Textbook of Pharmaceutics,* Rawlins, E. A., Ed., Bailliere Tindell, London, 1992.
18. Simpson, K. L. and Markham, A., Ofloxacin otic solution: a review of its use in management of ear infection, *Drugs,* 58, 509–531, 1999.
19. Thibodeau, G. A., and Patton, K. T., *Anatomy and Physiology,* 3rd ed., Mosby, St. Louis, MO, 1996.
20. U.S. Pharmacopoeia XX, N.F. XV, 1980.
21. Yang, C., Gao, H., Mitra A. K., Chemical stability, enzymatic hydrolysis, and nasal uptake of amino acid ester prodrugs of acyclovir, *Journal of Pharmaceutical Sciences,* 90, 617–624, 2001.

TUTORIAL

1. Which layer is the major rate limiting barrier for permeation of hydrophilic drugs across the cornea?
   (a) Endothelial layer
   (b) Stroma
   (c) Epithelial layer
   (d) (a) and (c)
2. What is the physiological pH of lachrymal fluid?
   (a) 6.4
   (b) 8.0
   (c) 7.4
   (d) None
3. Timolol has a pKa of 9.2. Maximum corneal penetration would be at pH:
   (a) 6.2
   (b) 7.5
   (c) 5.0
   (d) 6.8
4. Drugs are administered by the intravitreal route for treatment of:
   (a) Conjunctivitis
   (b) Glaucoma
   (c) Retinitis
   (d) Corneal infections
5. What are the important postcorneal factors to be considered in developing an effective topical formulation?
   (a) Tear drainage and pH
   (b) Conjunctival absorption and drug metabolism
   (c) Systemic drug absorption and drug binding to tear proteins
   (d) Melanin binding and drug metabolism
6. What are the advantages of nasal drug administration relative to oral drug administration?
7. The breakdown of a drug during absorption across the mucosal membrane constitutes which of the barriers to drug delivery?
   (a) Temporal barrier
   (b) Physical barrier
   (c) Enzymatic barrier
8. The rate of absorption of a drug across a mucosal membrane is expressed as its:
   (a) Partition coefficient
   (b) Permeability
   (c) Diffusion coefficient
   (d) Hydrodynamic radius
9. Write short notes on the suitability of the following dosage forms for nasal delivery:
   (a) Emulsions
   (b) Microspheres
   (c) Powders

10. Which of the following statements is not true about the lung?
    (a) It has a highly permeable membrane.
    (b) It has a membrane that provides an effective barrier to absorption.
    (c) It provides easy access to the bloodstream.
    (d) It has a high surface area available for absorption.
    (e) None of the above.
11. Traditional noninvasive systems such as pills and tablets fail to deliver macromolecules owing to:
    (a) Enzyme degradation
    (b) Degradation by acid
    (c) Poor absorption
    (d) All of the above
    (e) None of the above
12. The single anatomical unit in the lung where gases are exchanged is called the:
    (a) Trachea
    (b) Bronchiole
    (c) Alveoli
    (d) Terminal bronchi
    (e) None of the above
13. The systems generally used to deliver drugs are called:
    (a) Pressure cans
    (b) Aerosols
    (c) Spray devices
    (d) None of the above
    (e) All of the above
14. The rate of drug release from an aerosol depends on:
    (a) The power of a compressed gas to expel contents
    (b) The type of drug present
    (c) The type of container
    (d) All of the above
    (e) None of the above
15. Which statement is not true about aerosols?
    (a) Dry powders can be dispensed.
    (b) They avoid contamination of drug once a certain amount is dispensed.
    (c) Emulsions cannot be dispensed.
    (d) They result in more patient compliance than injectables.
    (e) None of the above.
16. In a dispersed system the stability decreases with:
    (a) Decreased agglomeration
    (b) Increased agglomeration
    (c) No agglomeration
    (d) All of the above
    (e) None of the above
17. The ideal particle size for aerosol systems to deliver drugs to the deep lung area is:
    (a) 100 to 1000 μm
    (b) 100 to 200 μm
    (c) 1 to 100 μm
    (d) 5 to 100 μm
    (e) 1 to 5 μm

18. The otic route is widely used as _______________ drug delivery.
    (a) Systemic
    (b) Local
    (c) (a) and (b)
19. Discuss the clinical relevance of the middle ear.
20. Name the commonly used agents in the removal of cerumen from ear.
21. Which aural disorder is termed otorrhea?
22. List the important characteristics of agents used in ear infections.

ANSWERS

1. (d); 2. (d); 3. (c); 4. (b); 5. (a); 7. (c); 8. (b); 10. (d); 11. (d); 12. (c); 13. (b); 14. (a); 15. (c); 16. (b); 17. (e); 18. (b)

HOMEWORK

1. Discuss the various precorneal factors affecting ocular bioavailability.
2. What are the different routes of drug administration to the eye?
3. Discuss the importance of pH in the formulation of ophthalmic medications.
4. What are ocuserts? Mention a drug delivered in the form of ocuserts.
5. Name two advanced ocular drug delivery systems.
6. For a weakly acidic drug, show that

$$k_{abs} = k_{HA}[H^+]/[H^+ + K_a]$$

given that $k_{abs} = k_{HA}[HA]$

7. Using the sodium chloride equivalent method, calculate the amount of sodium chloride to be added to make the following solution isotonic to tear fluids. Sodium chloride equivalent E value is 0.18 (Answer: 0.72 g)
    Tetracaine hydrochloride: 1 g
    Sodium chloride: x g
    Sterilized distilled water to make 100 ml
8. A novel compound is developed for the treatment of glaucoma. Frequent administration is required when administered as eye drops. Hence an ocusert device is developed to deliver the drug over a prolonged period of time. The diffusion coefficient of the novel drug in the ocusert device is 3.65×10^{-5} cm^2/h. The surface area is 0.80 cm^2. The partition coefficient between the ocusert and ocular fluids is 1.00, the thickness of the membrane is 0.008 cm, and the solubility in water is 2.80 mg/cm^3. Calculate the cumulative amount of drug released in 120 h.
    (a) 2.4 mg
    (b) 1.22 mg
    (c) 3.5 mg
    (d) 0.80 mg

ANSWERS

Hint: Apply principles from the Henderson–Hasselbach buffer equation.

7. First calculate the weight in grams of sodium chloride that is equivalent to the weight in grams of tetracaine hydrochloride = quantity of the drug × sodium equivalent = 1.0 g × 0.18 = 0.18 g. A total of 0.9% of sodium chloride is required for isotonicity if no other salts are present. Since tetracaine hydrochloride has contributed a weight of material osmotically equivalent to 0.18 g of NaCl, the amount of sodium chloride to be added is

$$0.90 - 0.18 = 0.72 \text{ g}$$

8. The release of sparingly soluble drugs from the ocusert can be calculated from the following equation:

$$M = \frac{SDKC_s}{h} t$$

where M is the accumulated amount released, t is time, S is the surface area of the device, D is the diffusion coefficient of the membrane, K is the liquid-liquid partition coefficient of the ocusert membrane and the eye fluids, C_s is the aqueous solubility of the drug, and h is the membrane thickness. On substituting the given values in the above equation, the cumulative amount released in 120 h is **1.22 mg**.

CASES

CASE I

In order to enhance nasal absorption of an antiviral compound, acyclovir (ACV), nasal absorption of the parent compound was compared with those of its amino acid ester prodrugs (i.e., Asp-ACV, Lys-ACV, and Phe-ACV) using a rat nasal perfusion model. All prodrugs were chemically stable. The first-order nasal perfusion rate constants (× 10^3 min^{-1}) of enzymatic hydrolysis and nasal absorption of the prodrugs are as follows:

Compound	Hydrolysis in Nasal Washings ($k_m \times 10^3$ min^{-1})	% Remaining after 90 min Perfusion	First-Order Tate Constant ($k \times 10^3$ min^{-1})
ACV	—	99.06	0.142
Asp-ACV	0.681 ± 0.072	86.72 ± 1.32	1.44
Lys-ACV	3.94 ± 0.86	69.34 ± 1.23	3.99
Phe-ACV	14.5 ± 1.8	29.06 ± 10.87	14.5

1. Determine the k_{abs} for acyclovir and its half-life ($t_{1/2}$) of nasal absorption.
2. Does prodrug derivatization improve the nasal absorption of acyclovir? If so, which prodrug is most suitable for nasal delivery of acyclovir?

CASE II

XT-302, a newly developed Naproxen analog has been found to show promise in the management of migraine. You have been asked to develop a formulation for the nasal administration of this compound. XT-302 is a weak acid and has a pK_a of 4.7. Nasal perfusion of the drug was performed at two pH values: 2.50 and 7.40. The time course of absorption was as follows:

Time (h)	Concentration of XT-302 Remaining in the Perfusate (μg/ml)	
	pH 2.5	pH 7.4
0	8.56	8.56
0.25	8.21	8.51
0.50	7.87	8.47
1.00	7.23	8.39
1.50	6.63	8.30
2.00	6.09	8.22
2.50	5.60	8.14
3.00	5.15	8.06

1. Calculate the overall rate constant of absorption at both pH values, assuming that there is no contribution of enzymatic or chemical degradation to the loss of drug in the perfusate. Calculate the half-life of absorption of the ionized species and the un-ionized species.
2. Estimate the overall rate of absorption of this drug at a pH of 4.5 and estimate the relative contributions of the ionized and un-ionized species to net transport.
3. With a view toward maximizing absorption, what pH would you use and why?

ANSWERS

Case 1

1. Recall that the overall rate constant (**k**) of loss of drug (or prodrug) from the nasal perfusate is the sum of the hydrolytic rate constant ($\mathbf{k_{hyd}}$), the metabolic rate constant ($\mathbf{k_m}$), and the absorption rate constant (k_{abs}). (Equation 16.3: $\mathbf{k = k_m + k_{hyd} + k_{abs}}$)

   Since acyclovir does not undergo any chemical or enzymatic hydrolysis, the rate constant of loss of drug from the perfusate (k) represents loss due to absorption alone. Therefore

$$k = k_{abs} = \mathbf{0.142 \times 10^{-3}\ min^{-1}}$$

   Therefore the $t_{1/2}$ of absorption of ACV = $0.693/0.142 \times 10^{-3}$ min^{-1} = **81.33 h**
2. Since the prodrugs are chemically stable, the k_{hyd} term drops out when calculating the overall rate of absorption of the prodrugs. However, the prodrugs do undergo enzymatic hydrolysis. k_{abs} may be obtained by rearranging Equation 16.3 as $k_{abs} = k - k_m$.

   k_{abs} for Asp-ACV = $1.44 \times 10^{-3}\ min^{-1} - 0.681 \times 10^{-3}\ min^{-1} = 0.759 \times 10^{-3}\ min^{-1}$. Therefore, the $t_{1/2}$ of absorption of Asp-ACV = $0.693/0.759 \times 10^{-3}\ min^{-1}$ = 15.21 h.

   A 5.4-fold increase in k_{abs} is seen for Asp-ACV over ACV, indicating that prodrug derivatization indeed increases the nasal absorption of ACV.

   However, the Lys-ACV and Phe-ACV esters are not metabolically stable and their overall rate constant of loss from the nasal perfusate (k) is equal to the rate constant of loss owing to enzymatic hydrolysis (k_m). This indicates that the k_{abs} term is almost equal to zero. Therefore Lys-ACV and Phe-ACV do not represent any advantage over ACV for nasal absorption.

Therefore Asp-ACV prodrug is most suitable for nasal delivery.

Case II

1. The first-order rate of nasal absorption at pH 2.5 and 7.4 is obtained by performing linear regression of the log concentration vs. time data provided. The k_{abs} at pH 2.5 and 7.4 are 0.170 h^{-1} and 0.02 h^{-1}, respectively.

   At pH 2.5, the drug is predominantly in the un-ionized form. The rate of absorption given by Equation 16.6 is

$$k_{abs} = k_{HA} (H^+/H^+ + K_a)$$

$$\text{Since pH} = 2.5, [H^+] = 3.16 \times 10^{-3}$$

$$\text{and } pK_a = 4.7, K_a = 1.99 \times 10^{-5}$$

   Substituting the above values in Equation 16.6 and solving for k_{HA}, we get a value of 0.168 h^{-1}, which is very close to the calculated rate of absorption at pH 2.5, indicating that the permeating species is almost entirely un-ionized. The half-life of absorption of the un-ionized species is given by 0.693/0.168 h^{-1} = **4.12 h.**

   Similarly, the rate of absorption of the ionized species is obtained at pH 7.4 using Equation 16.7:

$$k_{abs} = k_{A-} (K_a/H^+ + K_a)$$

   Since pH = 7.4, $[H^+] = 4.99 \times 10^{-8}$ and $K_a = 1.99 \times 10^{-5}$

   Substituting the above values in Equation 16.7 and solving for k_{A-} gives a value of 0.02 h^{-1}, which is very close to the calculated rate of absorption at pH 7.4, indicating that the permeating species is almost entirely ionized. The half-life of absorption of the ionized species is given by 0.693/0.02 h^{-1} = **34.65 h.**
2. At pH 4.5, the ratio of ionized species to un-ionized species is given by the expression pH = pKa + log [ionized]/[un-ionized]. Substituting the values of pH and pKa, the percentage of ionized species is 38.7 and the percentage of un-ionized species is 61.3. The overall rate constant of absorption at any pH is the sum of the rate constant of absorption of the un-ionized species times the percentage of un-ionized species at that pH and the rate constant of absorption of the ionized species times the percentage of ionized species. Therefore,

$$k_{abs} = [0.168\ h^{-1} \times (61.3/100)] + [0.02\ h^{-1} \times (38.7/100)]$$

$$= 0.0077 + 0.103 = 0.1107\ h^{-1}$$

   Therefore, the overall half-life of absorption at pH 4.5 = 0.693/0.1107 h^{-1} = **6.26 h.**
3. Although the rate of absorption at pH 2.5 is about 1.5-fold higher than at pH 4.5, such increase in absorption may be accompanied by a simultaneous loss of structural integrity of the nasal mucosa. Consequently, such low pH values in the formulation must be avoided. The use of pH 4.5 in the nasal formulation is optimal for absorption.

17 Delivery of Peptide and Protein Drugs

Emily Ha, Manjori Ganguly, Xiaoling Li, Bhaskara R. Jasti, and Uday B. Kompella

CONTENTS

0-4152-8863-0/05/$0.00+$1.50
© 2005 by CRC Press LLC

Learning Objectives

After studying this chapter, the student will have thorough knowledge of:

- Structure of proteins and peptides
- Stability of proteins and peptides
- Pharmacokinetics of biologicals
- Routes of administration of proteins and peptides
- Handling of protein and peptide drugs
- Regulatory aspects of biologicals

I. INTRODUCTION

Clinically available peptide and protein drugs have been made possible through advances in recombinant DNA technology and solid-phase peptide synthesis in the biotechnology industry. Few examples of clinically available biological pharmaceuticals are listed in Table 17.1. These protein pharmaceuticals provide life-saving replacement therapies and improve the comfort and quality of life for many patients. The therapeutic potential of protein and peptide drugs have prompted pharmaceutical companies to invest extensive effort and resources into research and development of these macromolecules as potential drug candidates. A number of peptide and protein drugs in development are listed in Table 17.2.

TABLE 17.1
Examples of Clinically Available Biologicals

Product	Brand (Manufacturer)	Indication
rFactor VIII	Kogenate (Miles)	Hemophilia A
Erythropoietin (epoetin alfa)	Epogen (Amgen)	Dialysis anemia
	Procrit (Ortho Biotech)	Chemotherapy-induced anemia
Tissue plasminogen activator (tPA)	Activase (Genentech)	Acute myocardial infarction
Somatotropin	Nutropin (Genentech) Humatrope (Eli Lilly) Protropin (Genentech)	Chronic renal insufficiency; human growth hormone (GH) deficiency in children
Hepatitis B vaccine	Engerix-B (SKB) Recombivax HB (Merck)	Hepatitis B prevention
rhDNase	Pulmozyme (Genentech)	Cystic cibrosis
Glucocerebrosidase	Cerezyme (Genzyme)	Gaucher's disease
Human insulin	Humulin (Eli Lilly)	Diabetes
Granulocyte/macrophage colony stimulating factor (GM-CSF; sargramostin)	Leukine (Immunex) Prokine (Hoechst-Roussel)	Bone marrow transplant; Neutropenia secondary to chemotherapy
Interferon beta-1b	Betaseron (Berlex/Chiron)	Multiple sclerosis
Interferon gamma-1b	Actimmune (Genentech)	Chronic granulomatous disease
Interferon alfa-2b	Alferon N (Interferon Sciences)	Genital warts
Interferon alfa-2a and 2b	Roferon-A (Roche); Intron-A (Schering)	Kaposi's sarcoma; Hairy cell leukemia
Muromonoab CD3	Orthoclone OKT3/Ortho	Heart, liver, and kidney transplant rejection
Staumonab pendetide	OncoScint (Cytogen)	Detection of colorectal and ovarian cancers

TABLE 17.2
Categories of Biotechnology Products Being Investigated

Tumor necrosis factor	Human growth hormones
Erythropoietines	Interferons
Dismutases	Growth factors
Tissue plasminogen activators	Interleukins
Clotting factor	Gene therapy
Recombinant soluble CD4s	
Colony stimulating factors	Vaccines
Antisense	Monoclonal antibodies
	Others

However, the large molecular weight along with the physical and chemical instability of proteins and peptides pose many challenges for pharmaceutical scientists. The primary challenge to successful protein development and formulation is preservation of the protein's native conformation during processing and storage. Changes to the three-dimensional structure of the protein caused by either chemical or physical degradation render the protein therapeutically inactive. Therefore, chemical instability, which includes reactions such as proteolysis, oxidation, reduction, deamidation, disulfide exchange and racemization, and physical instability, such as denaturation, surface adsorption, aggregation and precipitation, should be prevented or limited with a suitable formulation and proper handling. Research aimed at improving protein stability has successfully increased the shelf life of many protein drugs by optimizing the protein's immediate environment and promoting the conservation of the protein's native structure.

An understanding of the fundamentals of protein stability is required for the dispensing practitioner to ensure proper storage and effective delivery of these drug products. In this chapter, the structure, stability, disposition, routes of administration, and regulatory issues related to peptide and protein drugs will be discussed along with a presentation of three case studies of clinically available peptide and protein drug products.

II. STRUCTURE OF PEPTIDES AND PROTEINS

Peptides and proteins are made up of simple building blocks called amino acids. There are 20 naturally occurring amino acids that form the myriad proteins (Table 17.3). Specific functions of different proteins are attributed to the unique amino acid sequence and the resulting conformation that make up the protein. All amino acids contain an amino ($-NH_2$) and carboxyl (–COOH) group attached to the α-carbon, to which various side chains (–R) are connected. All of the amino acids except glycine contain an asymmetric center with four functional groups attached, resulting in a chiral molecule with the S-configuration (Figure 17.1).

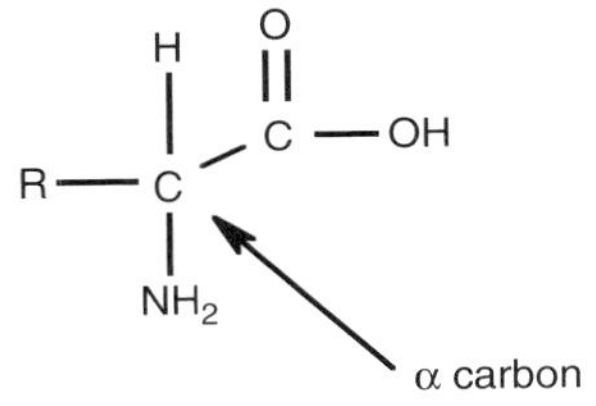

FIGURE 17.1 α carbon in an amino acid.

TABLE 17.3
Naturally Occurring Amino Acids

Amino Acid	Abbreviation	pKa (side-chain)	R Functional Group	Property
Aspartic acid	Asp, D	3.86	$–CH_2CO_2H$	Acidic
Glutamic acid	Glu, E	4.07	$–CH_2CO_2H$	Acidic
Arginine	Arg, R	12.48	$–CH_2CH_2CH_2NHC(NH)NH_2$	Basic
Histidine	His, H	6.10	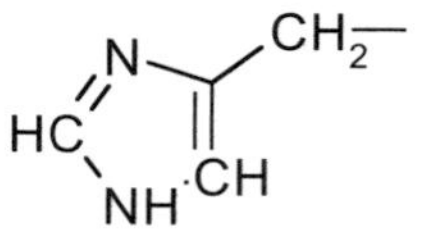	Basic
Lysine	Lys, K	10.53	$–CH_2CH_2CH_2CH_2NH_2$	Basic
Asparagine	Asn, N		$–CH_2CONH_2$	Basic
Cysteine	Cys, C	8.00	$–CH_2SH$	Polar
Glutamine	Gln, Q		$–CH_2CH_2CONH_2$	Polar
Serine	Ser, S		$–CH_2OH$	Polar
Threonine	Thr, T		$–CH(OH)CH_3$	Polar
Tryptophan	Trp, W		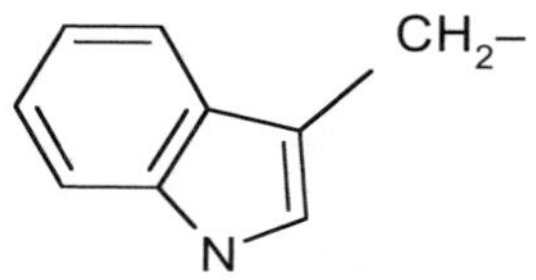	Polar
Tyrosine	Tyr, Y	10.07	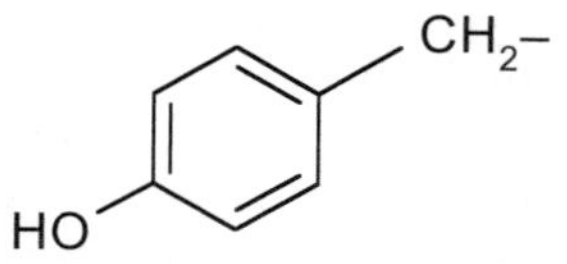	Polar
Alanine	Ala, A		$–CH_3$	Nonpolar
Glycine	Gly, G		$–H$	Nonpolar
Isoleucine	Ile, I		$–CH(CH_3)CH_2CH_3$	Nonpolar
Leucine	Leu, L		$–CH_2CH(CH_3)CH_3$	Nonpolar
Methionine	Met, M		$–CH_2CH_2SCH_3$	Nonpolar
Phenylalanine	Phe, F		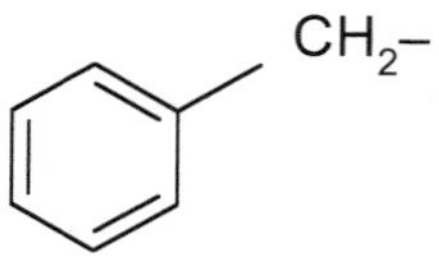	Nonpolar
Proline[a]	Pro, P		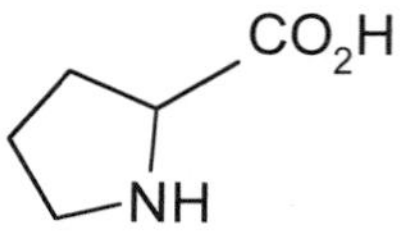	Nonpolar
Valine	Val, V		$–CH(CH_3)CH_3$	Nonpolar

[a] Full amino acid residue

Amino acids are usually classified by the acid/base or polar/nonpolar property of the side chain. Aspartic and glutamic amino acids are considered acidic because of the presence of ionizable carboxylic acid functional groups. Arginine, histidine, and lysine contain basic ionizable side chains and are referred to as the basic amino acids. Amino acids that do not have ionizable groups at physiological pH but can lose a proton at alkaline pH are called uncharged polar amino acids. Asparagine, cysteine, glutamine, serine, threonine, and tyrosine are included in this category. Examples of nonpolar amino acids include alanine, glycine, isoleucine, leucine, methionine, phenylalanine, proline, tryptophan, and valine.

The presence and position of an amino acid in the sequence determines whether that particular amino acid occupies the surface or the core of the protein. This influences the overall structure of the protein. The ionizable and polar amino acids usually occupy the surface when surrounded by an aqueous environment, while the nonpolar side chains cluster together to form the hydrophobic core.

Peptides and proteins are formed through peptide bonds, which are amide linkages between the $-NH_2$ and $-COOH$ groups of neighboring amino acids. For the purpose of this chapter, when more than 50 amino acids are joined by peptide bonds, it will be considered as a protein. Smaller sequences will be referred to as peptides.

Most proteins weigh between 5,500 to 220,000 Da and are made up of one or more polypeptide chains. The characteristically shorter, stronger and more rigid nature of the peptide bond is caused by its involvement in tautomerization. Along with bonds on either side, the amide bond makes up the protein backbone. Atoms that are associated with the peptide bond are inflexible and lie in the rigid amide plane. The C_α–C bonds in the peptide, however, can freely rotate, with degrees of rotational freedom limited only by the size and character of the R group. A combination of the rigid amide bond and the freely rotating C–C bonds determines the flexibility and the overall structural integrity of peptide backbone.

Large and various side chain functional groups cause the protein to fold into unique three-dimensional configurations. The four levels of structural organizations in proteins are primary, secondary, tertiary, and quaternary structures. The primary structure is the amino acid sequence and is the basis for the protein's folding and influences the protein's higher levels of structural organization.

Secondary structures are the conformation of the polypeptide backbone. The polypeptide chain can arrange itself into characteristic patterns, such as helical, pleated, and turn segments. The α-helix is the most common helical structure when the polypeptide chain is constructed from L-α-amino acids. Hydrogen bonds stabilize the carbonyl backbone between one amino acid and the NH of another amino acid four residues away. Throughout the helix, the amino and carboxyl groups are hydrogen bonded, while the side chains protrude outward creating a spiral arrangement. β-pleated sheets are another secondary structure. This structure is characterized by side-by-side hydrogen bonding either within the same chain (intrachain) or between two different chains (interchain). Alternating R groups point to opposite sides of the peptide plane, permitting close proximity of the peptide backbone for good β-strand hydrogen bonding. β-bends or (β-turns) on average contain four amino acids. Typically proline, glycine, and a polar amino acid are present at turn junctures. Proline creates a kink and reverses the direction of the polypeptide chain.

Tertiary structures are the three-dimensional shape of the protein caused by further folding of the secondary structures. Hydrogen bonding, ionic, and hydrophobic interactions are the types of interactions that are responsible for the stabilization of this higher order assembly. The globular protein's tertiary structures are made of domains, which are the functional units of a polypeptide. Each domain usually folds and maintains its conformation independently from other domains in the polypeptide chain. Within each domain, there is a hydrophobic core composed mainly of nonpolar amino acids. These hydrophobic cores have a very compact environment and are stabilized by Van der Waals forces and hydrophobic interactions. Water usually cannot penetrate this highly hydrophobic region of the protein. In contrast, polar amino acids are preferentially located at the protein's surface, increasing the domain's water solubility.

Quaternary structures are the highest level of protein organization and can be achieved by proteins that have more than one polypeptide chain. Each chain is a subunit, having its own primary, secondary, and tertiary structure. These subunits can associate to form dimers, trimers, and oligomers, which constitute the quaternary structure of a protein. Almost all proteins that are greater than 100 kDa have a quaternary structure.

III. PHYSICAL INSTABILITY

Physical instability is a concern with peptides and proteins and is generally caused by changes in the protein's conformation, referred to as denaturation. The possible results of protein and peptide instability are surface adsorption, aggregation, and precipitation.

A. Denaturation

Alterations in the tertiary, and frequently secondary, structure of globular proteins from the native conformation is termed denaturation. It is possible, however, for proteins to be denatured and retain some secondary structure. The simplest model for protein denaturation is a two-state model, in which the native state (N) of the protein exists in equilibrium with the unfolded denatured state (U):

$$N \leftrightarrow U \tag{17.1}$$

In solution, the protein's native state can fluctuate to form several different stable active states. Denatured state is actually a term used to describe an ensemble average of several nonnative protein conformations.

How easily a protein denatures depends on the forces that keep the protein in its native conformation. Of the various stabilizing forces, interactions between hydrophobic residues in the interior of the protein play a critical role in determining the physical stability of the protein. Changes in temperature and pH or the presence of organic solvents or denaturants can interfere with the bonding forces and can shift the equilibrium toward the unfolded denatured state. Upon restoration to native conditions, reversible denaturation processes reestablish the equilibrium, while irreversible denaturation will not. Irreversible denaturation or inactivation typically refers to a variety of processes. The protein may have undergone a conformational change in which renaturation cannot be achieved, or additional processes such as chemical or physical alterations have taken place. As the protein denatures, internal hydrophobic residues are often exposed to the solvent. These residues can then interact with nonpolar surfaces such as container surfaces, air interfaces, or other denatured proteins.

B. Surface Adsorption

One possible result of protein denaturation is surface interaction. Peptides and proteins are continually exposed to surfaces such as air, glass, rubber, plastic, and other synthetic materials during isolation, recovery, and formulation processes. Contact with these nonpolar surfaces can cause proteins to expose their hydrophobic interior, leading to adherence or adsorption to the surfaces of the containers. Once the adsorbed protein dissociates from the surface, its denatured conformation could result in aggregation.

The physical and chemical properties of both the molecule and the surface govern the type and extent of these interactions. Alterations in the pH and ionic strength of the media can significantly enhance or reduce the protein's tendency to adsorb. Owing to the polyelectrolyte nature of protein, adsorption to neutral or slightly charged surfaces is greatest near the protein's isoelectric point. Of the various types of surfaces studied, polyethylene oxide (Teflon) has been shown to be most effective at reducing protein adsorption and preventing surface denaturation. The extent and reversibility of these protein–surface interactions are dependent on time, temperature, and agitation. Often, prolonged exposure to surfaces, high temperatures, and agitation causes irreversible loss of the protein.

Surface adsorption is also determined by the available surface area. Once a closely packed monolayer is formed onto the surface, the adsorption process is saturated and further loss of protein usually occurs only by surface-induced denaturation of the bulk solution. Therefore, product loss through adsorption is negligible when large amounts of protein are present. However, at low concentrations, as in the case of insulin, loss from adsorption can be significant.

Several methods are available to reduce peptide and protein loss caused by surface adsorption. For therapeutic proteins, various excipients are usually added to the formulation to prolong stability. Incorporation of serum albumin into protein formulations prevents protein loss by competing for the surface active sites. Surfactants such as copolymers of ethylene oxide (Pluronics, Genapol) and polysorbates (Tween, Triton X-100) were also found to be effective for preventing adsorption. Addition of nonpolar solvents such as alcohol have also been used to prevent protein adsorption and denaturation.

C. Aggregation and Precipitation

Aggregation and precipitation is often the end product of protein instability and denaturation. It is a common problem encountered by peptides and proteins during processing, formulation, and delivery. Often, the resultant aggregates sustain the loss of bioactivity along with other possible deleterious effects such as increased immunogenicity, altered pharmacokinetics, pharmacology, and toxicology. When aggregation occurs on a macroscopic scale, precipitates that are formed can cause blockage in tubings and pumps used in drug delivery. Insoluble aggregates that do not sustain these consequences are simply undesirable because of the protein's altered physical appearance.

Protein aggregation and precipitation can occur during ordinary drug product handling. Contact with hydrophobic air interfaces, shearing and shaking during shipment, and passage through needles are common processes that result in denaturation and self-association. Although many physicochemical interactions can lead to protein aggregation, the hydrophobic interactions are believed to be the major cause. Interestingly, the same forces are believed to be responsible for protein refolding.

Some proteins reversibly self-associate to form dimers and higher oligomers of the native state. Insulin exists in several associated states that can fully revert back to the active monomer. At concentrations above 0.6 μg/ml, insulin exists as a dimer, and in the presence of zinc, insulin further associates into hexamers. Since association of insulin into the hexameric form is a noncovalent interaction, chelation of the zinc by ethylene diamine tetra acetic acid (EDTA) dissociates the insulin from the hexamer back to the dimer. Long-term storage, however, can potentially induce covalent transformations to form aggregates.

Moisture induced aggregation occurs in lyophilized formulations of proteins. Results of residual moisture level for several protein formulation studies showed that in order to maintain stability, an optimum residual moisture level is required. A fully hydrated monolayer of water surrounding the protein yields the greatest stability. Removal of this monolayer results in aggregation, but excess hydration results in loss of biological activity. The optimum amount of water needed is reduced when proteins are formulated in combination with glycine and mannitol as excipients in the case of human growth hormone. Such excipients, which can protect the protein from denaturation during freezing, are known as cryoprotectants. Several studies reported that proteins must remain hydrated either by having increased residual moisture levels or by adding water-substituting excipients. Once a critical moisture level is determined for a formulation, maintenance of this level becomes a challenge owing to evaporation of moisture from the product or absorption of moisture from the environment. The type of container stoppers and caps used is important in preventing moisture content exchange. Specially designed elastomeric caps may be used to reduce the moisture absorption in most lyophilized products.

IV. CHEMICAL DEGRADATION

A. Hydrolysis

Proteolysis is the hydrolysis of the peptide bond between amino acids in a peptide or a protein. Generally, the peptide bond is very stable at physiological pH, in the range of 6 to 8. At extremes of pH or temperature, the peptide bond can undergo rapid proteolysis resulting in protein degradation. The reaction proceeds much faster in acidic mediums and can be further increased by the proximity of a negatively charged side chain residue. However, the presence of a positively charged residue or bulky side chains such as leucine and valine retards the rate of peptide cleavage.

B. Oxidation

There are several reactive amino acids that may undergo oxidation in peptides and proteins. Methionine, cysteine, histidine, tryptophan, and tyrosine are among the amino acids susceptible to oxidation in both solution and lyophilized formulations (Table 17.4). Metal ions, light, base and free radical initiators are all catalysts for the oxidation of these amino acids. Even peroxides from excipients such as polysorbates, polyethylene glycol and gray silicone rubber stoppers can cause protein oxidative damage.

Several strategies are available to minimize protein oxidation. Refrigeration, protection from light, pH optimization, and reduction of oxygen exposure are simple measures that can be taken to reduce oxidative damage. More rigorous methods require the use of antioxidants. Antioxidants fall into one of three categories. Phenolic compounds such as butylated hydroxytoluene (BHT), butylated hydroxyanisole (BHA), propyl gallate, and vitamin E are classified under the true antioxidants. A second class of antioxidants is the reducing agents, which include compounds such as methionine, ascorbic acid, sodium sulfite, thioglycerol, and thioglycolic acid. They act as oxygen scavengers by becoming preferentially oxidized over the protein. A third class of antioxidants known as chelators includes EDTA, citric acid, and thioglycolic acid. These compounds complex trace metals that are responsible for initiating auto-oxidation reactions. The choice of antioxidants should be carefully considered since deleterious interactions with the protein or with excipients in the formulation may occur. Finally, lyophilization of a formulation may prevent protein oxidation.

C. Deamidation

Deamidation of glutamine (Gln) and asparagine (Asn) residues is one of the most common nonenzymatic reactions of protein degradation *in vivo* (Table 17.4). The reaction involves hydrolysis of the amide bond on the side chain with the release of an amine group. The product of deamidation is an aspartic acid (Asp) and a glutamic acid (Glu) residue. This reaction can take place rapidly at physiological conditions.

A global conformational change in the protein may not result from deamidation; however, localized changes in the vicinity of the deamidated site may occur owing to a change in charge at the site of modification. Loss in activity and elicited immunogenicity cannot be generalized for all deamidated proteins. Studies show that deamidated proteins are cleared faster from the body owing to their greater susceptibility to protease digestion, the first clearance mechanism in the body.

Several strategies can be employed to decrease the rate of deamidation. A pH of 6 afforded the greatest stability. The pH should be chosen to minimize all the various pathways of protein inactivation. One method is to develop a lyophilized formulation. Eliminating the use of aqueous solvent by freeze-drying is expected to reduce or eliminate deamidation, since hydrolysis of Asn and Gln requires the presence of water. However, residual moisture present in the lyophilized formulation can still allow deamidation reactions to take place. Protein engineering to replace the Asn residue with Ser can be an alternative method for proteins that retain conformation and biological activity after the mutation. Ser is an ideal replacement residue because of its similar size and potential hydrogen donor functional group.

D. β-Elimination and Disulfide Exchange

Thermal stress can destroy the disulfide bonds in proteins by β-elimination from the cystine residues. This can also occur at lower temperatures and high pH. At 100°C, several proteins undergo β-elimination at similar rates. However, it cannot be considered a general decomposition pathway for proteins. Other amino acids that are susceptible to β-elimination at alkaline conditions are Cys,

TABLE 17.4
Summary of Amino Acid Sequences and Corresponding Chemical Degradation Pathways

Hydrolysis	Oxidation	Deamidation	Racemization	Disulfide Formation
Asp-Pro	Met	Asn-Gly	Asp	Cys
Asp-X (X)	His	Asn-X	Ser-X (X ≠ Gly, Asp)	
X-Ser	Trp		Phe-X (X ≠ Gly, Asp)	
X-Thr	Tyr		Tyr-X (X ≠ Gly, Asp)	
X-X	Cys			

X, any amino acid unless otherwise specified

Ser, Thr, Phe, and Lys. The resulting thiols from β-elimination may contribute to further degradation through disulfide exchange. The reaction mechanism for disulfide depends on the pH of the media.

E. Racemization

Except for Gly, racemization can occur for all the chiral amino acids (Table 17.4). The reaction is base-catalyzed and generates D-enantiomers, which may have reduced bioactivity. However, D-enantiomers are often more resistant to proteolytic enzymes with improved stability of the resulting peptide.

F. Thermal Stability

Storage of peptides and proteins under normal refrigeration (4°C) may not be enough to prevent physical and chemical degradation. For instance, bovine insulin is recommended to be stored at –20°C. Kept at 4°C, insulin undergoes deamidation and polymerization reactions. Thermal stress often results in denaturated, deamidated, or aggregated products. To prevent thermal stress, lyoprotectants or surfactants are often included in protein formulations.

G. Multiple Pathways of Chemical Instability

It is rare for a protein to degrade by a single pathway. A major route of degradation may be apparent; however, peptide and protein breakdown typically involves numerous, simultaneous pathways of degradation. The two most common chemical degradation pathways are oxidation and deamidation.

V. PHARMACOKINETICS AND CLEARANCE MECHANISMS

Pharmacokinetics is defined as the study of the time course of drug absorption, distribution, metabolism, and excretion upon administration to a subject. For most protein and peptide drugs, its large molecular weight and degradation at site of administration prohibit its uptake into the circulatory system to any appreciable extent when given by an extravascular route. Once absorbed, however, the drug rapidly distributes to the different tissues in the body, as blood vessels are relatively permeable to peptides and proteins. Proteolytic enzymes in the tissues throughout the body continue to metabolize and clear the drug. The degradation of protein and peptide occurs throughout the drug's journey in the body, starting at the site of administration, during entry into the blood vessels, and in the blood through the liver and kidneys.

A. Absorption

Absorption is the first pharmacokinetic phase for any drug given extravascularly. Formulating a protein or peptide drug to be administered extravascularly has proven to be a formidable task because of poor absorption and thus, bioavailability. The reason is that protein drugs, in general, exhibit low permeability through the various exterior membranes such as nasal, oral, buccal, transdermal, ocular, rectal, vaginal, and upper branches of the pulmonary track. The membranes available for drug absorption differ significantly in the type of protective tissue layers that restrict the drug's passage into the circulatory system. Even the walls of blood vessels are layered with endothelial cells that can limit drug passage in and out of the blood vessels. Peptide drugs such as luteinizing hormone releasing hormone (LHRH) analogs (~1 kDa) given orally show negligible absorption. Nasal spray of an LHRH formulation exhibits a bioavailability of approximately 3%. Following subcutaneous administration, leuprolide, an LHRH agonist is absorbed to the extent of 94%. Thus, even peptide drugs are absorbed minimally from extravascular routes, making it currently impossible to deliver protein drugs to the systemic circulation by these routes.

Macromolecules can cross the epithelial cell layers two ways. Drug molecules can be transported either through the cell by the transcellular route or between the cells through the paracellular pathway. The junctions between the epithelial cells and the paracellular pathways are very closely packed, prohibiting the transport of macromolecules such as proteins and peptides. It is a more efficient route of transport for small hydrophilic drugs. The narrowness of the paracellular pathway in the various epithelia can be arranged in the order skin > cornea > colon = conjunctiva > small intestine ≈ nasal epithelium, with drug permeability being least for skin and highest for small intestinal and nasal epithelia. Several investigational penetration or absorption enhancers (chelating agents, surfactants, bile salts, fatty acids, and their derivatives) can help open the tight junctions and improve the permeablity and absorption of peptide and protein drugs.

B. Distribution

Peptide and protein drugs generally exhibit multiexponential disposition with distinct distribution and elimination phases. Their half-lives are relatively short, and drug analysis requires complete assay methods such as radioimmunoassays and enzyme-linked immunoassay. Because of their multiexponential disposition and short half-life, many, frequent plasma samples are required for a complete pharmacokinetic analysis.

C. Clearance

Enzyme-catalyzed degradation causes peptide and protein drugs to exhibit short plasma elmination half-lives, as these drugs are rapidly metabolized and excreted. Regular human insulin has an elimination half-life of 58 min, while the elimination half-life of native LHRH is 7.8 min. The mammalian body possesses numerous enzymes and enzyme systems that degrade proteins and peptides similarly to their endogenous macromolecular counterparts. Degradation of proteins and peptides *in vivo* usually involves either hydrolytic cleavage of peptide bonds by numerous proteases in the body. These enzymes are widely distributed throughout the body, ready to hydrolyze peptide and protein drugs for elimination. Like many enzymes, the catalytic activity of the peptidase enzymes depends on the pH, temperature, and concentrations of substrate and enzyme.

The liver plays a major role in metabolic clearing of drugs from circulation. It houses many of the drug-degrading enzymes, and much of a drug's metabolic biotransformation occurs inside this highly perfused organ. The kidneys also play a major role in the clearance of polypepetides and proteins through glomerular filtration. For polypeptides with molecular weight less than 30 kDa, glomerular filtration is estimated to clear as much as 20% of an administered dose in a single pass. Factors that affect drug excretion include size and charge. With increasing size and greater negative charge, the filtration rate is slower. This is due to filtration being a passive difussion and the glomerular filters being negatively charged. Polypeptides that are cleared by the kidneys include growth hormone, glucagon, and insulin.

VI. ROUTES OF ADMINISTRATION

A. Oral Route

Of the many ways in which a drug can be administered, oral delivery is still the most attractive and acceptable route of administration for therapeutically active compounds. This route is unsuitable for most protein and peptide drugs, owing mainly to their extensive degradation in the gastrointestinal tract. The highly acidic environment and proteolytic enzymes present such as pepsin, trypsin, and chymotrypsin immediately degrade the drug before it can be absorbed into the systemic circulation. In addition, the large molecular mass of peptides and proteins limits their ability to diffuse through the intestinal wall. There are, however, two clinically available small bioactive peptides that are marketed for oral use. One is DDAVP, a vasopressin analogue that is available as tablets and has comparable bioavailability to the intravenous dosage form. The other is cyclosporin, which is formulated in soft gelatin capsules and oral solutions. Because of this attractive route of delivery, continued effort is made to improve the bioavailability of macromolecules by using protease inhibitors, enhancers, and drug encapsulation and by chemically modifying the active compound into prodrugs.

B. Buccal Route

There are several advantages to delivering drug through the buccal mucosa. The rich vasculature embedded in the walls of the buccal membrane allows for noninvasive rapid absorption into the systemic circulation. Another significant advantage is the bypassing of the hepatic first pass metabolism. For peptides and proteins, this can significantly increase bioavailability. The easily accessible site of administration permits simple implantation and removal of adhesives and patches. A multitude of dosage forms have been explored for the delivery of large macromolecules. These include aqueous solutions, conventional buccal and sublingual tablets, adhesive gels, adhesive patches, and possibly devices mechanically attached to the teeth or implanted in the tooth enamel.

Like other routes of administration, there are obstacles that must be overcome before buccal delivery of macromolecules becomes possible. Lack of dosage form retention at the site of absorption is a major limitation. The patient can involuntarily swallow a large portion of the drug in the case of the solution dosage form. Tablets and capsules can be retained next to the buccal membrane longer; however, they can hinder speaking abilities. Adhesive gels and patches may offer a solution to this problem. Additionally, the presence of drug-metabolizing enzymes including oxidases, reductases, cyclooxygenases, and peptidases can limit the integrity of peptide and protein drugs at the site of administration.

C. Parenteral Routes

The more common intravenous, intramuscular, subcutaneous and the less popular intraperitoneal, intraspinal, intrathecal, intracerebroventricular, intraarterial, subarachnoid, and epidural methods of administration are all parenteral routes of administration. Aqueous solutions are the most prevalent dosage form available for parenteral delivery. Novel dosage forms for injections include suspensions of biodegradable microspheres, microemulsions, liposomes, and polymeric conjugates. This route is the preferred route of administration for drugs that have low bioavailability such as peptides and proteins and for emergency drugs requiring immediate action owing to the elimination of the absorption phase. Most clinically available peptides and proteins are delivered by the parenteral route as shown in Table 17.5. However, parenteral administration is invasive, causes discomfort, and can be traumatic when multiple injections are required. In addition, the possibility of severe immunogenic response, requirement of trained health professionals, needlestick injury,

TABLE 17.5
Examples of Marketed Macromolecular Drugs Intended for Parenteral Administration

Brand	Active Drug (Abbreviations)	No. of Amino Acids	Molecular Weight	Formulation[a]	Origin	Storage/ Stability
Prolastin (Miles)	α_{1}-Antitrypsin (ATT)	394	52 kDa	Recombinant product contains ATT, Na^+ (100–210 mEq/L), Cl^-(60–180 mEq/L), PEG (NMT 0.1%)	Pooled human plasma	2–8°C Avoid freezing
Miacalcin (Sandoz) Calcimar (RhonePoulenc, PA)	Calcitonin (CT)	32	3418 Da	Each milliliter contains 200 IU of protein with acetic acid 2.25 mg, sod. acetate 2 mg, NaCl 7.5 mg, and phenol 5 mg	Synthetic	2 to 8°C
Profasi (Serono Labs, MA)	Human chorionic gonadotropin (HCG)	—	—	Each vial contains 5 to 10,000 USP units of protein with sodium phosphate dibasic 16 mg, sodium phosphate monobasic 4 mg, and mannitol 100 mg	Human placenta/ pregnancy urine	15 to 30 °C; after reconstitution, refrigerate at 2 to 8°C and use within 30 days.
Neupogen (Amgen)	Granulocyte-colony-stimulating factor or Filgrastim (G-CSF)	175	19 kDa	Each vial contains 300-mcg proteins in a 10 mM sodium acetate buffer (pH 4.0) containing 50 mg mannitol, and 0.004% Tween 80	Recombinant	2 to –8°C; do not freeze or shake.
Epogen (Amgen) Procrit (Ortho Biotech)	Erythropoietin (EPO)	165	30.4 kDa	Single dose vial containing 2 to 10,000 U of protein with citrate buffer (pH 6.9), sodium chloride 0.58% and human serum albumin 0.25%. Also available as multidose vial preserved with 1% benzyl alcohol (pH 6.1)	Recombinant	2 to 8°C; do not freeze or shake.
Kogenate (Miles) Helixate (Armour/Miles)	Factor VIII antihemophilic factor (FVIII)	—	80 to 90 kDa	Final reconstituted product contains 100 IU/ml of protein with glycine 10 to 30 mg, imidazole NMT 500 µg, polysorbate 80[b] 600 µg, calcium chloride 2 to 5 mg mM, sodium 100 to 130 mq/l chloride 100 to 130 mEq/l, and human serum albumin 4 to 10 mg	Recombinant (human plasma products are also available)	2 to 8°C; avoid freezing; lyophilized powder may be stored up to 25°C for 3 months without loss of activity.
Glucagon for injection USP (Eli Lilly)	Glucagon	29	3483 Da	Each vial contains 1 or 10 mg of protein. Each 1-mg vial contains 49 mg of lactose. The diluting solution contains glycerin 1.6% and phenol 0.2%	Beef or pork pancreas	15 to 30°C; after reconstitution, 5°C up to 48 hr if diluting solution is used; use immediately if sterile WFI used.

Humatrope (Eli Lilly)	Human growth hormone or somatotropin (hGH)	191	22,125 kDa	Phosphate buffer (pH 7.5 after reconstitution) mannitol 25 mg, glycine 5 mg. diluting solution contains *m*-cresol 0.3% and glycerin 1.7%	Recombinant	2 to 8°C; use within 14 days after reconstitution.
Protropin (Genentech)	Methionine-human growth hormone or somatrem (Met-hGH)	192	22 kDa	Each vial contains 5 or 10 mg protein. The 10-mg vial contains mannitol 80 mg, monobasic sodium phosphate 0.2 mg, and dibasic sodium phosphate 3.2 mg	Recombinant	2 to 8°C; use within 14 days after reconstitution.
Wydase (Wyeth-Ayerst)	Hyaluronidase bovine testes			Lyophilized powder: Each vial contains 150 USP units of protein with 2.66 mg lactose and 0.075 mg thimerosal. Stabilized solution: Each milliliter contains 150 USP units of protein with 8.5 mg NaCl, 1 mg edetate disodium, 0.4 mg calcium chloride, monobasic sodium phosphate buffer, and not more than 0.1 mg thimerosal		Store below 30°C; use reconstituted solution within 14 days
Roferon-A (Roche)	Interferon alfa-2a (IFN-α-2a)	165	19 kDa	Each vial contains 3 to 36 million IU protein with 9 mg NaCl, 5 mg human serum albumin, and 3 mg phenol	Recombinant	2 to 8°C; do not shake or freeze.
Pitocin (Parke-Davis) Syntocinon (Sandoz)	Oxytocin	9	1,007	Each 1-ml ampule contains 10 IU of protein with acetic acid (qs to pH 4), alcohol (0.61%), chlorobutanol (0.5%), sodium acetate 1 mg, and sodium chloride 0.017 mg	Synthetic	15 to 25°C; protect from freezing.
Activase (Genentech, CA)	Tissue plasminogen activator or alteplase (t-PA)	527	59.04/63.0 kDa	Each vial contains 20, 50, or 100 mg protein. The 100-mg vial contains L-arginine 3.5 g, phosphoric acid 1.0 g, and less than 11 mg of polysorbate 80	Recombinant	
Abbokinase (Abbott)	Urokinase	267	—	Each vial contains 250,000 IU of protein with 25 mg of mannitol, 250 mg of albumin (human), and 50 mg of NaCl	Tissue culture of human kidney cells	2 to 8°C
Pitressin (Parke Davis)	Vasopressin or 8-arginine vasopressin (AVP)	9	1084	Acetate buffer chlorobutanol 0.5%	Synthetic	15 to 25°C

Abbreviations: Nacl, Sodium chloride; NMT, not more than

[a] All solutions are in water for injection USP; bacteriostatic water for injection (BWI) contains benzyl alcohol as a preservative

Modified from Banga, A., Therapeutic Peptides and Proteins, Formulation, Processing, and Delivery Systems, Technomic Publishing Company, Lancaster, PA, 1995.

TABLE 17.6
Commercially Available Peptide Nasal Sprays

Peptide	Therapeutic Indication	Inactive Ingredient
Lypressin nasal solution USP (Diapid, Sandoz)	Diabetes insipidus	Acetic acid and sodium acetate, chlorambutanol, citric acid, disodium phosphate, glycerin, methyl and propyl paraben, sodium chloride, sorbitol, and purified water
Desmopressin (DDAVP, Rhone-Poulenc Rorer), (Stiamte, Armour/Ferring)	Diabetes insipidus Primary nocturnal enuresis	Chlorobutanol, sodium chloride and purified water
Oxytocin nasal solution USP (Syntocinon, Sandoz) (Stimate, Armour/Ferring)	To assist initial postpartum milk ejection	Chlorobutanol, citric acid, sod. phosphate, glycerin, methyl and propylparaben, sodium chloride, sorbitol solution and purified water
Nafarelin (LHRH; agonist) (Synarel, Syntex)	Endometriosis	Benzalkonium chloride, glacial acetic acid, sorbitol, and purified water
Calcitonin-Salmon nasal spray (Miacalcin, Novartis)	Postmenopausal osteoporosis	Sodium chloride, benzalkonium chloride, nitrogen, hydrochloric acid (added as necessary to adjust pH) and purified water

Modified from Banga, A., Therapeutic Peptides and Proteins, Formulation, Processing, and Delivery Systems, Technomic Publishing Company, Lancaster, PA, 1995.

splash back of bodily fluids as well as cross contamination with blood-borne pathogens such as HIV and hepatitis B renders this route of administration the last alternative.

D. Nasal Delivery

The nasal route of delivery shows promise as a possible route for delivering macromolecules. Because of its rich capillary vasculature, the permeability for most drugs through the nasal mucosa is comparable to the small intestine. For some small drug molecules, this route offers rapid and complete absorption. Large molecular weight drugs such as peptides and proteins are absorbed to the extent of about 5% or less. Although bioavailability of proteins and peptides through the nasal mucosa is relatively low, it is still greater than oral absorption. Clinically, peptides have been successfully delivered by this method, as shown in Table 17.6. An added advantage to the nasal route of delivery is the avoidance of first pass metabolism. This noninvasive method of delivery is not ideal for macromolecular administration owing to low drug permeability and metabolism. Thus, enhancers and protease inhibitors have been used to increase absorption and decrease enzymatic degradation of proteins and peptides following intranasal administration. Other disadvantages include variable absorption and possible irritation and pathological changes to the nasal mucosa over prolonged administration.

E. Pulmonary Delivery

The greatest advantage for delivering drug through the pulmonary route is the very thin epithelium and large surface area (140 m^2) available for drug absorption. Administration by this route also avoids hepatic first pass effect. Rapid and complete absorption is possible for certain macromolecules when drug is deposited into the deep lungs. Thus, the pulmonary route offers another noninvasive method for delivering a wide variety of peptides and possibly proteins. However, depositing drug into the deep lungs is still a challenge with the existing delivery systems. Currently, new dry powder devices are in clinical trials for the systemic delivery of insulin via the pulmonary route. If this route proves clinically successful for insulin, it may become the preferred route for several protein and peptide drugs that are currently administered by the parenteral route. However, concerns with protein and peptide degradation by peptidases present in the lungs and possible immune response resulting from changes in pulmonary epithelial permeability will have to be addressed before the protein and peptide can be delivered by this route. Also, permeability changes in smokers complicates the dosing.

F. Transdermal Route

There is great potential for transdermally delivering peptide and protein drugs with the use of physical aids, such as sonophoresis, iontophoresis, and electroporation. Sonophoresis or phonophoresis facilitates the transport of macromolecules using ultrasound, while iontophoresis helps their transport under the influence of an electric field. Electroporation, a more recent approach for transdermal delivery, uses high voltage electrical pulses to create pores in lipid bilayers to lower the transdermal layer resistance. The main advantage of this route is the possibility of controlling and sustaining the release of drug into the systemic circulation. The efficiency and tolerability of these new modes of administration will have to be compared with other noninvasive modes of administration prior to the selection of this route for protein administration.

G. Ocular Delivery

The ocular membrane offers a route for delivery of protein and peptides to the eye for local and systemic effect. Absorption through conjunctiva was shown to be rapid, with direct deposition into the systemic circulation avoiding the hepatic first pass metabolism. Ocularly administered insulin has been shown to exert pharmacological effect, possibly through its absorption through conjunctiva as well as the nasal epithelium following nasolacrimal drainage of administered drops. Tear turnover, degradation by proteolytic enzymes, and protein binding may lead to a low bioavailability of drugs administered by this route. The main concern with this route of administration is low patient acceptance owing to sensitivity of the corneal epithelium to external irritation and to the innate aversion of instilling drugs to the eye.

H. Rectal Route

The rectal area has an extensive blood supply from the various rectal arteries available for the absorption of drugs. Some of the advantages to the rectal route include low levels of protease activity as well as avoidance of hepatic first pass metabolism. Patients suffering from nausea and vomiting benefit greatly from rectal administration. The disadvantages include low patient compliance, limited surface area available for drug absorption, and elimination of drug during bowel movements.

VII. PROPER HANDLING OF PEPTIDE AND PROTEINS

Healthcare professionals are expected to be knowledgeable about how to properly store, dispense, and administer peptide and protein pharmaceuticals. Appropriate usage of these biological drugs requires knowledge of the individual macromolecule. In hospital settings, biological drugs often require reconstitution with a suitable diluent before administering to patients. Water for injection is a commonly used solvent for dissolving the lyophilized drug. Bacteriostatic water for injection, which contains benzyl alcohol as a preservative may also be used for certain products. In some formulations, such as Humatrope (human growth hormone), a diluent is supplied with the lyophilized drug. The diluent in this case contains 0.3% *m*-cresol as a preservative and 1.7% glycerin as a stabilizer and tonicity agent. Incompatibilities of drug product with excipients in diluents should be noted and avoided. An example is interleukin-2 (IL-2) becoming incompatible with bacteriostatic agents in the dileunt 4 hours after reconstitution. Observation of incompatibility between valinomycin and granulocyte/macrophage colony stimulating factor were reported when the two drugs were administered through the same line. Owing to the potential problems that can arise from infusing aggregated or precipitated proteins directly into the systemic circulation, extra precautions should be taken when administering biological products through the parenteral route. It is advisable to check for particulates in all injectable drug products. Protein solutions suspected of containing any particulates should be discarded.

VIII. REGULATORY ASPECTS

CBER (Center for Biologics Evaluation and Research) is the regulatory agency responsible for ensuring the safety, effectiveness, and timely delivery to patients of biological and related products including blood, vaccines, tissue, allergenics, and biological therapeutics. CBER is the biological equivalent of CDER (the Center for Drug Evaluation and Research), whose mission is to regulate nonbiological therapeutics. Other responsibilities delegated to CBER are regulations for the following:

- The nation's entire blood supply and the products derived from it
- The production and approval of safe and effective childhood vaccines, including any future AIDS vaccines
- Human tissue for transplantation
- The adequate and safe supply of allergenic materials and antitoxins and biotechnology-derived products used to treat diseases such as cancer and AIDS

Approval for a new biological drug requires the application of a product license application to CBER. The product license application is equivalent to the new drug application for nonbiological drug molecules. The product license application requires approval of the specific series of production steps required to manufacture the biological and process control tests along with end product specifications to ensure uniformity of the product on a lot-to-lot basis.

After approval, the company must obtain two manufacturing licenses before production can be launched: an establishment license, which grants the company the right to manufacturer a biological at a particular site, and a production license, which establishes the right to manufacture a particular approved drug at a designated site.

CASES

1. Leuprolide

Leuprolide, a leuteinizing hormone releasing hormone (LHRH), is an endogenous hormone synthesized and stored in the neurons of the hypothalamus. It is responsible for stimulating the release of the luteinizing hormone and the follicle-stimulating hormone, which produces the trophic and steroidogenic effects on gonadal tissues. It is currently on the market for treatment of prostatic cancer and endometriosis.

2. Human Growth Hormone

Generic human growth hormone (hGH) comes in two forms: regular hGH, called somatropin, and a methionylated form called somatrem, with an extra methionine residue at the NH_2-terminus of the regular hGH. The two forms of growth hormone possess the same efficacy and safety profiles and undergo similar degradation pathways.

3. Insulin

Insulin is an important therapeutic agent used for the treatment of diabetes. It is essential for the treatment of Type II and Type I diabetic patients who have become insensitive to oral diabetic agents.

Review the literature and answer the following questions.

1. What is the structure of each of the proteins?
2. Based on the structures, what kind of instabilities do you anticipate?
3. What dosage forms are available in the market?
4. Are there any special diluents necessary for administration of their dosage forms?

ANSWERS

1. Leuprolide

Structure

The native structure is a decapeptide with a lactam at the NH_2-terminus and an amine group attached at the COOH end.

$$\text{(Pro)Glu-His-Trp-Ser-Tyr-Gly-Leu-Arg-Pro-Gly}NH_2$$

Studies show that the first three amino acids at the NH_2 terminus play a functional role in the biological activity. Substitution of these residues with amino acids that do not resemble Glu, His, and Trp can significantly decrease or abolish activity. However, replacement of residues 1 through 4 has been shown to improve the potency. Many analogues of LHRH have been successfully synthesized, some with increased activity 30 to 80 times over that of the native compound.

Stability Issues

Leuprolide undergoes two pathways of chemical modification. In acidic solutions, the proposed mechanism of degradation is the opening of the lactam ring followed by hydrolytic degradation of LHRH. At alkaline conditions, the principle reaction observed is racemization at the His and Ser amino acids. *In vivo*, LHRH is eliminated from the body mainly by enzymatic hydrolysis at the (pyro)Glu-His group as well as excretion of intact drug by the kidneys.

Delivery Systems

A marketed formulation of leuprolide is Lupron Depot manufactured by TAP Pharmaceuticals, a sustained-release intramuscular injectable that comes in two strengths, a 7.5-mg dosage indicated for prostatic cancer and a 3.75-mg dose used in endometriosis. The drug is entrapped inside the matrix of biodegradable microspheres made of poly(*d,l*-lactideglycolide. Drug is released when the matrix degrades to soluble lactate and glycolate. The sustained-release product consists of the following:

Leuprolide acetate: 11.25 mg
Polylactic acid/polyglycolide: 99.3 mg
D-mannitol: 19.45 mg
Vehicle for 2 ml reconstitution

The accompanying vehicle contains carboxymethylcellulose sodium (10 mg), D-mannitol (100 mg), polysorbate 80 (2 mg), water for injection, USP, and glacial acetic acid, USP to control pH.

The nasal and pulmonary routes of administration have been investigated for the delivery of leuprolide because of the inconvenience and pain of the currently marketed parenteral formulations. Unfortunately, absorption through the nasal mucosa was very low and erratic. Intersubject variation, approximately 90%, was also unacceptably high. Delivery through the lungs was more promising, showing significant absorption when drug was formulated in both solution and suspension aerosols.

2. Human Growth Hormone

Structure

The hGH is a single polypeptide chain made of 191 amino acids, internally cross-linked by two disulfide bonds. It has a wide range of growth-promoting effects in the body and is used clinically for hormone replacement therapy in the treatment of pediatric hypopituitary dwarfism and for children suffering from low levels of the hGH. It has also been used for and shown efficacy toward improving the recovery of burn patients.

Stability Issues

Of the 9 asparagine and 13 glutamine residues in the hGH that are available for deamidation, only Asn-149 was found to undergo deamidation to a significant extent. Minor amounts of deamidation can also take place at Asn-152 under alkaline conditions. Both aspartic acid and its iso-form were found as products of deamidation. The deamidated form of the hGH has essentially the same degree of potency as the native protein.

In the presence of peroxides (H_2O_2), the methionine residues at positions14 and 125 were oxidized to Met-sulfoxide, but Met-170 was spared, possibly because of steric hindrance, interaction with another side chain, or the fact that it is buried in the interior of the protein. The partially oxidized form (ox-Met-14, ox-Met-125) of the hGH exhibited full bioactivity, whereas the fully oxidized form (ox-Met-14, ox-Met-125, and ox-Met-170) possessed only about one fourth the activity of the unoxidized species.

Unlike many other proteins where intact and correct disulfide linkages are required for the native conformation and biological activity, reduction and alkylation of the four thiol groups in the cysteine residues from hGH conserved the tertiary structure as well as the biological activity. Therefore, when proteolytic cleavage of the hGH by proteases occurs, mostly between residues 133 and 149, the resultant two large fragments held together by disulfide linkage at Cys-53 and Cys-165 retain full growth-promoting activity.

Physical degradation such as denaturation and subsequent aggregation of the hGH can be caused by agitation of the solution vial. Particulates that are formed in solution may be the result of surface denaturation of the protein, followed by aggregation of the unfolded molecules. A number of excipients that have effectively prevented hGH aggregation include cyclodextrins, cellulose derivatives, and nonionic surfactants.

Delivery Systems

Both commercially available hGH products sold in the U.S. are parenteral formulations packaged as lyophilized products for reconstitution before administration. Solid-state stability studies show that the major routes of degradation upon storage in the freeze-dried formulation include deamidation, oxidation, and aggregation.

In the U.S., hGH is sold under the trade name Protropin (somatrem for injection) by Genentech, Inc. and Humatrope (somatropin for injection) by Eli Lilly and Co. The contents of the drug products are listed below.

Protropin[a]		Humatrope[b]	
Somatrem	5 mg	Somatropin	5 mg
Mannitol	40 mg	Mannitol	25 mg
Sodium phosphates	1.7 mg	Sodium phosphate	1.13 mg
Glycine	5 mg		

[a] The diluent is BWI, containing 0.9% benzyl alcohol.
[b] The diluent contains 0.3% *m*-cresol as a preservative and 1.7% glycerin for isotonicity.

3. Insulin

Structure

Insulin is the principal hormone required for proper glucose control in normal metabolic processes. This protein has 51 amino acid residue proteins made of a 21-residue A-chain and a 30-residue B-chain connected by two disulfide bonds (Figure 17.2). The A-chain also possesses a third disulfide bond, which connects the Cys-6 and Cys-11 residues. At very low concentrations (<0.6 μg/ml),

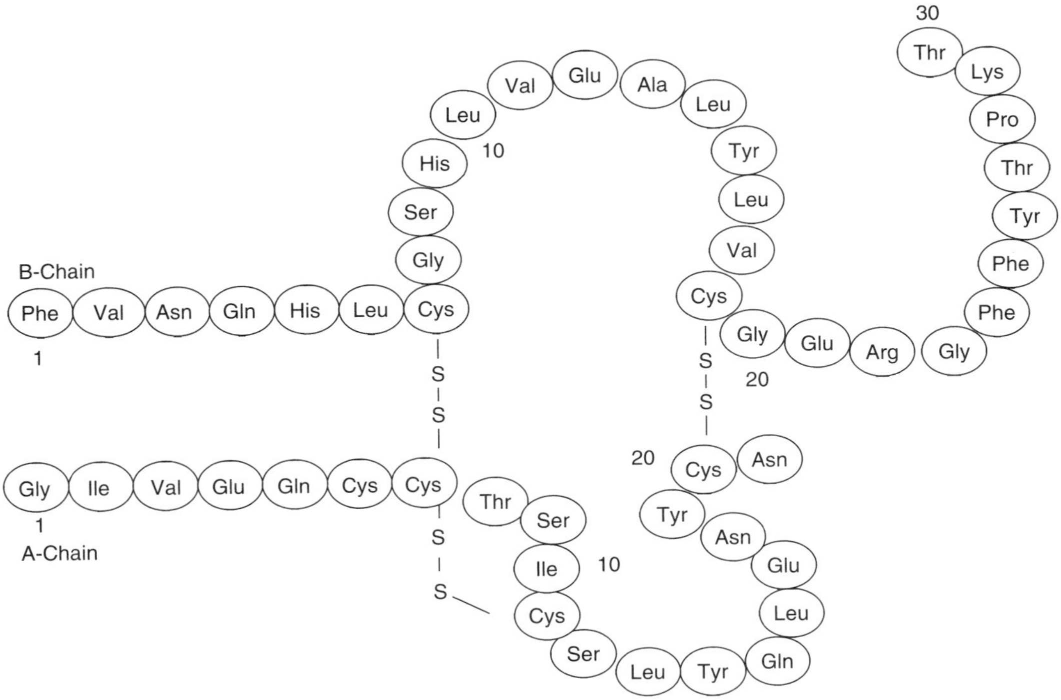

FIGURE 17.2 Structure of human insulin.

insulin exists as a monomer in solution. At pharmaceutically relevant concentrations or in the presence of Zn, insulin tends to associate in solution to form a noncovalent hexamer. The hexameric form of insulin is found in all available formulations on the market.

Stability Issues

The pharmaceutical significance of insulin has made it one of the most well characterized proteins. The chemical and physical breakdown of insulin during storage has been well documented. The most prevalent chemical route of degradation for insulin is deamidation at the possible Asn and Gln residues. Deamidation of Asn at A-21 is most extensive in acid conditions followed by deamidation at B-3 in neutral solutions at a much slower rate. The B-3 deamidation results in the formation of isoAsp as well as Asp derivatives.

Fibrils, aggregates, viscous gels, and insoluble precipitates are products of insulin's physical instability. They are the result of insulin's tendency to undergo conformational changes in solution. Fibrils are formed by the noncovalent association of intact proteins arranged into β-sheets. Insulin fibrils are often observed when infusion pumps are used to deliver insulin. Seen under the electron microscope, the fibrils look like long fibers with diameters of 10 to 50 nm in length. Studies have shown that the rate of fibril formation is dependent on the insulin concentration, pH, temperature, and exposure to hydrophobic surfaces such as air–water interfaces. The bonds between the subunits of the insulin fibrils are noncovalent and can be broken in the presence of strong denaturants and in extreme alkaline conditions ($pH > 11$) or strong organic acids.

Adsorption of insulin to hydrophobic surfaces such as glass is another physical instability and may be caused by a conformational change of the protein at the container surface. At concentrations of greater than 0.2 mg/ml, the amount of insulin loss is negligible. When working with insulin at lower concentrations, protein loss can be minimized by the addition of albumin to decrease surface adsorption.

Delivery Systems

The various insulin preparations available can be found in the Facts and Comparison.

TUTORIAL

1. The peptide bonds containing D-amino acids as opposed to L-amino acids are less stable in proteolytic environments.
   (a) True
   (b) False
2. The secondary structure of proteins refers to the conformation of the polypeptide backbone.
   (a) True
   (b) False
3. Oxidation of methionine to methionine sulfoxide can be reversed with a suitable reducing agent.
   (a) True
   (b) False

4. The peptide bonds between aspartic acid (Asp) and proline (Pro) are highly susceptible to hydrolysis at acidic pH.
   (a) True
   (b) False
5. The amide groups of asparaginyl (Asn) or glutaminyl (Gln) residues are labile at extremes of pH and may be hydrolyzed easily to leucine (Leu) and valine (Val), respectively.
   (a) True
   (b) False
6. Lyophilized protein in freeze-dried formulations require a large amount of residual moisture to retain stability.
   (a) True
   (b) False
7. Administration by the nasal route avoids the hepatic first pass metabolism.
   (a) True
   (b) False
8. The challenge to delivering peptides and proteins orally is the low permeability and degrading enzymes in the gastrointestinal tract.
   (a) True
   (b) False
9. The skin is the most permeable barrier among the noninvasive routes of administration for peptides and proteins.
   (a) True
   (b) False
10. Partial unfolding of proteins when exposed to hydrophobic environments may promote protein aggregation.
    (a) True
    (b) False
11. Which of the following statements is true regarding protein denaturation?
    I. Protein denaturation can be either reversible or irreversible.
    II. Protein denaturation can be caused by exposure to hydrophobic surfaces.
    III. Factors causing protein denaturation include thermal stress and extremes of pH.
       (a) I and II
       (b) II and III
       (c) III and I
       (d) I, II, and III
12. Which of the following statements regarding protein aggregation is true?
    I. In general, protein aggregation enhances activity.
    II. Aggregation always precedes precipitation.
    III. It is possible to have a soluble aggregate.
       (a) I only
       (b) II only
       (c) II and III
       (d) I and II
13. Which of the following routes of administration is likely to yield the highest systemic bioavailability of protein and peptide drugs?
    (a) Ocular
    (b) Buccal
    (c) Nasal
    (d) Pulmonary

14. Which of the following enzymes is most responsible for peptide and protein metabolism and degradation?
    (a) Kinases
    (b) Proteases
    (c) Oxidases
    (d) Phosphorylases
15. Which of the following are not methods used to increase permeability of macromolecules through the skin?
    (a) Sonophoresis
    (b) Iontophoresis
    (c) Electroporation
    (d) Dermaphoresis

Match the following amino acids and their common route of chemical degradation.

A. Oxidation
B. Deamidation
C. Hydrolysis
D. Disulfide exchange

16. (a) Methionine
    (b) Asparagine
    (c) Aspartic acid
    (d) Cysteine
    (e) Glutamine
    (f) Histidine

Respond to questions 17 to 20 based on the following information.

Intron A (Schering) is supplied as either solution or lyophilized powder. Each milligram of the solution contains the following ingredients. The lyophilized powder contains all the following ingredients except parabens.

Interferon alfa-2b (active ingredient)	10–25 million IU
Sodium phosphate dibasic	2.3 mg
Sodium phosphate monobasic	0.55 mg
Glycine	20 mg
Human serum albumin	1.0 mg
Methyl paraben	1.2 mg
Propyl paraben	0.12 mg

17. In intron A solution, the purpose of sodium phosphate (dibasic and monobasic) is:
    (a) Preservative
    (b) Cryoprotectant
    (c) Anti-adsorption agent
    (d) Buffering agent
18. In Intron A powder, the purpose of glycine is:
    (a) Preservative
    (b) Cryoprotectant
    (c) Anti-adsorption agent
    (d) Buffering agent

19. In Intron A solution, the purpose of human serum albumin is:
    (a) Preservative
    (b) Cryoprotectant
    (c) Anti-adsorption agent
    (d) Buffering agent
20. In Intron A solution, the purpose of parabens is:
    (a) Preservative
    (b) Cryoprotectant
    (c) Anti-adsorption agent
    (d) Buffering agent

Answers: 1. (b); 2. (a); 3. (a); 4. (a); 5. (b); 6. (b); 7. (a); 8. (a); 9. (b); 10. (a); 11. d; 12. c; 13. (d); 14. (b); 15. (e); 16a. (A); 16b. (B); 16c. (C); 16d. (D); 16e. (B); 16f. (A); 17. (d); 18. (b); 19. (c); 20. (a)

SELECTED REFERENCES

1. Banga, A., *Therapeutic Peptides and Proteins, Formulation, Processing, and Delivery Systems,* Technomic, Lancaster, PA, 1995.
2. Cleland, J.L. and Wang, D.I.C., Refolding and aggregation of bovine carbonic anhydrase B: quasi-elastic light scattering, *Biochemistry,* 29, 11072, 1990.
3. Fisher, B.V. and Porter, P.B., Stability of bovine insulin, *J. Pharm. Pharmacol.,* 33(4), 203–206, 1981.
4. Flatmark, T., *Acta. Chem. Scand.,* 20, 1487–1496, 1966.
5. Ikai, A., Tanka, S., and Noda, H., Reactive kinetics of guanidine denatured bovine carbonic anhydrase B, *Arch. Biochem. Biophys.,* 190, 39, 1978.
6. Kroon, D.J., Baldwin-Ferro, A., and Lalan, P., Identification of sites of degradation in a therapeutic monoclonal antibody by peptide mapping, *Pharm. Res.,* 9, 1386–1393, 1992.
7. Kompella, U.B., Protein and drug delivery, in *Biopharmaceutical Drug Design and Development,* Wu-pon, S., and Rojanasakul, Y., Eds., Humana Press, Totowa, NJ, 1999, 239–274.
8. Kompella, U.B. and Lee, V.H.L., *Pharmacokinetics of peptide and protein drugs,* in *Peptide and Protein Drug Delivery,* Lee, V.H.L., Ed., Marcel Dekker, New York, 391-484, 1991.
9. Lackowicz, J.R. and Weber, G., Quenching of protein fluorescence by oxygen: detection of structural fluctuations in proteins in the nanosecond timescale, *Biochemistry,*12, 1471, 1973.
10. Lee, V.H.L., Yamamoto, A., and Kompella, U.B., Mucosal penetration enhancer for facilitation of peptide and protein drug absorption, *CRC Crit. Rev. Drug Carrier Sys.,* 8, 91–192, 1991.
11. Liu, W.R., Langer, R., and Klibanov, A.M., Moisture-induced aggregation of lyophilized proteins in the solid state, *Biotech. Bioeng.,* 37, 177, 1991.
12. Lougheed, W.D., et al., Physical stability of insulin formulations, *Diabetes,* 32, 424, 1983.
13. Lukash, A.M., Pushkina, N.V., and Tsibulsky, I.E., Autoantigenic properties of deamidated serum albumin, *Immunologiya,* 68, 89, 1987.
14. Marcus, F., *Int. J. Peptide Protein Res.,* 25, 542–546, 1983.
15. Mitraki, A. and King, J., Protein folding intermediates and inclusion body formation, *Biotechnology,* 7, 690, 1989.
16. Nguyen, T.H., Oxidation degradation of protein pharmaceuticals, in *Formulation and Delivery of Proteins and Peptides,* Cleland, J.L. and Langer, R., Eds., American Chemical Society, Washington, DC, 1994, 59–71.
17. Pikal, M.J., Dellerman, K., and Roy, M.I., Formulation and stability of freeze-dried proteins: effect of moisture and oxygen on the stability of freeze-dried formulations of human growth hormone, *Dev. Biol. Stand.,* 74, 21, 1991.
18. Sadana, A., Protein adsorption and inactivation on surfaces. Influence of heterogenicities, *Chem. Rev.,* 1992. 92: p. 1799.
19. Schlicktkrull, J., et al., Monocomponent insulin and its clinical implications, *Horm. Metab. Res.,* 5, 134, 1974.
20. Zale, S.E. and Klibanov, A.M., *Biochemistry,* 25, 5432–5444, 1986.

Index

A

B

C

D

E

F

G

H

I

Q

R

S

T

U

V

W

Z